90 03

D1759890

WITHDRAWN
FROM
UNIVERSITY OF PLYMOUTH
LIBRARY SERVICES

This b

CH

Stable Isotopes

integration of biological,
ecological and geochemical
processes

ENVIRONMENTAL PLANT BIOLOGY series:

Editor: H. Griffiths
Department of Agricultural and Environmental Science, University of Newcastle, Newcastle upon Tyne NE1 7RU, UK

Abscisic Acid: physiology and biochemistry

Carbon Partitioning: within and in between organisms

Pests and Pathogens: plant responses to foliar attack

Water Deficits: plant responses from cell to community

Photoinhibition of Photosynthesis: from molecular mechanisms to the field

Environment and Plant Metabolism: flexibility and acclimation

Embryogenesis: the generation of a plant

Plant Cuticles: an integrated functional approach

Stable Isotopes: integration of biological, ecological and geochemical processes

Forthcoming titles include:

Biological Rhythms and Photoperiodism in Plants

H. GRIFFITHS
*Department of Agricultural and Environmental Science, University of Newcastle,
Newcastle upon Tyne NE1 7RU, UK*

Stable Isotopes

*integration of biological,
ecological and geochemical
processes*

βIOS
SCIENTIFIC
PUBLISHERS

© BIOS Scientific Publishers Limited, 1998

First published in 1998

All rights reserved. No part of this book may be reproduced or transmitted, in any form or by any means, without permission.

A CIP catalogue record for this book is available from the British Library.

ISBN 1 85996 135 5

BIOS Scientific Publishers Ltd
9 Newtec Place, Magdalen Road, Oxford OX4 1RE, UK.
Tel. +44 (0) 1865 726286. Fax. +44 (0) 1865 246823
World-Wide Web home page: http://www.Bookshop.co.uk/BIOS/

DISTRIBUTORS

Australia and New Zealand
    Blackwell Science Asia
    54 University Street
    Carlton, South Victoria 3053

Singapore and South East Asia
    Toppan Company (S) PTE Ltd
    38 Liu Fang Road, Jurong
    Singapore 2262

India
    Viva Books Private Limited
    4325/3 Ansari Road, Daryaganj
    New Delhi 110002

USA and Canada
    BIOS Scientific Publishers
    PO Box 605, Herndon,
    VA 20172-0605

UNIVERSITY OF PLYMOUTH

Item No. 900 346776 1

Date 1 9 MAR 1998

Class No. 581.028 STA

Contl. No. 1859961355

LIBRARY SERVICES

Production Editor: Andrea Bosher
Typeset by Saxon Graphics Ltd, Derby, UK.
Printed by Biddles Ltd, Guildford, UK.

Contents

Section 7. Palaeoclimatic reconstructions from Precambrian to Quaternary

Contributors

Beerling, D.J. Department of Animal and Plant Sciences, University of Sheffield, Sheffield, S10 2TN, UK

Borland, A.M. Department of Agricultural and Environmental Science, University of Newcastle upon Tyne, Ridley Building, Newcastle upon Tyne, NE1 7RU, UK

Brenna, J.T. Division of Nutritional Sciences, Cornell University, Savage Hall, Ithaca, NY 14853, USA

Broadmeadow, M.S.J. Forest Research, Alice Holt Lodge, Farnham, Surrey, UK

Brooks, J.R. Department of Biology, University of South Florida, Tampa, FL 33620-5150, USA

Brugnoli, E. CNR Instituto per l'Agroselvicoltura, Viale Marconi 2, I-05010 Porano, (TR), Italy

Buchmann, N. Lehrstuhl Pflanzenökologie, Universität Bayreuth, D-95440 Bayreuth, Germany

Cerling, T.E. Department of Geology and Geophysics, University of Utah, Salt Lake City, UT 84112, USA

Ciais, P. Laboratoire de Modélisation du Climat et de l'Environment, DSM/CE Saclay, 91191 Gif sur Yvette Cedex, France

Corso, T.N. Division of Nutritional Sciences, Cornell University, Savage Hall, Ithaca, NY 14853, USA

Dawson, T.E. Laboratory for Isotope Research and Analysis and the Section of Ecology and Systematics, Cornell University, Ithaca, NY 14853–2701, USA

Eglinton, G. Biogeochemistry Research Centre, Department of Geology, Wills Memorial Building, Queens Road, Bristol BS8 1RJ, UK

Ehleringer, J.R. Stable Isotope Facility for Environmental Research, Department of Biology, University of Utah, Salt Lake City, UT, 84112, USA

Evans, R.D. Department of Biological Sciences, University of Arkansas, Fayetteville, AR 72701, USA

Farquhar, G.D. Research School of Biological Sciences, Australian National University, GPO Box 475, Canberra, ACT 2601, Australia

Flanagan, L.B. Department of Biological Sciences, University of Lethbridge, Lethbridge, Alta. T1K 3M4, Canada

Fujiwara, H. National Agriculture Research Centre, Kannoudai 3–1–1, Tsukuba, Ibaraki 305, Japan

Gat, J.R. Department of Environmental Sciences and Energy Research, Weizmann Institute of Science, 76100 Rehovot, Israel

Gillon, J.S. Department of Agricultural and Environmental Science, University of Newcastle upon Tyne, Ridley Building, Newcastle upon Tyne, NE1 7RU, UK

Gleixner, G. Lehrstuhl für Allgemeine Chemie und Biochemie, Technische Universitat, München, Vöttingerstrasse 40, D-85350 Freising-Weihenstephan, Germany

Griffiths, H. Department of Agricultural and Environmental Science, University of Newcastle upon Tyne, Ridley Building, Newcastle upon Tyne, NE1 7RU, UK

Handley, L.L. Unit for Stable Isotopes Studies in Biology, Scottish Crop Research Institute, Invergowrie, Dundee, DD2 5DA, UK

Harris, J.M. Los Angeles County Museum of Natural History, Los Angeles, CA 90007, USA

Harwood, K.G. Department of Agricultural and Environmental Science, University of Newcastle upon Tyne, Ridley Building, Newcastle upon Tyne, NE1 7RU, UK

Hopkins, D.W. Department of Biological Sciences, University of Dundee, Dundee, DD1 4HN, UK

Huang, Y. Biogeochemistry Research Centre, Department of Geology, Wills Memorial Building, Queens Road, Bristol BS8 1RJ, UK. Current address: Department of Geosciences, Pennsylvania State University, University Park, PA 16802, USA

Johnston, A.M. Scottish Crop Research Institute, Invergowrie, Dundee, DD1 4HN, UK.

Kennedy, H. School of Ocean Sciences, University of Wales, Bangor, Menai Bridge, Anglesey, Gwynedd, LL59 5EY, UK

Lauteri, M. CNR Instituto per l'Agroselvicoltura, Viale Marconi 2, I-05010 Porano, (TR), Italy

Máguas, C. Ecologia, Biologia Vegetale, Universidade de Lisboa, Lisbon, Portugal

MacFadden, B.J. Florida Museum of Natural History, University of Florida, Gainesville, FL 32611, USA

Meijer, H.A.J. Centrum voor Isotopen Onderzoek, Nijenborgh 4, NL-9747, Groningen, Netherlands

Monteverdi, M.C. CNR Instituto per l'Agroselvicoltura, Viale Marconi 2, I-05010 Porano, (TR), Italy

Owens, N.J.P. Department of Marine Sciences and Coastal Management, University of Newcastle upon Tyne, Ridley Building, Newcastle upon Tyne, NE1 7RU, UK

Parker, H.M. Laboratory for Isotope Research and Analysis and the Section of Ecology and Systematics, Cornell University, Ithaca, NY 14853–2701, USA

Pausch, R.C. Laboratory for Isotope Research and Analysis and the Section of Ecology and Systematics at Cornell University, Ithaca, NY 14853–2701, USA

Perrott, R.A. Tropical Palaeoenvironments Research Group, Department of Geography, University of Wales Swansea, Swansea, SA2 8PP, UK

Pollard, A.M. Department of Archaeological Sciences, University of Bradford, Bradford, BD7 1DP, UK

Raven, J.A. Department of Biological Sciences, University of Dundee, Dundee, DD1 4HN, UK

Roberts, A. Department of Agricultural and Environmental Science, University of Newcastle upon Tyne, Ridley Building, Newcastle upon Tyne, NE1 7RU, UK

Robinson, D. Scottish Crop Research Institute, Invergowrie, Dundee, DD2 5DA, UK

Scartazza, A. CNR Instituto per l'Agroselvicoltura, Viale Marconi 2, I-05010 Porano, (TR), Italy

Schleucher, J.W. Department of Medical Biochemistry and Biophysics, Umeå University, S-90187 Umeå, Sweden

Schmidt H.-L. Lehrstuhl für Allgemeine Chemie und Biochemie, Technische Universität, München, Vöttingerstrasse 40, D-85350 Freising-Weihenstephan, Germany

Scrimgeour, C.M. Unit for Stable Isotopes Studies in Biology, Scottish Crop Research Institute, Invergowrie, Dundee, DD2 5DA, UK

Street-Perrott, F.A. Tropical Palaeoenvironments Research Group, Department of Geography, University of Wales Swansea, Swansea, SA2 8PP, UK

Switsur, R. Godwin Institute for Quaternary Research, University of Cambridge, Free School Lane, Cambridge, CB2 3RS, UK

Tobias, H.J. Division of Nutritional Sciences, Cornell University, Savage Hall, Ithaca, NY 14853, USA

Waterhouse, J.S. Environmental Science Research Centre, Anglia University, East Road, Cambridge, CB1 1PT, UK

Watts, L. Department of Marine Sciences and Coastal Management, University of Newcastle upon Tyne, Ridley Building, Newcastle upon Tyne, NE1 7RU, UK

Wheatley, R.E. Scottish Crop Research Institute, Invergowrie, Dundee, DD2 5DA, UK

Williams, D.G. School of Renewable Natural Resources, University of Arizona, Tucson, AZ 85721, USA

Wilson, J.M. Department of Agriculture and Environmental Science, University of Newcastle upon Tyne, NE1 7RU, UK

Woodward, F.I. Department of Animal and Plant Sciences, University of Sheffield, Sheffield, S10 2TN, UK

Yakir, D. Department of Environmental and Energy Research, Weizmann Institute of Science, Rehovot 76160, Israel

Yoneyama, T. Plant Nutrition Diagnosis Laboratory, National Agriculture Research Centre, Kannondai 3–1–1, Tsukuba, Ibaraki 305, Japan

Abbreviations and symbols

ϵ	fractionation during liquid vapour transition with subscripts eq – at equilibrium; k – kinetic fractionation.
ϵ_p	fractionation between source inorganic carbon and plant material
ϕ	bundle sheath leakiness in C_4 plants
ϕ	ratio of oxygenation: carboxylation
Γ	CO_2 compensation point (Γ^* in the absence of respiration)
ξ	ratio of CO_2 drawdown in the leaf cuvette to the ambient CO_2
α	fractionation factor
σ	pressure at altitude / surface pressure
β	proportion of carbon fixed by PEPC
$\delta^{13}C$	carbon isotope composition, vs (V)PDB, with subscripts a – atmospheric CO_2; e – CO_2 entering the leaf cuvette; f – photorespired CO_2; fs – photorespiratory substrate; o – CO_2 leaving the leaf cuvette
$\Delta^{13}C$	discrimination against ^{13}C in CO_2 ($^{13}C^{16}O^{16}O$) ‰ with subscripts obs – measured net discrimination; ps – measured gross photosynthetic discrimination; i – predicted from gas exchange; c – predicted from mesophyll conductance; e -ecosystem level; l – leaf level; sug – leaf soluble carbohydrates; 3 – sucrose p – leaf material; ps – photosynthetically-fixed carbon (gross); r – dark respired CO_2 in leaf; rs – dark respiratory substrate; R – soil or ecosystem respiration; T – troposphere CO_2; SOM – soluble organic matter; DIC – dissolved organic carbon; POM – particulate organic matter
$\delta^{15}N$	nitrogen isotope composition *vs.* atmospheric N_2 standard
$\Delta^{18}O$	discrimination against ^{18}O in CO_2 ($^{12}C^{18}O^{16}O$) with same subscripts defined as $\Delta^{13}C$
$\delta^{18}O$	oxygen isotope composition of CO_2 or H_2O with subscripts (see for $\delta^{13}C$), with VPDB, VPDB – CO_2 or VSMOW as standard
δD	hydrogen isotope composition, with VSMOW as standard
(V)PDB	(Vienna) Pee Dee Belemnite standard; CO_2 standard based on NBS – 19 calcite used to calibrate $^{13}C/^{12}C$ composition
(V)PDB-CO_2	CO_2 standard based on NBS-19 calcite used to calibrate ^{18}O composition of CO_2
(V)SMOW	Vienna Standard Mean Oceanic Water used to calibrate ^{18}O and D in water
$[CO_{2(aq)}]_{equil}$	aqueous CO_2 concentration in equilibrium and bulk air CO_2
‰	per mille (or parts per thousand)
a	fractionation against ^{13}C for CO_2 diffusion through air, with subscript *i* - dissolution and diffusion through water.

a	fractionation against $^{12}C^{18}O^{16}O$ during diffusion
A	net CO_2 assimilation rate
A/Pa	ratio of assimilation to external partial pressure of CO_2
AM	abuscular mycorrhiza
b	fractionation against ^{13}C during carboxylation with subscripts 3 – during CO_2 fixation by RUBISCO alone; 4 – during CO_2 fixation by PEPC alone; ′ – during CO_2 fixation by RUBISCO and PEPC
b	fractionation against $^{12}C^{18}O^{16}O$ during carboxylation defined for CO_2, O_2 and H_2O
BP	^{14}C dated years before present
C	concentration of CO_2, with subscripts a – in the atmosphere; i – in the sub-stomatal cavity; c – at the chloroplast; e, o – in air entering and leaving the cuvette
C	concentration of 15N label
C_3	photosynthetic pathway with primary 3-carbon products
C_4	photosynthetic pathway with primary 4-carbon products
CA	carbonic anhydrase
CAM	crassulacean acid metabolism
CBL	convective boundary layer
CCM	CO_2 concentrating mechanism
CENTURY	biogeochemistry decomposition model
CF	continuous flow
CG	Craig and Gordon equation for evaporative enrichment
CIO	Centrum voor Isotoepen Onderzoek
CPS	carbonyl phosphate synthase
CSIA	compound specific isotope analysis
CTF	cool temperate forest
C_W	molar concentration of water
D	diffusion coefficient
DHAP	dihydroxyacetone phosphate
DI	dual inlet
DIC	dissolved inorganic carbon
e	fractionation against ^{13}C during respiration
E	evaporative flux
e_i/e_a	ratio of internal : external vapour pressure in leaf
ET	evapotranspiration
f	fractionation against ^{13}C during photorespiration
F	rate of photorespiratory CO_2 production
f	fraction of tissue water
f	proportion of new production: total production in phytoplankton
$F_{(in)}$	inflow of H_2O
$F_{(out)}$	outflow of H_2O
FG	forest–grassland mosaic
F_{oi}	one way turbulent flux of CO_2 into canopy
GC	gas chromatograph
GCC	gas chromatograph-combustion
GCM	General Circulation Model
g_i	internal leaf CO_2 transfer conductance

GMWL	global meteoric water line
GOGAT	glutamate synthase
GPP	gross primary productivity
GS	glutamine synthetase
h	relative humidity (RH)
IRGA	infra-red gas analyzer
IRMS	isotope ratio mass spectrometry
$J_e, J_o,$	the molar flux of CO_2 entering and leaving the cuvette
k	carboxylation efficiency
K	ratio between diffusion/backdiffusion and photosynthesis for CO_2 exchange on leaves
$K_{1/2}$	half-saturation concentration
K_m	Michaelis Menten constant (O_2, CO_2)
L	mixing pathlength
LAI	leaf area index
LC	liquid chromatography
LGM	last glacial maximum
LRF	lowland rainforest
m/z	mass: charge ratio of ion
M_i	molar density of air
MRF	montane rainforest
NALMA	North American Land Mammal Age
NEP	net ecosystem productivity
N_f	concentration of particulate N
NMR	nuclear magnetic resonance
NPP	net primary productivity
P	atmospheric pressure
$\mathscr{P}$	péclet number
PAL	phenylalanine ammonia lyase (Chapter 2)
PAL	present atmospheric level (Chapter 20)
PBL	planetary boundary layers
PC	pyruvate carboxylase
PCR	photosynthetic carbon reduction (cycle)
PDH	pyruvate dehydrogenase
PEPC	phospho-*enol*pyruvate carboxylase
PEPCK	phospho-*enol*pyruvate carboxykinase
PFD	photon flux density
PGA	phosphoglycerate
PPC	pentose phosphate cycle
R	molar isotope ratio for $^{13}C/^{12}C$ (subcripts as for $\delta^{13}C$)
R	molar isotope ratio $^{18}O/^{16}O$ with subscripts: L – liquid or leaf water; in – input water; out – evaporative flux; E – evaporated water; V – water vapour; c – chloroplast; a – ambient air; l – input water; ss – steady state isotope composition during evaporation
r	resistance to diffusion, with subscript s – stomatal; bl – boundary layer
R	respiration rate
RPI	relative preference index
RUBISCO	ribulose-1,5-*bis*phosphate carboxylase/oxygenase

RuBP	ribulose 1,5 *bis*phosphate
SAM	s-adenosylmethionine
sd	stomatal density, where subscript *max* = maximum
SHMT	serine hydroxymethyl transferase
SiB	simple biosphere model
SOC	soil organic carbon
SPAC	soil–plant–atmosphere continuum
S_{rel}	RUBISCO selectivity factor
SST	sea-surface temperature
τ	ratio of carboxylation of RUBISCO: rate of hydration by CA
τ	equilibration time for CO_2 and H_2O
THF	tetrahydrafolic acid
TOC	total organic carbon
TPP	thiamine pyrophosphate
u	molar flow of CO_2 through leaf cuvette
UV-B	ultra violet B radiation
V	leaf water volume
V_m	maximum velocity
VPD	vapour pressure difference (leaf–air)
WTF	warm temperate forest
WUE	water efficiency (g $DW.g^{-1}H_2O$), with subscripts i – photosynthetic, instantaneous WUE during gas exchange; s – seasonal integrated WUE

Introduction

This volume represents the synthesis of a Stable Isotope Symposium, held at Newcastle upon Tyne in July 1996. Stable isotopes have provided a powerful adjunct to environmental studies, since the majority of elements have a higher natural abundance of at least one (non-radioactive) isotope. The majority of interest to date has been devoted to carbon ($^{13}C/^{12}C$), oxygen ($^{18}O/^{16}O$), hydrogen/deuterium ($^{2}H(D)/H$) and nitrogen ($^{29}N/^{28}N$) which leave signals in organic material (plant, animal) and inorganic constituents (carbon dioxide, water, soils, sediments, fossils and rocks). The heavier isotopes are present at such low abundances (generally less than 1%), and generally diffuse and react more slowly during transformations, which leads either to enrichment (e.g. ^{18}O and D, as evaporation of water favours the lighter isotopes) or depletion (^{13}C during photosynthesis). Thus, the relative abundance of the isotopes can be related through the theoretical fractionation or discrimination processes to conditions operating at the time of deposition, or serve as a marker for ecosystem transformations.

The meeting provided a theoretical and practical framework to the analysis and interpretation of stable isotope signals for a range of applications. Latest developments in the analysis and sample preparation for carbon, oxygen, hydrogen and nitrogen (and sulphur) were set against the theory underlying fractionation and implications for primary production of organic material. The transformations associated with the hydrological cycle, soils and vegetation were compared for a range of ecosystems, so as to compare terrestrial and aquatic models of resource acquisition and turnover. The importance of stable isotopes for validating our understanding of extant biological systems could then be set in the context of improving palaeoclimatic reconstruction for the past and making more reliable predictions for future climate change.

The scope of the volume reflects closely the enormous breadth of the meeting. The book starts with the recent developments in mass spectrometer technology and interpretation of organic signals, which have been stimulated by increasingly specialist, compound-specific analyses, while high sample throughput has been associated with continuous-flow sample preparation. Having established a progression from cell through plant to terrestrial and aquatic ecosystems, it was then possible to explore the broader implications for reconstructing past climates, and modelling climate change processes. In editing the volume, I have tried where possible to allow each author to retain their favoured nomenclature. While the isotope compositions are measured as a differential (δ), compared against a defined standard (arbitrarily set at 0‰), processes may be defined in terms of fractionation factors or isotopic effects during transformations (α and ϵ), or more commonly, in terms of the discrimination (Δ) between in organic source material and subsequent primary product. Additionally, while many processes lead to the depletion of the heavy isotope in organic material, real-time measurements of gas exchange now allow us to detect the concomitant enrichment in residual bulk air, which, by isotopic mass balance, provides an instantaneous measure of discrimination processes. Accordingly, I have attempted to simplify terms used generally for more than one isotope (e.g. δ in Craig–Gordon Model for ^{18}O and D in water), while at the same time explicitly defining similar terms when applied to different isotopes (e.g. $\delta^{13}C_c$ or $\delta^{18}O_c$ for signals associated with CO_2 in the chloroplast).

Finally, I must gratefully acknowledge the huge amount of work and support provided by a host of helpers, both in organising the original meeting and editing the subsequent volume: firstly the contributors and delegates, who believed in the meeting and came to Newcastle; the financial support provided by the Association of Applied Biologists, HRI, Wellesbourne, UK, the British Ecological Society, the Society for Experimental Biology and mass spectrometer manufacturers; the organisational skills and laughter of Carol Millman at the AAB; the publishers, BIOS, but particularly Jonathan Ray and Andrea Bosher for their patient support; my entire research group (The Potter Laboratory) who have all been tremendous, but particularly Jackie, Kirsty and Jim; colleagues within the Agricultural and Environmental Science Department, but in particular Isabel Blackburn who has provided monumental support in processing and reprocessing edited manuscripts, sorting correspondence and who has even tried to get me organised; and finally, the support from the most important team out at Ingoe in Northumberland, where Anna, Jemma, David and now young Jared are still waiting patiently for me to finish just one task on time.

Howard Griffiths (*Newcastle upon Tyne*)

High-precision deuterium and ^{13}C measurement by continuous flow-IRMS: organic and position-specific isotope analysis

J. Thomas Brenna, Herbert J. Tobias, Thomas N. Corso

1.1 Introduction

The 1990s have witnessed a dramatic increase in research on instrumentation and methods in high-precision isotope ratio mass spectrometry (IRMS) for determination of isotopes of organic elements. Prior to 1990, the dual-inlet (DI) instrument developed in the 1940s (McKinney *et al.*, 1950; Murphy, 1947), was the standard for high-precision measurements, requiring microgrammes (μg) of sample and 10–20 minutes for analysis. The continuous flow (CF) approach to isotope ratio monitoring was introduced in the late 1970s (Matthews and Hayes, 1978; Sano *et al.*, 1976), consisting of the coupling of a gas chromatograph (GC) with mass spectrometry (MS) via a combustion furnace, followed by the first high-precision system for ^{15}N (Preston and Owens, 1983) and ^{13}C (Preston and Owens, 1985) with an elemental analyser interfaced with IRMS in the mid-1980s. The first demonstration of high-precision compound-specific IRMS with a multicollector mass spectrometer utilized GC coupled with IRMS and also appeared in the mid-1980s (Barrie *et al.*, 1984). The success of these developments reduced sample sizes to 10 ng and analysis times to less than 1 minute. Both approaches became commonplace with commercial introduction around 1990. The instruments are now found in hundreds of laboratories worldwide in fields as diverse as ecology, geochemistry, and biomedicine.

The DI approach remains the technique of choice for highest precision analyses. However, the sample must be admitted to the instrument as a specific gas, CO_2, H_2, N_2, or SO_2, requiring conversion to one of these gases while maintaining isotopic representation of the original sample. Besides those which exist naturally in these gaseous

Stable Isotopes, edited by H. Griffiths.
© 1998 BIOS Scientific Publishers Ltd, Oxford.

forms, such as atmospheric CO_2, chemically complex samples such as whole plants or animals, and whole petroleum residues or fractions thereof were natural and popular subjects of research. Analysis of individual chemical compounds required cumbersome preparative separation from mixtures on a vacuum line followed by conversion to the appropriate gas. Uncontrolled isotopic fractionation and contamination is an ever present issue in these procedures. Rarely attempted was the even more cumbersome analysis of intramolecular isotope ratios, requiring isolation of individual positions in a molecule following chemical isolation, all in isotopically representative and uncontaminated form (Abelson and Hoering, 1961; DeNiro and Epstein, 1978).

The advent of GC-combustion-IRMS (GCC-IRMS) using online separation for automated compound-specific isotope analysis (CSIA), also served to reduce contamination dramatically and produce isotopic fractionation which can be characterized and accounted for with relative ease (Goodman and Brenna, 1994; Ricci et al.,1994). The GCC-IRMS technique, first developed for $^{13}C/^{12}C$ analysis (Barrie et al., 1984) was recently extended to $^{15}N/^{14}N$ (Merritt and Hayes, 1994; Preston and Slater, 1994). The technique has also been extended to liquid chromatography (LC) for carbon analysis (Caimi and Brenna, 1993; Caimi et al., 1994; Teffera et al., 1996) although commercial LC/IRMS systems have not yet appeared.

CF techniques have been applied to improve the sensitivity and speed of non-CSIA analysis other than elemental analyzers. For instance, the Pt equilibration technique, introduced by Horita et al. (1989) has been refined and incorporated into a CF system for HD/H_2 measurements and is commercially available (Prosser and Scrimgeour, 1995). Recent advances in automated trace concentrators have resulted in high-precision carbon isotope analysis of CH_4 in 100 ml of ambient air (Brand, 1995).

Perhaps the most challenging element for CF analysis has been hydrogen. Recent work has now demonstrated methods for HD/H_2 analysis, along with continuing challenges for routine implementation. In addition, the levels at which isotope ratios can be measured have progressed along with analytical instrumentation, first from bulk mixture analysis using the DI, to CSIA using GC or LC. The final step in this progression is automated position-specific isotopic analysis (PSIA), or the measurement of intramolecular isotope ratios (see Chapter 4 for analyses using NMR). Here we review the current state of the art for hydrogen analysis and describe in detail our work on CF-IRMS applied to organic hydrogen. We also discuss progress toward automated PSIA for C analysis.

1.2 CF-IRMS analysis of HD/H_2

1.2.1 Static and Dynamic Techniques for Dual-Inlet Analysis

Traditionally, static DI analysis is used to determine the hydrogen isotope ratio of pure hydrogen gas derived from water and organic compounds. The conversion of the analyte to isotopically representative hydrogen gas can be undertaken in one of many ways. In the chemical technique (Wong and Klein, 1986), an organic sample is combusted to CO_2 and H_2O using cupric oxide, either by a static method (Schimmelmann and DeNiro, 1993; Sofer, 1980) where the sample is oxidized in an evacuated sealed tube (Boutton et al., 1983) containing copper oxide at >700 °C, or by a dynamic method (Craig, 1953), where combustion is performed in a vacuum line in the presence of excess oxygen. Next, CO_2 is cryogenically separated from the water, and the water can then be reduced by zinc at >450 °C in a sealed tube (Coleman *et al.*, 1982) or dynamically reduced over hot uranium (Frazer, 1962) or zinc (Friedman, 1962). These techniques are laborious and time-consuming, although development of batch processing (Kendal and Coplen, 1985) for water analysis helps reduce the inconvenience of the methodology. Nevertheless, they often require large sample quantities for precise analysis and problems of isotopic fractionation and sample memory effects (Wong *et al.*, 1984) result from the difficulty in working with water vapour. The need for handling water vapour in a vacuum line can be eliminated by directly converting organic hydrogen to hydrogen gas using online pyrolysis at >900 °C (Sofer, 1986a,b).

The advent of chemical equilibration methods for water analysis (Coplen *et al.*, 1991; Horita *et al.*, 1989) using platinum catalysts reduce sample preparation requirements. The isotope ratio of the hydrogen gas obtained upon equilibration of water with hydrogen gas is very temperature sensitive (6‰/°C) over short time periods (<1 hour) in the original systems, although Pt reagents have been found which require up to 3 days (Scrimgeour *et al.*, 1993) but dramatically reduce this sensitivity. In addition, these methods produce hydrogen gas with a severe depletion in D/H ratio that makes precise measurement more problematic by reducing the abundance of the minor isotope (Scrimgeour *et al.*, 1993).

In the early 1980s, an online static system became commercially available that allowed the simultaneous determination of D/H and $^{18}O/^{16}O$ in aqueous samples (Wong *et al.*, 1984). This consisted of an on-line uranium reduction furnace preceding a dual mass spectrometer for HD/H_2 and $H_2^{18}O/H_2^{16}O$ detection. Unfortunately, this system suffered from severe memory effects and reportedly required a washout procedure consisting of a least 6 injections of a sample between analytes of different enrichment before accurate data could be obtained.

1.2.2 *Obstacles to CF-IRMS of hydrogen*

CF techniques for C, N, O, and S have been developed mostly due to the successful advancement of online sample chemistry via microreactors and traps. Although helpful for hydrogen isotopic analysis, obstacles to continuous monitoring of hydrogen using IRMS were significant and had to be addressed at the start.

First of all, the continuous-flow techniques benefit from the ability to couple GC with IRMS, allowing compound-specific analysis of complex mixtures. Such a system requires an inert carrier gas to sweep the analytes of interest into the mass spectrometer. The carrier of choice is pure He gas. For the analysis of hydrogen isotopes, one monitors m/z 2 for H_2 and m/z 3 for HD to measure the D/H ratio of a sample; however, He interferes with HD measurement. Standard lab grade He gas contains ~1 ppm ^{3}He isotope, although its small contribution can be corrected for by background subtraction. More importantly, the abundance sensitivity of conventional hydrogen IRMS is not sufficient to resolve HD from the tail of a considerable $^4He^+$ beam, resulting in a saturated m/z 3 detector. A logical step would be to choose another carrier, however obvious choices such Ne and Xe, are very expensive, and all have higher densities and low ionization potentials compared to He, thereby complicating analysis. For instance, Ar was tested as a potential carrier with some promising results (Tobias *et al.*, 1995). In the tight IRMS ion source, Ar has a high level of ionization and the stray unfocused beam sputters the metal of the ion lenses leading to electronic problems.

Two satisfactory resolutions to the carrier problem have been presented. Prosser and Scrimgeour (1995) introduced a novel mass spectrometer with an extra collector spur with a larger radius. This achieves a high dispersion of m/z 3 from m/z 4 yielding remarkably high abundance sensitivity, preventing interference from ^{4}He. An alternative approach developed at the same time by Tobias *et al.* (1995) is a Pd filter system (PFS) in which carrier does not enter the IRMS ion source. A palladium membrane held at elevated temperatures is used to selectively admit only hydrogen and deuterium to pass into the mass spectrometer while diverting the carrier gas to waste. This membrane consists of pure palladium metal or a palladium alloy for more durability and permeability, such as Pd, 25 wt% Ag. A vanadium alloy, with a thin plate of Pd on both sides, can also be used to facilitate even more permeability (Tobias and Brenna, 1996). Both these systems permit the use of He as a carrier gas for continuous-flow hydrogen isotope analysis.

Another consideration for hydrogen analysis is the false contributions to the m/z 3 signal due to effects in the tight IRMS ion source that change with varying analyte quantity. This is important because in most cases CF analyte signals change continuously, in contrast to dual-inlet analysis where sample and standard pressures are always equalized. First, pressure-dependent mass selection (Kirshenbaum, 1951) is known to produce isotope ratios that depend on source pressure with selectivity toward the ions with higher mass depending on the interaction of the fields from the collimating magnets on ion acceleration. Second, a much greater effect well known for hydrogen analysis, is the production of H_3^+ in an ion-molecule reaction. The intensity of H_3^+ depends on hydrogen–hydrogen collision frequencies and it is not mass-resolved from HD in commercial IRMS instruments, including the high-dispersion variety. Isotope ratios are therefore measured from continuously varying analyte levels that also have correspondingly varying contaminant levels. These effects have been evaluated and correction procedures have been proposed for over a 2-fold signal intensity range by Prosser and Scrimgeour (1995) and a much larger range, up to 22-fold, by Tobias *et al.* (1995). Ultimately, this is important for chromatographic analysis, where peak intensities usually cannot be matched and vary over a large range. Tobias *et al.* (1995) demonstrated that linear algorithms over a small signal range may be adequate. However, over larger ranges these effects become nonlinear and require higher order equations for unbiased correction. This nonlinear response is in part due to specific instrument tuning parameters and unmatched m/z 2 and 3 detector responses. The corrections presented thus far are of a 'peakwise' nature where characterization and correction of ion source effects are conducted after hydrogen peak detection and isotope ratio calculation, that is the whole peak is corrected at once. Ultimately, a 'pointwise' correction would be more desirable where characterization and correction of ion source effects are conducted before hydrogen peak detection and isotope ratio calculation. This goal is complicated by a slower m/z 3 detector response. As a result, a continually changing H_3^+ contribution, as a peak rises and falls, cannot be accurately represented by analyte intensity at each point. A method using a decay rate deconvolution is currently under development in our lab and early data show some promising results (Tobias and Brenna, 1996a).

1.2.3. *Compound specific isotope analysis for deuterium via CF-IRMS*

Direct hydrogen gas analysis using CF-IRMS has been demonstrated via the PFS, using pure palladium (Tobias *et al.*, 1995), with precisions of SD(δD) <6‰ and the high-dispersion instrument (Prosser and Scrimgeour, 1995) with precisions of ±4‰ using water equilibrated hydrogen. The first online water reduction system for hydrogen isotope analysis via CF-IRMS was demonstrated by Tobias *et al.*, (1995). The interface consists of a reduction microreactor, filled with pure Ni metal held at >850 °C, for conversion of water to hydrogen gas, and the sample size was 100–400 nl (Tobias and Brenna, 1996b). Due to non-quantitative conversion, a water trap was added in-line after the reactor and before the Pd filter assembly to remove any residual unreduced water vapour that degrades analytical performance. No memory effects were observed over a water enrichment range of –55 to 4700 ‰ (δD_{SMOW}). This online reduction scheme was later adapted by Begley and Scrimgeour (1996) to facilitate direct water analysis using the high-dispersion instrument. In addition, they used nickelized carbon for water reduction to facilitate analysis of $^{18}O/^{16}O$ from CO using the same instrumental setup, making it convenient for analysis of isotopes of O and H from water samples, for the doubly-labelled water technique.

Analysis for hydrogen isotope ratios of individual organic compounds from a mixture of organics was first demonstrated by Tobias *et al.* (1995) using the PFS. This system consists of a GC interfaced with a modified Pd filter assembly (Pd/Ag alloy) via a combustion reactor filled with CuO held at 850 °C, and a reduction reactor filled with Ni metal held at 950 °C. This system facilitates online organic mixture analysis subsequent to chromatographic separation with precisions of SD (δD) <5‰. It was evaluated using benzene as an internal standard in a mixture of ethyl benzene and cyclohexanone in hexane solvent with analyte measured at quantities of <3 ng (<300 pg H).

Long capillary columns are often required when analysing complex mixtures of organics; the resulting low flow rate and pressure at the Pd filter results in less H transport into the vacuum system. A pressure unit was inserted between the water trap and the Pd filter to increase the pressure with which the carrier and analyte impinge upon the palladium surface. This facilitates a dramatic increase in sensitivity. No memory is observed between peaks of differently enriched analytes within the same chromatogram. However, a small dependence of isotope ratio on palladium membrane temperature is demonstrated over a range of 4 °C. Figure 1.1 demonstrates

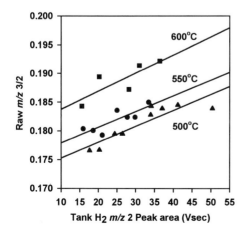

Figure 1.1. Continuous-flow peaks of tank hydrogen gas measured via the PFS, using a V/Nb alloy coated with Pd metal, held at 500, 550 and 600 °C. The gas was analysed over a range of injection quantities. Data acquired using a Finnigan MAT delta S IRMS.

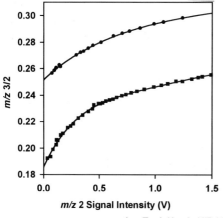

Figure 1.2. Static measurement of tank H_2 gas measured over a range of m/z 2 signal intensities. The gas is measured either directly from a dual-inlet bellows, or through a V/Nb/Pd membrane at ~550 °C. Data acquired using a Finnigan MAT delta S IRMS. The membrane admits gas that is significantly enriched in D compared to direct admission across the crimp.

the temperature effect in a V/Nb membrane coated with Pd. With increasing temperature, more deuterium is detected in relation to protium, resulting in an apparent increase in enrichment. These data indicate that tighter control of palladium temperature would facilitate improved precision.

Tank H_2 is more enriched in deuterium when measured as a static signal through the V/Nb/Pd filter than when measured directly via the DI bellows, as shown in Figure 1.2, indicating D transport through the foil is more efficient than protium. This effect is easily calibrated and can be beneficial for equilibration techniques, where the severe depletion of the hydrogen gas analyte reduces the abundance of the minor isotope resulting in lower counts and further limiting precision (Prosser and Scrimgeour, 1995; Scrimgeour *et al.*,1993). The PFS offers the advantage that co-eluting contaminants do not enter the vacuum system along with analyte hydrogen. The importance of this feature has been discussed elsewhere (Kirshenbaum, 1951). Figure 1.3 demonstrates that a

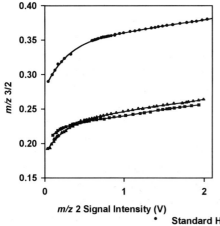

Figure 1.3. Statically measured standard H_2, tank H_2, and tank H_2 contaminated with N_2 gas at different m/z 2 signal intensities. Data acquired using a Finnigan MAT delta S IRMS. Levels of the linear sections of the curves are different, indicating significant degradation of accuracy when N_2 is admitted along with analyte H.

co-admitted contaminant, N_2, produces statistically difference tank H_2 isotope ratio than for pure tank H_2. This can be important in organic analysis using chromatography, where different compounds will be reduced to hydrogen and different gas contaminants. For example, nitrogen-containing compounds (e.g. amino acids) will be converted to H_2, CO, and N_2, while hydrocarbons will be reduced only to H_2 and CO. Similar effects might be expected with other common gases, such as CO.

In summary, CF-IRMS of hydrogen using current technologies analysed with well characterised, well-behaved mass spectrometers facilitates an efficient alternative to DI analysis of hydrogen isotope ratios from water and organic compounds. For the latter, it will be most useful for circumstances where isotope ratios are sought from a limited number of compounds for which peak size can be controlled.

1.3 Carbon isotopes: automated position-specific isotope analysis (PSIA)

Position-specific measurements for analysis of natural variability have been published but are generally limited only to the most important problems because of laborious chemical degradation and separation steps required prior to isotope ratio mass spectrometry (IRMS) analysis. The earliest position-specific work is that of Abelson and Hoering (1961), who isolated the carboxyl position manually via the ninhydrin reaction, and showed it to be enriched relative to the rest of the molecule. They further showed indirect evidence that glycolysis proceeds without isotope discrimination. Almost a decade passed before further work appeared. A classic paper on isotope fractionation established the precise enzymatic step at which acetate carbon is fractionated, resulting in the well known low enrichment of ^{13}C in lipids (DeNiro and Epstein, 1977). Pyruvate oxidation to acetate was accomplished by tedious procedures to isolate the carboxyl and methyl carbons of acetate for conventional bulk analysis. Results showed that the isotopic composition of these two positions were very different and that a kinetic isotope effect that discriminates against ^{13}C at the acetate carbon is the depleted site. These results predict alternating ^{13}C abundances in fatty acids, which have been verified (Melzer and Schmidt, 1987; Monson and Hayes, 1982; see Chapter 2). In related work, Hayes and coworkers (Vogler and Hayes, 1990) analysed the ^{13}C content of the carboxyl and olefinic C in monounsaturated fatty acids from microorganisms (*E. coli* and *S. cerevisiae*) and refined models of fatty acid metabolism based on their position-specific isotopic results (Monson and Hayes, 1980; Monson and Hayes, 1982). These studies required Schmidt decarboxylation and oxidative ozonolysis to isolate specific carbon positions prior to high-precision determination. A representative review of work up until 1985 can be found (Galimov, 1985) including a discussion of Galimov theory of biological isotopic fractionation. More recently, Ivlev and coworkers have discussed the relationship between isotopic fractionation in photosynthetic cells and the order of amino acid synthesis (Ivlev, 1986). In total, the body of scientific work on the intramolecular distribution of carbon isotopes clearly indicates that they are indicative of a variety of important physiological processes, and would be routinely analysed if an instrument were available to do so conveniently. Such an instrument would also facilitate a wide variety of tracer experiments which are presently accomplished with high enrichments and analysed using either NMR or organic MS. Presently, high-precision position-specific measurements are at a similar stage as was CSIA prior to the mid-1970s, requiring manual isolation of carbon positions within a molecule prior to analysis, which is labour intensive and practically impossible for determining many internal positions (Ivlev, 1991).

The bulk of recent work on pyrolysis-GC has applied the analytical capabilities of GC to non-volatile, usually intractable, solids, such as high molecular weight polymer resins, keragens (Eglinton, 1994), and bacteria. Previously, several groups reported vapour phase pyrolysis of small organic molecules for qualitative analysis as a possible alternative for GC-MS-based analysis (Dhont, 1961, 1963; Walker and Wolf, 1968). These early studies showed that organic molecules of biological and commercial interest fragmented in a useful and characteristic way, and did so according to predictions of simple free radical fragmentation mechanisms worked out decades earlier for hydrocarbons (Fabuss et al., 1964; Kossiakoff and Rice, 1943). Although pyrolysis-GC systems never overshadowed GC-MS, they suggest a convenient and powerful means for fragmenting organic molecules (see Chapter 22), prior to separation and analysis by IRMS, thereby providing an online strategy for determination of isotope ratios from groups or individual positions within molecules.

Our strategy for PSIA is to (1) fragment molecules pyrolytically; (2) measure fragment isotope ratios, and (3) calculate isotope ratios for specific positions and/or for moieties. Although there is considerable work involving the generation of isotope ratios from pyrolysis products (Mycke et al., 1994), there are no data on the prospects for generating reproducible isotope ratios from on-line pyrolysis in a microreactor. We have constructed test systems to evaluate this question for pentane and methyl palmitate (Me16: 0). The system for pentane is shown in Figure 1.4.

About ten µl of headspace pentane (<300 ng), was injected into the pyrolysis chamber at temperatures ranging from 100 °C to 500 °C. Pyrograms and results are shown in Figure 1.5 and Table 1.1. Below about 380 °C, no fragmentation is observed. At 420 °C, partial pyrolysis is observed and products are strong reproducible peaks, as shown in Figure 1.5(a). Figure 1.5(b) shows results at 480 °C, where pentane is almost quantitatively pyrolyzed and the products are the major peaks in the spectrum. Assuming no rearrangements, the four fragments correspond to C-1, C-2, C-3, and C-4, eluting in that order.

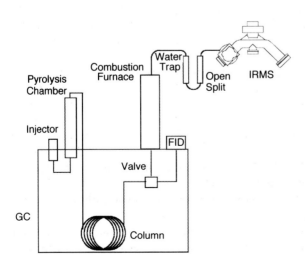

Figure 1.4. *Diagram of the pyrolysis-GCC-IRMS/ prototype system used to evaluate precision and performance of pentane fragmentation. Sample is injected into a ceramic furnace for pyrolysis, fragments separated by GC, and analysed by IRMS for C isotope ratio.*

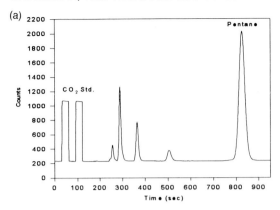

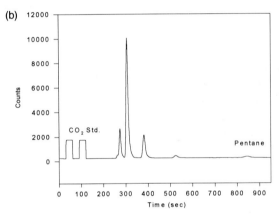

Figure 1.5. Results of pentane pyrolysis at two temperatures. (a) 420 °C, where partial pyrolysis into four fragments, C_{1-4}, is observed, (b) 480 °C where pyrolysis is near complete.

Table 1.1 presents the results of replicate analyses of the highly reproducible pyrolysis fragment peaks at 420 °C. These data show online pyrolysis to be highly reproducible and the subsequent chromatography of sufficient quality for IRMS. The average precision is SD(δ^{13}C) <1‰ and is therefore sufficiently reproducible to follow isotope ratio changes relevant in nature, even with this preliminary system. The mean isotope ratios are statistically different, indicating that the system is sensitive to changes in isotope ratio. Methane is often observed to be the isotopically lightest hydrocarbon in nature. Also, progressively larger hydrocarbon chains increase in ^{13}C content until about pentane, (see Chapter 2).

Since online pyrolysis is reproducible, we constructed a second system, not shown, similar to that of Levy and Paul (1967) in a tandem GC configuration. The system includes an organic MS for structure analysis of pyrolysis products to verify structures predicted by theory and to perform labelling experiments to rule out rearrangement. Me16: 0 was purified in a first GC, fragmented pyrolytically, fragments separated in the second, and analysed by either an ion trap MS, or an IRMS for high-precision isotope ratio analysis. Pyrolysis of Me16: 0 yields two series of fragments differing by 1 C in length up to the C_{17} parent. Data from labelling studies using [1-^{13}C], [Me-^{13}C], [16-^{13}C], and [Me, 16-^{13}C]-Me16: 0 indicate no rearrangement of carbon due to the pyrolysis process. The difference in ^{13}C atom fraction between consecutive fragments corresponds to the ^{13}C atom fraction of the additional carbon position in the larger molecule, and can be calculated as a weighted difference between the two fragments.

Table 1.1. Precision of carbon isotope analysis for pentane pyrolysis fragments

Run	C_1	C_2	C_3	C_4	Pentane
1	−36.15	−31.39	−29.02	−28.15	−28.32
2	−35.38	−30.15	−28.08	−26.18	−27.07
3	−35.34	−29.99	−27.7	−25.25	−27.01
4	−34.38	−29.77	−26.67	−26.67	−27.25
Mean	−35.31	−30.33	−27.87	−26.56	−27.41
S.D.	0.72	0.73	0.97	1.21	0.61

Data obtained for fragmentation at differing temperatures showed pyrolysis-induced fractionation to be nearly insensitive to temperature and that pyrolysis conducted at a single temperature results in fragments with isotope ratios that are readily calibrated against standards. Results thus far indicate that controlled pyrolysis can be conducted on-line and produce single bond breakage, stabilization, and isotopically representative fragments of carbon. The absence of measurable rearrangement or major fractionation shows the approach to be applicable to position-specific studies of carbon at natural abundance or very low enrichment.

References

Abelson, P.H. and Hoering, T.C. (1961) Carbon isotope fractionation in formation of amino acids by photosynthetic organisms. *Proc. Natl. Acad. Sci.* **47**, 623–632.

Barrie, A., Bricout, J. and Koziet, J. (1984) Gas chromatography-stable isotope ratio analysis at natural abundance levels. *Biomed. Mass Spectrom.* **11**, 583–588.

Begley, I.S. and Scrimgeour, C.M. (1996) On-line Reduction of H_2O for delta-^2H and delta-^{18}O Measurement by Continuous-flow Isotope Ration Mass Spectometry. *Rap. Comm. Mass. Spectrom.* **10**, 969–973.

Boutton, T.W., Wong, W.W,. Hachey, D.L., Lee, S.L., Cabrera, M.P. and Klein, P.D. (1983) Comparison of Quartz and Pyrex Tubes for Combustion of Organic Samples for Stable Carbon Isotope Analysis. *Anal. Chem.* **55**, 1832–1833.

Brand, W.A. (1995) Precon: A fully automated interface for the pre-GC concentration of trace gases in air for isotopic analysis. *Isotopes Environ. Health Stud.* **31**, 277–284.

Caimi, R.J. and Brenna, J.T. (1993) High Precision Liquid Chromatography-Combustion Isotope Ratio Mass Spectrometry (LCC-IRMS). *Anal. Chem.* **65**, 3497–3500.

Caimi, R.J., Houghton, L.H. and Brenna, J.T. (1994). Condensed phase standards for high precision compound specific isotope analysis. *Anal. Chem.* **66**, 2989–2991.

Coleman, M.L., Sheperd, T.J., Durham, J.J., Rouse, J.E. and Moore, G.R. (1982) Reduction of Water with Zinc for Hydrogen Isotope Analysis. *Anal. Chem.* **54**, 993–995.

Coplen, T.B., Wildman, J.D. and Chen, J. (1991) Improvements in the Gaseous Hydrogen-Water Equilibration Technique for Hydrogen Isotope Ratio Analysis. *Anal. Chem.* **63**, 910–912.

Craig, H. (1953) The geochemistry of the stable carbon isotopes. *Geochim. Cosmochim. Acta.* **3**, 53–92.

DeNiro, M.J. and Epstein, S. (1977) Mechanism of Carbon Isotope Fractionation Associated with Lipid Synthesis. *Science.* **197**, 261–263.

DeNiro, M.J. and Epstein, S. (1978) Influence of diet on the distribution of carbon isotopes in animals. *Geochim. Cosmochim. Acta.* **42**, 485–506.

Dhont, J.H. (1961) Pyrolysis and gas chromatography for the detection of the benzene ring in organic compounds. *Nature.* **192**, 747–748.

Dhont, J.H. (1963) Identification of organic compounds from food odours by pyrolysis. *Nature* **200**, 882.

Eglinton, T.I. (1994) Carbon isotopic evidence for the origin of macromolecular aliphatic structures in keragens. *Org. Geochem.* **21**, 721–735.

Fabuss, B.M., Smith, J.O. and Satterfield, C.N. (1964) Thermal cracking of pure saturated hydrocarbons, in *Advances in Petroleum Chemistry and Refining,* (ed J. J. McKetta), Wiley and Sons, NY, pp. 157–201.

Frazer, J.W. (1962) Simultaneous Determination of Carbon, Hydrogen and Nitrogen. Part II. *Mikrochim. Acta.* **6**, 993–999.

Friedman, I. (1953) Deuterium content of natural waters and other substances. *Geochim. Cosmochim. Acta.* **4**, 89–103.

Galimov, E.M. (1985) *The Biological Fractionation of Isotopes.* Academic Press, New York.

Goodman, K.J., and Brenna, J.T. (1994) Curve-fitting for resolution of overlapping peaks in high precision gas chromatography-combustion isotope ratio mass spectrometry. *Anal. Chem.* **66**, 1294–1301.

Horita, J., Ueda, A., Mizukami, K. and Takatori, I. (1989) Automatic D and ^{18}O analysis of multi-water samples using H_2-and CO_2 water equilibration methods with common equilibration methods with a common equilibration set-up. *Appl. Radiat. Isot.* **40**, 801–805.

Ivlev, A.A. (1986) Distribution of carbon isotopes in amino acids of protein fraction of microorganisms as a means of studying the mechanisms of their biosynthesis in the cell. *Biokhimiya* (English Translation) **50**, 1605–1615.

Ivlev, A.A. (1991) Distribution of carbon isotopes ($^{13}C/^{12}C$) in the cell and the temporal organization of cellular processes. *Biophysics* **36**, 1078–1087.

Kendall, C. and Coplen, T.B. (1985) Multisample Conversion of Water to Hydrogen by Zinc for Stable Isotope Determination. *Anal. Chem.* **57**, 1437–1440.

Kirshenbaum, I. (1951) *Physical Properties and Analysis of Heavy Water.* National Nuclear Energy Series, (ed H.C. Urey and G.M. Murphy), Vol. 4A, McGraw-Hill, New York.

Kossiakoff, A. and Rice, F.O. (1943) Thermal decomposition of hydrocarbons, resonance stabilization and isomerization of free radicals. *J. Amer. Chem. Soc.* **65**, 590–595.

Levy, E.J. and Paul, D.G. (1967) The application of controlled gas phase thermolytic dissociation to the identification of gas chromatographic effluents. *J. Gas Chromatogr.* **5**, 136–145.

Matthews, D.E. and Hayes, J.M. (1978) Isotope-ratio-monitoring gas chromatography-mass spectrometry. *Anal. Chem.* **50**, 1465–1473.

McKinney, C.R., McCrea, J.M., Epstein, S., Allen, H.A. and Urey, H.C. (1950) Improvements in mass spectrometers for the measurement of small differences in isotope abundance ratios. *Rev Sci Instrum.* **21**, 724–730.

Melzer, E. and Schmidt, H.-L. (1987) Carbon Isotope Effects on the Pyruvate Dehydrogenase Reaction and Their Importance for Relative Carbon-13 Depletion in Lipids. *J. Biol. Chem.* **262**, 8159–8164.

Merritt, D.A. and Hayes, J.M. (1994) Nitrogen isotopic analysis by isotope-ratio-monitoring gas chromatography/mass spectrometry. *J. Am. Soc. Mass Spectrom.* **5**. 387–397.

Monson, K.D. and Hayes, J.M. (1980) Biosynthetic control of the natural abundance of carbon 13 at specific positions within fatty acids in *Escherichia coli. J. Biol. Chem.* **255**, 11435–11441.

Monson, K.D. and Hayes, J.M. (1982a) Biosynthetic control of the natural abundance of carbon 13 at specific positions within fatty acids in *Saccharomyces cerevisiae. J. Biol. Chem.* **257**, 5568–5575.

Monson, K.D. and Hayes, J.M. (1982b) Carbon isotopic fractionation in the biosynthesis of bacterial fatty acids. Ozonolysis of unsaturated fatty acids as a means of determining the intramolecular distribution of carbon isotopes. *Geochim. Cosmochim. Acta.* **46**, 139–149.

Murphy, B.F. (1947) The temperature variation of the thermal diffusion factors for binary mixtures of hydrogen, deuterium and helium. *Phys Rev.* **72**, 834–837.

Mycke, B., Hall, K. and Leplat, P. (1994) Carbon isotopic composition of individual hydrocarbons and associated gases evolved from microscaled sealed vessel (MSSV) pyrolysis of high molecular weight organic material. *Org. Geochem.* **21**, 787–800.

Preston, T. and Slater, C. (1994) Mass spectrometric analysis of stable-isotope-labelled amino acid tracers. *Proc. Nutr. Soc.* **53**, 363–372.

Preston, T. and Owens, N.J.P. (1983) Interfacing and automatic elemental analyser with an isotope ratio mass spectrometer: the potential for fully automated total nitrogen and nitrogen-15 analysis. *Analyst* **108**, 971–977.

Preston, T. and Owens, N.J.P. (1985) Preliminary ^{13}C measurements using a gas chromatograph interfaced to an isotope ratio mass spectrometer. *Biomed. Mass Spectrom.* **12**, 510–513.

Prosser, S.J. and Scrimgeour, C.M. (1995) High-precision determination of ^{2}H/^{1}H in H_2 and H_2O by continuous-flow isotope ratio mass spectrometry. *Anal. Chem.* **67**, 1992–1997.

Ricci, M.P., Merritt, D.A., Freeman, K.H. and Hayes, J.M. (1994) Acquisition and processing of data for isotope-ratio-monitoring mass spectrometry. *Org. Geochem.* **21**, 561–571.

Sano, M., Yotsui, Y., Abe, H. and Sasaki, S. (1976) A new technique for the detection of metabolites labelled by the isotope 13C using mass fragmentography. *Biomed. Mass Spectrom.* **3**, 1–3.

Schimmelmann, A. and DeNiro, M.J. (1993) Preparation of organic and water hydrogen for stable isotope analysis: Effects due to reaction vessel and zinc reagent. *Anal. Chem.* **65**, 789–792.

Scrimgeour, C.M., Rollo, M.M., Mudambo, S.M.K.T,. Handley, L.L. and Prosser, S.J. (1993) A simplified method for deuterium/hydrogen isotope ratio measurements on water samples of biological origin. *Biol. Mass Spec.* **22**, 383–387.

Sofer, Z. (1980) Preparation of carbon dioxide for stable carbon isotope analysis of petroleum fractions. *Anal. Chem.* **52**, 1389–1931.

Sofer, Z. (1986a) Chemistry of hydrogen gas preparation by pyrolysis for the measurement of isotope ratios in hydrocarbons. *Anal. Chem.* **58**, 2029–2032.

Sofer, Z., and Schiefelbein, C.F. (1986b) Hydrogen isotope ratio determinations in hydrocarbons using the pyrolysis preparation technique. *Anal. Chem.* **58**, 2033–2036.

Teffera, Y., Kusmierz, J.J. and Abramson, F.P. (1996) Continuous-flow isotope ratio mass spectrometry using the chemical reaction interface with either gas or liquid chromatographic introduction. *Anal. Chem.* **68**, 1888–1894.

Tobias, H.J. and Brenna, J.T. (1996a) Correction of ion source nonlinearities over a wide signal range in continuous-flow isotope ratio mass spectrometry of water-derived hydrogen. *Anal. Chem.* **68**, 2281–2286.

Tobias, H.J. and Brenna, J.T. (1996b) High-precision D/H measurement from organic mixtures by gas chromatography continuous flow isotope ratio mass spectrometry using a palladium filter. *Anal. Chem.* **68**, 3002–3007.

Tobias, H.J., Goodman, K.J., Blacken, C.E. and Brenna, J.T. (1995) Determination of D/H ratios from H_2 and H_2O by continuous flow isotope ratio mass spectrometry. *Anal. Chem.* **67**, 2486–2492.

Vogler, E.A. and Hayes, J.M. (1980) Carbon isotopic compositions of carboxyl groups of biosynthesized fatty acids, in *Physics and Chemistry of the Earth*, (eds A.G. Douglas and J.R. Maxwell), Pergamon, London, pp. 697–704.

Walker, J.Q. and Wolf, C.J. (1968) Complete identification of chromatographic effluents using interrupted elution and pyrolysis-gas chromatography. *Anal. Chem.* **40**, 711–714.

Wong, W.W. and Klein, P.D. (1986) A review of techniques for the preparation of biological samples for mass-spectrometric measurements of hydrogen-2/hydrogen-1 and oxygen-18/oxygen-16 isotope ratios. *Mass Spec. Rev.* **5**, 313–342.

Wong, W.W., Cabrera, M.P. and Klein, P.D. (1984) Evaluation of a dual mass spectrometer system for rapid simultaneous determination of hydrogen-2/hydrogen-1 and oxygen-18/oxygen-16 ratios in aqueous samples. *Anal. Chem.* **56**, 1852–1858.

Carbon isotope effects on key reactions in plant metabolism and ^{13}C-patterns in natural compounds

H.-L. Schmidt and G. Gleixner

2.1 Introduction

In 1961 Abelson and Hoering detected that the carboxylic group of many amino acids is enriched in ^{13}C relative to their global ^{13}C-content. Meinschein *et al.* (1974) and Rinaldi *et al.* (1974) found a non-statistical ^{13}C-distribution in acetic acid and acetoin, and the latter group postulated and found a relative alternating ^{13}C-abundance in the hydrocarbon chain of some fatty acids which they attributed to an isotope effect on the pyruvate dehydrogenase reaction (Monson and Hayes, 1982). While Ivlev *et al.* (1987) elucidated the partial pattern of natural glucose, we elaborated the complete pattern of this primary compound (Roßmann *et al.*, 1991). This was the initial impact for a systematic study combining results of *in-vitro* measurements of kinetic isotope effects and pattern elucidations of natural products (Schmidt *et al.*, 1995).

For the development of a general interpretation and theory of isotopic discriminations in a closed system like a plant, besides the above-mentioned involvement of isotope effects, the existence of pools and metabolic fluxes in different directions has to be taken into account. This was expressed by S. Epstein in 1968 by the statement: 'If the carbon of lipids is light, conservation laws require that some other portion of carbon is heavier. Or stated in another way, a change in isotopic ratios implies the splitting of a population of molecules into two or more classes and streams.'

A first attempt to find a general explanation and principle for intermolecular and intramolecular non-statistical isotope distributions was performed by E. Galimov. In his book on 'The Biological Fractionation of Isotopes', he stressed mainly thermodynamic reasons (β-factors) for the explanation of the patterns known at that time. Since then, several *in-vitro* kinetic isotope effects on enzyme catalysed reactions have been determined, some by our group, and the patterns of several natural products have

Stable Isotopes, edited by H. Griffiths.
© 1998 BIOS Scientific Publishers Ltd, Oxford.

been elucidated. The philosophy of our work is to correlate the ^{13}C-enrichments or -
depletions in positions of natural products to influences of kinetic isotope effects in
the course of their biosynthesis, or vice versa, to deduce kinetic isotope effects on
given biosynthetic reaction sequences from non-statistical ^{13}C-distributions in natural
products. It is quite clear that the corresponding correlations can only become evident
in a closed system like a plant, when we presume the existence of several pools of
intermediates connected to a branching of metabolic pathways and fluxes of metabolites, involving isotope effects in the various directions (Figure 2.1).

The first parameter influencing the isotopic composition of organic matter is the isotope abundance of the primary source materials, CO_2, H_2O and N_2, which occur in infinite pools with consistent δ-values. However, depending on local conditions and plant
type, carbon may be assimilated as CO_2 or HCO_3^-, having different $\delta^{13}C$-values; the leaf
water, origin of organic hydrogen and oxygen, may have different [2]H(D)- and ^{18}O-con-
tents, and apart from N_2, other nitrogen compounds may be primary N-sources. The
detailed variations in these source isotope compositions are discussed subsequently: see
Chapters 8, 19 (^{13}C); Chapters 10, 12 (^{18}O); Chapters 4, 11 (D) and Chapters 5, 6 and 7
(^{15}N). Kinetic isotope effects accompanying the photosynthetic assimilations are well-
known to determine the δ-values of the primary compounds and of the bulk of the biomass, however, various isotope effects are also implied in the formation of secondary
products. Deposition of polymers eliminates some of the organic matter from the continuous turnover, and partial or final degradations contribute to local formations of new
starting material with altered isotopic abundances, which may also be recycled. In general,
secondary products are depleted in ^{13}C relative to primary compounds (carbohydrates),
and catabolic reactions prefer the 'light' molecules, while the 'heavy' ones are involved in

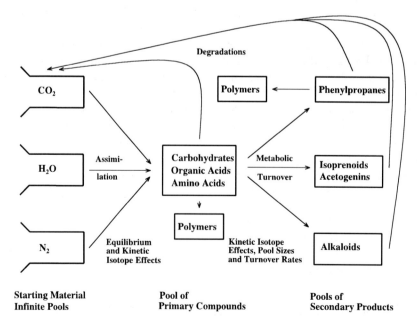

Figure 2.1. Control and locations of isotopic discrimination in biological systems, indicating
main pools and metabolic fluxes.

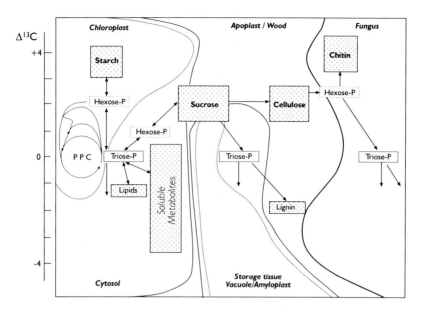

Figure 2.2. Δ¹³C-values [‰ vs. PDB] of metabolites in different compartments of plant cells and fungi (adapted from Gleixner et al., 1993). PPC = Pentose phosphate cycle

biomass formation, e.g. polymerization. As an example, dealing with the formation and the degradation of biomass we have studied the isotopic correlations in the system wood (cellulose, lignin) and chitin of decomposing fungi, which confirmed the generally accepted results mentioned before (Figure 2.2, Gleixner *et al.*, 1993). Cellulose as a polymer of primary products (carbohydrates) is ¹³C-enriched relative to lignin, a secondary product. However, chitin, the main polymer of the fungal cell walls, is even 'heavier' than the cellulose. Probably the fungi metabolize the 'lighter' intermediates for energy production and polymerize the 'heavier' ones. This is obviously a general principle which may also predominate in the further formation of fossil material.

Entering more into the biochemical reasons for isotopic patterns of secondary products, it is obvious that they are linked to kinetic isotope effects on defined enzymatic reactions. The relative location of these reactions in the secondary metabolism and their corresponding metabolite pools are displayed in Figure 2.3, and the enzymes catalysing these main reactions with kinetic isotope effects identified so far are compiled in Table 2.1. The catalysed reactions are mostly of the lyase type, forming or breaking C–C bonds, however, even those reactions belonging to other groups in the enzyme classification, involve corresponding steps. In the following the effects of these reactions are treated, not necessarily always according to the reaction type, but also taking into account their position and importance in the biosynthesis of secondary products (see Figure 2.3).

2.2 Influence of carboxylations on the initial ¹³C-distribution pattern

The reason for the typical ¹³C-depletion in C_3- and C_4-plants, relative to the bound CO_2 and HCO_3^-, are kinetic isotope effects on ribulose-1,5-biphosphate carboxylase

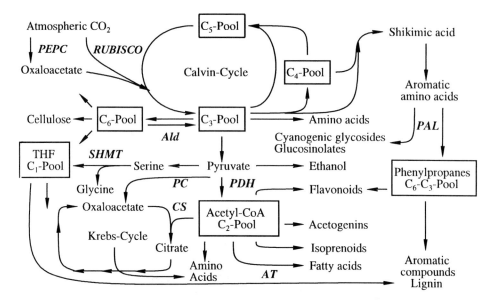

Figure 2.3. *Sites of carbon isotope fractionation and metabolite pattern formation in plant metabolism. Key reactions with isotope effects are associated with metabolic branching. Abbreviations for enzymes (PEPC, Phosphoenol pyruvate carboxylase; RUBISCO, Ribulose bisphosphate carboxylase-oxygenase; PAL, Phenylalanine ammonia lyase; PDH, Pyruvate dehydrogenase; CS, Citrate synthase; PC, Pyruvate carboxylase; SHMT, Serine hydroxymethyl transferase; Ald, Aldolase; AT, Acetyl-CoA-acetyl transferase)*

oxygenase (RUBISCO) and the phophoenol pyruvate carboxylase (PEPC) reactions, respectively (Farquhar *et al.*, 1982; Winkler *et al.*, 1982). These reactions are discussed in more detail in Chapters 8 and 9, together with the variations in ^{13}C-abundances of bulk-biomass, CO_2 diffusion processes and photorespiration. Kinetic isotope effects on reactions during secondary metabolism, implying isotope discriminations, may not severely influence the bulk $\delta^{13}C$-value of the biomass, but cause differences between groups of compounds and, more importantly, between positions in these compounds (intermolecular and intramolecular patterns). As already mentioned, secondary products in a plant (aromatics, proteins, isoprenoids) are depleted by 3–6‰ relative to primary compounds (carbohydrates) from the same source. As the kinetic isotope effects on the reactions responsible are associated with metabolic branching, the extent of the observed discriminations is always dependent on pool-sizes of intermediates and on metabolic fluxes in different directions, leading to depletions in one direction and enrichments in the other (isotopic balance, see Epstein, 1968). A carboxylation preferring 'heavy' CO_2 is the pyruvate carboxylase reaction, leading to C_4-dicarboxylic acids with enriched carboxyl groups (Melzer and O'Leary, 1987); this should be remembered in the context of the pattern associated with these intermediates.

2.3 Influence of aldol reactions and decarboxylations on secondary distribution patterns

There are various reactions in the Calvin cycle which could imply isotopic effects, for

Table 2.1. Enzymatic reactions involved in isotopic discriminations in plant metabolism and consequences of isotope effects on the catalysed reactions (for abbreviations see Figure 2.3)

Reaction with identified kinetic isotope effect	Enzyme (trivial name) and EC-No.	Consequence on product pattern or δ^{13}C-value
Ru1,5-BP+CO_2→2 PGA	RUBISCO [4.1.1.39]	General depletion of C_3 plants relative to ambient CO_2
CH_2=CHP-COOH+ HCO_3^-→HOOC-CH_2-CO-COOH+P	PEPC [4.1.1.31]	General depletion in C_4 plants, relative enrichment in position C-4
H_3C-CO-COOH+CO_2 →HOOC-CH_2-CO-COOH	Pyruvate-carboxylase, PC [6.4.1.1]	Relative enrichment in position C-4 of dicarboxylic acids
Fru-1,6-BP↔GAP+ DHAP	Aldolase, Ald [4.1.2.1.3]	Relative enrichment in pos. C-3 and C-4 of carbohydrates
H_3C-CO-COOH+ HSCoA→H_3C-CO-SCoA+CO_2+2 [H]	Pyruvate-dehydrogenase, PHD [1.2.4.1]	Relative depletion in CO-group of acetyl-CoA, general depletion of fatty acids and isoprenoids, alternating abundance
Oxaloacetate+Acetyl -CoA→Citrate+CoA	Citrate-synthase, CS [4.1.3.7]	Depletion methylene groups, general effect on corresponding groups by other aldol reactions
2 AcetylCoA→ AcetoacetylCoA+HSCoA	AcetylCoA-acetyl transferase, AT [2.3.1.9]	Positional ^{13}C-abundance in fatty acids and isoprenoids
Phenylalanine→Cinnamic acid+NH_3	Phenylalanine-ammonia-lyase, PAL [4.3.1.5]	Relative depletion in position 2 of side chain of phenylpropanes, enrichment in cyanogenic glucosides
Serine+THF↔Glycine +THF-CH_2O	Serine-aldolase, SHMT [2.1.2.1]	Formation C1-pool, discrimination on C-3
R-CR^1O+R^2-CHO→ R-CR^1HO-CR^2O	Acyloin-condensation catalysing [4.1.n.n]	Enrichment of the C-atom of the transferred active aldehyde

example the aldolase -, transaldolase - and transketolase reactions (Roßmann *et al.*, 1991). Although the turnover and the direction of these reactions depend on the actual metabolic situation of the plant (assimilation or dissimilation) or on the different tasks of the compartments in which they occur (chloroplast or mitochondrion), carbohydrates seem to have a reproducible non-statistical ^{13}C-pattern, showing relative enrichments in positions C-3 and C-4 and depletions in positions C-1 and C-6 (Figure 2.4). We have recently been able to connect the enrichment in position C-3 and C-4 of glucose from different origins to an equilibrium isotope effect on the aldolase reaction (Gleixner and Schmidt, 1997). Similarly, isotope effects on other reactions of the pentose phosphate cycle and substrate fluctuations in either direction may be the reason for the occurrence of the pattern observed in other positions.

This pattern of glucose is certainly transferred to descendants, provided that these are the sole or main products of an irreversible turnover of the precursor. Thus, lactic

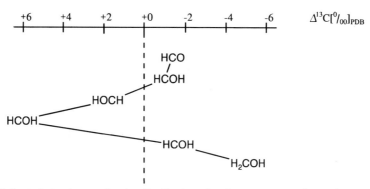

Figure 2.4. *Relative intramolecular distribution of carbon isotopes in glucose from C$_3$ and C$_4$ plants (adapted from Roβmann et al., 1991).*

acid provided by a fermentation of glucose is enriched in the carboxylic group (originating from C-3 and C-4 of the precursor) and depleted in the methyl group (C-1 and C-6 of glucose). Correspondingly, the carbon dioxide evolved in the alcoholic fermentation is enriched, while ethanol and natural acetic acid are depleted as compared to the glucose from which they originate, yet still preserve the pattern between C-1 and C-2 corresponding to the mean values of the positions C-2/C-5 and C-1/C-6 of the precursor.

Provided a metabolite pool is the starting point for different pathways, and one of these includes a kinetic isotope effect, the isotopic balance requires that respective abundances change in both directions, particularly if one of the pathways represents only a small flux, the corresponding isotopic shift will be dramatic. This is the case with glycerol, showing a depletion in position C-1 by –25‰, as compensation for a relatively small enrichment in the corresponding position of the main descendants of dihydroxyacetone phosphate (Weber *et al.*, 1997). We have observed a similar effect with ascorbic acid, where this descendant of glucose is enriched in position C-1 as compared to carbohydrates from the same origin.

Quite distinct effects on the ^{13}C-patterns of secondary products have been predicted from the isotope effect on the pyruvate dehydrogenase reaction (Melzer and Schmidt, 1987). The thiamine pyrophosphate (TPP) catalysed fission of C–C-bonds adjacent to carbonyl groups prefers the conversion of pyruvate with 'light' carbon atoms in position C-1 and C-2, to form 'active acetaldehyde' (Melzer and Schmidt, 1987). This is not only thought to be the main reason for the well-known relative depletion of lipids, but should also lead to an alternating ^{13}C-abundance in all products formed from the acetyl unit (Monson and Hayes, 1982). Moreover, this isotopic effect should, in a given pool of pyruvate after a partial turnover, change dramatically the pattern of that part which has not been converted to acetyl-CoA. Depending on the turnover of an individual pool, this would result in a completely different pattern of this intermediate (as well as of the acetyl moiety, Figure 2.5). In consequence, the pyruvate carboxylase reaction enriches the second carboxyl group (Melzer and O'Leary, 1987), and one then has to expect a pattern of dicarboxylic acids with a remarkably depleted methylene group. This has in principle been found – quite naturally with individual variations – in malic acid isolated from many samples of different fruit juices (Gensler *et al.*, 1995).

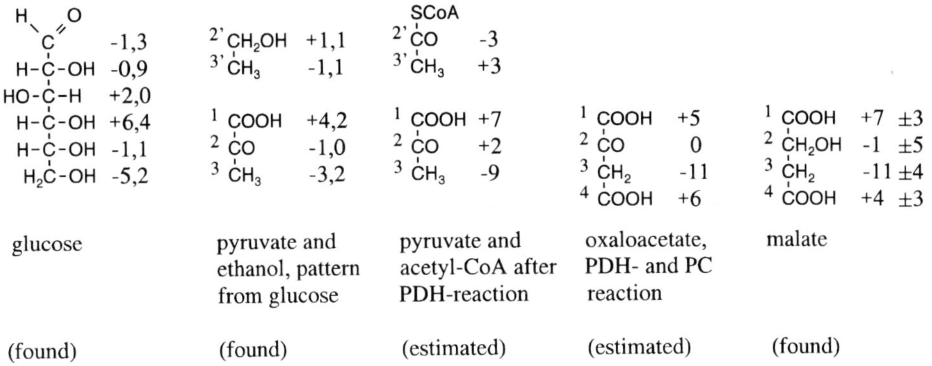

Figure 2.5. *Intramolecular distribution of carbon isotopes (Δ¹³C-values) of descendants of glucose with and without an arbitrary partial turnover of pyruvate via the pyruvate dehydrogenase reaction. The pattern of malate is indicated with individual variations, as found from various sources.*

2.4 Ester and aldol reactions during secondary metabolism

C_4-dicarboxylic acids, as well as the acetyl group from acetyl-CoA, are essential starting points for many secondary products. Citric acid from a given source should have a pattern composed from that of its moieties starting from the same origin. However, for citric acid from fruit juices, where the pattern of the oxaloacetic acid moiety should be identical to that of malic acid from the same source, only a part of the skeleton was similar (Figure 2.6). The isotopic content of those atoms involved in the aldol addition of the citrate synthase reaction had been changed: we deduced from this that the reaction includes an isotope effect, leading to a large depletion of the methylene group introduced from the acetyl precursor. We are actually measuring this isotope effect *in vitro*. There is evidence that corresponding isotope effects accompany most aldol reactions of acetyl-CoA involved in the biosynthesis of natural compounds, as found for sinigrin, a glucosinolate from black mustard, synthezised via homomethionine (Butzenlechner *et al.*, 1996).

Therefore, this should be the case for the synthesis of mevalonic acid and its descendants, the isoprenoids. However, the shifts in ¹³C-abundance in these compounds

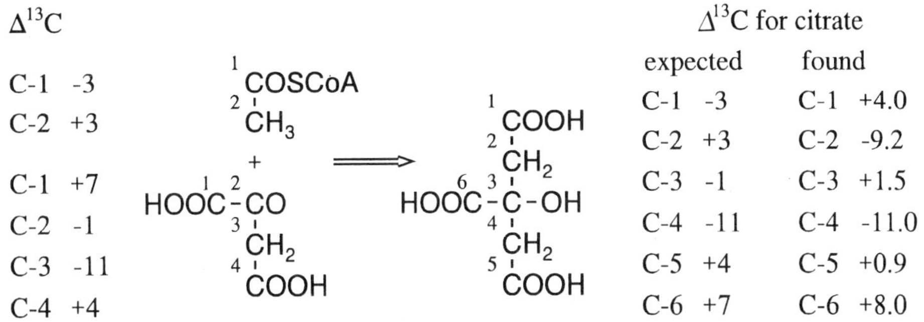

Figure 2.6. *Intramolecular distribution of carbon isotopes in citric acid in relation to the precursors oxaloacetate and acetyl-CoA as expected from precursors and as found.*

must be caused by isotope effects acting on several steps which overlap, and their relative influences must depend on pools and turnover rates in particular directions. We have found individual δ^{13}C-values for fatty acids (δ^{13}C = −30.6‰) and for isoprenoids from different pools, e. g. for limonene −26.6‰, sitosterol −26.0‰, violaxanthene −31.3‰ in extracts from orange peels. Correspondingly, positional δ^{13}C-values of terpenes from various origins never fit into a general pattern. This confirms the influence of reactions depends on the individual conditions of biosynthesis; even different pathways of isoprenoid synthesis may be involved (Hano et al., 1994; Rohmer et al., 1993).

2.5 Effect of other lyase reactions

The possible influence of additional lyase reactions, such as the transketolase reaction, has already been mentioned. While the isotope effect on this reaction has not yet been investigated, the influence of other TPP-dependent enzymatic reactions transferring 'active aldehydes' have already been studied. Model systems include the transfer of 'active aldehydes' to carbonyl groups, leading to 'acyloin' structures, as in the biosynthesis of the branched chain amino acids (valine, isoleucine). Rinaldi et al. (1974) described an extreme ^{13}C-enrichment in the carbonyl group of acetoine, formed by the transfer of 'active acetaldehyde' to acetaldehyde (or pyruvate). We have found a corresponding enrichment in this position in natural ephedrine, biosynthetic L-phenylacetylcarbinol and in angelic acid, a descendant of isoleucine (Ihle et al., unpublished data; Figure 2.7).

Aromatic compounds, among which lignin and catechins are most important in respect of biomass, are synthesized via the shikimic acid pathway from erythrose and phosphoenolpyruvate (see Figure 2.3). The side chain of the aromatic amino acids, central intermediates in this pathway, seems to preserve the pattern of the precursor pyruvate (Butzenlechner et al., 1996). However, derivatives of these amino acids, in which the original carboxylic group has been lost (cyanogenic glycosides and glucosinolates), demonstrate a dramatic ^{13}C-enrichment in the original α-position, the former asymmetric C-atom bearing the amino group. We suggest that this is an expression of a carbon isotope effect on the phenylalanine ammonia lyase (PAL) reaction. Obviously, the isotopic shift is easily detected in these minor products, while it is relatively small in the main products (Figure 2.7; Butzenlechner et al., 1996).

2.6 Reactions involving C$_1$-metabolism

Many aromatic compounds have O-methyl groups. It has been observed (Galimov et al., 1976) that they are depleted in ^{13}C relative to the rest of the molecules in question. Any methyl groups in natural products originate from S-adenosylmethionine (SAM), which in turn receives the C_1-unit from a pool of tetrahydrofolic acid (THF)-bound 'active formaldehyde' and 'active formiate'. The primary origin of all these C_1-units is the atom C-3 of serine, which is transferred to THF in the serine aldolase reaction (SHMT, serine hydroxymethyl transferase). We have measured the in vitro kinetic isotope effect acting on this reaction (Gleixner, Schmidt, Schirch and Matthews, unpublished data) and found that it is practically identical to unity. The δ^{13}C-value of the 'active formaldehyde' formed initially will therefore be identical to that in the corresponding position in serine, and hence related to C-1 and C-6 of glucose from the same source.

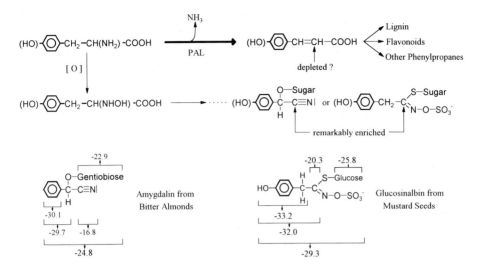

Figure 2.7. *Possible influence of a carbon isotope effect on the PAL- reaction and intramolecular distribution in phenylpropanes and cyanogenic glycosides or glucosinolates from aromatic amino acids. In the latter product the δ-value of the original α-C-atom shows a dramatic ^{13}C-enrichment (Butzenlechner et al., 1996).*

Caffeine is a secondary natural compound synthesized mainly from C_1-units of several oxidation states, hence originating from the THF- and the SAM-pools. We have investigated the ^{13}C-pattern of natural caffeine (Figure 2.8) and found that the C_1-units originating from the THF-pool are relatively enriched, while those from the SAM-pool are relatively depleted in respect to the mean $\delta^{13}C$-value of the alkaloid. This demonstrates a dramatic isotope abundance difference between the two C_1-pools, and we postulate that the C_1-transformation from the oxidised THF- to the reduced SAM-pool leads to a large isotopic discrimination. As obviously the 'light' C_1-units are preferably reduced, the THF-pool is enriched in 'heavy' C_1-units. This phenomenon corresponds to that discussed for the pyruvate pool after partial turnover of the compound into acetyl groups, or to the dramatic ^{13}C-depletion in position C-1 of glycerol accompanying the turnover of trioses at a branching point (see above). It also proves the general

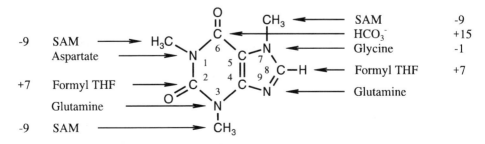

Figure 2.8. *Intramolecular distribution of carbon isotopes (Δ^{13}C-values) in caffeine as elucidated by chemical degradation (adapted from Weilacher et al., 1996).*

rule that in a closed system within a plant, isotope effects only become manifested when they are located at metabolic branching points.

2.7 Isotope effects on reactions introducing hydrogen, oxygen and nitrogen in biological compounds

The discrimination of hydrogen isotopes in nature starts with thermodynamic isotope effects on the evaporation and condensation of water and on the transpiration by plants (White 1989; see Chapters 4 and 11). The primary step of hydrogen introduction into organic binding and subsequent hydrogenations or dehydrogenations, as well as reactions with protons imply kinetic isotope effects, leading to isotopic patterns of the non-exchangeable (carbon-bound) hydrogen atoms of organic molecules. The patterns of some natural products have been investigated by NMR (Martin, 1995) and found to be characteristic for the kind and origin of plants as well as for the type of (bio)-synthesis. Indeed, evidence for a direct correlation of this pattern to *in vitro* kinetic hydrogen isotope effects on defined enzyme-catalysed reactions has only recently been established (see Chapter 4).

With the exception of carbohydrates, most organic molecules have only one or a few oxygen atoms. The oxygen isotopic abundance is primarily determined by the element's origin, namely from CO_2, H_2O or O_2. Due to isotope effects in the oxygen cycle, these three sources are quite different as to their $\delta^{18}O$-value. Leaf water may have values between $+12‰$ and $0‰$, the CO_2 in equilibrium is relatively enriched to this water by some $40‰$ to $45‰$, and atmospheric O_2 ($\delta^{18}O = +22.4‰$), being the source for hydroxylations, may also be mixed with oxygen produced by photosynthesis within the plant (see Chapter 10). In fact, carbohydrates, being primary products metabolically closest to CO_2, have the highest enrichment in ^{18}O. Secondary products are less enriched, due to the increasing influence of reactions introducing oxygen from other sources during biosynthesis. The influence of the primary source is, in addition, accompanied by an oxygen atom exchange with the surrounding water, especially with carbonyl groups or corresponding reaction intermediates. In addition kinetic isotope effects are of importance in the determination of the oxygen isotope abundances in given positions, for example there is evidence that they contribute to the $\delta^{18}O$-value of esters (Werner and Schmidt, unpublished data).

There are many reports on fractionations of isotopes in the nitrogen cycle, however, mostly dealing with global isotopic abundances in biomass or fertilizer pools (see Chapters 5, 6, 7 and 14). Some time ago we showed that the heavy oxygen as well as the nitrogen isotopic abundances of nitrate are determined by those of the given elemental sources of this anion, and that both are enhanced for a given NO_3^--pool after partial reduction, in agreement with N- and O- isotope effects on the nitrate reductase reaction (Schmidt *et al.*, 1992). Most organic compounds, with the exception of heteroaromatic bases, contain only one nitrogen atom, and the mean $\delta^{15}N$-value of these atoms is determined by that of the primary source, such as the N- fertilizer. However, in metabolic chains, for instance in the biosynthesis of amino acids, the nitrogen donor, mostly glutamate, seems to have the most positive $\delta^{15}N$-value, while subsequent acceptors are relatively depleted (Hare and Estep, 1983). This is expressed most clearly in excretion products like urea (Medina and Schmidt, 1982). Correspondingly, a continuous global ^{15}N-enrichment is observed in food chains (Tieszen and Boutton,

1989). It has been suggested that this phenomenon is due to kinetic isotope effects on transamination and deamination reactions, as well as on reactions of the ornithine cycle, although there have been no detailed isotopic studies of defined reactions to date. For a sample of a bacterial arginine in which the α-amino group had a $\delta^{15}N$-value of 0‰, we have found that the guanidino group was extremely depleted, attaining −37‰, but we have no information about the biosynthesis of this product. Certainly, a non-statistical ^{15}N-distribution will exist within organic molecules of biological origin, yet correlations with kinetic isotope effects on defined reactions are rarely available (for example see Yoneyama et al., 1993).

2.8 Conclusion

It is clear that, particularly for carbon and hydrogen, intermolecular and intramolecular non-statistical distributions of isotopes exist in all biological systems. For primary natural compounds, they are consistently expressed, while the pattern of secondary products reflects more the specific conditions during synthesis. Although shifts of global δ-values occur in the course of diagenesis (Galimov, 1995), the molecules must preserve their typical pattern, and elucidation of this will, in our opinion, become a valuable tool for distinguishing authentic material and artifacts in archeological and geochemical questions (Balzer et al., 1996, Schidlowski et al., 1983 see Chapter 17). Furthermore, knowledge of the influences of isotopic effects, pools and metabolic fluxes on shifts of δ-values in the course of metabolic turnover and transfer of these shifts to reactions in diagenetic and post-biotic transformations will help clarify metabolic correlations in this area, for instance in humification processes. Finally, we suggest that the interaction between isotope effects and turnover rates will be important for the study of metabolic fluxes in plants and other problems in plant physiology (see Chapters 3, 8 and 10).

References

Abelson, P.H. and Hoering, T.C. (1961) Carbon isotope fractionation of amino acids by photosynthetic organisms. *Proc. Natl. Acad. Sci. USA* **47**, 623–632.

Balzer, A., Gleixner, G., Grupe, G., Schmidt, H.-L., Schramm, S. and Turban-Just, S. (1997) In vitro decomposition of bone collagen by soil bacteria. Consequences for stable isotope analysis in archaeometry. *Archaeometry*, **39**, 415–429.

Butzenlechner, M., Thimet, S., Kempe, K., Kexel, H. and Schmidt, H.-L. (1996) Inter- and intramolecular isotopic correlations in some cyanogenic glycosides and glucosinolates and their practical importance. *Phytochem.*, **43**, 585–592.

Epstein, S. (1968) Distribution of carbon isotopes and their biochemical and geochemical significance. In *Proc. Symp. on CO_2 Chemical, Biochemical and Physiological Aspects* (eds Forster et al.) NASA Sp-188, Haverford PA, pp. 5–14.

Farquhar, G.D., O'Leary, M.H. and Berry, J.A. (1982) On the relationship between carbon isotope discrimination and the intercellular carbon dioxide concentration in leaves. *Aust. J. Plant Physiol.* **9**, 121–137.

Galimov, E. (1995) Fractionation of Carbon Isotopes on the Way from Living to Fossile Organic Matter. In *Stable Isotopes in the Biosphere* (eds E. Wada, Y. Eitaro, T. Yoneyama, M. Minagawa, T. Ando and B.D. Fry). Kyoto University Press, pp. 133–170.

Galimov, E.M., Kodina, L.A. and Generalova, V.N. (1976) Experimental investigation of intra- and intermolecular isotope effects in biogenic aromatic compounds. *Geochem. Int.* **1**, 9–13.

Gensler, M., Roßmann, A. and Schmidt, H.-L. (1995) Detection of Added L-Ascorbic Acid in Fruit Juices by Isotope Ratio Mass Spectrometry. *J. Agr. Food Chem.* **43**, 2662.

Gleixner, G., Danier, H-J., Werner, R.A. and Schmidt, H.-L. (1993) Correlations between the ^{13}C-content in different cell compartments and that in decomposing basidiomycetes, *Plant Physiol.* **102**, 1287–1290.

Gleixner, G. and Schmidt, H.-L. (1996) Carbon isotope effects on the fructose-1,6-bisphospate aldolase reaction. Origin for non-statistical ^{13}C-distribution in carbohydrates, *J. Biol. Chem.* **272**, 5382–5387.

Hano, Y., Ayukawa, A., Nomura, T. and Ueda, S. (1994) Origin of the Acetate Units Composing the Hemiterpene Moieties of Chalcomoranin in Morus alba Cell Cultures. *J. Am. Chem. Soc.* **116**, 4189–4193.

Hare, P.E. and Estep, M.L.F. (1983) Carbon and Nitrogen Isotopic Composition of Amino Acids in Modern and Fossil Collagens. *Carn. Inst. Wash. Yearbook* **82**, 410–413.

Ivlev, A.A., Apin, A.V. and Brizanova, L.Y. (1987) Distribution of carbon isotopes in the glucose of maize starch. *Fiziologica Rastenij* (Sov. plant physiol. New York) **34**, 493–498.

Martin, G.J. (1995) Inference of Metabolic and Environmental Effects from the NMR Determination of Natural Deuterium Isotopomers. In *Stable Isotopes in the Biosphere* (eds E. Wada, Y. Eitaro, T. Yoneyama, M. Minagawa, T. Ando and B.D. Fry). Kyoto University Press, pp. 36–45.

Medina, R. and Schmidt, H.-L. (1982) Nitrogen isotope ratio variations in biological material, indicator for metabolic correlations? In *Stable Isotopes* (eds. H.-L. Schmidt, H. Förstel and K. Heinzinger) Elsevier, Amsterdam, pp. 465–473.

Meinschein, W.G., Rinaldi, G.G., Hayes, J.M. and Schoeller, D.A. (1974) Intramolecular isotopic order in biologically produced acetic acid. *Biomed. Mass Spec.* **1**, 172–174.

Melzer, E. and O'Leary, M.H. (1987) Anaplerotic CO_2 fixation by Phosphoenolpyruvate carboxylase in C-3 plants. *Plant Physiol.* **84**, 58–60.

Melzer, E. and Schmidt, H.-L. (1987) Carbon isotope effects on the pyruvate-dehydrogenase reaction and their importance for relative carbon ^{13}C-depletions in lipids. *J. Biol. Chem.* **262**, 8159–8164.

Monson K.D. and Hayes J.M. (1982) Carbon isotope fractionation in the biosynthesis of bacterial fatty acids. Ozonolysis of unsaturated fatty acids as a means of determining the intramolecular distribution of carbon isotopes. *Geochim. Cosmochim. Acta* **46**, 139–149.

Rinaldi, G.G., Meinschein, W.G. and Hayes, J.M. (1974) Intramolecular carbon isotope distribution in biologically produced acetoin. *Biomed. Mass Spect.* **1**, 415–417.

Rohmer, M., Seemann, M., Horbach, S., Bringermeyer, S. and Sahm, H. (1996) Glyceraldehyde 3-phosphate and Pyruvate as Precursors of Isoprenic units in an alternative non-mevalonate pathway for terpenoid biosynthesis. *J. Am. Chem. Soc.* **118**, 2564–2566.

Roßmann, A., Butzenlechner, M. and Schmidt, H.-L. (1991) Evidence for a non-statistical carbon isotope distribution in natural glucose. *Plant Physiol.* **96**, 609–614.

Schidlowski, M., Hayes, J.M. and Kaplan, I.R. (1983) Isotopic Inferences of Ancient Biochemistries: Carbon, Sulfur, Hydrogen, and Nitrogen. In *Earth's earliest Biosphere: Its Origin and Evolution* (ed J.W. Schopf) Princeton University Press, Princeton, NY, pp.149–186.

Schmidt, H.-L., Kexel, H., Butzenlechner, M., Schwarz, S., Gleixner, G., Thimet, S., Werner, R.A. and Gensler, M. (1995) Non Statistical Isotope Distribution in Natural Compounds: Mirror of their Biosynthesis and Key of their Origin Assignment. In *Stable Isotopes in the Biosphere* (eds E. Wada, Y. Eitaro, T. Yoneyama, M. Minagawa, T. Ando and B.D. Fry). Kyoto University Press, pp. 17–35.

Schmidt, H.-L., Voerkelius, S. and Amberger, A. (1992) Nitrogen and Oxygen Isotopes as Indicators for Nitrification and Denitrification. I: *Progress in Hydrogeochemistry* (eds G. Matthess, F.H. Frimmel, P. Hirsch, H.D. Schulz and E. Usdowski). Springer Verlag, Berlin, pp. 212–218.

Tieszen, L.L. and Boutton, T.W. (1989) In: *Stable Carbon Isotopes in Terrestrial Ecosystem Research* (eds P.W. Rundel, J.R. Ehleringer and K.A. Nagy). Springer Verlag, Berlin, pp. 167–195.

Weber, D., Kexel, H. and Schmidt, H.-L. (1997) ^{13}C-pattern of natural glycerol: origin and practical importance. *J. Agr. Food Chem.* **45**, 2042–2046.

Weilacher, T., Gleixner, G. and Schmidt, H.-L. (1996) Carbon isotope pattern in purine alkaloids, a key to isotope discrimination in C_1-compounds, *Phytochem.* **41**, 1073–1077.

White, J.W.C. (1989) Stable Hydrogen Isotope Ratios in Plants: A Review of Current Theory and Current Potential Applications. In *Stable Isotopes in Ecological Research* (eds P.W. Rundel, J.R. Ehleringer and KA. Nagy). Springer Verlag, Berlin, pp.142–162.

Winkler, F.J., Kexel H., Kranz C., Schmidt, H.-L. (1982) Parameters affecting the $^{13}CO_2/^{12}CO_2$ isotope discrimination of the ribulose-1,5-bisphosphate carboxylase reaction. In *Stable isotopes* (eds H.-L. Schmidt, H. Förstel and K.Heinzinger). Elsevier, Amsterdam, pp. 83–89.

Yoneyama, T., Kamachi, K., Yamaya, T. and Mae, T. (1993) Fractionation of nitrogen isotopes by glutamine synthase isolated from spinach leaves. *Plant Cell Physiol.* **34**, 489–491.

Interpretation of oxygen isotope composition of leaf material

G.D. Farquhar, M.M. Barbour and B.K. Henry

3.1 Why are we interested in the oxygen isotope ratio of organic matter?

Measurements of the oxygen isotope ratio of plant organic matter (typically cellulose in tree rings) assist in reconstruction of palaeoclimatic data. For example, they help in establishing time-series for oxygen isotope composition ($\delta^{18}O$) of the rain water supplying moisture to the tree, which in turn can be used as a palaeothermometer (see Chapter 18). If, however, the $\delta^{18}O$ of rainwater is already known from independent measurements, the isotopic composition of organic matter can reveal information about the relative humidity of the air during the time of growth (for example, Yakir et al., 1994; Aucour et al., 1996; see Chapter 10). The $\delta^{18}O$ of organic matter aids identification of the origin of the material, because the $\delta^{18}O$ of rain water varies strongly with latitude and other factors (see Chapter 23). This has implications for forensic and other studies of the origins of paper, cloth (DeNiro et al., 1988) and other products (see Chapter 17). As we discuss below, there are also physiological and genetic influences on the $\delta^{18}O$ of plant organic matter, and these should prove useful in, for example, understanding differences in yield among lines of cereals (Farquhar et al., unpublished data). Measurement of $\delta^{18}O$ also aids interpretation of differences in carbon isotope discrimination among individuals growing in the same environment (see Chapters 8 and 9). The aim of this chapter is to provide theoretical background to, and analytical procedures for, ^{18}O composition in organic material in order to provide a new tool for the study of plant-environment interactions.

3.2 Oxygen isotopes in plant organic matter: a historical perspective

The earliest references to measurements of $\delta^{18}O$ in organic matter, of which we are aware are those of Libby (1972), Libby and Pandolfi (1973), Hardcastle and Friedman

(1974) on sugar from cane and beet, Bricout *et al.* (1975), Ferhi *et al.* (1975), Libby *et al.* (1976), and Gray and Thompson (1976). Apart from the study of Libby and Pandolfi (1973), which was on algae, these studies predominantly analysed wood, and with the aim of using tree rings as 'isotopic tree thermometers'. The main issues, then and perhaps even now, were the origins of the oxygen atoms, and how $\delta^{18}O$ varied between compounds (cellulose and lignin), and between early- and late-wood deposited in the growing season.

Lignin was found to be 10.5‰ less enriched than cellulose (Gray and Thompson, 1977) while Burk (1979) showed that cellulose was enriched compared to whole wood by ca. 4.5‰. Gray and Thompson (1976) thought it seemed 'fair to assume' that the cellulose oxygen is derived from CO_2, but that isotope exchange could occur between CO_2 and water before photosynthesis occurs. However they noted that the overall temperature dependence was four times greater than that for the isotopic composition of precipitation in their area. The differences between early- and late-wood drew attention to the possible effects of evaporative enrichment in leaves (Wilson, 1978). As often happens, laboratories were not always aware of work done elsewhere, and in this case Ferhi and Letolle (1977) had already shown that humidity affected the $\delta^{18}O$ in organic matter of bean leaves, with decreasing enrichment as humidity increased.

Epstein *et al.* (1977) found that the cellulose of aquatic plants was enriched in ^{18}O by a factor of about 1.027 compared to the water in which they grew. They suggested as a possible explanation that the ribulose bisphosphate carboxylation reaction involves a CO_2 molecule and an H_2O molecule so that 2/3 of the O atoms could come from CO_2 (which, at equilibrium, is enriched compared to water by about 41‰: Epstein *et al.*, 1977) and 1/3 from H_2O, making an enrichment of 27‰ compared with H_2O. In this context, it is interesting that, in the same year, Ferhi and Letolle (1977) found that organic matter of bean leaves was enriched in ^{18}O by 20‰ compared to that measured in leaf water. In retrospect this raises two issues, one again being the differences between organic matter, cellulose and water and the other being the extent to which whole-leaf-water reflects the water at the site of synthesis. Ferhi and Letolle, (1977), like Bricout (1978) also believed that $\delta^{18}O$ was greater in organic matter than in water because of the isotopic enrichment in CO_2.

Berry *et al.* (1978) fed labelled oxygen to leaves and noted that its presence in 3-PGA indicated that it probably enters the photosynthetic carbon reduction (PCR) cycle. While they did not pursue this, they noted that there was no evidence that the ^{18}O was *retained* in the cycle. DeNiro and Epstein (1979) confirmed that it is the water in the plant that determines the ^{18}O signature of cellulose. They ruled out the 2/3: 1/3 hypothesis on the grounds that one of the oxygens from fixed CO_2 is actually lost in the PCR cycle, and also suggested that there could be exchange between the oxygen of glyceraldehyde phosphate and the oxygen of water. The difference between the $\delta^{18}O$ of cellulose and that of the water distilled from their wheat plants was 28.2‰.

Sternberg and DeNiro (1983) suggested that the carbonyl oxygens of dihydroxy-acetone phosphate (DHAP) in the PCR cycle exchange with water (with a half-time of less than 10 sec: Reynolds *et al.*, 1971), and they demonstrated, using acetone as a model system, a 27‰ enrichment of the carbonyl O compared with water. More recently, there has been increasing evidence for exchange of O during carbohydrate metabolism and cellulose synthesis. Sternberg *et al.* (1986) showed that 45% of the oxygens in sugar were lost (i.e. exchanged with water) before cellulose is formed. DeNiro and Cooper (1989) suggested that the oxygen in cellulose formed from starch may have passed through triose phosphates, allowing greater exchange of O with water. Hill *et al.* (1995) provided evidence that in tree rings there is a rapid cycle between hexose monophosphates and triose phosphates during cellulose synthesis.

DeNiro and Cooper (1989) argued that there will be difficulties in interpreting fossil cellulose data: the fractionation between water and cellulose in aquatic plants varies somewhat (Cooper and DeNiro, 1989), and they could see no reason why this should not also be the case for terrestrial plants. Nevertheless, Sternberg (1989) reviewed the literature and concluded that the $\delta^{18}O$ in cellulose is $27\pm3‰$ enriched compared to the water at the site of synthesis. While the above conclusions form the basis of our current understanding, the work of Cooper and DeNiro (1989) must be borne in mind. They found from experiments with aquatic plants that they could not account for all the oxygen sources contributing to the $\delta^{18}O$ in cellulose. They did show that it was not O_2 directly, by feeding a highly enriched source of the gas. Although they suggested that recycling of organic matter might account for the results, they also noted that terrestrial plants seemed to have a similar problem of interpretation.

3.3 On the enrichment of ^{18}O in water within the plant

From the above it is apparent that a clear understanding is needed of the $\delta^{18}O$ in water at a number of sites: in the chloroplast, where the PCR cycle takes place, in the cytosol, where further exchange could occur, and at the sites of cellulose synthesis, where there is more exchange. Further complications apply for other compounds, but cellulose is the major interest in this Chapter, although the implications for photosynthetic gas exchange and scaling to global climate models are explored, respectively, in Chapters 10, 12 and 24.

Gonfiantini *et al.* (1965) were the first to demonstrate that the water in leaves is enriched compared to that supplied from the soil. Craig and Gordon (1965) developed an expression for enrichment of a free water surface, where the water vapour transport was by turbulent diffusion (see Chapter 23). The first applications for the oxygen isotope were by Lesaint *et al.* (1974), Dongmann *et al.* (1974) and then by Farris and Strain (1978). The kinetic isotope effect, ϵ_k, was a fitted value, and typically involved a value lower than that accepted today, more appropriate to the boundary layer of a leaf, than for diffusion through stomata. This value ignored the large amount of work done on gas exchange of leaves, where techniques had been developed to estimate independently the stomatal and boundary layer resistances to diffusion. Farquhar *et al.* (1989) used this plant physiological theory, to be applied with an experimentally determined value for the ratio of diffusivities of $H_2^{18}O$ and $H_2^{16}O$ in air, (1.028; Merlivat, 1978). It applies to the enrichment, $\Delta^{18}O_e$, in ^{18}O of liquid water at the sites of evaporation into the intercellular spaces within a leaf, compared to source water, modifying the notation of Farquhar and Lloyd (1993):

$$\Delta^{18}O_e = R_e/R_s - 1 \qquad (3.1)$$

where R_e and R_s are the $^{18}O/^{16}O$ ratios of liquid water at the evaporating sites, and of the source water, respectively. The equation is

$$\Delta^{18}O_v = \epsilon^* + \epsilon_k + (\Delta_v - \epsilon_k)\, e_a/e_i \qquad (3.2)$$

where e_a and e_i are the vapour pressures in the atmosphere and intercellular spaces, ϵ^*

is the proportional depression of equilibrium vapour pressure by the heavier molecule compared to $H_2{}^{16}O$ (9.2‰ at 25°C: Bottinga and Craig, 1969), ϵ_k is the kinetic fractionation factor. ϵ_k has a value of 28‰ for diffusion in air through the stomatal resistance (r_s) (Merlivat, 1978) and 19‰ in the boundary layer resistance, r_b (assuming a 2/3 power effect according to the Pohlhausen analysis: Kays, 1966), and are related by

$$\epsilon_k = (28r_s + 19r_b)/(r_s + r_b)\text{‰} \tag{3.3}$$

The isotopic composition of water vapour in the air, R_v, is again expressed relative to that of source water by

$$\Delta^{18}O_v = (R_v/R_s) - 1 \tag{3.4}$$

As noted by Farquhar and Lloyd, Equation 3.2 is an approximation to a more precise formulation (White, 1989; Flanagan et al., 1991) written in terms of isotope ratios, R, which is derived as follows: the isotopic ratio, R_E, of transpired water is $^{18}E/^{16}E$, where E denotes transpiration rate, and the superscripts refer to ^{18}O and ^{16}O, respectively. In turn,

$$E = g \times (e_i - e_a)/P, \tag{3.5}$$

where g is the conductance to water vapour (stomata and boundary layer in series, as cuticular losses are ignored), and P is the atmospheric pressure. Equation 3.5 applies directly to ^{16}E. For ^{18}E, g is replaced by $g/(1+\epsilon_k)$, e_i by $R_e e_i/(1+\epsilon^*)$ and e_a by $R_v e_a$ so that

$$^{18}E = [g/(1+\epsilon_k)]\{R_e e_i/(1+\epsilon^*) - R_v e_a\}/P \tag{3.6}$$

and then

$$R_E = \{R_e e_i/(1+\epsilon^*) - R_v e_a\}/\{(1+\epsilon_k)(e_i - e_a)\}. \tag{3.7}$$

In the steady state, R_E must be equal to R_s, and using the definitions of Equations 3.1 and 3.4 one obtains (Farquhar and Lloyd, 1993):

$$\Delta_e = (1+\epsilon^*)[1 + \epsilon_k + (\Delta_v - \epsilon_k) e_a/e_i] - 1 \tag{3.8}$$

for which Equation 3.2 is a good approximation (within 0.1‰) and has the advantage of being amenable to mental arithmetic. There are theoretical uncertainties involved here, such as the effects of the exchange from CO_2, which are probably of the same order as 0.1‰.

Equations 3.2, 3.7 and 3.8 apply only to steady-state conditions. The important effects of non-steady-state have been examined by Dongmann et al. (1974), Zundel et al. (1978), Farris and Strain (1978). Subsequently Harwood et al. (1998) have noted that since Equation 3.7 applies in the non-steady state, R_e can be estimated even during varying conditions, provided R_E is measured. They then demonstrated its use in field-based measurements. As an aside, we note that Equations 3.2, 3.7 and 3.8 apply in the non-steady state if the definitions of Equations 3.1 and 3.4 are modified such that the reference is now R_E, rather than R_s.

3.4 Variation within the leaf

While the above equations may apply at the sites of evaporation, they need not represent leaf water as a whole. The isotopic composition of water in the petiole and veins of a leaf is very close to that of source water (Bariac *et al.*, 1994), so that gradients must occur within the leaf. Some must occur over a fine spatial scale, for example from a fine vein to the cells adjacent to a substomatal cavity. There will also be variation on the scale associated with stomatal heterogeneity of the kind described by Terashima *et al.* (1988). On a larger scale, Wang and Yakir (1995) showed increasing enrichment towards the tips and edges of leaves, with variation of up to 13‰ (see Chapter 10).

We suggest that the net result of the fine scale gradient should be that whole leaf-water is less enriched than one calculates for the sites of evaporation. This is in fact what is observed (see review by Flanagan, 1993). Leaney *et al.* (1985), White (1989), and Yakir *et al.* (1989) discussed compartmentation at various scales, to explain the observations. Farquhar and Lloyd (1993) suggested that the effects were associated with convection and diffusion of the isotopes within the leaf. Their theory suggested the disparity between $\delta^{18}O$ and whole leaf water $\delta^{18}O$ should increase as transpiration rate increases.

3.4.1 *The Péclet effect*

With water at the sites of evaporation becoming enriched in ^{18}O, $H_2^{18}O$ will tend to diffuse away from these sites in the liquid phase. This effect, in turn, is opposed by convection of source water to the sites. Taking a simple, one-dimensional analogue, Farquhar and Lloyd (unpublished data) showed that diffusion and convection opposed across a small element yields, in the steady-state, a value that decreases from that at the site of evaporation. For the enrichment, compared to water source at a distance l (m) from the site of evaporation, the decrease depends on the dimensionless term $El/(C_W D)$, where E (mol m^{-2} s^{-1}) is the rate of water movement, C_W (mol m^{-3}) is the concentration of water, and D (m^2 s^{-1}) is the diffusivity of $H_2^{18}O$ in water and the dimensionless term is known as the Péclet number, $\mathscr{P}$, (Farquhar and Lloyd, 1993).

Zimmermann *et al.* (1967), who discussed the Péclet effect on isotopes in water moving through and evaporating from soil, were apparently the first to note that the large value of $\mathscr{P}$, in the stem explains why transpirational enrichment does not usually influence the soil water. Some data are consistent with the Farquhar and Lloyd hypothesis (White, 1989; Walker *et al.*, 1989; Flanagan *et al.*, 1991). Walker *et al.* (1989) interpreted their data in terms of a model assuming that a fraction, **f**, of the tissue water was at isotope ratio R_e while the rest, (1-f), was at R_s (Leaney *et al.*, 1985). This is equivalent to having one pool (f) at $\mathscr{P}$, = 0, and (1-f) at infinite $\mathscr{P}$. It seems more reasonable that there should be a continuum in $\mathscr{P}$. So, when water is expressed from a leaf under pressure, the isotope ratio of the emerging water should change from R_s to R_e, with the rate of change reflecting local velocities and cross sectional areas.

A complication arises if evaporation occurs from a file of cells and not just from the terminal cell. Farquhar and Lloyd (unpublished data) have showed analytically that the average enrichment will still be given by the Craig equation, with greater enrichment at the terminus and less where evaporation first occurs. This phenomenon has been reported by Wang and Yakir (1995), who also suggested the analogy of a string of interconnected evaporating lakes (see Chapter 10). The mathematics of such strings have been described, and have related properties.

3.4.2 *On the $^{18}O/^{16}O$ ratio of metabolic water in the chloroplast*

Ignoring for the moment the evaporation from a file of cells and considering only a uniformly evaporating leaf, we would expect the isotopic composition of chloroplast water to be close to that of the evaporating cell walls, but somewhat towards that of source water. The quantitative nature of that term 'somewhat' is a matter of debate and uncertainty at the present.

Yakir *et al.* (1994) described experiments where the $^{18}O/^{16}O$ ratios in evolved oxygen were measured as a function of relative humidity, and where, in separate runs, the $^{18}O/^{16}O$ ratios of respired and photorespired CO_2 were also examined. The calculated isotopic composition of 'metabolic water' was similar using both techniques, and less than $\delta^{18}O_e$. The oxygen evolution was carried out in a He/CO_2 mixture so that the effective value for ϵ_k in such a system is about 9‰ compared to 27‰ in air. When this is allowed for, the oxygen data lie half way between calculated $\delta^{18}O_e$ and source water. However, some experiments examining the ^{18}O composition of CO_2 exchanging with leaves suggest that the chloroplastic water has an isotopic composition rather closer to that calculated for the sites of evaporation (Farquhar *et al.*, 1993; Flanagan *et al.*, 1994 and Chapter 12). Harwood *et al.* (1998), using the latter technique under field conditions, found the chloroplastic water $\delta^{18}O$ to be sometimes near $\delta^{18}O_e$ while at other times to be like the observations of Yakir *et al.* (1994). They cautioned against overinterpretation of their results because of some uncertainty, under field conditions, in the determination of the CO_2 concentration in the chloroplast. Further discussion of all these data demands an understanding of the calculations involved in the latter experiments (see next section).

3.4.3 *Effects of leaf water isotopic composition on discrimination against $C^{18}O^{16}O$ during photosynthesis*

The enrichment in $^{18}O/^{16}O$ in water in the chloroplast is passed to the CO_2 there *via* the action of carbonic anhydrase (CA). This can contribute to enrichment in ^{18}O of CO_2 passing over a leaf. The extent of this effect depends on rates of air flow and of assimilation of CO_2 and is analogous to effects on ^{13}C enrichment in CO_2 (Evans *et al.* 1986: see Chapter 8). The latter are now routinely scaled by ξ, the ratio of the concentration of CO_2 in air entering (before) a gas exchange chamber, C_e, to the depletion in the well-mixed cuvette, C_e-C_o, where C_o is the $[CO_2]$ in the air leaving the chamber. If the CO_2 in the chamber is enriched in ^{13}C by a small amount, $(-\Delta^{13}C_a)$ then the discrimination by the leaf, $\Delta^{13}C$, against $^{13}CO_2$ during photosynthesis is approximately ξ times as much. $\Delta^{13}C$ is the $^{13}C/^{12}C$ ratio of CO_2 in the air divided by that of the CO_2 taken up by the plant, minus one (Farquhar and Richards, 1984; for derivation, see Chapter 8). The same idea can be applied to the enrichment in $^{18}O/^{16}O$ to calculate $\Delta^{18}O_A$, the discrimination against ^{18}O during CO_2 assimilation (Farquhar *et al.*, 1993) defined as $R_a/R_A - 1$, where R_a is the oxygen isotope ratio of CO_2 in the air, and R_A is the oxygen isotope ratio of the flux of CO_2 into the leaf.

The enrichment of ^{18}O in the chloroplast water, in evaporating leaves, equilibrates with CO_2 *via* CA, which affects the isotopic fluxes of $C^{16}O^{16}O$ and $C^{16}O^{18}O$ (Farquhar *et al.*, 1993; for detailed analysis, see Chapter 24). Farquhar *et al.* (1993) assumed that the action of CA was complete and analysed the resulting effects on global atmospheric CO_2. The equations to describe this effect at the leaf level have been used to 'back-calculate' the isotopic composition of water in the chloroplast (Farquhar *et al.*, 1993; Flanagan *et al.*, 1994; Harwood *et al.*, 1998). However, this full equilibrium between CO_2 and water in the chloroplast is unlikely to be precisely the case and Farquhar and Lloyd (1993) gave a more complete description taking into account a non-zero ratio, τ, of the rate of carboxylation by Rubisco to the rate of hydration of CO_2 by CA.

We know that τ must be greater than zero. From concurrent measurements of Rubisco and carbonic anhydrase activities in wheat (Evans, 1983), a value of 0.05 can be calculated, which is in line with theoretical estimates of Cowan (1986). Nevertheless, Farquhar et al. (1993) noted that measurements with wheat gave results similar to those of their tree species, for example, isotopic composition of chloroplast CO_2, $\delta^{18}O_c$, appeared to be equal to, or perhaps even slightly greater than, the full equilibrium value. The higher values may be caused by greater exchange with parts of a leaf where there has been enrichment along a file of cells (see above).

Working with common bean (*Phaseolus vulgaris*), Flanagan et al. (1994) measured the average isotopic composition of bulk leaf water concurrently with $C^{18}O^{16}O$ discrimination over a range of vapour pressure deficits. Back-calculation of the effective $\delta^{18}O_c$, from measurements of $\Delta^{18}O_A$ also showed it to be very close to equilibrium with water at the leaf evaporating surface, even though the bulk leaf water was less enriched, and they estimated a value for τ of 0.013 (see Chapter 12). Despite the progress outlined above, we cannot explain the discrepancy between the data typified by those of Yakir et al. (1994) and Farquhar et al. (1993; Section 3.4.2), and clearly more work is required.

3.4.4 *On the relationship between Δ_A and the $^{18}O/^{16}O$ ratio in organic matter*

Unlike $^{13}CO_2$, the $C^{18}O^{16}O$ discrimination described here is not reflected in the isotopic composition of organic matter. Firstly, the proportion of oxygen (in CO_2) initially fixed (by Rubisco into phosphoglycerate) with a gaseous isotopic signature will normally be small. In fact, it is about 3τ of the chloroplast-water-related signal (Farquhar and Lloyd, unpublished data) and τ is normally only a few percent. After that there are many opportunities for isotopic exchange, as discussed below.

3.5 Isotopic exchange of oxygen during metabolism

3.5.1 *Carbonyl oxygen exchange with water*

Organic molecules often reflect the oxygen isotope ratio of water in which they formed due to isotopic exchange of oxygen between water and oxygen in carbonyl groups (Samuel and Silver, 1965). Exchange is via formation of a *gem*-diol:

$$H_2O^{\bullet} + \begin{array}{c} R \\ \diagdown \\ R \end{array} C = O \rightleftharpoons \begin{array}{c} R \diagup OH \\ C \\ R \diagdown O^{\bullet}H \end{array} \rightleftharpoons \begin{array}{c} R \\ \diagdown \\ R \end{array} C = O^{\bullet} + H_2O \tag{3.9}$$

Oxygens in other functional groups, such as hydroxyl, carboxyl and phosphate are not exchangeable under temperature and pH ranges found in plants (Sternberg et al., 1986). The fractionation factor of carbonyl oxygen exchange with water has been measured in acetone to be 1.028 at 25°C (Sternberg and DeNiro, 1983). The authors ascribed no significance to any temperature effect, but there is an indication in their data that the effect diminishes with increasing temperature. It has been assumed that carbonyl oxygens in all molecules have the same fractionation factor, as cellulose from aquatic plants and animals was found to be $27\pm3‰$ more enriched than the water in which the organisms grew (DeNiro and Epstein, 1981). Measurements of fractionation factors in other organic molecules to confirm this assumption would be valuable, as discussed in Section 3.2.

3.5.2 Isotopic equilibrium

Many intermediates leading to cellulose synthesis contain carbonyl oxygen, so that the exchange reaction becomes important in determining the isotopic signature of plant organic material. The rate of exchange of carbonyl oxygen varies considerably between molecules. Acetone, CH_3COCH_3, has a half-time for equilibration with water of about 10 minutes, compared to 166 and 29.5 min for fructose 6-phosphate and fructose 1,6-bisphosphate respectively (Model et al., 1968). The two fructose phosphates are the only two molecules relevant to pathways in plants for which the half-time to equilibration with water has been measured, of which the authors are aware. Other intermediates within the PCR cycle, sucrose and cellulose synthesis pathways that contain a carbonyl group and therefore need to be looked at are: ribulose 1,5-bisphosphate, glyceraldehyde 3-phosphate, dihydroxyacetone 3-phosphate, erythrose 4-phosphate, xylulose 5-phosphate, sedoheptulose 1,7-phosphate, sedoheptulose 7-phosphate, ribose 5-phosphate, ribulose 5-phosphate and glucose 1-phosphate. We further note that the half-times described above refer to non-enzymatic exchange. In vivo the times are likely to be shorter because of such enzymes as aldolase (Model et al., 1968).

Sternberg et al. (1986) identified the importance of the carbonyl oxygens in triose phosphates, the molecules exported from the chloroplast for sucrose synthesis. Two of the three oxygens bound to carbons in triose phosphates are in carbonyl groups (see Scheme 1, Appendix 1), and so the rate of exchange of these oxygens is vital to the isotopic signature of sucrose, and ultimately cellulose. The pool size of triose phosphates in leaves of Phaseolus vulgaris is fairly constant over a wide range of photosynthetic rates, at about 57 μmol m^{-2} (Badger et al., 1984). This implies that the residence time of these molecules must vary considerably with assimilation rate.

The residence time of triose phosphates at a given assimilation rate (A) may be calculated as follows:

$$\text{triose phosphate flux} = (2 + 1.5\phi)V_c \tag{3.10}$$

where V_c is the rate of carboxylation and ϕ is the ratio of oxygenation to carboxylation (see Figure 3.1). Following the notation developed by Farquhar and von Caemmerer (1982),

$$\phi = \frac{2\Gamma^*}{C_c} \tag{3.11}$$

where $\Gamma^* = CO_2$ compensation point
$\quad\quad C_c$ = chloroplastic CO_2 concentration.

$$A = V_c(1 - 0.5\phi) - R_d \tag{3.12}$$

Ignoring dark respiration (R_d)

$$V_c = \frac{A}{1 - \Gamma^*/C_c} \tag{3.13}$$

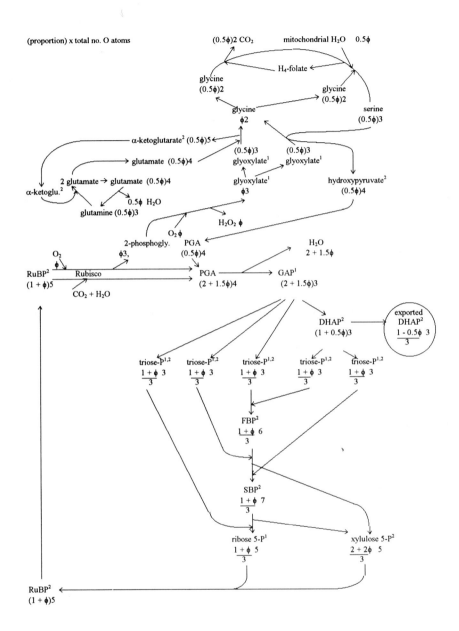

Figure 3.1. The proportional pathway of oxygen in the PCR and Photorespiratory cycles. All fluxes are in proportion to the rate of carboxylation. Super-scripted members refer to the carbon to which the exchangeable oxygen is bound. Adapted from Farquhar and von Caemmerer (1982); Farquhar et al. (1983).

Therefore triose phosphate flux

$$
\begin{aligned}
&= \frac{2C_c + 3\Gamma^*}{C_c} \times \frac{A}{1 - \Gamma^* / C_c} \\
&= \frac{2C_c + 3\Gamma^*}{C_c - \Gamma^*} \times A
\end{aligned}
\tag{3.14}
$$

So if $\Gamma^* = 35$ µmol mol^{-1} and $C_c = 250$ µmol mol^{-1}, then at an assimilation rate of 25 µmol m^{-2} s^{-1} the residence time is 0.8 seconds, while if the assimilation rate is as low as 5 µmol m^{-2}s^{-1}, then the residence time is 4.0 seconds. As noted earlier, the exchange rate of oxygen in triose phophates is expected to be fast, Sternberg et al. (1986) suggesting a half-time to equilibration of less than 10 seconds, but residence times like those calculated above would suggest incomplete isotopic exchange. Sternberg et al. (1986) found some evidence of incomplete exchange between cellulose formed from sucrose and the water in which it formed. A fractionation factor of 16.3‰ was measured rather than the predicted 27‰.

3.6 Isotopic history of oxygen in organic molecules

By following the history of each oxygen atom through the biochemical pathways we are able to highlight steps that are important in determining the isotopic signal of cellulose. Schemes 1–5, in Appendix 1, represent current understanding of the pathways involved. The only reactions for which the outcome, in terms of movement of oxygen atoms, is uncertain are phosphatase reactions. However, the action of phosphatase has been described for phosphoglycolate phosphatase by Christeller and Tolbert (1978), who suggest the bond between the bridging oxygen and the phosphate group is broken. This reaction is shown in step two, scheme one. Benkovic and deMaine (1982) observed that during steady-state hydrolysis of fructose-1,6-bisphosphate by the phosphatase in H$_2$^{18}O, one oxygen of the inorganic phosphate formed is isotopically equilibrated with water. This suggests that fructose-1,6-bisphosphatase also breaks the oxygen to phosphate group bond, employing the same mechanism of action as phosphoglycolate phosphatase. If we assume that all phosphatases break this bond, then predictions about the history of each oxygen in sucrose and cellulose can be made.

3.6.1 Isotopic history of oxygen in PCR cycle intermediates

Evaporative enrichment of leaf water is passed on to chloroplastic CO$_2$ via the action of CA with CO$_2$ being enriched in ^{18}O by 41.2‰ compared to water at equilibrium (O'Neil et al., 1975). The CO$_2$ then either diffuses back out of the leaf, or is fixed by carboxylation. Assimilated CO$_2$ has mostly become equilibrated with water, but a small proportion, maybe 2% is not equilibrated, and so reflects the isotopic signature of atmospheric CO$_2$. During carboxylation CO$_2$ and water are added to RuBP to form two 3-phosphoglycerate molecules. The 'upper' and 'lower' molecules are distinguished by the fact that the 'upper' molecule contains carbon and oxygen from CO$_2$. The 'lower' molecule contains an oxygen from water (see Figure 3.2). However, differences in isotopic signature between 'upper' and 'lower' molecules become masked due to isotopic exchange with water as the PCR cycle continues.

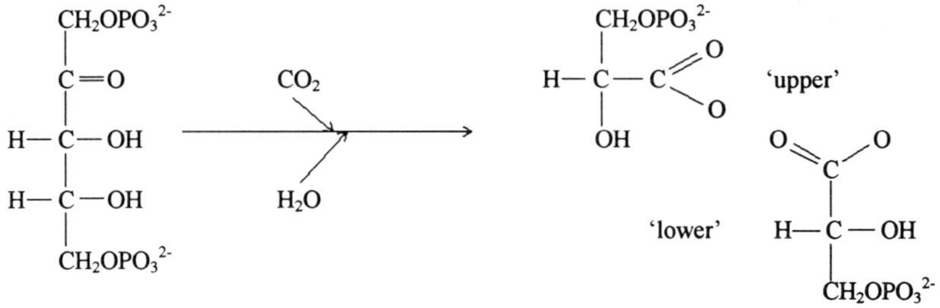

Figure 3.2. *The carboxylation of RuBP to form two 3-phosphoglycerate (Taiz and Zeiger, 1991).*

Photorespiration also feeds PGA into the pool to be subsequently exported as triose phosphates for sucrose synthesis, or continue around the cycle, reforming RuBP. The PGA formed directly by oxygenation will be the same as the 'lower' PGA formed by carboxylation (see Figure 3.3), while that formed via the glycolate pathway after oxygenation will have a carboxyl oxygen from O_2, and another from O_2 of RuBP. Berry *et al.* (1978) reported that labelled O_2, fixed into 2-phosphoglycolate by Rubisco, was found in PGA. They found no evidence of doubly-labelled PGA, which would indicate that the isotopic signal of molecular oxygen was retained through the PCR cycle: that is there was complete equilibration with cellular water. However, errors in GCMS measurements were relatively high so that small levels of signal that may have been retained might have been within the measurement error. The experiment of Cooper and DeNiro (1989), in which highly enriched O_2 was fed to submerged aquatic plants but left no signal, may not necessarily translate to a similar result in terrestrial plants, if the aquatic plants had some sort of CO_2 concentrating mechanism (Badger and Price, 1994).

PGA from both carboxylation and oxygenation continues around the PCR cycle, to be converted into glyceraldehyde 3-phosphate (G3P). G3P is reversibly converted to DHAP via the enzyme triose phosphate isomerase. The equilibrium constant for this reaction is 5×10^{-2}, so we expect the ratio of DHAP to G3P to be close to 20: 1 (Rawn, 1989). The high activity of triose phosphate isomerase means that 95% of triose phosphates are in the DHAP form, which has an important implication for the exchange of oxygen in the carbonyl group of G3P. This oxygen is derived, in part, from the newly fixed CO_2, O_2 and H_2O. Even at low photosynthetic rates (say 5 μmol

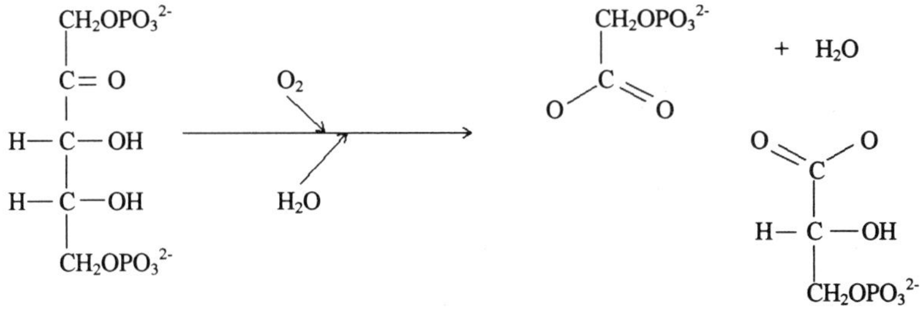

Figure 3.3. *The oxygenation of RuBP to form 2-phosphoglycolate and 3-phosphoglycerate.*

$m^{-2} s^{-1}$) the residence time of G3P will be just 0.2 seconds, hardly long enough for any isotopic exchange to occur.

The carbonyl oxygen in G3P, O-1, becomes non-exchangeable in fructose bisphosphate and remains to form four out of 11 oxygens in sucrose and 4 out of 10 oxygens in cellulose. The difference in isotopic signal between CO_2 and O_2 fixed by Rubisco is expected to be large. The isotope ratio of oxygen in PGA from O_2 will be close to that of oxygen in the atmosphere (23.5‰ on the SMOW (standard mean ocean water) scale), minus discrimination by Rubisco (discrimination; b_{O2} = 21‰ (Guy et al., 1987), giving 2.5‰), whereas chloroplastic CO_2 will be enriched due to exchange with water via carbonic anhydrase. If leaf water is about 20‰ (SMOW), then CO_2 equilibrated with this water will be 61‰. Discrimination against $C^{16}O^{18}O$ by Rubisco (b_{CO2}) is expected to be small (Farquhar and Lloyd, 1993) but has not been measured. During both carboxylation and oxygenation Rubisco fixes an oxygen from H_2O into PGA. Discrimination against $H_2^{18}O$ (b_{H2O}) has not yet been measured, and could be large. The potentially large differences in isotopic ratio between sources of oxygen forming O-1 of triose phosphate, and the short time for equilibration with water while in GAP, lead us to suggest that sucrose may not be in isotopic equilibrium with cellular water, and that differences in ϕ (the ratio of oxygenation to carboxylation, Farquhar and von Caemmerer, 1982) may show up in cellulose $\delta^{18}O$.

The PGA formed from the glycolate pathway after oxygenation will have a carbon-to-phosphate bridging oxygen derived from mitochondrial water (see step six, Scheme 2, Appendix 1). This is an important point as mitochondrial water may be less enriched than chloroplast water due to the Péclet effect. Phosphate-bridging oxygens are protected from exchange with water by the phosphate group, and assuming all phosphatases break the oxygen to phosphate group bond, the signal from mitochondrial water should show up in sucrose and cellulose.

3.6.2 *Isotopic history of oxygen in sucrose*

While plant organic material largely reflects the oxygen isotope composition of water involved in its synthesis, the $\delta^{18}O$ of water varies throughout the plant. If oxygen becomes non-exchangeable in biochemical intermediates its isotope ratio may reflect that of water from some other part of the plant. For example, if oxygen in the intermediates leading to sucrose exchanged with leaf water but became non-exchangeable in sucrose and remained non-exchangeable during cellulose synthesis then cellulose laid down from that sucrose in woody tissue would reflect, in part, the isotope composition of leaf water.

The reactions of the sucrose synthesis pathway are well understood and are outlined in Scheme 3. Assuming all phosphatases break the O – P bond, we predict that three out of 11 oxygens in sucrose are the same as those in the RuBP that was carboxylated or oxygenated to form PGA (oxygens a, h and k, Figure 3.4). A small proportion of two of these three/11 oxygens comes via the glycolate pathway, so is derived from mitochondrial water (oxygens a and h). Of the remaining eight out of 11 oxygens, four out of 11 reflect the isotopic signal of chloroplastic CO_2, O_2 and H_2O, though slightly masked due to exchange with chloroplastic and cytosolic water for a short time before becoming non-exchangeable (oxygens d, e, i and j). The remaining four/11 oxygens have exchanged with chloroplastic water, but also exchanged for a longer time with cytosolic water when in carbonyl groups of hexose phosphates, before becoming non-exchangeable in sucrose (oxygens b, c, f and g).

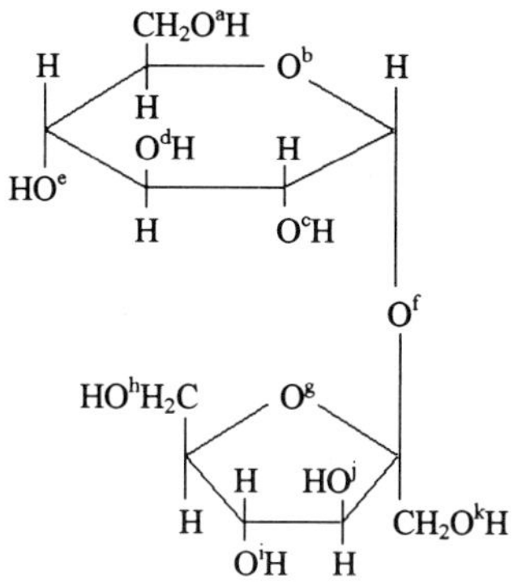

Figure 3.4. *Molecular structure of sucrose.*

3.6.3 *Isotopic history of oxygen in cellulose*

The steps from sucrose to cellulose are not as well understood as the sucrose synthesis pathway, although it is known that both UTP-glucose and GTP-glucose are intermediates, the ratio of which depends on the type of cellulose that is being laid down (Tarchevskey and Marchenko, 1991). Again, assuming that phosphatases break the oxygen to phosphate bond, the proportion of exchangeable oxygens can be estimated. Sternberg *et al.* (1986) predicted that five/11 oxygens in cellulose formed from sucrose are exchangeable, two carbonyl positions from each fructose and glucose (assuming a rapid equilibrium between the two) and one from the dissociation of fructose and glucose. The predicted proportion, 45%, was very close to the 47% exchangeable oxygens observed (Sternberg *et al.*, 1986).

Recent work by Hill *et al.* (1995) shows a futile recycling of hexose phosphates through triose phosphates before cellulose synthesis. This is shown in Scheme 5, Appendix 1. The effect of this recycling will be to exchange two out of three oxygen that go through triose phosphates. In light of this new work, data presented by Sternberg *et al.* (1986) need to be re-interpreted.

By following the biochemical pathway outlined in Scheme 5, Appendix 1, predictions about the isotope ratio of individual oxygens in cellulose may be made. In every cellobiose unit, two out of 10 oxygens have exchanged in glucose and fructose, not five out of 11 suggested by Sternberg *et al.* (1986), as the oxygen attached to carbon 1 remains with U(G)DP when the glucose residue is added to the growing cellulose polymer. The oxygen forming the β1–4 link between glucose residues is therefore always that attached to carbon 4 of the non-reducing end of the polysaccharide (W. York personal communication). From Scheme 5, Appendix 1, it can be seen that if there is recycling through triose phosphates then six out of 10 oxygens in a cellobiose unit will have exchanged in triose phosphates. Thus the proportion of exchangeable oxygen in a cellobiose unit becomes the proportion that are exchangeable in triose phosphates (six/10) multiplied by the proportion of hexose phosphates that go through the triose phosphate pathway (y), plus the proportion that exchange in hexose phosphates (two/10).

$$\text{Exchangeable O} = y\, 6/10 + 2/10. \qquad (3.15)$$

The data presented by Sternberg *et al.* (1986) can be recalculated as follows;

$$0.47 = y\ 6/10 + 2/10,\ \text{so}\ y = 0.45$$

That is, 45% of hexose phosphates are broken down into triose phosphates before being incorporated into cellulose. This compares to 20–25% redistribution of ^{14}C from C-1 to C-6, or 40–50% of hexose phosphates cycling through triose phosphates, calculated by Hill *et al.* (1995) in oak stem tissue. Other estimates of fluxes from hexose phosphates through triose phosphates are 28% in ripening banana (Hill and ap Rees, 1994), 30–40% in developing wheat grain (Keeling *et al.*, 1988), 22–39% in potato tubers (Viola *et al.*, 1991), and 60% in maize endosperm (Hatzfeld and Stitt, 1990).

By following the biochemical pathways outlined in Schemes 1 – 5, Appendix 1, we are able to predict the isotopic history of each oxygen in a cellobiose unit. Oxygens -2 and O–7 (Figure 3.5) reflect the isotopic ratio of oxygens bound to C-1 and C-5 of RuBP, with a small proportion from mitochondrial water via the photorespiratory cycle. Oxygen -4 and O–9 (Figure 3.5) exchanged with cellular water while in fructose and in DHAP, so are expected to largely reflect this exchange, but some signal from chloroplastic and mitochondrial water may remain. Oxygens -1, O–5, O–6 and O–10 will reflect the isotope ratio of CO_2, H_2O and O_2 originally fixed by carboxylation and oxygenation, with some masking after exchange with chloroplastic and cytosolic water during sucrose synthesis. A proportion have gone through triose phosphates during cellulose synthesis, so have exchanged in G3P. This exchange is likely to be incomplete as G3P forms only 5% of triose phosphates and is expected to exchange more slowly than DHAP. Oxygen -3 and O–8 exchanged with chloroplastic and cytosolic water when in DHAP and a proportion, dependent on the percent of hexose to triose phosphate recycling, also exchanged with cellular water in DHAP. If 45% recycling through triose phosphate occurs (as calculated above), then these oxygens will reflect 45% cellular water and 55% chloroplastic and cytosolic water.

3.6.4 *Isotopic history of oxygen in plant compounds other than cellulose*

Little is known about oxygen isotope effects in secondary metabolism. The lignin fraction of white spruce tree rings was found to be 10.5 ± 1.7‰ less enriched than cellulose (Gray and Thompson, 1977), while fatty acids were also found to be less enriched than cellulose (Bricout, 1978). Farquhar *et al.* (1997b) reported that cellulose of *Banksia prionotes* was up to 9.2‰ more enriched when compared to whole leaf tissue.

3.6.5 *Summary of metabolic effects*

Until now it has been assumed that there is complete equilibration of oxygen in carbonyl groups with water, and that oxygen in cellulose formed from newly synthesised

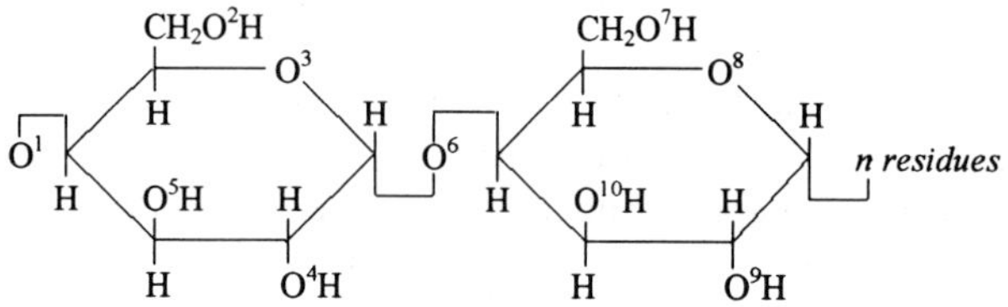

Figure 3.5. *Molecular structure of a cellobiose unit (Tarchevskey and Marchenko, 1991).*

sucrose is expected to preserve about half the signal caused by variation in composition at the sites of evaporation in the leaf. This is because 45% of oxygens in cellulose are expected to exchange with water in the expanding cell when cellulose is synthesised from sucrose (Sternberg *et al.*, 1986, Saurer *et al.*, 1997). Given that chloroplast, mitochondria and cytosol water may be slightly closer in composition to that of the source water (compared to water at the sites of evaporation), due to the Péclet effect (Farquhar and Lloyd, 1993; see also Chapter 10), this estimate could decrease.

By following the biochemical steps leading to sucrose and cellulose synthesis, we suggest the possibility that complete equilibration between water and all carbonyl oxygens may not occur. As pool sizes of G3P are lower than that of DHAP by about 20: 1, the carbonyl oxygen in G3P (O-1) will have substantially less time for isotopic exchange with water. We suggest that the signal from CO_2 and O_2 fixed during carboxylation and oxygenation may not be completely lost by isotopic exchange. This implies that sucrose may not be in isotopic equilibrium with cellular water and that differences in ϕ (the ratio of oxygenation to carboxylation) may be evident in $\delta^{18}O$ of cellulose. Experiments are needed to test the extent of equilibration.

3.7 The potential for oxygen isotopes in evaluating plant water use

3.7.1 *Variation in stomatal conductance revealed in organic oxygen isotope composition*

From the foregoing we conclude that factors that affect the isotopic composition of leaf water may be revealed in leaf organic matter. Farquhar *et al.* (1989) suggested that it might be possible to recover leaf–air humidity differences to augment the information in carbon isotope composition that is used in water-use efficiency studies (see Chapter 9). Since greater stomatal conductance cools the leaf, reducing the intercellular vapour pressure, e_i, it seemed feasible to us that organic $\delta^{18}O$ might be a useful tool to pick up differences in stomatal conductance. This would enable one to distinguish the effects of greater conductance from lower photosynthetic capacity, in material with increased carbon isotope discrimination, thereby removing considerable ambiguity. We have demonstrated this to be the case (Farquhar *et al.*, unpublished data). For cotton plants, where the effects of humidity on $\Delta^{18}O$ are, we suggest, combinations of effects on Δ_e and on $\mathscr{P}$. If these are combined, the overall sensitivity of organic material $\Delta^{18}O$ to e_a/e_i was -11.5‰, that being 0.12‰ per 1% change in relative humidity. This compares well with tree ring data showing a slope for $\delta^{18}O_{cellulose}$ *versus* relative humidity of -0.12‰, which is one third of the slope for the relationship Δ_e *versus* e_a/e_i see Chapter 18. Stomatal conductance, when varied by edaphic effects (soil compaction and water content), hormonal means (abscisic acid), and genetic effects (genotypes of cotton and of wheat), can affect $\delta^{18}O$. We showed that the signal is recoverable in leaf organic matter, and quickly, using the technique described in Section 3.8. One exciting result was found among lines of wheat released over 26 years from the wheat breeding centre, CIMMYT, which were grown in one site for comparative purposes. Improvement in yield with year of release has correlated with increase of stomatal conductance (Fischer *et al.*, 1997), and reduction of $\delta^{18}O$ in organic matter. These trials were for yield under well-watered conditions. Thus, stable isotopes appear to be useful for studies of both water-use efficiency in water-limited environments (via carbon isotope discrimination – see Farquhar and

Richards, 1984; Farquhar *et al.*, 1989; see Chapter 9), and, now, yield potential in well watered conditions (Farquhar *et al.*, 1994) as determined via $\delta^{18}O$.

3.7.2 *Relationship between carbon and oxygen isotopic composition*

The inter-relationships between changing stomatal conductance and internal mesophyll limitations to photosynthetic carbon isotope discrimination are discussed in detail in Chapters 8 and 9. In addition, the interplay between leaf-water ^{18}O signal and coupling of gas exchange is explored in Chapters 10, 12 and 24. It is evident that the extent of coupling between discrimination (measured as $\delta^{13}C$ or $\delta^{18}O$ of organic matter minus that of source) is dependent on a variety of external and internal factors. In this chapter, however, we present the theory to aid integration of the ^{13}C and ^{18}O signal remaining in leaf organic material. When humidity is the underlying source of variation among a set of samples, as humidity increases both $\delta^{13}C$ and $\delta^{18}O$ will decrease, with $\delta^{13}C_{PDB}$ becoming more negative and $\delta^{18}O_{SMOW}$ becoming less positive. That is, the isotopic composition of C and O should then be positively related. This should also be the case when the source of variation is stomatal conductance. A positive relation has been observed by Saurer *et al.* (1997). However, when the source of variation is increasing photosynthetic capacity, which will tend to draw down internal CO_2 concentration carbon isotope discrimination will diminish (see Chapters 8 and 9). Thus, if stomata do not respond, when $\delta^{13}C$ becomes less negative, $\delta^{18}O$ is unaffected. Generally, an increase in capacity is accompanied by *some* increase in conductance, so that $\delta^{18}O$ should decrease, if anything. Under these circumstances the relationship between ^{13}C and ^{18}O discrimination should be non-existent or show negative co-variance.

3.8 Oxygen isotope composition of organic matter: methodology and analysis

As part of the background to the theory developed above, the practice of analysing ^{18}O in organic material has also been the subject of considerable development, as recently reviewed by Farquhar *et al.* (1997). Traditionally, analysis of oxygen isotopes has involved decomposition of the sample by pyrolysis and conversion of the oxygen-containing products to carbon dioxide. The focus has been on CO_2 as the target gas because it is readily formed by chemical action and is easy to trap and purify. The Schutze–Unterzaucher procedure (Unterzaucher, 1952) is well established as a technique for oxygen elemental analysis (for example, Doering and Dorfman, 1953; Kirsten, 1978; Gygli, 1993) and, more recently, for isotope analysis (Santrock and Hayes, 1987). The sample is pyrolysed over nickelised carbon at 1060°C to convert all oxygen products to carbon monoxide, which is then oxidised to CO_2 with iodine pentoxide. As well as correcting for any contamination of the oxygen pool in CO_2 by an oxygen blank (associated with the cups used for containing the sample for pyrolysis) and memory (oxygen from previous samples), it is necessary to correct for the oxygen contributed by iodine pentoxide to the CO_2.

Other methods used include those of Rittenberg and Pontecorvo (1956) and Thompson and Gray (1977). The first, involving pyrolysing the sample in the presence of mercuric chloride at around 500°C, has been improved more recently by Schimmelmann and DeNiro (1985). A mixture of CO and CO_2 is formed and these two species have been found to differ in isotopic composition. Epstein *et al.* (1977)

reported that 5 – 20% of the oxygen from cellulose was not recovered as CO or CO_2 using this method, but stated that this did not affect the accuracy or precision of their measurements. Thompson and Gray (1977) describe a method in which the sample is pyrolysed in an evacuated nickel vessel and the CO formed is converted to CO_2 in a discharge. During the electrical discharge process nitrogen oxides may be formed making it unsuitable for samples containing nitrogen such as organic material since $^{14}N^{16}O_2$ (mass 46) interferes with measurement of $^{12}C^{18}O^{16}O$ shifting $\delta^{18}O$ by up to 10‰ for only 50ppm NO_2. Schimmelmann and DeNiro (1985) determined oxygen isotope ratios of compounds containing nitrogen after quantitatively disproportion-ating the CO to CO_2 on metallic nickel. Whilst these methods avoid oxidising the sample with an outside source, they are time-consuming, and the number of steps involved increases the potential for contamination. The isotopic ratio of oxygen in water has been measured by variations of the pyrolysis/disproportionation technique (Brenninkmeijer and Mook, 1981; Fehri et al., 1983), reaction with guanidine hydrochloride to produce CO_2 (Dugan et al., 1985) and also by the CO_2 equilibration method (Epstein and Mayeda, 1953; Talesk and Daugherty, 1993). The last method gives good precision (0.05‰) and is available commercially as a fully automated sys-tem but has the disadvantages of requiring large sample size and being time-consum-ing (Scrimgeour et al., 1993; Scrimgeour, 1995).

We have been working on a new method since the 1980s and recently published a description (Farquhar et al., 1997) that enables organic samples (solid or liquid, and with or without nitrogen) to be analysed. The technique involves pyrolysis at 1080°C to 1100°C using a modified Unterzaucher procedure. The approach was used by Brand et al. (1994) to analyse the isotopic composition of water. Begley and Scrimgeour (1996, 1997) applied this technique to both water and volatile organic compounds. Werner et al. (1996) and Koziet (1997) recently reported analyses of solid organic mat-ter using a similar method, but did not explore the modifications required to cope with the presence of N in the material. We have minimised contact between the hot (1100°C) carbon and the quartz walls of the pyrolysis reactor using a nickel foil sleeve (Santrock and Hayes, 1987), and extended the reacting region. Following pyrolysis of the sample any N_2 produced (which has the same molecular weight as CO) is sepa-rated from CO using a gas chromatograph column (Molecular Sieve 5A at 50°C and 120 ml min^{-1} ultra-high purity He carrier gas), allowing application of the technique to nitrogen-containing compounds. The CO is then introduced into the inlet of an Isochrom mass spectrometer (Micromass UK Ltd). We obtain a precision of 0.2‰, and the technique takes 7 min per sample. The effects of aging of the pyrolysis column and the efficiency of pyrolysis are currently the focus of our attention.

In conclusion, we hope that the combination of theory and practice outlined in this Chapter will contribute to the study of the ^{18}O signal in plant material, with implica-tions for other studies reported subsequently in this volume.

Acknowledgements

We wish to acknowledge John Andrews, A.N.U., for advice and the Australian Rural Industries Research & Development Corporation, the Cotton Research & Development Corporation, the National Greenhouse Gas Inventory Committee and Micromass UK Ltd. for their support.

References

Aucour, A.-M., Hillaire-Marcel, C., Bonnefille, R. (1996) Oxygen isotopes in cellulose from modern and quaternary intertropical peatbags: implications for palaeohydrology. *Chem. Geol.* **129**, 341–359.

Badger, M.R. and Price, G.D. (1994) The role of carbonic anhydrase in photosynthesis. *Ann. Rev. Plant Physiol. Plant Mol. Biol.* **45**, 369–392.

Badger, M.R., Sharkey, T.D. and von Caemmerer S. (1984) The relationship between steady-state gas exchange of bean leaves and the levels of carbon-reduction-cycle intermediates. *Planta* **160**, 305–313.

Bariac, T., Gonzalez-Diuna, J., Kataerji, N., Bethenod, O., Bertrolini, J.M. and Mariotti, A. (1994) Spatial variation of the isotopic composition of water (^{18}O, ^2H) in organs of aerophytic plants. 1. Assessment under laboratory conditions. *Chem. Geol. including Isotope Geosci.* **115**, 307–315.

Begley, I.S. and Scrimgeour, C.M. (1996) Online reduction of H_2O for delta – ^2H and delta – ^{18}O Measurement by Continuous-flow isotope ration mass spectrometry. *Rap.Comm.Mass. Spectrom.* **10**, 969–973.

Begley, I.S. and Scrimgeour, C.M. (1977) High precision δ^2H and $\delta^{18}O$ measurement for water and volatile organic compounds by continuous flow pyrolysis isotope ratio mass spectrometry. *Anal. Chem.* **69**, 1530–1535.

Benkovic, S. J. and deMaine, M.M. (1982) Mechanism of action of fructose 1,6-bisphosphatase. *Advan. Enzymol.* **53**, 45–82.

Berry, J.A., Osmond, C.B. and Lorimer, G.H. (1978) Fixation of $^{18}O_2$ during photorespiration. *Plant Physiol.* **62**, 954–967.

Bottinga, Y. and Craig, H. (1969) Oxygen isotope fractionation between CO_2 and water, and the isotopic composition of marine atmospheric CO_2. *Earth Plan. Sci. Lett.* **5**, 285–295.

Brand, W.A., Tegtmeyer, A.R. and Hilkert, A. (1994) Compound-specific isotope analysis: extending toward $^{15}N/^{14}N$ and $^{18}O/^{16}O$. *Org. Geochem.* **21**, 586–594.

Brenninkmeijer, C.A. and Mook, W.G. (1981) A batch process for direct conversion of organic oxygen and water to CO^2 for $^{18}O/^{16}O$ analysis. *Int. J. Appl. Radiation and Isotopes* **32**, 137–141.

Bricout, J. (1978) Recherches sur la fractionnement des isotopes stable de l'hydrogène et de l'oxygène dans quelques végétaux. *Revue de Cytologie, Biologie et Vegetable – Botansie* **I**, 133–209.

Bricout, J., Fontes, J.Ch., Letolle, R., Mariotti, A. and Merlivat, L. (1975) Essai de caractérisation de certaines substances organiques et minérales par leur composition en isotopes stables. In: *Isotopes Ratios as Pollutant Source and Behaviour Indicators*, proceedings of a symposium, Vienna.

Burk, R.L. (1979) Factors affecting oxygen-18/oxygen-16 ratios in cellulose. Ph.D. Dissertation, VMI Ann. Arbor.

Christeller, J. T. and Tolbert, N.E. (1978) Mechanism of phosphoglycolate phosphatase. *J. Biol. Chem.* **253**, 1791–1798.

Cooper, L.W. and De Niro, M.J. (1989) Covariance of oxygen and hydrogen isotopic composition in plant water: species effects. *Ecology* **70**, 1619–1628.

Cowan, I.R. (1986) Economics of carbon fixation in higher plants. In: *On the economy of plant form and function.* (ed. T.J. Givnish) Cambridge University Press, Cambridge, pp. 133–170.

Craig, H. and Gordon, L.I. (1965) Deuterium and oxygen-18 variations in the ocean and the marine atmosphere. In: Proceedings of a Conference on Stable Isotopes in Oceanographic Studies and Palaeotemperatures (ed. T. Tongiorgi). Spoleto, Italy, pp. 9–130.

De Niro, M.J. and Epstein, S. (1979) Relationship between oxygen isotope ratios of terrestrial plant cellulose, carbon dioxide and water. *Science,* **204**, 51–53.

DeNiro, M.J. and Epstein, S. (1981) Isotopic composition of cellulose from aquatic organisms. *Geochim. et Cosmochim. Acta* **45**, 1885–1894.

De Niro, M.J., Sternberg, L.D., Marino, B.D. and Druzik, J.R. (1988) Relation between D/H ratios and $^{18}O/^{16}O$ ratios in cellulose from linen and maize – Implications for paleoclimatology and for sindonology. *Geochim. et Cosmochim. Acta* 52, 2189–2196.

DeNiro M.J. and Cooper L.W. (1989) Post-photosynthetic modification of oxygen isotope ratios of carbohydrates in the potato: Implications for palaeoclimatic reconstruction based upon isotopic analysis of wood cellulose. *Geochim. Cosmochim. Acta* 53, 2573–2580.

Doering, W. von and Dorfman, E. (1953) Mechanism of the peracid ketone-ester conversion. Analysis of organic compounds for oxygen-18. *J. Am. Chem. Soc.* 75, 5595–5598.

Dongmann, G., Nurnberg, M.W., Forstel, M. and Wagner, R.K. (1974) On the enrichment of $H_2^{18}O$ in the leaves of transpiring plants. *Radiation Environ. Biophys.* 11, 41–52.

Dugan, J.P.J., Borthwick, J., Harmon, R.S., Gagnier, M.A., Glahn, J.E., Kinsel, E.P., MacLeod, S. and Viglino, J.A. (1985) Guanidine hydrochloride method for determination of water oxygen isotope ratios and the oxygen-18 fractionation between carbon dioxide and water at 25°C. *Anal. Chem.* 57, 1734–1736.

Epstein, S., Thompson, P. and Yapp, C.J. (1977) Oxygen and hydrogen isotopic ratios in plant cellulose. *Science* 198, 1209.

Epstein, S. and Mayeda, T.K. (1953) Variation of ^{18}O content of water from natural sources. *Geochim. et Cosmochim. Acta* 4, 213–224.

Evans, J.R. (1983) *Photosynthesis and Nitrogen Partitioning in leaves of* Triticum aestivum L. *and related species.* Ph.D. thesis, Australian National University, Canberra.

Evans, J.R., Sharkey, T.,D., Berry, J.A. and Farquhar, G.D. (1986) Carbon isotope discrimination measured concurrently with gas exchange to investigate CO_2 diffusion in leaves of higher plants. *Aust. J. Plant Physiol.* 13, 281–292.

Farquhar, G.D. and von Caemmerer, S. (1982) Modelling of photosynthetic response to environmental conditions. In: *Encyclopedia of Plant Physiology New Series vol. 12b.* (eds. O.L.Lange, P.S.Nobel, C.B.Osmond and H.Ziegler) Springer, Heidelberg, pp. 549–587.

Farquhar, G.D. and Richards, R.A. (1984) Isotopic composition of plant carbon correlates with water use efficiency of wheat genotypes. *Aust J. Plant Physiol.*

Farquhar, G.D., Ehleringer, J.R. and Hubick, K.T. (1989) Carbon isotope discrimination in photosynthesis. *Ann. Rev. Plant. Physiol Mol. Biol.* 40, 503–537.

Farquhar, G.D. and Lloyd, J. (1993) Carbon and oxygen isotope effects in the exchange of carbon dioxide between terrestrial plants and the atmosphere. In: *Stable Isotopes and Plant Carbon-Water Relations.* (eds. J.R. Ehleringer, A.E. Hall and G. D. Farquhar.) Academic Press, San Diego, pp. 47–70

Farquhar, G.D., Lloyd, J. Talor, J.A., Flanagan, L.B., Syversten, J.P., Hubick, K.T. Chin Wong, S.C. and Ehleringer, J.R. (1993) Vegetation effects on the isotope composition of oxygen in atmospheric CO_2. *Nature* 363, 439–443.

Farquhar, G.D., Condon, A.G. and Masle, J. (1994) On the use of carbon and oxygen isotope composition and mineral ash content in breeding for improved rice production under favourable, irrigated conditions. In: *Breaking the Yield Barrier.* (ed. K.G. Cassman). International Rice Research Institute, Manilla, pp. 95–101.

Farquhar, G.D., Henry, B.K. and Styles, J.M. (1997) A rapid on-line technique for determination of oxygen isotope composition of nitrogen-containing organic matter and water. *Rapid Communications in Mass Spectrometry* 11, 1554–1560.

Farris, F. and Strain, B.R. (1978) The effects of water stress on leaf $H_2^{18}O$ enrichment. *Radiation and Environ. Biophys.* 15, 167–202.

Ferhi, A.M., Lerman, J.C. and Letolle, R. (1975) Oxygen isotope ratios of organic matter analysis of natural composition. Second International Conference on Stable Isotopes, Oakbrook, USA.

Fehri, A. and Lotelle, R. (1977) Transpiration and evaporation as the principal factors in oxygen isotopes variations of organic matter in land plants. *Physiol. Veg.* 15, 363–370.

Fehri, A., Bariac, T., Jusserand, C. and Letolle, R. (1983) An integrated method for isotopic analysis of oxygen from organic compounds, air, water vapour and leaf water. *Int. J. Appl. Radiation and Isotopes* **34**, 1451–1457.

Fischer, R.A., Rees, D., Sayre, K.D., Lu, Z., Condon, A.G., Larquesaavedra, A. and Zeiger, E. (1997) Wheat yield progress is associated with higher stomatal conductance, higher photosynthetic rate and cooler canopies. *Crop Sci.* in press.

Flanagan, L.B., Comstock, J.P. and Ehleringer, J.R. (1991) Comparison of modeled and observed environmental influences on the stable oxygen and hydrogen isotope composition of leaf water in *Phaseolus vulgaris* L. *Plant Physiol.* **96**, 588–596.

Flanagan, L.B. (1993) Environmental and biological influences on the stable oxygen and hydrogen isotopic composition of leaf water. In: *Stable Isotopes and Plant Carbon-Water Relations* (eds. J.R.Ehleringer, A.E.Hall and G.D.Farquhar), Academic Press, London. pp. 71–90.

Flanagan, L.B., Marshall, J.D. and Ehleringer, J.R. (1993) Photosynthetic gas exchange and the stable isotope composition of leaf water: comparison of a xylem-tapping mistletoe and its host. *Plant, Cell Environ.* **16**, 623–631.

Flanagan, L.B., Philips, S.L., Ehleringer, J.R., Lloyd, J. and Farquhar, G.D. (1994) Effect of changes in leaf water oxygen isotopic composition on discrimination against $C^{18}O^{16}O$ during photosynthetic gas exchange. *Aust. J. Plant Physiol.* **21**, 221–234.

Gonfiantini, R., Gratziu, S. and Tongiorgi, E. (1965) Oxygen isotopic composition of water in leaves. In: *Isotope Atomic Energy Commission*, Vienna. pp 405–410.

Gray, J. and Thompson, P. (1976) Climatic information from $^{18}O/^{16}O$ ratios of cellulose in tree rings. *Nature* **262**, 481–482.

Gray, J. and Thompson, P. (1977) Climatic information from $^{18}O/^{16}O$ analysis of cellulose, lignin and whole wood from tree-rings. *Nature* **93**, 325–332.

Guy, R.D., Fogel, M.F. Berry, J.A. Hoering, T.C. (1987) Isotope fractionation during oxygen production and consumption by plants. In: *Progress in Photosynthesis Research.* (ed. J. Biggins). Martinus Nijhoff, Dordrecht, pp. 597–600.

Gygli, A. (1993) Microdetermination of oxygen in organic compounds using a glassy carbon pyrolysis tube and nondispersive infrared detection. *Mikrochimica Acta* **111**, 37–43.

Hardcastle, K.G. and Friedman, I. (1974) A method for oxygen isotope analysis of organic material. *Geophys. Res. Lett.* **1**, 165–167.

Harwood, K.G., Gillon, J.S., Griffiths, H. and Broadmeadow, M.S.J. (1998) Diurnal variation of $\Delta^{13}CO_2$, $\Delta C^{18}O^{16}O$ and evaporative site enrichment of $\delta H_2^{18}O$ in *Piper aduncum* under field conditions in Trinidad. *Plant, Cell Environ.*, in press.

Hatzfeld, W.-D. and Stitt, M. (1990) A study of the rate of recycling of triose phosphates in heterotrophic *Chenopodium rubrum* cells, potato tubers, and maize endosperm. *Planta* **180**, 198–204.

Hill, S.A. and ap Rees, T. (1994) Fluxes of carbohydrate metabolism in ripening bananas. *Planta* **192**, 52–60.

Hill, S.A., Waterhouse, J.S., Field, E.M., Switsur, V.R. and ap Rees, T. (1995) Rapid recycling of triose phosphates in oak stem tissue. *Plant, Cell Environ.* **18**, 931–936.

Kays, W.M. (1966) Convective Heat and Mass Transfer. McGraw-Hill, New York.

Keeling, P.L., Wood, J.R., Tyson, R.H. and Bridges, I.G. (1988) Starch biosynthesis in developing wheat grain. Evidence against direct involvement in the metabolic pathway. *Plant Physiol.* **87**, 311–319.

Kirsten, W.J. (1978) Micro and trace determination of oxygen in organic compounds. *Anal. Chim. Acta.* **100**, 2799–288.

Koziet, J. (1997) Isotope ratio mass spectrometric method for the on-line determination of oxygen-18 in organic matter. *J. Mass Spectrom.* **32**, 103–108.

Leaney, F.W., Osmond, C.B., Allison, G.B. and Ziegler, H. (1985) Hydrogen – isotope composition of leaf water in C_3 and C_4 plants: its relationship to the hydrogen-isotope composition of dry matter. *Planta* **164**, 215–220.

Lesaint, C., Mercivat, L. Bricout, J, Fontes, J.C. and Gautheret, R. (1974) Sur la composition en isotopes stable de l'eau de la tomate et du mais. *C.R. Acad.Sci. Paris Ser. D.* **278**, 2925–1930.

Libby, L.M. (1972) Multiple thermometry in paleoclimate and historic climate. *J. Geophys. Res.* 77, 4310–4317.

Libby L.M. and Pandolfi L.J. (1973) Temperature dependence of isotopic ratios in tree rings. *Proc. Natl. Acad. Sci.* 71, 2482 – 2486.

Libby, L.M., Pandolfi, L.J., Payton, P.H., Marshall, J. III, Becker, B. and Giertz-Sienbenlist, V. (1976) Isotopic tree thermometers. *Nature* 261, 284–288.

Merlivant, L. (1978) Molecular diffusivities of $H_2^{18}O$ in gases. *J. Chemical Physics* 69, 2864–2871.

Model, P., Ponticorvo, L., and Rittenberg, D. (1968) Catalysis of an oxygen-exchange reaction of fructose-1,6-diphosphate and fructose-1-phosphate with water by rabbit muscle aldolase. *Biochemistry* 7, 1339–1347.

O'Neil, J.R., Adami, L.H. and Epstein S. (1975) Revised value for ^{18}O fractionation between CO_2 and H_2O at 25°C. *Journal of the Research of the U.S. Geological Survey* 3, 623–624.

Rawn, J.D. (ed.) (1989) 'Biochemistry'. Neil Patterson Publishers, Burlington, NC.

Reynolds, S.J., Yates, D.W. and Pogson, C.I. (1971) Dihydroxyacetone phosphate. *Biochem. J.* 122, 285–297.

Rittenberg, D. and Ponticorvo, L. (1956) A method for the determination of the ^{18}O concentration of oxygen and organic compounds. *Int. J. Appl. Radiation and Isotopes* 1, 208–214.

Samuel, D. and Silver, B.L. (1965) Oxygen isotope exchange reactions of organic compounds. *Advances Phys. Organ. Chem.* 3, 1885–1895.

Santrock, J. and Hayes, J.M. (1987) Adaption of the Unterzaucher procedure for determination of oxygen-18 in organic substances. *Anal. Chem.* 59, 119–127.

Saurer, M., Aellen, K. and Siegwolf, R. (1997) Correlating $\delta^{13}C$ and $\delta^{18}O$ in cellulose of trees. *Plant, Cell Environ.* (in press).

Schimmelmann, A. and DeNiro, M.J. (1985) Determination of oxygen stable isotope ratios in organic matter containing carbon, hydrogen, oxygen and nitrogen. *Anal. Chem.* 57, 2644–2646.

Scrimgeour, C.M. (1995) Measurement of plant and soil water isotope composition by direct equilibration methods. *J. Hydrology,* 172, 261–274.

Scrimgeour, C.M., Rollo, M.M., Mudambo, S.M.K.T., Handley, L.L. and Prosser, S.J. (1993) A simplified method for deuterium hydrogen isotope ratio measurements on water samples of biological origin. *Biol. Mass Spec.* 22, 383–387.

Sternberg, L. and DeNiro, M. (1983) Biogeochemical implications of the isotopic equilibrium fractionation factor between oxygen atoms of acetone and water. *Geochim. et Cosmochim. Acta* 47, 2271–2274

Sternberg, L., DeNiro, M. and Savidge, R. (1986) Oxygen isotope exchange between metabolites and water during biochemical reactions leading to cellulose synthesis. *Plant Physiol.* 82, 423–427.

Sternberg, L.S.L. (1989) Oxygen and hydrogen isotope ratios in plant cellulose: mechanisms and applications. In: Stable Isotopes in Ecological Research (eds. P.W.Rundel, J.R. Ehleringer, and K.A.Nagy). Springer Verlag, New York, pp. 124–141.

Taiz, L., and Zeiger, E. (eds.) (1991) *Plant Physiology.* (Benjamin/Cummings Publishing Company, Redwood City, CA.

Talesk, R.T. and Daugherty, K.E. (1993) Analysis of carbon monoxide by molecular sieve tapping. *J. Chromatogr.* 639, 221–226

Tarchevsky, I.A. and Marchenko, G.N. (Eds.) (1991) Cellulose; Biosynthesis and Structure. Springer, Berlin.

Terashima, I., Wong, S-C., Osmond,C.B. and Farquhar, G.D. (1988) Characterisation of non-uniform photosynthesis induced by abscisic acid in leaves having different mesophyll anatomies. *Plant Cell Physiol.* 29, 385–394.

Thompson, P. and Gray, J. (1977) Determination of the $^{18}O/^{16}O$ ratios in compounds containing C, H and O. *Int. J. Appl. Radiation and Isotopes* 28, 411 – 415.

Unterzaucher, J. (1952) The direct micro-determination of oxygen in organic substances. *Int. Congress on Anal. Chem.* **77**, 584–595.

Viola, R., Davies, H.V. and Chudeck, A.R. (1991) Pathways of starch and sucrose biosynthesis in developing tubers of potato (*Solanum tuberosum* L.) and seeds of faba bean (*Vicia faba* L.). *Planta* **183**, 202–208.

Walker, C.D., Leaney, F.W. Dighton, J.C. and Allison, G.B. (1989) The influence of transpiration on the equilibration of leaf water with atmospheric water vapour. *Plant, Cell Environ.* **12**, 221–234.

Wang, X.F. and Yakir, D. (1995) Temperal and spatial variations in the oxygen – 18 content of leaf water in different plant species. *Plant, Cell Environ.* **18**, 1377–1385.

Werner, R.A., Kornexl, A., Roßmann, A. and Schmidt, H.-L. (1996) On-line determination of $\delta^{18}O$-values of organic substances. *Anal. Chim. Acta.* **319**, 159–164.

White, J.W.C. (1989) Stable hydrogen isotope ratios in plants: a review of current theory and some potential applications. In: *Stable Isotopes in Ecological Research* (eds. P.W.Rundel, J.R. Ehleringer and K.A. Nagy), Springer, Berlin, pp. 142–162.

Wilson, A.T. (1978) A reply to 'Climatic interpretation of $\delta^{18}O$ and $\delta\Delta$ in tree rings' by Wigley, T.M.L, Gray, B.M. and Kelly, P.M. *Nature* **271**, 92–93.

Yakir, D., De Niro, M.J. and Rundel, P.W. (1989) Isotopic inhomogeneity of leaf water: evidence and implications for the use of isotopic signals transduced by plants. *Geochim. Cosmochim. Acta* **53**, 2769 – 2773.

Yakir, D., Berry, J.A., Giles, L and Osmond C.B. (1994) Isotopic heterogeneity of water in transpiring leaves: identification of the component that controls the ^{18}O of atmospheric O_2 and CO_2. *Plant, Cell Environ.* **17**, 73–80.

Zimmerman, U., Ehhalt, D. and Munnich, K.O. (1967) Soil water movement and evaporation: changes in the isotopic composition of water. In: *Proc Symp Isotopes in Hydrology*. International Atomic Energy Agency, Vienna. pp. 567–584.

Zundel, G. Miekeleyn, Grisi, B.M. and Förstel, H. (1978) In $H_2^{18}O$ enrichment in the leaf water of tropic trees: Comparison of Species from the tropical rain forest and Semi-Arid region in Brazil. *Radiation and Environ. Biophys.* **15**, 203–212.

Appendix 1

A.1. *Biochemical pathways leading to cellulose synthesis*

The oxygen atoms in the molecules of the following pathways are numbered according to the number of the carbon atom to which they are bonded. The numbers on the right of reaction follow those in the starting molecules so that each oxygen may be traced, but labels revert to the numbering system at the beginning of every new reaction.

Scheme 1. *The Calvin Cycle (Robinson and Walker, 1981)*

1. *ribulose 1,5-bisphosphate + $C^{6,7}O_2$ + $H_2^8O \to 2(3$-phosphoglycerate) + 2H^+*

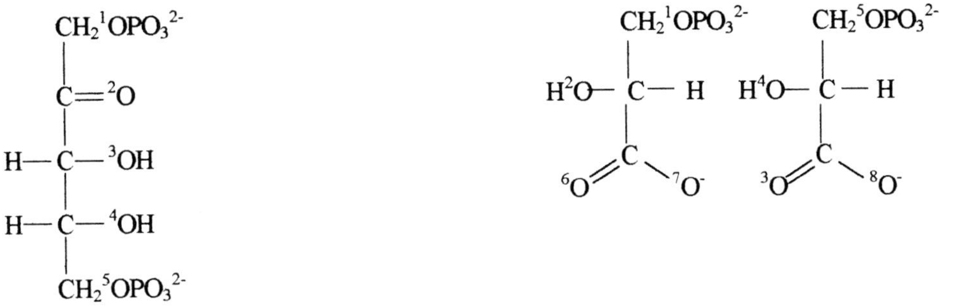

$$CH_2{}^1OPO_3{}^{2-}$$
$$|$$
$$C={}^2O$$
$$|$$
$$H-C-{}^3OH$$
$$|$$
$$H-C-{}^4OH$$
$$|$$
$$CH_2{}^5OPO_3{}^{2-}$$

$$CH_2{}^1OPO_3{}^{2-} \qquad CH_2{}^5OPO_3{}^{2-}$$
$$H^2O-C-H \qquad H^4O-C-H$$
$$\quad\; {}^6O{=}C{-}{}^7O^- \qquad {}^3O{=}C{-}{}^8O^-$$

2. *3-phosphoglycerate + ATP $\longrightarrow$ 1,3-bisphosphoglycerate + ADP*

$$CH_2{}^3OPO_3{}^{2-}$$
$$H^2O-C-H$$
$${}^{1a}O{=}C{-}{}^{1b}O^-$$

$$CH_2{}^3OPO_3{}^{2-}$$
$$H^2O-C-H$$
$${}^{1a}O{=}C{-}{}^{1b}OPO_3{}^{2-}$$

3. *1,3-bisphosphoglycerate + NADPH + H^+*
gives glyceraldehyde 3-phosphate + NADP^+ + HOPO_3^{2-}

$$CH_2{}^3OPO_3{}^{2-}$$
$$H^2O-C-H$$
$${}^{1a}O{=}C{-}{}^{1b}OPO_3{}^{2-}$$

$$CH_2{}^3OPO_3{}^{2-}$$
$$H^2O-C-H$$
$${}^{1a,b}O{=}C{-}H$$

4. *glyceraldehyde 3-phosphate $\longrightarrow$ dihydroxyacetone 3-phosphate*

$$CH_2{}^3OPO_3{}^{2-}$$
$$H^2O-C-H$$
$${}^1O{=}C{-}H$$

$$CH_2{}^3OPO_3{}^{2-}$$
$$|$$
$$C={}^2O$$
$$|$$
$$CH_2{}^1OH$$

5. *glyceraldehyde 3-phosphate + dihydroxyacetone 3-phosphate gives fructose-1,6-bisphosphate*

$$CH_2{}^3OPO_3{}^{2-}$$
$$H^2O-\overset{\displaystyle|}{C}-H$$
$$\underset{{}^1O}{\overset{\displaystyle|}{C}}{\diagdown}_H$$

$$CH_2{}^4OH$$
$$\overset{\displaystyle|}{C}={}^5O$$
$$\underset{CH_2{}^6OPO_3{}^{2-}}{\overset{\displaystyle|}{}}$$

${}^{-2}O_3P^3OH_2C$ — ${}^2O$ — ${}^5OH$, H, H, $CH_2{}^6OPO_3{}^{2-}$, ${}^1OH$ ${}^4OH$

6. *fructose 1,6-bisphosphate + H_2O* $\longrightarrow$ *fructose 6-phosphate + $HOPO_3{}^{2-}$*

${}^{-2}O_3P^6OH_2C$ — ${}^5O$ — ${}^2OH$, H, H, $CH_2{}^1OPO_3{}^{2-}$, H^4O ${}^3OH$

${}^{-2}O_3P^6OH_2C$ — ${}^5O$ — ${}^2OH$, H, H, $CH_2{}^1OH$, H^4O ${}^3OH$

7. *fructose 6-phosphate + glyceraldehyde 3-phosphate*

${}^{-2}O_3P^6OH_2C$ — ${}^5O$ — ${}^2OH$, H, H, $CH_2{}^1OH$, H^4O ${}^3OH$

$$CH_2{}^9OPO_3{}^{2-}$$
$$H^8O-\overset{\displaystyle|}{C}-H$$
$$\underset{{}^7O}{\overset{\displaystyle|}{C}}{\diagdown}_H$$

gives erythrose 4-phosphate + xylulose 5-phosphate

$$H{\diagdown}\overset{{}^3O}{C}$$
$$H-\overset{\displaystyle|}{C}-{}^4OH$$
$$H-\overset{\displaystyle|}{C}-{}^5OH$$
$$\underset{CH_2{}^6OPO_3{}^{2-}}{}$$

$$CH_2{}^1OH$$
$$\overset{\displaystyle|}{C}={}^2O$$
$$H^7O-\overset{\displaystyle|}{C}-H$$
$$H-\overset{\displaystyle|}{C}-{}^8OH$$
$$\underset{CH_2{}^9OPO_3{}^{2-}}{}$$

8. *erythrose 4-phosphate + dihydroxyacetone 3-phosphate gives sedoheptulose 1,7-phosphate*

$$H{\diagdown}\overset{{}^1O}{C}$$
$$H-\overset{\displaystyle|}{C}-{}^2OH$$
$$H-\overset{\displaystyle|}{C}-{}^3OH$$
$$\underset{CH_2{}^4OPO_3{}^{2-}}{}$$

$$CH_2{}^5OH$$
$$\overset{\displaystyle|}{C}={}^6O$$
$$\underset{CH_2{}^7OPO_3{}^{2-}}{}$$

$CH_2{}^4OPO_3{}^{2-}$, H, H, ${}^3O$ ${}^6OH$, H^2O H, H^5O $CH_2{}^7OPO_3{}^{2-}$, ${}^1OH$ H

9. *sedoheptulose 1,7-bisphosphate + H₂O* $\longrightarrow$ *sedoheptulose 7-phosphate + HOPO₃²⁻*

9. *sedoheptulose 1,7-bisphosphate + H_2O* $\longrightarrow$ *sedoheptulose 7-phosphate + $HOPO_3^{2-}$*

$CH_2^{7}OPO_3^{2-}$

$CH_2^{1}OPO_3^{2-}$

$CH_2^{1}OH$

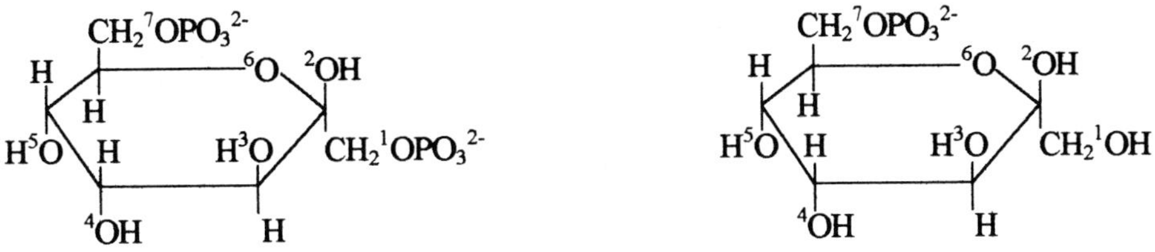

10. *sedoheptulose 7-phosphate + glyceraldehyde 3-phosphate*

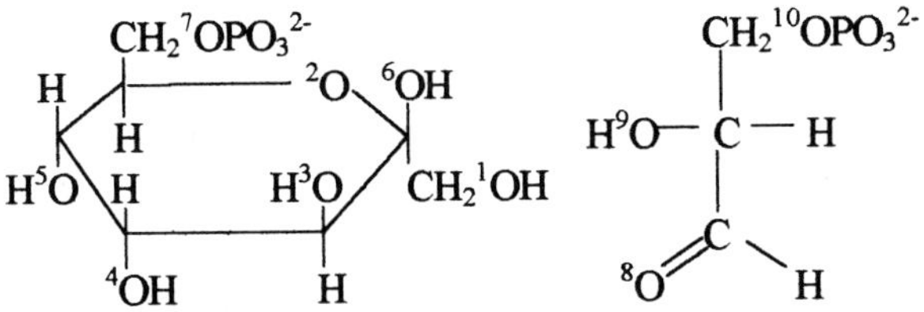

$CH_2^{7}OPO_3^{2-}$

$CH_2^{1}OH$

$CH_2^{10}OPO_3^{2-}$

H^9O-C-H

gives ribose 5-phosphate + xylulose 5-phosphate

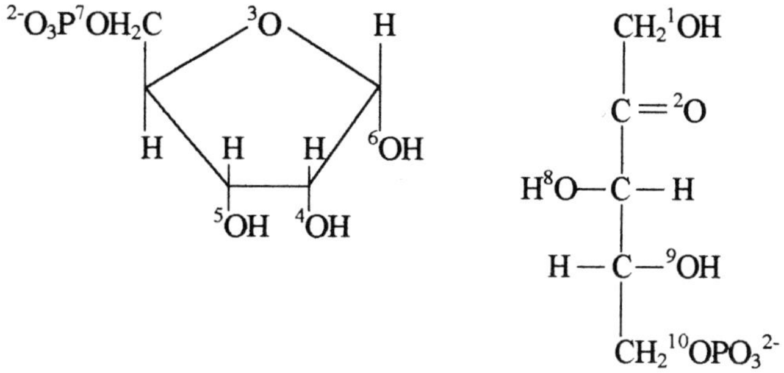

$^{2-}O_3P^7OH_2C$

$CH_2^{1}OH$

$C=^2O$

H^8O-C-H

$H-C-^9OH$

$CH_2^{10}OPO_3^{2-}$

11. *xylulose 5-phosphate* $\longrightarrow$ *ribulose 5-phosphate*

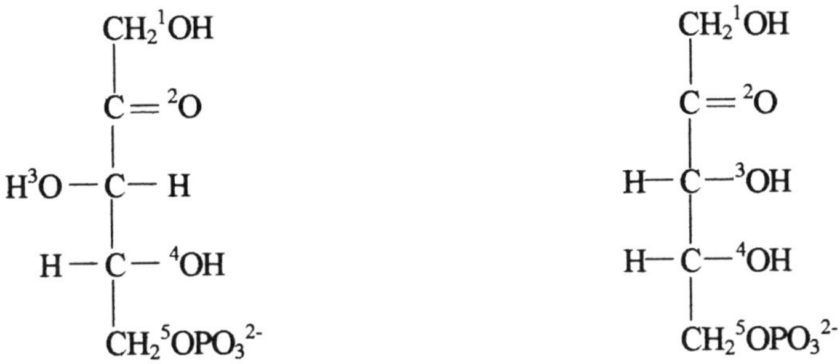

$CH_2^{1}OH$

$C=^2O$

H^3O-C-H

$H-C-^4OH$

$CH_2^{5}OPO_3^{2-}$

$CH_2^{1}OH$

$C=^2O$

$H-C-^3OH$

$H-C-^4OH$

$CH_2^{5}OPO_3^{2-}$

12. *ribose 5-phosphate* ⟶ *ribulose 5-phosphate*

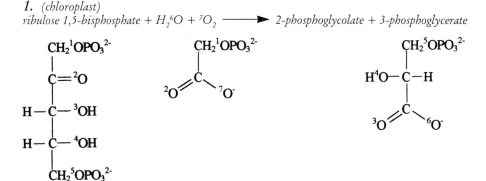

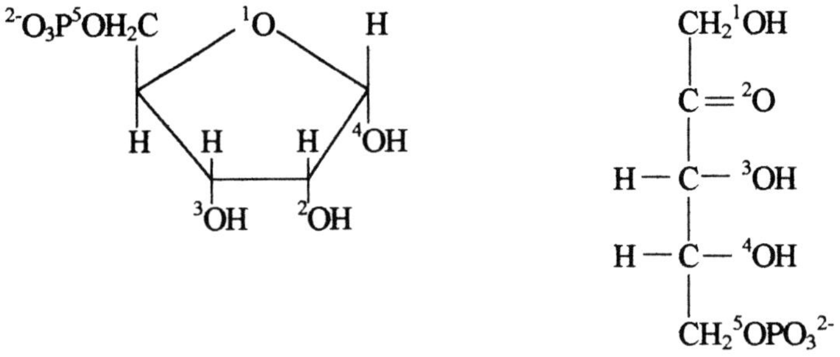

13. *ribulose 5-phosphate + ATP* ⟶ *ribulose 1,5-bisphosphate + ADP + H⁺*

Structures:

$CH_2{}^1OH$
$C={}^2O$
$H-C-{}^3OH$
$H-C-{}^4OH$
$CH_2{}^5OPO_3{}^{2-}$

⟶

$CH_2{}^1OPO_3{}^{2-}$
$C={}^2O$
$H-C-{}^3OH$
$H-C-{}^4OH$
$CH_2{}^5OPO_3{}^{2-}$

Scheme 2. The photorespiratory cycle (Farquhar et al, 1983)

1. *(chloroplast)*
ribulose 1,5-bisphosphate + $H_2{}^6O$ + ${}^7O_2$ ⟶ *2-phosphoglycolate + 3-phosphoglycerate*

2. *(chloroplast)*
phosphoglycolate + H_2O ⟶ *glycolate + $HOPO_3{}^{2-}$*

3. *(peroxisome)*
glycolate + O_2 ⟶ glyoxylate + H_2O_2

$$CH_2{}^2OH$$
$$^{1a}O=C-C-{}^{1b}O^-$$

$$H-C={}^2O$$
$$^{1a}O=C-{}^{1b}O^-$$

4. $2H_2O_2$ ⟶ $2H_2O + O_2$

5. *(peroxisome)*
glyoxylate + glutamate ⟶ glycine + α-ketoglutarate

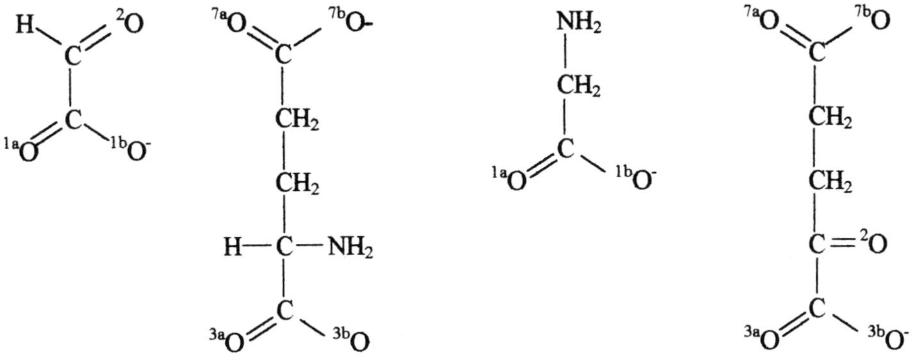

$$H-C={}^2O$$
$$^{1a}O=C-{}^{1b}O^-$$

$$^{7a}O=C-{}^{7b}O^-$$
$$CH_2$$
$$CH_2$$
$$H-C-NH_2$$
$$^{3a}O=C-{}^{3b}O$$

$$NH_2$$
$$CH_2$$
$$^{1a}O=C-{}^{1b}O^-$$

$$^{7a}O=C-{}^{7b}O$$
$$CH_2$$
$$CH_2$$
$$C={}^2O$$
$$^{3a}O=C-{}^{3b}O^-$$

6. *(mitochondrion)*
glycine + NAD+ + H_2-folate ⟶ NADH + H+ + CO_2 + NH_3

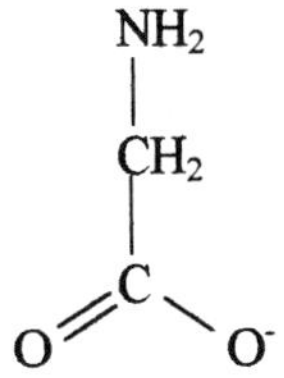

$$NH_2$$
$$CH_2$$
$$O=C-O^-$$

7. *(mitochondrion)*
methylene H_4-folate + H_2O + glycine ⟶ serine + H_4-folate

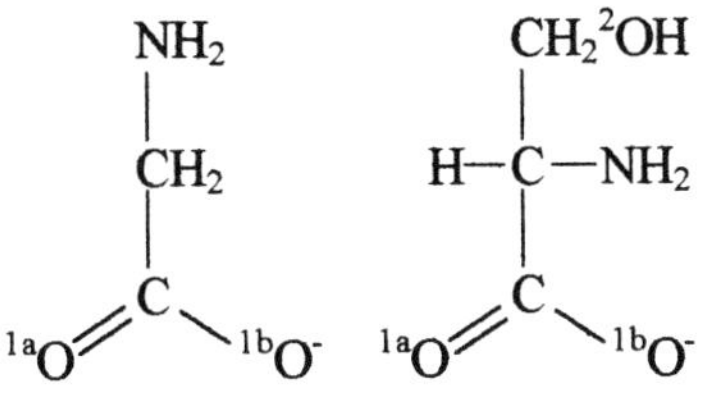

$$NH_2$$
$$CH_2$$
$$^{1a}O=C-{}^{1b}O^-$$

$$CH_2{}^2OH$$
$$H-C-NH_2$$
$$^{1a}O=C-{}^{1b}O^-$$

8. *(peroxisome)*
serine + glyoxylate $\longrightarrow$ *hydroxypyruvate + glycine*

$$\begin{array}{c} CH_2{}^3OH \\ | \\ H-C-NH_2 \\ | \\ {}^{1a}O{=}C{-}{}^{1b}O^- \end{array} \qquad \begin{array}{c} H{-}C{=}{}^5O^- \\ | \\ {}^{4a}O{=}C{-}{}^{4b}O^- \end{array} \qquad \begin{array}{c} CH_2{}^3OH \\ | \\ C{=}{}^5O \\ | \\ {}^{1a}O{=}C{-}{}^{1b}O^- \end{array} \qquad \begin{array}{c} NH_2 \\ | \\ CH_2 \\ | \\ {}^{4a}O{=}C{-}{}^{4b}O^- \end{array}$$

9. *(peroxisome)*
hydroxypyruvate + NADH + H+ $\longrightarrow$ *glycerate + NAD+*

$$\begin{array}{c} CH_2{}^3OH \\ | \\ C{=}{}^2O \\ | \\ {}^{1a}O{=}C{-}{}^{1b}O^- \end{array} \qquad\qquad \begin{array}{c} CH_2{}^3OH \\ | \\ H^2O{-}C{-}H \\ | \\ {}^{1a}O{=}C{-}{}^{1b}O^- \end{array}$$

10. *(chloroplast)*
glycerate + ATP $\longrightarrow$ *3-phosphoglycerate + ADP + H+*

$$\begin{array}{c} CH_2{}^3OH \\ | \\ H^2O{-}C{-}H \\ | \\ {}^{1a}O{=}C{-}{}^{1b}O^- \end{array} \qquad\qquad \begin{array}{c} CH_2{}^3OPO_3{}^{2-} \\ | \\ H^2O{-}C{-}H \\ | \\ {}^{1a}O{=}C{-}{}^{1b}O^- \end{array}$$

Scheme 3. *Sucrose synthesis (Taiz and Zeiger, 1991)*

1. *dihydroxyacetone 3-phosphate* $\longrightarrow$ *glyceraldehyde 3-phosphate*

$$\begin{array}{c} CH_2{}^1OH \\ | \\ C{=}{}^2O \\ | \\ CH_2{}^3OPO_3{}^{2-} \end{array} \qquad\qquad \begin{array}{c} {}^1O{=}C{-}H \\ | \\ H^2O{-}C{-}H \\ | \\ CH_2{}^3OPO_3{}^{2-} \end{array}$$

2. *dihydroxyacetone 3-phosphate + glyceraldehyde 3-phosphate*

gives fructose 1,6-bisphosphate

$$\begin{array}{c} CH_2{}^1OH \\ | \\ C{=}{}^2O \\ | \\ CH_2{}^3OPO_3{}^{2-} \end{array} \qquad \begin{array}{c} {}^4O{=}C{-}H \\ | \\ H^5O{-}C{-}H \\ | \\ CH_2{}^6OPO_3{}^{2-} \end{array} \qquad \begin{array}{c} {}^{2-}O_3P^6OH_2C \quad {}^5O \quad {}^2OH \\ \diagup \qquad \diagdown \\ H \quad H^1O \\ H \qquad CH_2{}^3OPO_3{}^{2-} \\ {}^4OH \quad H \end{array}$$

3. *fructose 1,6-bisphosphate + H₂O* ⟶ *fructose 6-phosphate + HOPO₃⁻²*

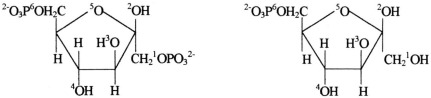

4. *fructose 6-phosphate* ⟶ *glucose 6-phosphate*

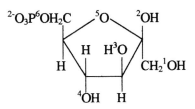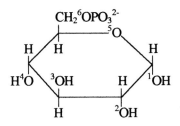

5. *glucose 6-phosphate* ⟶ *glucose 1-phosphate*

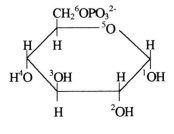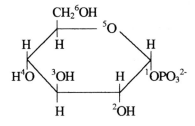

6. *glucose 1-phosphate + UTP* ⟶ *UDP-glucose + H₂P₂O₉²⁻*

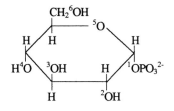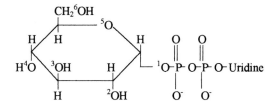

7. *UDP-glucose* + *fructose 6-phosphate*

gives UDP + sucrose 6-phosphate

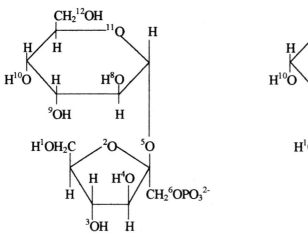

8. *sucrose 6-phosphate + H_2O*   ⟶   *sucrose + $HOPO_3^{2-}$*

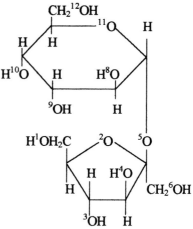

Scheme 4. *Starch synthesis (Taiz and Zeiger, 1991)*

1. *dihydroxyacetone 3-phosphate + glyceraldehyde 3-phosphate*

                                           gives fructose 1,6-bisphosphate

2. *fructose 1,6-bisphosphate + H_2O* $\longrightarrow$ *fructose 6-phosphate + $HOPO_3^{-2}$*

3. *fructose 6-phosphate* $\longrightarrow$ *glucose 6-phosphate*

4. *glucose 6-phosphate* $\longrightarrow$ *glucose 1-phosphate*

5. *glucose 1-phosphate + ATP* $\longrightarrow$ *ADP-glucose + $H_2P_2O_7^{2-}$*

7. *ADP-glucose* $+$ *$(1,4$-α-D-glucosyl$)_n$*

gives $ADP + (1,4\text{-}\alpha\text{-}D\text{-}glucosyl)_{n+1}$

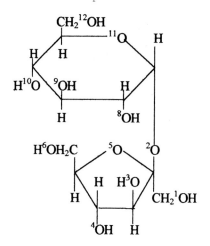

Scheme 5. *Cellulose synthesis (Tarchevskey and Marchenko, 1991, Hill et al., 1995)*

1. *sucrose* $+ H_2^{13}O$ *gives glucose*

and fructose

1b. *glucose 6P*   ⇌   *fructose 6P*

Exchangeable oxygens; 1,2

2. *A proportion of hexose phosphate cycles through triose phosphate*

Fructose 1,6-bisphosphate ⟶ *DHAP and G3P*

H^6OH_2C $\quad^5O\quad$ ^2OH

H H^{3}O

H $\quad CH_2{}^1OH$

^4OH H

$CH_2{}^3OH$

$C = {}^2O$

$CH_2{}^1OPO_3{}^{2-}$

^4O H
$\diagdown C \diagup$

$H^5O - C - H$

$CH_2{}^6OPO_3{}^{2-}$

Exchangeable oxygens: 2,3,4 and 5, as there is a rapid equilibrium between triose-phosphates

DHAP   and   G3P   *Fructose 1,6-bisphosphate*

$CH_2{}^1OH$

$C = {}^2O$

$CH_2{}^3OPO_3{}^{2-}$

^4O H
$\diagdown C \diagup$

$H^5O - C - H$

$CH_2{}^6OPO_3{}^{2-}$

$^{2-}O_3P^6OH_2C$ $\quad^5O\quad$ ^2OH

H H^{1}O

H $\quad CH_2{}^3OPO_3{}^{2-}$

^4OH H

Oxygen exchanged in FBP: 2,1,4 and 5, oxygen 2 is still able to exchange

3. *glucose-1-P + UTP or GTP*

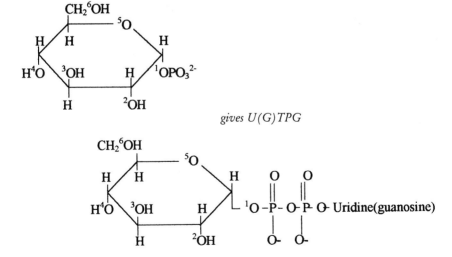

gives U(G)TPG

Exchanged oxygen:
while in DHAP, fructose, F6P and FBP: 2;
while in G3P: 4,3;
while in DHAP: 5;
while in glucose: 1;
not exchanged: 6.

4. U(G)DPG

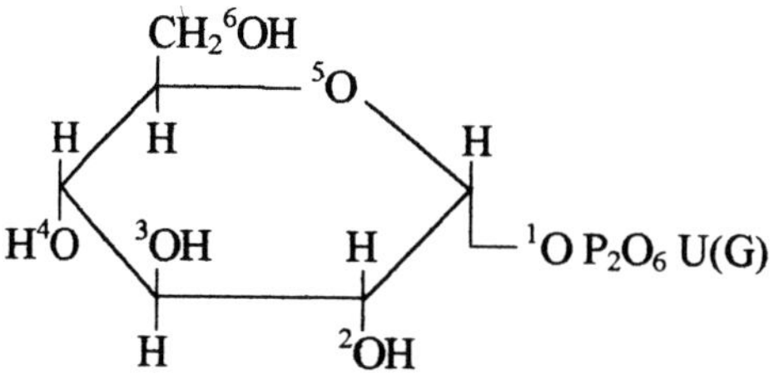

gives cellobiose

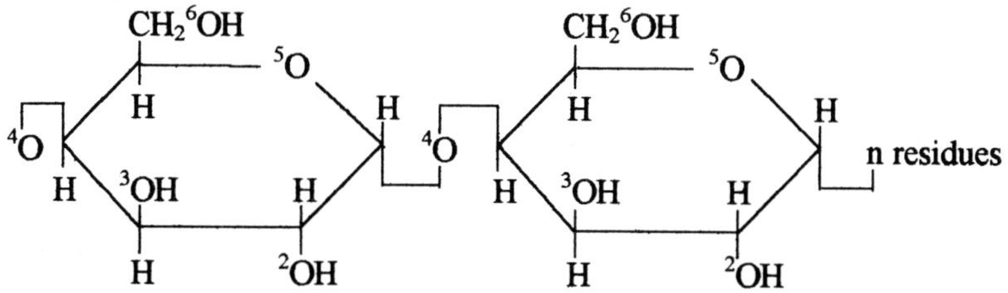

n residues

Appendix 2

Approximations inherent in the Craig–Gordon Model for ^{18}O when applied to terrestrial plants

In section 3.3 we made the comment that Equations 3.2 and 3.8 are only an approximation. We now show that the isotopic composition of source water, δ_S, does not precisely equal that of transpired water, (δ_E), in contrast to the normal assumption made for the application of the Craig–Gordon model under isotopic steady-state conditions.

Consider the conservation of mass for oxygen in the leaf. The flux of water from the roots is J, bringing $R_S J$ atoms of ^{18}O. CO_2 enters the leaf and exchanges its oxygens and either is carboxylated or leaves the leaf. The flux of CO_2 into the leaf, A, is $g(c_a - c_i)$, but each molecule has two O atoms, and the flux of ^{18}O is $(2g/\alpha_{kc})$ $(R_{ac}c_a - \alpha_{BC}R_e c_i)$ where the factor α_{kc} allows for kinetic effects, R_{ac} is the isotope ratio of oxygen in CO_2 in the air, and α_{BC} is the Bottinga–Craig factor desccribing equilibration between CO_2 and water (approximately 1.041). Organic O is formed at the rate A (assuming a C:O ratio of 1), and for O^{18} at $AR_e\alpha_m$, where α_m is the enrichment. The rate of loss of O_2, the Hill reaction, is denoted H and the corresponding flux is $2R_e H$, since there is no fractionation here. In a C_3 plant the rate of entry of O_2 associated with photorespiration is $1.5V_o$ where V_o is the rate of oxygenation, and so is doubled for atoms, with a corresponding flux for ^{18}O of $3R_o V_o/\alpha_u$, where R_o is the isotope ratio in atmospheric oxygen, and α_u is the weighted fractionation during uptake by Rubisco and glycolate oxidase (see Guy et al., 1987). Respiration takes up O_2 at rate R (we here assume a respiratory quotient of 1), with a corresponding ^{18}O flux of $2R_o R/\alpha_r$ (Guy et al., 1987) where α_r is the fractionation during respiration. We ignore the water which flows with exported sucrose. It is a potentially significant source of error, unless it returns in the xylem thereby cancelling its effects. Again the transpiration is E and of heavy water is $R_E E$.

For conservation of mass,

$$E = J + 2g(c_a - c_i) - A - 2(H - 1.5V_o - R) = J - A \qquad 3.A1$$

since

$$H - 1.5V_o - R = A$$

and

$$R_E E = R_s J + (2g/\alpha_{kc})(R_{ac} c_a - R_e \alpha_{BC} c_i) - R_e A \alpha_m - 2(R_e H - 1.5 R_o V_o(\alpha_u - R_o R/\alpha_r) \qquad 3.A2$$

Substituting Equation 3.A1 in 3.A2

$$R_E = R_s + \{R_s A + 2g(R_{ac} c_a - R_e \alpha_{BC} c_i)/\alpha_{kc} - R_e A \alpha_m - 2(R_e H - 1.5 R_o V_o(\alpha_u - R_o R/\alpha_r)\}/E \qquad 3.A3$$

The bracketed term (divided by E) represents the error involved in equating R_E and R_s in the steady state. In the non-steady state Equation 3.A1 is replaced by

$$J = E + A + dW/dt$$

where W is the leaf water content, and $d(R_s W)/dt$ is added to the right hand side of Equation 3.A2, so that within the bracketed term of 3.A3, $(R_s - R_e) dW/dt - W dR_e/dt$ must be added.

Now the value for R_e depends on $(1 - e_a/e_i) R_E$ [Similarly to a good approximation, increasing the concentration of the root solution by, say, 10‰ should increase that of leaf water and organic matter by $10(1 - e_a/e_i)$‰].

Thus from Equation 3.7, it follows that

$$R_e = R_{ecg} + \alpha^* \alpha_k (1 - e_a/e_i)\{ \}/E$$

where R_{ecg} is the Craig–Gordon estimate of R_e, given by $\alpha^*[R_E \alpha_k(e_a/e_i) + R_v e_a]/e_i$ and the term $\{ \}$ is that from Equation 3.A3.

Solving 3.A4 for R_e and moving to the δ notation, and using $\alpha^* = 1 + \epsilon^*$, etc. we obtain

$$\delta_e = (\delta_{ecg} + \epsilon_n)/(1 + \epsilon_d) \qquad 3.A5$$

where ϵ_n and ϵ_d in the numerator and denominator, respectively, are given by:

$$\epsilon_n = \{(A + dW/dt) \delta_s + 2g c_a(\delta_s - \epsilon_{kc}) - 2g c_i(\alpha_{BC}/\alpha_{kc} - 1) - A\epsilon_m + 3V_o(\delta_o - \epsilon_u) + 2R(\delta_o - \epsilon_r) - W d\delta_e/dt\} \alpha^* \alpha_k P/(1.6g e_i) \qquad 3.A6$$

and

$$\epsilon_d = \{2g \alpha_{BC} c_i/\alpha_{kc} + A\alpha_m + 2H + dW/dt\} \alpha^* \alpha_k P/(1.6g e_i) \qquad 3.A7$$

We have used $E = 1.6g (1 - e_a/e_i) e_i/P$ where P is atmospheric pressure. We note that, strictly, 1.6 should be replaced by $(1.37 r_b + 1.6 r_s)/(r_b + r_s)$.

Assuming $c_i/c_a \approx 0.7$, Equation 3.A6 is well approximated by

$$\epsilon_n = [0.3 \delta_s + 2(\delta_{ac} - \epsilon_{kc}) - 1.4(\epsilon_{BC} - \epsilon_{kc}) - 0.3\epsilon_m] C_a/(1.6g e_i) + \epsilon_n' \qquad 3.A8$$

where $C_a = c_a P$ and is the partial pressure of CO_2 in the atmosphere and

$$\epsilon_n' = [\delta_s dW/dt + 3V_o(\delta_o - \epsilon_u) + 2R(\delta_o - \epsilon_r) - W d\delta_e/dt] P/(1.6g e_i) \qquad 3.A9$$

and Equation 3.A7 by

$$\epsilon_d = 1.7 \, C_a/e_i + dW/dt/(1.6g \, e_i/P) \qquad\qquad\qquad\qquad 3.A10$$

Now using the SMOW scale, $\delta_o = 23.5‰$, which is close to the values of ϵ_u and ϵ_r (Guy et al., 1987), so in the case of normal atmospheric oxygen Equation 3.A9 becomes:

$$\epsilon_n' = [\delta_s \, dW/dt - W \, d\delta_e/dt] \, P/(1.6g \, e_i) \qquad\qquad\qquad 3.A11$$

which is only non-zero in the non-steady state

For normal conditions ($C_a = 36$Pa, $e_i \approx 3$kPa, $\delta_{ac} = 41‰$) the corrections are small. We note, however, that an increase in δ_{ac} of 7‰ causes a change in δ_e of 0.1‰. The corrections are thus relevant in the artificial conditions in a gas exchange cuvette.

Some support for these conclusions comes from the experiment of DeNiro and Epstein (1979). They imposed an increase in δ_{ac} of 980‰ and reported an increase of 14‰ in the water distilled from the leaves. This perfect agreement with our calculations is, however, fortuitous, as we take no account of the changes in concentration and composition in their cuvette.

It is also worth noting that, under normal conditions, a leaf with $c_i/c_a = 0.3$ (a C_4 species, say) should have δ_e more than 0.1‰ greater than for a C_3 leaf with $c_i/c_a = 0.7$.

Intramolecular deuterium distributions and plant growth conditions

J. Schleucher

4.1 Introduction

The dependence of the deuterium abundance (expressed as the hydrogen isotope ratio δD) in precipitation on climate parameters ('meteoric water line') is well-established (Dansgaard 1964). In spite of this, and in spite of the fact that all hydrogen in organic plant material is derived from water taken up by plants, δD of organic plant material is rarely used for palaeontological studies. This may be because the incorporation of hydrogen isotopes into plant material is influenced by a large number of interacting parameters. The first group influences the δD of the water that is taken up by the plant. These parameters are the δD of the precipitation, which itself is influenced by climate, and the plant-water source (e.g. groundwater or surface water, Dawson, 1993 see Chapter 11). Transpiration processes of the plant enrich D in leaf water, so the water entering photosynthesis is isotopically distinct from the water taken up by the plant.

Besides the δD of the water entering metabolism, isotope discriminations of bio-chemical reactions influence the isotopic composition of the metabolites. Kinetic hydrogen isotope effects, as they occur in biochemical reactions, can be as great as a factor of ten, that is, if a C–H bond is broken in a chemical (enzymatic) reaction, the deuterated molecules with a C–D bond react up to ten times more slowly than the pro-tonated analogue. Because the isotope effect is characteristic of each enzyme, different carbon-bound hydrogen atoms in a metabolite experience different isotope discrimina-tions, and a non-random intramolecular isotope distribution can be expected which has been observed in a number of metabolites (Martin and Martin, 1990). To illustrate this, Figure 4.1 shows a schematic drawing of the net reaction of photosynthesis, assuming glucose to be the final product. Water with a given δD enters photosynthesis, and glu-cose bearing seven non-equivalent carbon-bound hydrogen atoms is formed. Each of these hydrogen atoms has a distinct biochemical history and experiences a distinct D

Stable Isotopes, edited by H. Griffiths.
© 1998 BIOS Scientific Publishers Ltd, Oxford.

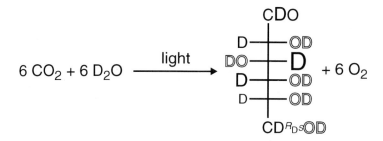

Figure 4.1. *Scheme depicting hydrogen isotope discrimination during the net reaction of photosynthesis. The letter 'D' does not stand for 100% deuteration; the different font sizes indicate the abundance of D in the respective position, the overall D content is close to natural abundance. Water with some δD enters photosynthesis, and glucose exhibiting a non-random intramolecular D distribution is formed. The exchangeable OD hydroxy groups are drawn in an outline font, because they play no role in this discussion.*

discrimination, as indicated by the different font sizes of the letters 'D' in the glucose molecule.

Transpirational isotope discrimination causes isotopic heterogeneity of plant water within leaves and between leaves and stem (Luo and Sternberg, 1992; Yakir *et al.*, 1989, 1994; Yakir, 1992; see Chapter 10). Isotopic heterogeneity combined with biochemical reactions could cause non-random intramolecular D distributions, even in the absence of isotope effects. If a metabolite is synthesized in one plant part (e.g. triose phosphates in chloroplasts) containing water of a certain δD, some of the carbon-bound hydrogens can exchange with the water in enzyme-catalysed reactions. After transport to another tissue (e.g. transport of sucrose to tubers), some hydrogen atoms of the metabolite can exchange with water again (e.g. during synthesis of starch from sucrose in the tubers). This can lead to a non-random intramolecular D distribution, because the leaf water is enriched in D by transpiration, while the water in the tubers is not.

If intramolecular D distributions of plant metabolites were non-random but constant, the δD of plant material would still faithfully track the δD of metabolic water. However, this will not be the case if changes in the metabolism induce changes in the intramolecular D distribution. The full kinetic isotope effect of a reaction is only observed if the back-reaction does not take place, that is in irreversible reactions or in reactions far from equilibrium. Depending on the metabolic state of an organism, any enzymatic reaction will fulfill this condition to a varying degree, leading to a varying isotope discrimination. Therefore, intramolecular D distributions can also be expected to vary depending on the physiological state of an organism.

As a fictitious example of differing intramolecular D distributions, the D distributions of two samples of glucose are displayed as bar diagrams in Figure 4.2. The samples differ in their patterns of D incorporation, in this case indicating that a change has occurred in the biochemical reaction incorporating the hydrogen atom of the C(4) position of glucose.

Underneath the bar diagrams of Figure 4.2, the δD values of both glucose samples is given in arbitrary units. Although the δD values differ, it is likely that the δD of the water entering photosynthesis was the same in both cases, and a direct interpretation of the δD values of the glucose samples in terms of the δD of the water entering

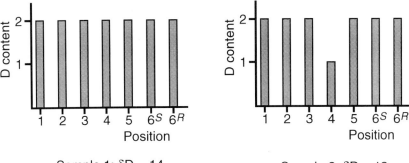

Figure 4.2. *Schematic display of the influence of variable intramolecular D distributions on δD. In this example, the overall difference in δD between samples 1 and 2 (14 vs. 13) results from a 50% decrease in the D abundance in the C(4) position of sample 2.*

photosynthesis would be misleading. In contrast, the knowledge of the δD and of the intramolecular D distribution would allow the conclusion that the water δD was the same, but that the plants producing the glucose samples differed in their metabolism.

Although the prediction of non-random intramolecular isotope distributions holds for any metabolite, only glucose is considered in this chapter because of its central role in plant metabolism. Sucrose is the end product of photosynthesis in most plants, so the intramolecular D distribution of the glucose fragment of sucrose is likely to contain information about the physiological state (e.g. rate limitation) of photosynthesis. Glucose is also the monomeric building block of starch, the main reserve carbohydrate in plants, and cellulose, the main structural material of plants. Because of their stability and ubiquity, starch and cellulose also are the best candidates for palaeontological studies.

It will be shown in this chapter that the intramolecular D distribution in photosynthetic glucose is not constant, and that this distribution depends on growth conditions. Because climate influences the physiology of plants, the influences of climate (δD of precipitation) and physiology (intramolecular D distribution) on the isotopic composition of plant material are likely to interact. Therefore, dependences of the intramolecular D distribution on physiology are likely to influence correlations of δD with climate parameters in an unpredictable way. Measurements of the intramolecular D distribution and of δD in laboratory experiments should help to separate climatic and physiological influences. Combined measurements of intramolecular D distributions and δD of ancient material could then be used to obtain information about climate and physiological responses of plants in past times.

4.2 Methodology

In the nuclear magnetic resonance (NMR) spectrum of a compound like glucose, one resonance line is observed for each carbon-bound hydrogen atom (compare Figures 4.2 and 4.3). Because hydroxy hydrogens exchange with water, only carbon-bound hydrogen atoms of organic compounds are useful for the NMR analysis, and care is taken to avoid interference between the signals of the hydroxy-hydrogens (not visible

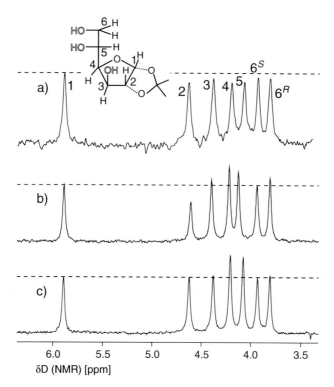

Figure 4.3. *Deuterium NMR spectra of the glucose derivative 1,2-O-isopropyliden-α-D-glucofuranose and chemical structure of this derivative (insert). Glucose was isolated from a) corn starch (natural abundance of D), b) starch from spinach leaves (deuterium enriched) and c) sucrose from spinach leaves (deuterium enriched). The dotted lines are drawn for easier comparison of intensities. The labelling of the peaks refers to the numbering of the carbons in glucose, the indices R and S distinguish between the diastereotopic methylene positions of C(6). Slight differences in the positions of the signals are due to minor differences in sample composition. The axis description 'δD (NMR) [ppm]' refers to the positions of the resonance lines, not to the isotope ratio.*

in Figure 4.3) and the carbon-bound hydrogens. NMR spectra can be obtained from H and D nuclei, and a large number of NMR experiments exist that allow each resonance line to assign to the hydrogen atom it represents, so it is easy to tell which resonance line in the D spectrum represents the D abundance in each position. Knowledge of the biochemical pathway of sucrose or starch synthesis allows each hydrogen atom to be distinguished and which biochemical reactions might have influenced its D abundance. D NMR spectra can be recorded under conditions that assure that the intensity of the lines is proportional to the D abundance in each position. From the spectra, the intensities of the resonances can be obtained by a line-shape fit.

In practice, sucrose, starch, or cellulose (for which the methodology does not yet exist) are isolated from plant material and are broken down to glucose. Glucose is converted into a derivative that gives D NMR spectra with sufficient resolution, such as 1,2-O-isopropyliden-α-D-glucofuranose (see insert in Figure 4.3 for structural formula). In order to increase the sensitivity and hence the accuracy of the measurements, the spectra displayed in this chapter (except Figure 4.3(a)) were obtained from deuterium-labelled glucose (plants were watered with water containing 1.5% D). All results were confirmed qualitatively by measurements at the natural abundance level of deuterium, using new isolation and measurement procedures (Schleucher, Markley and Sharkey unpublished data).

The absolute intensity of an NMR spectrum is a complicated function of the sample and the performance of the NMR instrument. While this does not interfere with the measurement of the intramolecular D distribution, it renders the determination of absolute intensities difficult. The only way to obtain δD from D NMR spectra would be to add a known amount of a compound of known δD to the sample or to introduce a molecular fragment of known δD into the glucose derivative as an intramolecular reference.

Isotope ratio mass spectrometry (IRMS) is a very sensitive technique for measuring δD. However, to derive intramolecular D distributions in this way would require measurement of the isotope ratios of a set of mass spectrometric molecular fragments (e.g. one fragment for each carbon-bound hydrogen, Guo et al., 1992), which would require all isotope discriminations during the mass spectrometric fragmentation to be constant and known. It appears to be harder still to measure D abundances at individual sites in a methylene group. In contrast, this is an easy task for NMR, because the hydrogen atoms in methylene groups of biochemical metabolites normally give separate signals. The resolved signals labelled 6^S and 6^R in Figure 4.3 represent the signals of the D atoms of the C(6) methylene group of glucose. It is important to be able to measure the D intensities in the sites of methylene groups individually, because such hydrogen atoms generally have distinct biochemical histories. It is apparent therefore, that NMR and IRMS complement each other, NMR being used to obtain intramolecular D distributions, and IRMS to measure δD.

In this chapter, C(1)-D denotes the D atom in position 1 of glucose. If the intramolecular D distribution of a molecule and the δD are known, absolute intramolecular δD values can be calculated. Because this is not the aim of the present discussion, intramolecular D intensity ratios (arbitrarily given in units of the well-resolved resonance of C(1)-D, setting the intensity of C(1)-D = 1) will be used throughout this chapter. The use of intensity ratios seems at first to preclude the identification of the positions whose intensities change. However, most results obtained so far can be interpreted as a change in the D content of individual positions.

4.3 Variation of intramolecular D distributions

4.3.1 Photosynthetic pathways

Organic material from C_3 and C_4 plants is known to differ in $\delta^{13}C$. The difference in $\delta^{13}C$ can be understood in terms of the different carbon isotope discrimination of PEPC and RUBISCO (Farquhar et al., 1989; see Chapters 8 and 9). There is evidence for a difference in δD of C_3 and C_4 cellulose of about 20‰ (Ziegler, 1989), but this seems not to be observed in all cases (Sternberg, 1989). Differences in the intramolecular D distribution between C_3 and C_4 plants (potato and corn starch) have been described previously (Martin et al., 1986).

In Figure 4.3, D NMR spectra of the glucose derivative 1,2-O-isopropyliden-α-D-glucofuranose isolated from corn starch (Figure 4(a) C_4 metabolism), spinach leaf starch (Figure 4.3(b), C_3 metabolism), and spinach leaf sucrose (Figure 4.3(c), C_3

metabolism) are compared. Differences in the D abundance of several hundred ‰ are observed. The similarity of the spectra in Figure 4.3(b) and 4.3(c) in all positions but C(2) and the big difference to the spectrum in Figure 4.3(a) suggest that C_3 and C_4 metabolism induce characteristic patterns in the intramolecular D distribution of glucose. On an empirical level, the intramolecular D distributions might provide an independent method to distinguish both photosynthetic pathways. Assuming that similar differences in the intramolecular D distribution exist between C_3 and C_4 cellulose as well, the small difference in δD between C_3 and C_4 cellulose must be considered to be the result of a large difference between the intramolecular D distributions. It is therefore unlikely that the difference in δD can be understood without an understanding of the intramolecular D distribution generated by both pathways.

In addition to plants that constitively operate by a C_3 or C_4 mechanism, C_3/C_4 intermediate plants are interesting as study objects. Similarly, the alga *Sargassum muticum* can fix CO_2 via RUBISCO or phosphoenolpyruvate carboxykinase (PEPCK), depending on the photoperiod. It was shown by D NMR that the intramolecular D distribution in mannitol isolated from this species correlates with the contribution of PEPCK to photosynthesis (Sancho *et al.*, 1989).

4.3.2 *Different metabolites*

A comparison of Figure 4.3(b) and 4.3(c) shows that intramolecular D distributions differ also between glucose isolated from the starch or sucrose of the spinach leaf. The biggest difference relative to the intensity of C(1)-D is observed in the intensity of C(2)-D. In this case, a conclusion about the absolute H/D ratios of the positions is justified on grounds of the intramolecular D distribution alone: either the C(2) position of glucose from starch is depleted in D, or all other positions in glucose from sucrose are enriched in D. Given that starch and sucrose have chloroplastic triose phosphates as common biochemical precursors, a depletion of C(2)-D in starch is likely to be the correct interpretation. This conclusion is in agreement with the observed negative δD of starch (Luo and Sternberg, 1991). Yakir and DeNiro (1990) propose that the D depletion of starch is caused by the incorporation of deuterium-depleted hydrogen from NADPH into carbohydrates in the glyceraldehyde-3-phosphate dehydrogenase reaction in the chloroplast. In this case, the depletion would be expected to be most pronounced in the hydrogen at the C(4) position of glucose, which originates from NADPH, however, the NMR results do not show a depletion in this position. The depletion observed in the C(2) position can be explained by a kinetic isotope effect of glucose-6-phosphate isomerase (Schleucher, Markley, Sharkey, unpublished data). Strikingly, the very low D incorporation into C(2)-D is not observed in starch isolated from corn kernels (C_4 metabolism, Figure 4.3(a)) or wheat kernels (C_3 metabolism, not shown). This suggests that the glucose-6-phosphate isomerase reaction is displaced from equilibrium in the chloroplast, but not in the cytosol or in non-photosynthetic plastids. A displacement of the glucose-6-phosphate isomerase reaction from equilibrium in chloroplasts was found in several studies (Dietz 1985, Gerhardt *et al.*, 1987, Kruckeberg *et al.*, 1989).

4.3.3 *Growth conditions*

The difference in the intramolecular D distributions of glucose isolated from sucrose and starch does not contradict the hypothesis that the D distributions of metabolites are non-random but constant. To test this hypothesis, spinach plants were grown in two light intensities (300 and 600 μmol photons s^{-1} m^{-2}) and two CO_2 mixing ratios in the air (C_a = 350 and 700 μmol mol^{-1}). Glucose from sucrose was isolated from the high-light-grown plants. For the low-light-grown plants, glucose from starch was analysed because the plants did not contain enough sucrose for analysis. Table 4.1 gives intensity ratios for C(6)-D^R/C(1)-D. The C(6)-D^R/C(1)-D ratio differs significantly (p = 0.003) between glucose isolated from plants grown under the different CO_2 levels.

In C_3 plants, increasing the CO_2 concentration during growth reduces photorespiration (see Chapter 8). If the C(6)-D^R/C(1)-D ratio reflected the amount of photorespiration, a reduction of the O_2 concentration should have the same effect as an increase in the CO_2 concentration, and the C(6)-D^R/C(1)-D ratio would be expected to approach a limiting value at high CO_2. Two experiments were carried out to test this hypothesis: spinach plants grown in 350 μmol mol^{-1} CO_2 were allowed to photosynthesize in air with a reduced O_2 concentration. The result of this experiment is given in the last row of Table 4.1. In a second experiment, spinach plants grown in 350 μmol mol^{-1} CO_2 were allowed to photosynthesize in CO_2 concentrations up to 2000 μmol mol^{-1}. The result of this experiment is displayed in Figure 4.4.

The results of both experiments support the hypothesis that the C(6)-D^R/C(1)-D ratio increases with decreasing photorespiration (decreasing ratio of oxygenation to carboxylation of ribulose-1,5-bisphosphate). At high CO_2 concentration (C_a), carboxylation dominates over oxygenation, and therefore the C(6)-D^R/C(1)-D ratio approaches a limiting value at high C_a (Figure 4.4). Because the ratio of carboxylation to oxygenation depends on the concentration ratio of CO_2 and O_2 inside the leaf, a reduction of the O_2 concentration has the same effect as an increase in C_a. The C(6)-D^R/C(6)-D^S ratio changes in parallel with the C(6)-D^R/C(1)-D ratio. (It should be noted that, as mentioned above, it would have been especially difficult to measure the D abundance in the two C(6) positions by mass spectrometry. D NMR allows to distinguish between the non-equivalent methylene protons, which appear as well-separated peaks.) This suggests that the change in these ratios is caused by a variable D content in the C(6)-D^R position.

Table 4.1. CO_2 dependence of the intramolecular D distribution in glucose isolated from spinach leaves. The values are the ratio of the intensities C(6)-D^R/C(1)-D, and in brackets standard deviation and sample size.

	Intensity ratio C(6)-D^R/C(1)-D	
Growth conditions	350 μmol mol^{-1} CO_2	700 μmol mol^{-1} CO_2
300 μmol photons m^{-2} s^{-1} [a]	1.081 (0.013, 2)	1.130 (0.038, 2)
600 μmol photons m^{-2} s^{-1} [b]	0.991 (0.016, 3)	1.110 (0.033, 3)
600 μmol photons m^{-2} s^{-1}, 12% O_2 [c]	1.102 (1)	N/A

[a] Glucose was isolated from starch, because plants gave too small a sucrose yield to be analysed.
[b] Glucose was isolated from sucrose.
[c] Plants (grown in 350 μmol mol^{-1} CO_2) were brought into darkness on the evening of day 1, kept in darkness during day 2, and allowed to photosynthesise in an O_2-depleted atmosphere during day 3 before harvest.

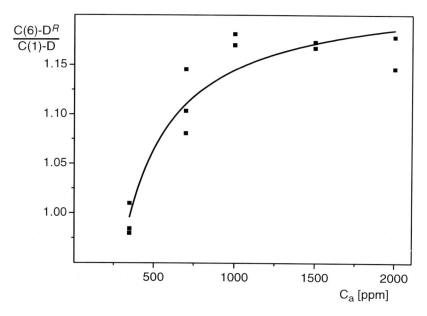

Figure 4.4. *Intensity ratio C(6)-D^R/C(1)-D in sucrose-derived spinach glucose as a function of the CO$_2$ concentration C$_a$ during photosynthesis. The values for 350 and 700 μmol mol^{-1} CO$_2$ were measured on material from plants grown from seeds in the respective atmosphere, the values for 1000, 1500 and 2000 μmol mol^{-1} were measured on material obtained from plants (grown in 350 μmol mol^{-1} CO$_2$) that had been kept in the dark for one day before photosynthesis at the respective CO$_2$ level.*

An examination of the reaction sequence from phosphoglycerate (PGA) to glucose in the photosynthetic carbon reduction cycle (PCR cycle, Calvin–Benson cycle), and of the catalytic properties of the enzymes involved shows that the C(3) methylene group of PGA is converted into the C(6) methylene group of glucose without the possibility of isotope exchange. This suggests that a change in the intensity ratio of C(6)-D^R/C(1)-D in glucose reflects a change in the isotopic composition of the C(3) methylene group of PGA.

Two, not mutually exclusive, explanations for the dependence of C(6)-D^R on photorespiration are: (1) The two PGA molecules formed in the carboxylation of ribulose-1,5-bisphosphate (RuBP) differ in their isotopic composition at C(3). In contrast to the carboxylation of RuBP, after oxygenation of RuBP, only the PGA molecule originating from C(3)–C(5) of RuBP enters the PCR cycle directly. A change in the ratio of oxygenation to carboxylation would therefore change the average isotopic composition of the PGA used in the synthesis of glucose and this would translate into a variable D content in the C(6)-D^R position of glucose. (2) The glyoxylate formed upon oxygenation of RuBP is recycled in the photosynthetic carbon oxidation cycle to PGA, which again enters the PCR cycle. This PGA is subject to different hydrogen isotope effects, so its intramolecular D distribution should differ. Although these hypotheses are speculative at present, experiments can be designed to investigate this question.

4.4 Discussion and outlook

Experiments have been conducted under controlled conditions to correlate changes in intramolecular D distributions to the metabolic conditions of plants. The intramolecular D distribution in plant glucose is neither random nor constant, but depends on (1) the photosynthetic pathway of the plant (C_3 or C_4), (2) the metabolite analysed (sucrose, leaf starch or storage tissues starch) and (3) the growth conditions of the plant (CO_2 concentration). The second and third effect can be localized to individual positions of the glucose molecule, and biochemical and physiological interpretations have been given for these changes in the intramolecular D distribution. Because hydrogen atoms are incorporated by a large number of biochemical reactions, intramolecular D distributions promise to be responsive to other physiological parameters besides the CO_2 concentration. It is likely that intramolecular D distributions of other metabolites show similar dependences on metabolic parameters.

At a biochemical level, intramolecular D distributions may give information about the kinetic operating conditions of biochemical reactions (rate-limiting or not). Suppose a hydrogen atom is incorporated into a metabolite by a particular biochemical reaction. If that reaction is rate-limiting for the biosynthetic pathway, each substrate molecule encounters the enzyme only once. In this case, the full kinetic isotope effect of the enzyme is operative, and the product of the pathway would be depleted in D in the position activated by the enzyme. If the reaction is at equilibrium, each substrate encounters the enzyme many times, only the much smaller equilibrium isotope effect operates, and the respective position of the product would be depleted in D to a smaller degree. Following this line of reasoning, intramolecular D distributions might be used as a tool to gauge the control that an enzyme exerts over the flux through a pathway. The explanation of the difference in the D NMR spectra of spinach sucrose and starch is a first example of this principle.

The results support the conclusion that the ratio of $C(6)\text{-}D^R$ to $C(1)\text{-}D$ reflects photorespiratory activity. As discussed above, a likely explanation for this is that the isotopic composition of $C(3)$ of PGA depends on whether it is formed by carboxylation of RuBP or by oxygenation. D NMR could therefore be used to obtain time-integrated measurements of photorespiration.

These observations can be generalized: any metabolite that can be formed via competing biochemical pathways will differ in its intramolecular isotope distribution depending on the contribution of each pathway, because the molecules are subject to different isotope discriminations during the respective pathways. These differing 'isotopic fingerprints' can be used to gauge the contribution of each pathway, provided the difference in the intramolecular isotope distribution is large enough to be measured.

With respect to isotope discriminations during photosynthesis, there are important differences between hydrogen and carbon. First, the $\delta^{13}C$ of atmospheric CO_2 is largely constant, while the δD of water is highly variable. Second, only one (or two) enzymatic reactions are responsible for the primary incorporation of carbon into photosynthate, while there are a large number of reactions incorporating hydrogen from water into organic material. Therefore, the $\delta^{13}C$ of organic material is largely determined by the carbon isotope discrimination of RUBISCO and PEPC, and $\delta^{13}C$ of organic material can be interpreted to a good approximation in terms of the carbon isotope discrimination of these enzymes. Other enzymes also show carbon isotope discrimination, and non-statistical intramolecular ^{13}C distributions have been observed (Monson and Hayes, 1982; Caer et al., 1991; Rossmann et al., 1991; see Chapter 2). However, non-statistical carbon distributions in metabolites do not blur the signature of RUBISCO and PEPC, because all carbon atoms go through these enzymatic steps, so $\delta^{13}C$ still provides a useful picture. In contrast, because of the large number of independent enzymatic reactions incorporating hydrogen atoms and the variability of the δD of source water, δD of plant metabolites does not provide a complete description of the hydrogen isotope discrimination.

The magnitude of the observed variability of the intramolecular D distribution of glucose suggests that changes in the intramolecular D distribution in the glucose fragments of cellulose should have a significant effect on δD. In spinach sucrose, the intensity of $C(6)$-D^R differs by 12% between plants grown in 350 and 700 μmol mol^{-1} CO_2. Accordingly, the rise in the atmospheric CO_2 concentration since industrialization from 280 to 360 μmol mol^{-1} should have caused an increase in the D incorporation into the $C(6)$-D^R position of glucose of about 3%, corresponding to a 4‰ change in δD. However, it remains to be seen whether the CO_2 dependence of the intramolecular D distribution is a general phenomenon of C_3 plants, and whether this dependence is also observed in storage starch and cellulose. If this was the case, changing environmental conditions would be expected to influence plant δD both directly (via the changing δD of precipitation) and indirectly (via changes in the intramolecular D distribution caused by differing physiological states of the plant).

Whatever the interaction of climatic and physiological influences on δD might be, the intramolecular D distribution should faithfully represent the plant's physiological status, because the isotope discriminations during metabolism do not depend on the δD of the water taken up by the plant. However, any interaction between climatic and physiological influences on δD should complicate the development of correlations between δD and climate parameters. On an empirical level, D NMR could be used to detect deviations from a standardized intramolecular D distribution and to correct δD values for the influence of changes in the D distribution. More interestingly, if physiological interpretations of dependences of intramolecular D distributions on environmental parameters are available, the combined use of IRMS and D NMR should help to separate climatic and physiological influences on δD and to obtain both climatic and physiological information from hydrogen isotope studies.

Acknowledgements

It is a pleasure to thank John Markley (Dept. of Biochemistry) and Tom Sharkey (Dept. of Botany) for the opportunity to carry out this research in their laboratories. This study made use of the National Magnetic Resonance Facility at Madison, WI, which is supported by the National Institutes of Health, the University of Wisconsin, the National Science Foundation and the U.S. Department of Agriculture. I thank the 'Deutsche Forschungsgemeinschaft' for a scholarship.

References

Caer, V., Trierweiler, M., Martin, G.J. and Martin, M.L. (1991) Determination of site-specific carbon isotope ratios at natural abundance by carbon-13 nuclear magnetic resonance spectroscopy. *Anal. Chem.* **63**, 2306–2313.

Dansgaard, W. (1964) Stable isotopes in precipitation. *Tellus* **16**, 436–468.

Dawson, T.E. (1993) Water sources of plants as determined from xylem-water isotopic composition: perspectives on plant competition, distribution, and water relations. In: *Stable Isotopes and Plant Carbon–Water Relations* (eds J.R. Ehleringer, A.E. Hall and G.D. Farquhar), Academic Press, San Diego, CA, pp. 465–496.

Dietz, K.J. (1985) A possible rate-limiting function of chloroplast hexosemonophosphate isomerase in starch synthesis of leaves. *Biochim. Biophys. Acta* **839**, 240–248.

Farquhar, G.D., Ehleringer, J.R. and Hubick, K.T. (1989) Carbon isotope discrimination and photosynthesis. *Ann. Rev. Plant Physiol., Plant Mol. Biol.* **40**, 503–537.

Gerhardt, R. Stitt, M. and Heldt, H.W. (1987) Subcellular metabolite levels in spinach leaves. *Plant Phys.* **83**, 399–407.

Guo, Z.K., Lee, W.N.P., Katz, J. and Bergner, A.E. (1992) Quantitation of positional isomers of deuterium-labelled glucose by gas chromatography/mass spectrometry. *Anal. Biochem.* **204**, 273–282.

Kruckeberg, A.L., Neuhaus, H.E., Feil, R., Gottlieb, L.D. and Stitt, M. (1989) Decreased-activity mutants of phosphoglucose isomerase in the cytosol and chloroplast of *Clarkia xantiana*. *Biochem. J.* **261**, 457–467.

Luo, Y. and Sternberg, L. (1991) Deuterium heterogeneity in starch and cellulose nitrate of CAM and C$_3$ plants. *Phytochemistry* **30**, 1095–1098.

Luo, Y.H., and Sternberg, L. (1992) Spatial D/H heterogeneity in leaf water. *Plant Physiol.* **99**, 348–350.

Martin, M.L. and Martin, G.J. (1990) Deuterium NMR in the study of site-specific natural isotope fractionation (SNIF-NMR). In: *NMR Basic Principles and Progress* (eds P. Diehl, E. Fluck, H. Günther, R. Kosfeld and J. Seelig), Vol. 23, Springer Verlag, Berlin, pp. 1–61.

Martin, G. Zhang, B.L., Naulet, N. and Martin M.L. (1986) Deuterium transfer in the bioconversion of glucose to ethanol studied by specific isotope labelling at the natural abundance level. *J. Am. Chem. Soc.* **108**, 5116–5122.

Monson, K.D. and Hayes, J.M. (1982) Carbon isotope fractionation in the biosynthesis of bacterial fatty acids. Ozonolysis of unsaturated fatty acids as a means of determining the intramolecular distribution of carbon isotopes. *Geochim. Cosmochim. Acta* **46**, 139–149.

Rossmann, A., Butzenlechner, M. and Schmidt, H.L. (1991) Evidence for a nonstatistical carbon isotope distribution in natural glucose. *Plant Physiol.* **96**, 609–614.

Sancho, A., Combaut, G., Jupin, H., Giraud, G., Quemerais, B., Naulet, N. and Martin, G. (1989) Evidence of variation in the CO$_2$ assimilating metabolism of Phaeophyceae by the SNIF-NMR method. *Plant Physiol. Biochem.* **27**, 537–544.

Sternberg, L.S.L. (1989) Oxygen and hydrogen isotope ratios in plant cellulose: mechanisms and applications. In: *Stable Isotopes in Ecological Research* (eds P.W. Rundel, J.R. Ehleringer and K.A. Nagy), Springer, New York, pp. 124–141.

Yakir, D., DeNiro, M.J. and Rundel, P.W. (1989) Isotopic inhomogeneity of leaf water: Evidence and implications for the use of isotopic signals transduced by plants. *Geochim. Cosmochim. Acta* **53**, 2769–2773.

Yakir, D. and DeNiro, M. (1990) Oxygen and hydrogen isotope fractionation during cellulose metabolism of *Lemna gibba* L. *Plant Physiol.* **93**, 325–332.

Yakir, D. (1992) Variations in the natural abundance of oxygen-18 and deuterium in plant carbohydrates, *Plant Cell Environ.* **15**, 1005–1020.

Yakir, D., Berry, J.A., Giles, L. and Osmond, C.B. (1994) Isotopic heterogeneity of water in transpiring leaves: identification of the component that controls the $\delta^{18}O$ of atmospheric O$_2$ and CO$_2$. *Plant Cell Environ.* **17**, 73–80.

Ziegler, H. (1989) Hydrogen isotope fractionation in plant tissues. In: *Stable Isotopes in Ecological Research* (eds P.W. Rundel, J.R. Ehleringer and K.A. Nagy), Springer, New York, pp. 105–123.

Stable isotope studies of soil nitrogen

D.W. Hopkins, R.E. Wheatley and D. Robinson

5.1 Introduction

It is probably true to say that without the stable isotope ^{15}N, we would know little about the mechanisms of the major transformations of N in soil. Some would argue that even with ^{15}N our knowledge is hardly profound. Sound generalisations and predictions of soil processes are difficult to make with confidence because of the chemical, physical and biological variations that occur in soil at all scales in both space and time. This applies particularly to the N cycle, as it involves a vast range of organisms, different chemical and physical processes, and the many interactions between them. Natural abundance studies have been used in a wide range of soil-N investigations and have provided some insights to the complexity of soil N, but at the same time the complexity of the soil often confounds interpretation in stable isotope studies. In principle, natural abundances can be used either as tracers, in which the ^{15}N natural abundance of one source is sufficiently different from other sources to allow it to be traced, or to report on a particular transformation during which isotopic fractionation occurs. We present examples of both approaches, but neither ought to be regarded as an off-the-shelf technique for field ecological investigations, particularly in soils.

In this chapter we outline the microbial processes of N transformation, the distribution of N and ^{15}N in soils in general and the variability in space and time of N and ^{15}N. Then we discuss some of the more recent insights to soil N transformations that have been gained from $\delta^{15}N$ measurements (either alone or in conjunction with other isotopes. Although natural abundance is the theme of this volume, it would be inappropriate for us to discuss soil N without mentioning briefly some of the other ^{15}N approaches that have contributed substantially to our present understanding (Powlson and Barraclough, 1993; see Chapter 16).

Addition of a ^{15}N enriched (or depleted) source, such as an inorganic fertilizer or organic residue, allows the ^{15}N to be used as an isotopically distinct but otherwise indistinguishable tracer for the added N. The approach is, of course, invasive, but this is particularly applicable for agricultural investigations where normal soil management is invasive. How isotopically distinct the tracer needs to be depends on a number of

factors, including the method of detection, the period of the experiment, the N transformation of interest, but at the very least, the isotopic enrichment (or depletion) of
the tracer needs to be greater than the range of variation in its natural abundance.

Advances in NMR spectroscopy have offered an alternative means of detecting ^{15}N
and provide information about the distribution of ^{15}N between different functional
groups. The combination of the relative insensitivity of ^{15}N NMR and the low natural
abundance of ^{15}N mean that the approach works best at high enrichments (Benzing-
Purdie *et al.*, 1992; Clinton *et al.*, 1995) and particular insights are likely to follow from
dual-tracer (^{13}C and ^{15}N) NMR experiments (Benzing-Purdie *et al.*, 1992; Hopkins
and Chudek, 1997).

Rather than adding a ^{15}N-enriched source, the sink for a particular reaction can be
isotopically enriched and the rate of ^{15}N dilution by ^{14}N can be used to determine the
gross rather than the net rates of microbial-N transformations. The dilution of the
isotopic pulse by unlabelled sources in the soil provides a means of estimating gross
process rates where the product is transient (Barraclough, 1991a, 1991b; Davidson *et
al.*, 1991).

5.2 Transformations of soil nitrogen

The transformations of N which occur in soil are a major component of the N cycle
(Figure 5.1). The particular processes on which we will focus are outlined below.

5.2.1 *Organic N transformations in soils*

N mineralization and immobilization are intimately associated with C transformations and decomposition in soils. N mineralization is the release of inorganic N from
the decomposing organic remains of plants, animals and microorganisms usually as
NH_4^+, N mineralization is also referred to as ammonification. This occurs as a result
of excretion by decomposer organisms and during extracellular breakdown (deamination) of organic compounds. N mineralization is a complex set of processes because
of the huge diversity and mixed origin of organic N compounds in soil (Ladd and
Jackson, 1982). N immobilization is the uptake and assimilation of N from the soil

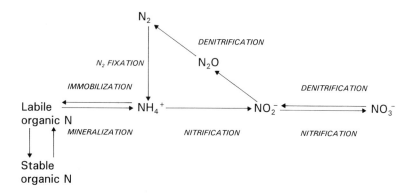

Figure 5.1. *Microbial transformations of soils N.*

NH_4^+ or NO_3^- pools by soil microorganisms, involving one of several pathways (see Paul and Clark, 1989). N immobilization is widely regarded as the opposite of N mineralization and in most studies it is the balance between N mineralization and N immobilization, that is, net N mineralization (or immobilization) rather than gross mineralization (or immobilization), which is measured. ^{15}N pool-dilution methods have proved effective for determining the rate of gross N mineralization during decomposition of crop residues (Watkins and Barraclough, 1996) and for distinguishing between different routes of N mineralization during decomposition of organic compounds in soil (Barraclough, 1997).

5.3.2 *Inorganic N transformations in soils*

Nitrification is the formation of NO_3^- by NH_4^+ oxidation by chemoautotrophic (nitrifying) bacteria (see Paul and Clark, 1989), during which some N_2O may be produced (Blackmer *et al.*, 1980). Heterotrophic organisms may also make a contribution to NO_3^- production, particularly in acidic soils, by the oxidation of either organic N or NH_4^+ (Killham, 1987). An important indirect consequence of nitrification is that the N is converted to a form susceptible to loss from the soil as a result of denitrification and/or leaching. Denitrification is the dissimilatory reduction of NO_3^- and NO_2^- to N_2O and N_2 by a wide range of facultatively anaerobic bacteria (see Paul and Clark, 1989), some fungi (Schoun *et al.*, 1992) and possibly also by some bacteria under aerobic conditions (Lloyd, 1993). Although regulated by different environmental factors and for the most part carried out by different groups of microorganisms, the processes of nitrification and denitrification share several common metabolites, and soil microsite variability often permits both processes to occur simultaneously (Webster and Hopkins, 1996b).

5.3.3 *Biological N_2 fixation in soils*

Reduction of atmospheric N_2 to NH_3, from whence it is assimilated into organic N, is carried out by some bacteria, which may live in symbiosis or in a looser association with plants or may be free-living in the soil (see Paul and Clark, 1989). N_2 fixation is the main natural route of N into the biosphere and, therefore, into soils. Although widely referred to as nitrogenase, there are at least three different enzyme complexes responsible for N_2 fixation, which based on their co-factor requirements are termed Mo-nitrogenase, V-nitrogenase and nitrogenase-3 (or Fe only-nitrogenase) (Pau, 1991). Most attention has been focused on Mo-nitrogenase, but there are important differences in the substrate specificities and affinities between the nitrogenases and there have been few studies to establish the controlling environmental factors on the different enzymes amongst soil microorganisms.

5.4 Distribution of soil nitrogen

The amount of N in soils ranges from 2 t N ha^{-1} in warm deserts, to 16 t N ha^{-1} in subtropical wet forests and 20 t N ha^{-1} in rain tundra (Post *et al.*, 1985). Typically, soils used for intensive agriculture contain between 2 and 6 t N ha^{-1} (Powlson, 1993). Virtually all the N in soils is in organic compounds (Stevenson, 1982). Despite generations of effort and the belief that knowledge of the chemical structures of organic N compounds holds the key to understanding soil N cycling, the chemical identities of most organic N compounds in soils have proved particularly difficult to elucidate. The classical methods of organic N fractionation, based on acid hydrolysis and steam distillation, incompletely divide the soil N into poorly-defined fractions only a few of which can be regarded as either biologically- or chemically-meaningful, and several of which contain N of largely unknown identity (Table 5.1).

The $\delta^{15}N$ of total soil N is usually slightly greater than that of atmospheric N_2 (Table 5.1). Although the ^{15}N natural abundance may be lower where the soil organic matter is dominated by a large amount of recently added litter (Natelhoffer and Fry, 1988). The considerable difficulty in assigning a $\delta^{15}N$ value to a particular organic N fraction is well-illustrated by one of the earlier reports (Cheng *et al.*, 1964) of variations in $\delta^{15}N$ of soil N (Table 5.1). The range $\delta^{15}N$ values for a particular fraction is quite large and it does not always show the same relationship to the total soil $\delta^{15}N$ across all soils. This, together with the uncertainty over their identity makes assigning source values for N transformations doubly difficult. As observed in Chapter 6, the complexity of soil N transformations means that despite large differences in the rates of particular processes between soils, coincidentally similar $\delta^{15}N$ values for the total soil N may occur. This, together with the fact that only a small fraction of the soil N may actually be involved in cycling mean that total soil $\delta^{15}N$ values of soil with contrasting rates of N transformation may by very similar. For example, the $\delta^{15}N$ values of total soil N from two experimental plots on an upland, high organic matter and high N containing soil were +4.6‰ (s.d. = 1.42‰, $n = 6$) and +5.4‰ (s.d. = 0.97‰, $n = 8$) (D.W. Hopkins, unpublished data). Yet as a result of lime application to the former plot some 10 years earlier, higher soil pH values (from 3.9 to 6.8), greater amount of microbial biomass N (6- fold) and higher rates of N_2-fixation (acetylene reduction; 11- fold), denitrification (14- fold), organic matter decomposition (1.4- fold) and net N mineralization/immobilization

Table 5.1. Variations in the $\delta^{15}N$ of soil N and different organic fractions of soil N for five different soils, and the typical percentage represented by each fraction. Based on Cheng et al., (1964), Sowden et al., (1977) and Stevenson (1982)

| Form of N | Percent of total soil N | $\delta^{15}N$ (‰) | | | | |
		Soil 1	Soil 2	Soil 3	Soil 4	Soil 5
Total N	100	+16	+7	+ 5	+ 3	+ 2
Total hydrolysable N[a]	84–89	+18	+10	+7	+ 5	+4
NH$_3$[a]	18–32	+7	+7	+3	+6	+5
Amino sugar N[a]	7–7	+25	+8	0	+2	+8
α-amino acid N[a]	33–42	+16	+11	+8	+7	+3
Hydrolysable unidentified N[a]	16–18	−3	−2	−1	0	−4
Mineralized N[b]	<1	+6	+2	+1	+1	+1
Fixed NH$_3$[c]	5–10	+6	+6	+4	+2	0

[a] N fractionation was by acid hydrolysis and steam distillation (see Stevenson, 1982).
[b] Mineralized N was ($NH_4^+ + NO_3^-$) produced by incubation of the soil at 30 °C and at 50% water capacity for 14 days.
[c] Fixed ammonia is ammonium intercalated in the clay crystal and not biologically fixed N.

(2-fold) were recorded in the limed compared with the unlimed soil (Hopkins, 1997; Isabella and Hopkins, 1994). For this normally acidic, cold, wet, high N soil, it is estimated even with the major increase in the N cycling rate, that less than about 2% of the total soil N had actually been mobilized over the period since lime application (Isabella and Hopkins, 1994). This contrasts with arid soil where the small amount of soil N is concentrated at the surface and is highly active, and in which differences in N cycling rates were reflected by much larger differences total soil $\delta^{15}N$ values (Evans and Ehleringer, 1993; see Chapter 14).

Mineralized N is susceptible to plant uptake and loss from the soil because of nitrification and/or denitrification. Fractionation during these processes (Table 5.2; Mariotti et al., 1981) accounts for the ^{15}N enrichment of the soil N and for the fact that plants often have $\delta^{15}N$ values slightly lower than the soil N (Broadbent et al., 1980; Cheng et al., 1964; Karamanos et al., 1981; Sutherland et al., 1993; van Kessel et al., 1994). Several researchers have reported an increase in $\delta^{15}N$ of the total soil N with depth in the soil profile (Ledgard et al., 1984; Mariotti et al., 1980; Natelhoffer and Fry, 1988 e.g. Riga et al., 1971); observations consistent with a general increase in age of the soil N with depth. However, caution is required in the interpretation of such data because of N redistibution in the profile and bioturbation (Handley and Scrimgeour, 1996). The greater $\delta^{15}N$ of N associated with clay-sized soil particles compared to that associated with the larger silt- and sand-sized particles (Tiessen et al., 1984) is presumably a reflection of the ^{15}N enrichment in the residual N after fractionation during loss processes, since reduction in particle size usually accompanies microbial degradation of organic matter.

5.5 Spatial variability of soil N

The tremendous spatial variability of soils, which can occur on scales measured in μm through to km and, unlike the oceans and atmosphere, the virtual absence of any mixing are constant factors in soil investigations. At the larger scale, soils vary in space

Table 5.2. Maximum $^{15/14}N$ discriminations, $\Delta^{15}N^{14}N$ (‰), for N transformation processes, based on data collated by Handley and Raven (1992). $\Delta^{15}N^{14}N$ (‰) = 1000 × [(k$_{14}$/k$_{15}$)-1], where k$_{14}$/k$_{15}$ is the ratio of the transformation rates for ^{14}N and that for ^{15}N. Each has a minimum Δ of zero (i.e., no $^{15/14}N$ discrimination)

Transformation	$\Delta^{15}N^{14}N$(‰)
NO_3^- or NH_4^+ diffusion in solution	0
Adsorption/desorption of NH_4^+	1
N_2 fixation	9
Nitrification	9
NH_4^+ assimilation by plants	8
Mineralisation	20
NO_3^- assimilation by plants	20
Denitrification	33

according to parent material, profile depth, topography, distribution of organisms and climate (Stark, 1994). At the smaller scale, the variability of porosity (connectivity and size distribution; both of which influence gas and liquid movement) and the variability in the presence and activity of many microorganisms contribute further to spatial heterogeneity.

Spatial variability in soil is best handled using geostatistical techniques (Oliver and Webster, 1991), but this has been rarely used to quantify isotope distributions *per se*. Most relevant geostatistical information exists for NO_3^- and NH_4^+ concentrations or the rates of their transformations. For example, Jackson and Caldwell (1993a) found that sites of net N mineralization existed within 12.5 cm of others where N immobilization was the predominant process. At the same site, the concentrations of both NO_3^- and NH_4^+ varied by between two and three orders of magnitude at m distances apart (Jackson and Caldwell, 1993b). At the dm scale, relative variations of these ions were up to 12 fold, falling to 3 fold at cm scales. The factors controlling such variation are not always obvious. However, Goovaerts and Chiang (1993) showed that net N mineralization was related to the amount of oxidizable C when a field was sampled at 10 m intervals and that the spatial patterns persisted between seasons. There have been several other investigations of the spatial and temporal variability of N transformations in soil (*e.g.* Lechowicz and Bell, 1991; Robertson *et al.*, 1988; van Kessel *et al.*, 1993).

Given the well-acknowledged spatial variability of soils, corresponding variability in ^{15}N natural abundances might be expected. The earlier soil $\delta^{15}N$ work was done with manual dual-inlet mass spectrometers, which although very precise did not usually allow enough samples to be processed for geostatistical analyses. To date, there have still been insufficient measurements of $\delta^{15}N$ of soil NO_3^- and NH_4^+ to provide the answer, but some recent sets of total soil $\delta^{15}N$ data reveal informative patterns. Sutherland *et al.* (1993) sampled soil and wheat at 10 m intervals and found that the $\delta^{15}N$ of the soil and the wheat varied predictably with landscape features. The variability could be explained in terms of denitrification flux since the total soil $\delta^{15}N$ and the plant $\delta^{15}N$ were both greater at lower-lying depressions which received run-off and the wheat $\delta^{15}N$ was inversely related to redox potential so that large $\delta^{15}N$ values were associated with soil conditions likely to lead to large, ^{15}N-depleted N losses by denitrification. In contrast, at a drier site nearby, van Kessel *et al.* (1994) found that the $\delta^{15}N$ of soil N and crop residues varied widely and apparently randomly at ten intervals, whereas corresponding $\delta^{13}C$ values had a strong dependence on topography or hydrology. These observations indicate firstly, that spatial (and, presumably temporal) uniformity in ^{15}N abundances in soil cannot be taken for granted and that to account for variability, each field site should be surveyed using geostatistical techniques.

5.6 Use of ^{15}N natural abundance to investigate soil processes

5.6.1 *Organic N transformations in soil*

Studies of aquatic ecosystems have shown that $\delta^{15}N$ is relatively poor indicator of diet because, unlike $\delta^{13}C$, it increases with successive trophic level in a food chain since nitrogenous excretory wastes are usually depleted in ^{15}N. However, this means that $\delta^{15}N$ may act as a guide to an organism's trophic position (*e.g.* Owens, 1987; Wada *et al.*, 1993). N excreted by soil animals involved in the decomposer food chain, be they protozoa, nematodes, annelids, molluscs or mammals, has an important role in N cycling in soil (Anderson, 1987; Griffiths, 1986) and in the absence of large fertilizer inputs may make a substantial contribution to the N supply to plants (Robinson *et al.*, 1989). Schmidt *et al.* (1997) have recently provided the first indication that the $\delta^{15}N$ relations with trophic position may hold for soil communities in a similar fashion to those observed in aquatic systems. In a grassland soil, the $\delta^{15}N$ of detritivores (earthworms and slugs) was up to 2‰ greater than either the living plant material or the total soil N and the $\delta^{15}N$ of carnivores (spiders and beetles) was up to 3‰ greater than that of the detritivores (R. Neilson, personal communication). Whether this approach can be further exploited in soil biology investigations will depend on the development of analytical procedures for determining $\delta^{15}N$ on very many small samples.

5.6.2 *Inorganic N transformations in soil*

Nitrate sources. NO_3^- in soil arises from several sources including coupled mineralization–nitrification, inorganic fertilizer, industrial effluent and atmospheric deposition. Establishing the fractional contributions from these sources has recently acquired considerable importance because of threats to drinking water from NO_3^- contamination. Kohl *et al.* (1971) were the first to attempt to use $\delta^{15}N$ to establish the origin of NO_3^- in surface water draining from agricultural land. They were unsuccessful because of isotopic fractionations during soil N transformations obscured the source $\delta^{15}N$ (Blackmer and Bremner, 1977; Hauck *et al.*, 1972). However, the dual isotope approach using both $\delta^{15}N$ and $\delta^{18}O$ offers greater potential, despite the difficulties in determining the $\delta^{18}O$ of NO_3^- (Amberger and Schmidt, 1987; Böttcher *et al.*, 1990; Durka *et al.*, 1994; Wassenaar, 1995). Wassenaar (1995) has conducted the most comprehensive study to date. The $\delta^{15}N$ and $\delta^{18}O$ of NO_3^- from 117 wells in an aquifer, many of which contained water with NO_3^- concentrations in excess of the acceptable limit, were determined. None of the NO_3^- samples had $\delta^{18}O$ values consistent with the NO_3^- originating from either precipitation or the nitrification of NH_4^+-based fertilizers (Figure 5.2). Most, however, had $\delta^{15}N$ values consistent with the NO_3^- originating from the mineralization of organic N and nitrification of NH_4^+ in poultry manure and its subsequent transport through the soil profile by mass flow. The isotopic data also indicated that no NO_3^- was being removed from the aquifer by denitrification.

Nitrous oxide sources. The dual isotope (^{15}N and ^{18}O) approach has also been useful in tracking the sources of N_2O in the atmosphere (Kim and Craig, 1990, 1993). The hypothesis that tropospheric N_2O is a mixture from oceanic and terrestrial emissions did not agree with measurements of $\delta^{15}N$ and $\delta^{18}O$ of tropospheric N_2O (Kim and Craig, 1993). The latter fell some way from a mixing line connecting the mean $\delta^{15}N$ and $\delta^{18}O$ of soil N_2O and of N_2O produced at the surface of oceans (Figure 5.3). This was strong evidence for a third, and previously unsuspected, source of N_2O contributing to the tropospheric N_2O mixture. Stratospheric N_2O has another, distinct combination of $\delta^{15}N$ and $\delta^{18}O$ (Figure 5.3). A three-source mixing model that included N_2O recycled into the lower atmosphere from the stratosphere, as well as from oceanic and soil sources accounted for the isotopic composition of tropospheric N_2O (Figure 5.3).

Soils are a major contributory source to increase in atmospheric N_2O (Bouwman, 1990; Weiss, 1981), but the contributions from different microbial processes is not

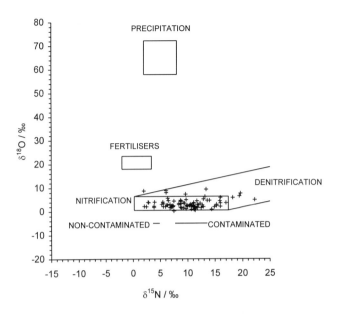

Figure 5.2. *$\delta^{15}N$ and $\delta^{18}O$ in NO_3 sampled from the Abbotsford aquifer in British Columbia (+) (Wassenaar, 1995). Rectangles enclose ranges of $\delta^{15}N$ and $\delta^{18}O$ in precipitation, NH_4^+-based fertilizers and from nitrification of soil N. Sloping lines enclose the likely trend in $\delta^{15}N$ and $\delta^{18}O$ in residual NO_3^- undergoing denitrification. Ranges of $\delta^{15}N$ in NO_3^- from soil contaminated by poultry manure and from beneath non-contaminated soil are shown as horizontal lines.*

well-understood (Robertson and Tiedje, 1987). The values of $\delta^{15}N$ and $\delta^{18}O$ in N_2O produced by nitrifying bacteria and denitrifying bacteria differ (Wahlen and Yoshinari, 1985; Webster and Hopkins, 1996a; Yoshida, 1988) The isotopic composition of N_2O from soils is closer to that produced during denitrification than from nitrification (Figure 5.3). This type of information may prove particularly valuable because the estimates of N_2O partitioning between different metabolic processes obtained from investigations using metabolic inhibitors are highly variable (Davidson *et al.*, 1993; Mummey *et al.*, 1994; Parton *et al.*, 1988; Robertson and Tiedje, 1987; Webster and Hopkins, 1996) and there is reasonable uncertainty about the selectivity and, in wet soils, the efficacy of the inhibitors (Robertson and Tiedje, 1987; Webster and Hopkins, 1996b).

Much of the N_2O produced by nitrifying bacteria actually arises during denitrification, that is denitrifying nitrifiers, in which N_2O is the main product (Poth and Focht, 1985; Poth, 1986). If similar denitrification pathways operate in both denitrifying and nitrifying bacteria, differences in isotopic composition between N_2O produced by the two groups of bacteria may not be due to different fractionations during contrasting processes of production. Fractionation during N_2O reduction to N_2 was used to explain the increase (less negative) in $\delta^{15}N$ of N_2O emitted from soil under conditions of increased wetness and organic C supply (Webster and Hopkins, 1996a, 1996b).

5.6.3 *Biological N fixation in soils*

The difference in $\delta^{15}N$ between atmospheric N_2 and plant-available soil N has been widely used as the basis of estimating the relative contribution of atmospheric N_2 to N_2-fixing organisms (Shearer and Kohl, 1986, 1989). Although the technique gives results comparable with other methods (*e.g.* acetylene reduction, yield comparison of fixing and non-fixing strains and isotope pool dilution) (Shearer and Kohl, 1993), its main value probably lies in natural ecosystems where these other methods cannot usually be applied (and are not, therefore, available for verification).

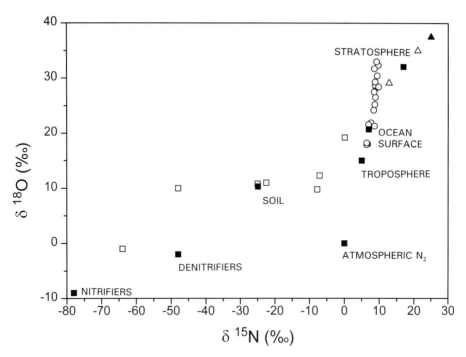

Figure 5.3. $\delta^{15}N$ and $\delta^{18}N$ in N_2O sampled from different environments based on Figure 1 in Kim and Craig (1993): Surface of Pacific Ocean (○, Kim and Craig 1990); soils (□ Kim and Craig 1993; Webster and Hopkins 1996a); Stratosphere (△, Kim and Craig 1993). Filled squares are mean values for the soil, ocean surface, stratosphere, and tropospheric measurements and for N_2O produced by nitrifiers and denitrifiers (Kim and Craig 1993; Webster and Hopkins, 1996a). The filled triangle is the maximum $\delta^{15}N$ and $\delta^{18}O$ value for stratospheric N_2O (Kim and Craig 1993).

Successful application of the approach relies on there being a genuine, robust and measurable difference between the two N sources (soil N and atmospheric N_2) (Shearer and Kohl, 1993), and in order to satisfy these criteria a number of precautions are required (Bremner and van Kessel, 1990; Shearer and Kohl, 1993). Determining the $\delta^{15}N$ of the plant-available soil N is usually achieved by analysis of non-N_2-fixing plants rather than analysis of extractable (mineral) soil N. This is because the former should integrate any fractionation during uptake and assimilation as well as avoiding uncertainty about fractionation during loss of soil mineral N prior to plant uptake. Careful selection of suitable non-fixing reference plants is essential, since they should be ecologically and physiologically indistinguishable from the putative N_2-fixer in all ways apart from receiving a supply of N via fixation; this is a fairly tall order. Handley and Raven (1992) comment on the possibility that some reference plants may be unsuspected symbiotic or associative N_2-fixers. The $\delta^{15}N$ of reference plants may also be affected by mycorrhizas (Handley *et al.*, 1993; Ibijbijen *et al.*, 1996). The spatial and temporal variability in N mineralization and plant-available soil N concentrations (Bremner and van Kessel, 1990) and different isotopic partitioning between roots, shoots, fruits *etc.* (Bergersen *et al.*, 1988; Handley and Scrimgeour, 1997) are both relevant considerations in sampling the reference plant. Finally, although the $\delta^{15}N$ of atmospheric N_2 is usually very close to that of the fixed

N, any small differences in isotopic composition arising from fractionation during uptake and assimilation (see Chapter 6) must be accounted for and Shearer and Kohl (1993) recommend growing the plants with N_2 as the sole N source and for a sufficiently long period for the seed N to be adequately diluted. Given these constraints, the natural abundance of ^{15}N should best be regarded as a usually reliable, although not precise, indicator of N_2 fixation.

5.7 Conclusions

The principal advantage of ^{15}N natural abundance studies of soil N is that it is non-invasive and that it may provide a record of previous transformations. However, the benefits of non-invasiveness particularly in studies of those soil systems in which normal management is invasive needs to be offset against the difficulties of accurately determining $\delta^{15}N$ of soil N pools which may be either transient, unextractable or spatially variable. By far the most informative natural abundance studies to date have involved $\delta^{15}N$ measurements alongside the natural abundances of other isotopes. Despite this, we feel it appropriate to repeat our introductory comment that natural abundance approaches ought not to be regarded as off-the-shelf techniques for field ecological investigations. For $\delta^{15}N$ measurements really to work for soil N studies, reliable information about the sizes and isotopic compositions all the potential N sources is required. As with investigative journalism, verifiable sources are essential.

Acknowledgements

DWH is grateful to the UK Natural Environment Research Council's Terrestrial Initiative in Global Environmental Research programme (GST/02/618). The Scottish Crop Research Institute is funded by the Scottish Office Agriculture, Environment and Fisheries Department. We are grateful to L.L. Handley and J.I. Sprent for useful discussions during the preparation of this chapter.

References

Amberger, A. and Schmidt, H-L. (1987) Naturliche Isotopengehalte von Nitrat als Indikatoren für dessen Herkunft. *Geochim. Cosmochim. Acta* **51**, 2699–2705.

Anderson, J.M. (1987) Interactions between invertebrates and microorganisms: noise or necessity for soil processes? In: *Ecology of Microbial Communities* (eds M. Fletcher, T.R.G. Gray and J.G. Jones). Cambridge University Press, Cambridge, pp. 125–145.

Barraclough, D. (1991a) The use of mean pool abundances to interpret ^{15}N tracer experiments. II. Theory. *Plant Soil* **131**, 89–96.

Barraclough, D. (1991b) The use of mean pool abundances to interpret ^{15}N tracer experiments. II. Application. *Plant Soil* **131**, 89–96.

Barraclough, D. (1997) The direct or MIT route for nitrogen immobilization: A ^{15}N mirror image study with leucine and glycine. *Soil Biol. Biochem.* **28**, 101–108.

Benzing-Purdie, L., Cheshire, M.V., Williams, B.L., Ratcliffe, C.I., Ripmeester, J.A. and Goodman, B.A. (1992) Interactions between peat and sodium acetate, ammonium sulphate, urea of wheat straw during incubation studied by ^{13}C and ^{15}N NMR spectroscopy. *J. Soil Sci.* **43**, 113–125.

Bergersen, F.J., Peoples, M.B. and Turner, G.L. (1988) Isotopic discriminations during the accumulation of nitrogen by soybeans. *Aust. J. Plant Physiol.* **15**, 407–420.

Blackmer, A.M. and Bremner, J.M. (1977) Nitrogen isotope discrimination in denitrification of nitrate in soils. *Soil Biol. Biochem.* 9, 73–77.

Blackmer, A.M., Bremner, J.M. and Schmidt, E.L. (1980) Production of nitrous oxide by ammonium-oxidizing chemoautotrophic microorganisms in soil. *Appl. Environ. Microbiol.* 40, 1060–1066.

Böttcher, J., Strebel, O., Voerkelius, S. and Schmidt, H-L. (1990) Using isotope fractionations of nitrate-nitrogen and nitrate-oxygen for evaluation of microbial denitrification in a sandy aquifer. *J. Hydrol.* 114, 413–424.

Bouwman, A.F. (1990) *Soils and the Greenhouse Effect.* Wiley, New York. Bremner, E, and van Kessel, C. (1990) Appraisal of the nitrogen-15 natural abundance method for quantifying dinitrogen fixation. *Soil Sci. Soc. Am. J.* 56, 1141–1146.

Broadbent, F.E., Rauschkolb, R.S., Lewis, K.A. and Chang, G.Y. (1980) Spatial variability of nitrogen-15 and total nitrogen in some virgin and cultivated soils. *Soil Sci. Soc. Am. J.* 44, 524–527.

Cheng, H.H., Bremner, J.M. and Edwards, A.P. (1964) Variations of ^{15}N abundance in soils. *Science* 146, 1574–1575.

Clinton, P.W., Newman, R.H. and Allen, R.B. (1995) Immobilization of ^{15}N in forest litter studied by ^{15}N CPMAS NMR spectroscopy. *Eur. J. Soil Sci.* 46, 551–556.

Davidson, E.A., Hart, S.C., Shanks, C.A. and Firestone, M.K. (1991) Measuring gross nitrogen mineralization, immobilization and nitrification by ^{15}N isotopic pool dilution in intact soil cores. *J. Soil Sci.* 42, 335–349.

Davidson, E.A., Matson, P.A., Vitousek, P.M., Riley, R., Dunkin, K., Garcia-Mendez, G. and Maas, J.M. (1993) Factors regulating soil emissions of NO and N_2O in a seasonally dry tropical forest. *Ecology*, 74, 130–139.

Durka, W., Schulze, E-D., Gebauer, G. and Voerkelius, S. (1994) Effects of forest decline on uptake and leaching of nitrate determined from ^{15}N and ^{18}O measurements. *Nature* 372, 765–767.

Evans, R.D. and Ehleringer, J.R. (1993) A break in the nitrogen cycle in aridlands? Evidence from $\delta^{15}N$ of soils. *Oecologia* 94, 314–317.

Griffiths, B.S. (1986) Mineralisation of nitrogen and phosphorus by a mixed of the ciliate protozoan *Colpoda steinii*, the nematode *Rhabditis* sp. and the bacterium *Pseudomonas fluorescens*. *Soil Biol. Biochem.* 18, 637–641.

Goovaerts, P. and Chiang, C.N. (1993) Temporal persistence and spatial patterns for mineralizable nitrogen and selected soil properties. *Soil Sci. Soc. Am. J.* 57, 372–381.

Handley, L.L. and Raven, J.A. (1992) The use of natural abundance of nitrogen isotopes in plant physiology and ecology. *Plant Cell Environ.* 15, 965–985.

Handley, L.L., Daft, M.J., Wilson, J., Scrimgeour, C.M., Ingleby, K. and Sattar, M.A. (1993) Effects of the ecto- and VA-mycorrhizal fungi *Hydnagium carneum* and *Glomus clarum* on the $\delta^{15}N$ and $\delta^{13}C$ values of *Eucalyptus globulus* and *Ricinus communis*. *Plant Cell Environ.* 16, 375–382.

Handley, L.L. and Scrimgeour, C.M. (1997) Terrestrial plant ecology and ^{15}N natural abundance; the present limits to interpretation for uncultivated systems with original data from a Scottish old-field. *Adv. Ecol. Res.* (in press).

Hauck, R.D., Bartholomew, W.V., Bremner, J.M., Broadbent, F.E., Cheng, H.H., Edwards, A.P., Keeney, D.R., Legg, J.O., Olsen, S.R. and Porter, L.K. (1972) Use of variations in nitrogen isotope abundance for environmental studies: A questionable approach. *Science* 177, 453–454.

Hopkins, D.W. and Chudek, J.A. (1997) Solid-state NMR investigations of organic transformations during the decomposition of plant material in soil. In: *Driven by Nature: Plant Litter Quality and Decomposition* (eds G. Cadisch and K.E. Giller), CAB International, Wallingford, pp. 85–94.

Hopkins, D.W. (1997) Decomposition in a peaty soil improved for pastoral agriculture. *Soil Use Man.* 13, 104–106.

Ibijbijen, J., Urquiaga, S., Ismail, M., Alves, J.R. and Boddey, R.M. (1996) Effect of arbuscular mycorrhizas on uptake of nitrogen by *Briachia arrecta* and *Sorghum vulgare* from soils labelled for several years with ^{15}N. *New Phytol.* **133**, 487–494.

Isabella, B.L. and Hopkins, D.W. (1994) Nitrogen transformations in a peaty soil improved for pastoral agriculture. *Soil Use Man.* **10**, 107–111.

Jackson, R.B. and Caldwell, M.M. (1993a) The scale of nutrient heterogeneity around individual plants and its quantification with geostatistics. *Ecology* **74**, 612–614.

Jackson, R.B. and Caldwell, M.M. (1993b) Geostatistical patterns of soil heterogeneity around individual perennial plants. *J. Ecol.* **81**, 683–692.

Karamanos, R.E., Voroney, R.P. and Rennie, D.A. (1981) Variations in natural N-15 abundance of central Saskatchewan soils. *Soil Sci. Soc. Am. J.* **45**, 826–828.

Killham, K. (1987) A perfusion system for the measurement and characterisation of potential rates of nitrification. *Plant Soil* **97**, 267–272.

Kim, K-R. and Craig, H. (1990) Two-isotope characterization of N_2O in the Pacific Ocean and constraints on its origin in deep water. *Nature* **347**, 58–61.

Kim, K-R. and Craig, H. (1993) Nitrogen-15 and oxygen-18 characteristics of nitrous oxide: a global perspective. *Science* **262**, 1855–1857.

Kohl, D.H., Shearer, G.B. and Commoner, B. (1971) Fertilizer nitrogen: Contribution to nitrate in surface water in a corn belt watershed. *Science* **174**, 1331–1334.

Ladd, J.N. and Jackson, R.B. (1982) Biochemistry of ammonification. In: *Nitrogen in Agricultural Soils* (ed F.J. Stevenson) American Society of Agronomy/Crop Science Society of America/Soil Science Society of America, Madison WI, pp. 173–228.

Lechowicz, M.J. and Bell, G. (1991) The ecology and genetics of fitness in forest plants. II. Microspatial heterogeneity of the edaphic environment. *J. Ecol.* **79**, 687–696.

Ledgard, S.F., Freney, J.R. and Simpson, J.R. (1984) Variations in natural enrichment of ^{15}N in the profiles of some Australian pasture soils. *Aust. J. Soils Res.* **22**, 155–164.

Lloyd, D. (1993) Aerobic denitrification in soils and sediments: from fallacies to facts. *TREE* **8**, 352–355.

Mariotti, A., Pierre, D., Vedy, J.C., Bruckert, S. and Guillemot, J. (1980) The abundance of natural nitrogen-15 in the organic matter of soils along an altitudinal gradient (Chablais, Haute Savoie, France). *Catena* **7**, 293–300.

Mariotti, A., Germon, J.C., Hubert, P., Kaiser, P., Letolle, R., Tardieux, A. and Tardieux, P. (1981) Experimental determinations of nitrogen kinetic isotope fractionation: Some principles: Illustration for the denitrification and nitrification process. *Plant Soil* **62**, 413–430.

Mummey, D.L., Smith, J.L. and Bolton, H. (1994) Nitrous oxide flux from a shrub-steppe ecosystem: Sources and regulation. *Soil Biol. Biochem.* **26**, 279–286.

Natelhoffer, K.J. and Fry, B. (1988) Controls on natural nitrogen-15 and carbon-13 abundances in forest soil organic matter. *Soil Sci. Soc. Am. J.* **52**, 1633–1640.

Oliver, M.A. and Webster, R. (1991) How geostatistics can help you. *Soil Use Man.* **7**, 206–217.

Owens, N.J.P. (1987) Natural variations in ^{15}N in the marine environment. *Adv. Mar. Biol.* **24**, 389–451.

Parton, W.J., Mosier, A.R. and Schimel, D.S. (1988) Rates and pathways of nitrous oxide production in a shortgrass steppe. *Biogeochem.* **6**, 45–58.

Pau, R.N. (1991) The alternative nitrogenases. In: *Biology and Biochemistry of Nitrogen Fixation.* (eds M.J. Dilworth and A.R. Glenn), Elsevier Science Publishers, Amsterdam, pp. 37–57.

Paul, E.A. and Clark, F.E. (1989) *Soil Microbiology and Biochemistry.* Academic Press, San Diego, CA.

Post, W.M., Pastor, J., Zinke, P.J. and Stangenberger, A.G. (1985) Global patterns of soil nitrogen storage. *Nature* **317**, 613–616.

Poth, M. (1986) Dinitrogen production from nitrite by a *Nitrosomonas* isolate. *Appl. Environ. Microbiol.* **52**, 957–959.

Poth, M. and Focht, D.D. (1985) ^{15}N kinetic analysis of N_2O production by *Nitrosomonas europaea:* An examination of nitrifier denitrification. *Appl. Environ. Microbiol.* **49**, 1134–1141.

Powlson, D.S. (1993) Understanding the soil nitrogen cycle. *Soil Use Man.* **9**, 86–94.

Powlson, D.S. and Barraclough, D. (1993) Mineralization and assimilation on soil-plant systems. In: *Nitrogen Isotope Techniques* (eds R. Knowles and T.H. Blackburn), Academic Press, San Diego, CA, pp. 209–242.

Robertson, G.P. and Tiedje, J.M. (1987) Nitrous oxide sources in aerobic soils; nitrification, denitrification and other biological processes. *Soil Biol. Biochem.* **19**, 187–193.

Robertson, G.P., Huston, M.A., Evans, F.C. and Tiedje, J.M. (1988) Spatial variability in a successional plant community: patterns of nitrogen availability. *Ecology* **69**, 1517–1524.

Robinson, D., Griffiths, B.S., Ritz, K. and Wheatley, R.E. (1989) Root-induced nitrogen mineralisation: a theoretical analysis. *Plant Soil* **117**, 185–193.

Riga, A., van Praag, H.J. and Brigode, N. (1971) Rapport isotopique naturel de l'azote dans quelques sols forestiers et agricoles de Belgique soumis à divers traitements cultureax. *Geoderma* **6**, 213–222.

Schmidt, O., Scrimgeour, C.M. and Handley, L.L. (1997) Natural abundance of ^{15}N and ^{13}C in earthworms from a wheat and a wheat–clover field. *Soil Biol. Biochem.*, **29**, 1301–1308.

Schoun, H., Kim, D., Uchiyama, H. and Ugiyama, J. (1992) Denitrification by fungi. *FEMS Microbiol. Lett.* **94**, 277–282.

Shearer, G. and Kohl, D.H. (1986) N_2 fixation in field settings: estimates based on natural abundance. *Aust. J. Plant Physiol.* **16**, 305–313.

Shearer, G. and Kohl, D.H. (1989) Estimates of N_2 fixation in ecosystems: the need for and the basis of the ^{15}N natural abundance method. In: *Stable Isotopes in Ecological Research. Ecological Studies 68.* (eds P.W. Rundel, J.R. Ehleringer and K.A. Nagy), Springer Verlag, New York, pp. 342–374.

Shearer, G. and Kohl, D.H. (1993) Natural abundance of ^{15}N: Fractional contribution of two sources to a common sink and use of isotope discrimination. In: *Nitrogen Isotope Techniques* (eds R. Knowles and T.H. Blackburn), Academic Press, San Diego, CA, pp. 89–125.

Sowden, F.J., Chen, Y. and Schnitzer, M. (1977) The nitrogen distribution in soils formed under widely differing climatic conditions. *Geochim. Cosmochim. Acta* **41**, 1524–1526.

Stark, J.M. (1994) Causes of soil nutrient heterogeneity at different scales. In: *Exploitation of Environmental Heterogeneity by Plants* (eds M.M. Caldwell and R.W. Pearcy). Academic Press San Diego, pp. 255–284.

Stevenson, F.J. (1982) Organic forms of soil nitrogen. In: *Nitrogen in Agricultural* Soils (ed F.J. Stevenson) American Society of Agronomy/Crop Science Society of America/Soil Science Society of America, Madison WI, pp. 67–122.

Sutherland, R.A., van Kessel, C., Farrell, R.E. and Pennock, D.J. (1993) Landscape-scale variations in plant and soil nitrogen-15 natural abundance. *Soil Sci. Soc. Am. J.* **57**, 169–178.

Tiessen, H., Karamanos, R.E., Stewart, J.W.B. and Selles, F. (1984) Natural nitrogen-15 abundance as an indicator of soil organic matter transformations in native and cultivated soils. *Soil Sci. Soc. Am. J.* **55**, 312–315.

van Kessel, C., Pennock, D.J. and Farrell, R.E. (1993) Seasonal variations in denitrification and nitrous oxide evolution at the land-scape scale. *Soil Sci. Soc. Am. J.* **57**, 988–995.

van Kessel, C., Farrell, R.E. and Pennock, D.J. (1994) Carbon-13 and nitrogen-15 natural abundance in crop residues and soil organic matter. *Soil Sci. Soc. Am. J.* **58**, 382–389.

Wada, E., Kabaya, Y. and Kurihara, Y. (1993) Stable isotope structure of aquatic ecosystems. *J. Biosci.* **18**, 483–499.

Wahlen, M. and Yoshinari, T. (1985) Oxygen isotope ratios from nitrification at a wastewater treatment facility. *Nature* **317**, 349–350.

Wassenaar, L.I. (1995) Evaluation of the origin and fate of nitrate in the Abbotsford aquifer using the isotopes of ^{15}N and ^{18}O in NO_3^-. *Appl. Geochem.* **10**, 391–405.

Watkins, N. and Barraclough, D (1996) Gross rates of N mineralization associated with the decomposition of plant residues. *Soil Biol. Biochem.* **28**, 169–175.

Webster, E.A. and Hopkins, D.W. (1996a) Nitrogen and oxygen isotope ratios of nitrous oxide emitted from soil and produced by nitrifying and denitrifying bacteria. *Biol. Fertil. Soils* **22**, 326–330.

Webster, E.A. and Hopkins, D.W. (1996b) Contributions from different microbial processes to N_2O emission from soil under different water regimes. *Biol. Fertil. Soils* **22**, 331–335.

Weiss, R.J. (1981) The temporal and spatial distribution of tropospheric nitrous oxide. *J. Geophys. Res.* **86**, 7185–7195.

Yoshida, N. (1988) ^{15}N-depleted N_2O as a product of nitrification. *Nature* **335**, 528–529.

^{15}N at natural abundance levels in terrestrial vascular plants: a précis

Linda L. Handley, C.M. Scrimgeour and J.A. Raven

6.1 Introduction

When measured against atmospheric N_2 as a reference, the naturally occurring isotope ratios of $^{15}N/^{14}N$, are termed $\delta^{15}N$. $\delta^{15}N$ and ^{15}N-enrichment are quite different approaches to studying plant N relations. Enriched ^{15}N labelling has been used extensively to trace N source(s) and/or the amounts of N which move from a source to a sink, hence the term 'tracer.' N which is ^{15}N-enriched can be used in this way (because the $^{15}N/^{14}N$ isotope ratio is much larger than the background variations) to create a strong ^{15}N signal (see Chapter 16).

$\delta^{15}N$ values are very small compared to those of ^{15}N-enriched tracers. In air, ^{15}N comprises only 0.3663% of total N_2. ^{15}N-enriched tracers may contain > 99% ^{15}N. A difference in ^{15}N-enrichment of 1 atom percent (e.g. between 5 atom % and 6 atom %) is equivalent to a change of 3,046‰ using $\delta^{15}N$ terminology, which is derived as follows:

$$\delta^{15}N \text{ in ‰} = [(^{15}N/^{14}N\text{sample} - {}^{15}N/^{14}N\text{standard})/\ {}^{15}N/^{14}N\text{standard}] \times 10^3,$$

where the internationally accepted standard is N_2 of air and is defined at $\delta^{15}N_{air}$ = 0‰. Discrimination (Δ), as used here, is the difference in $\delta^{15}N$ between source N and sink N. Fractionation is a change in the ratio of $^{15}N/^{14}N$ but not necessarily implying that the absolute values of source or sink are known.

Repeatedly identifiable patterns have been found in the $\delta^{15}N$ values of plants, soils, and of individual chemical molecules within plants (see Chapters 5 and 7). The largest goal of current $\delta^{15}N$ research is to understand the processes causing these observed patterns. It is hoped, that when these patterns are better understood, useful mechanisms related to $\delta^{15}N$ will emerge and will perhaps be equivalent to that for plant $\delta^{13}C$ (e.g. Farquhar et al., 1982; Farquhar and Richards, 1984; O'Leary, 1988). Meanwhile, new patterns continue to be discovered as continuous-flow isotope ratio mass spectrometry (CF-IRMS) makes data gathering much easier than was previously the case,

Stable Isotopes, edited by H. Griffiths.
© 1998 BIOS Scientific Publishers Ltd, Oxford.

but care should be taken to achieve the precision associated with traditional dual-inlet mass spectrometer systems (Handley and Scrimgeour, 1997).

To date, studies have suggested that plant $\delta^{15}N$ may not simply be related to instantaneous source N, but also to plant physiological processes, rapid changes in soil-N signal and symbiotic associations (Azcón, Handley and Scrimgeour, unpublished data; Marriott et al., 1997; Pate et al., 1993; Shearer and Kohl, 1986).

6.2 What are the patterns in the $\delta^{15}N$ signal?

Relevant to terrestrial, vascular plants, $\delta^{15}N$ patterns have been found in individual N positions on molecules, in different molecules (see Chapter 2), in the different organs of plants (see Chapter 7), between taxa, in the different soil pools of plant-available N, in the products of various microbial processes, and in whole soil $\delta^{15}N$ over the growing season (Chapters 5 and 7; Handley and Scrimgeour 1997; Marriott et al., 1997; Yoneyama, 1995). The present discussion begins with what is known about $\delta^{15}N$ partitioning within plants and moves to whole plants growing in soil, soil N pools, then to a discussion of the aims and limits to using various kinds of experimental systems for studying vascular-plant $\delta^{15}N$. This discussion is limited to vascular plants and their soil N sources. Atmospheric sources of N are discussed *inter alia* in Handley and Scrimgeour (1997).

6.2.1 *Inside plants*

Changes of $\delta^{15}N$ (fractionations) are known to occur within plants because of enzymically mediated reactions (assimilation and transaminations, mainly, see Chapter 7). Enzymatic Δ, which have been measured *in vitro* and *in vivo* (see Handley and Raven 1992, as well as Chapter 5) range between about 0‰ to 40‰; they are large relative to the range of fractionations normally measured in a single ecosystem (- 12‰ maximally, but more frequently 3‰ to 5‰). Estimates for the discrimination caused by the assimilation of NH_4^+-N by glutamine synthetase in plants range from 9‰ for *Oryza sativa, in vivo* at an external [NH_4^+-N] of 7.1 mol m^{-3} N (Yoneyama et al., 1991) to 20‰ for *Skeletonema costatum, in vivo* at 0.05 mol m^{-3} N (Pennock et al ., 1988). The equivalent range of discriminations attributed to NO_3^- N assimilation by nitrate reductase are 3‰ for *Pennisetum americanum* seedlings, *in vivo* at 12 mol m^{-3} N (Marriotti et al., 1980) to 23‰ for *Phaeodactylum tricornatum, in vivo* at 10 mol m^{-3} N (Wada and Hattori 1978). There is no evidence for fractionation of $^{15}N/ {}^{14}N$ by physical transport of any kind, and this includes uptake, active and passive membrane transport and translocation within the plant (Handley and Raven, 1992; Handley and Scrimgeour, 1997; Yoneyama, 1995; see Chapter 7). The fractionations associated with internal transport are thought to occur only at the enzymes involved.

Enzymes discriminate against the heavier isotope by a constant instantaneous fractionation factor. However, enzyme activity and external concentration of N can affect the net fractionation. In general, discrimination relative to source N increases when supply substantially exceeds enzymic demand (Cleland, 1977). If all of a given N pool is converted from source to product, then no net fractionation occurs; this, however, is seldom the case in nature, and is more laboratory construct than a biological reality.

When NH_4^+-N enters a plant root, almost all of it is immediately assimilated by glutamine synthetase (GS) (see also Chapter 7). NH_4^+-N is also generated in the leaf by photorespiration and reassimilated with a Δ measured as -16.5‰ (Yoneyama et al., 1993) for isolated chloroplasts of spinach leaves (in vitro). When NH_4^+-N is the sole N source there may be whole plant Δ against ^{15}N. Yoneyama et al (1991) found whole plant Δ relative to source N of -7.5‰ and -7.9‰ in Oryza sativa L. Grown at 1.4 and 7.1 mol m^{-3} external NH_4^+-N, there was no significant difference in the amount of whole plant Δ over this range of concentrations at 20 days growth. At 32 days growth, whole plant Δ varied with external NH_4^+-N concentrations (Δ = -4.1‰ for plants grown at 1.4 mol m^{-3} and Δ = -12.6‰ for plants grown at 7.1 mol m^{-3}). Yoneyama (1995) and Chapter 7 provide a detailed discussion of plant δ^{15}N versus N sources.

The assimilation of NO_3^--N is conceptually more complicated than the assimilation of NH_4^+-N. Plant assimilation of NO_3^--N implies at least three isotopic sources, (i) external soil solution NO_3^--N, (ii) cytoplasmic NO_3^--N and (iii) vacuolar NO_3^--N (the plant storage pool in which NO_3^--N can be in quite high concentrations (Zhen et al., 1996). The concentration relevant to net Δ is the concentration at the site of the enzyme, nitrase reductase, not the N concentration external to the root. When NO_3^--N is the plant-available form of N, the soil NO_3^--N pool is not the sole, or even possibly the main, isotopic source. External NO_3^--N concentrations, over a wide range, appear to have little effect on whole plant δ^{15}N. Whole plant δ^{15}N did not differ greatly from the externally supplied NO_3^--N at external concentrations ranging from 0.2 mol m^{-3} to 12 mol m^{-3} (Yoneyama and Kaneko, 1989). However, root–shoot differences occurred because of partial assimilation in the root. Partial assimilation by nitrate reductase preferentially reduces the lighter isotope (Cleland, 1977) leaving a ^{15}N-enriched NO_3^--N pool for further assimilation in the shoots (Yoneyama and Kaneko, 1989). Thus, barring other later translocations of N, shoots should be more ^{15}N-enriched than roots under NO_3^--N nutrition.

At an external concentration of 5 mol m^{-3} NO_3^--N, shoots of hydroponically-grown wild and cultivated barley varieties were shown to express different shoot δ^{15}N values as a function of genotype, salt treatment and genotype x treatment interactions (Handley et al., 1997). The ranges of variation were as large as for concurrently measured δ^{13}C, which is routinely used to measure integrated genotypic variation for carbon discrimination (e.g. Zhang et al., 1993). Roots could not be harvested, but previous results by others (see discussion above) suggest that these were variations caused by plant N metabolism rather than whole plant Δ relative to external source.

Most of the information which exists for fractionations of δ^{15}N results from NO_3^--N or NH_4^+-N being the sole source of plant N. Plants are generally thought not to specialise in N sources to this extent. We are aware of only one paper (Steele, 1983) in which both NH_4^+-N and NO_3^--N were simultaneously supplied. Most agricultural plants use both forms of N to some extent and grasses are not generally selective for NO_3^--N or NH_4^+-N. Also, there is a growing body of evidence (see for example Chapin et al., 1993, Kielland, 1994, 1997) that many plants in the wild use NO_3^--N, NH_4^+-N and yet-unidentified organic N pools. For perennial plants, there is the additional isotopic source of internally stored N which is remobilised at various times of year (e.g., Millard, 1988).

6.2.2 Whole plants and soil

There is evidence that the δ^{15}N of plant leaves and shoots (Handley and Scrimgeour, 1997), of whole soil δ^{15}N (Marriott et al., 1997) and the δ^{15}N of soil NO_3^--N and

NH_4^+-N change during the growing season (see also Chapter 7). Additionally, the δ^{15}N of soil N changes more in the presence of plants than in the absence of plants (e.g. Farrell et al., 1996; Kreitler, 1971; Turner et al., 1987). However, values for soil mineral N must be treated with caution, as discussed later under soil N pools.

We propose that soil-N pools change isotopic composition by analogy with changing isotopic composition of CO_2 in the leaf substomatal chamber (see Chapter 9). Thus the N isotope pools which are measured in the soil (= rhizosphere) are analogous with the δ^{13}C of the substomatal cavity for CO_2 and the δ^{15}N of bulk soil N is roughly analogous with the δ^{13}C of atmospheric CO_2. However, this must remain a qualitative analogy, because the assimilation of NO_3^--N, especially, does not emulate that of CO_2. Hence, the model (Farquhar et al., 1982; Farquhar and Richards 1984) for carbon discrimination cannot provide a direct basis for modelling N assimilation: CO_2 has a single external isotopic source (air CO_2 at ~ -8‰) and a single, but variable, internal isotopic source (substomatal chamber CO_2) as well as the mesophyll conductance component (Chapters 8 and 9). NO_3^--N has at least three additional internal sources, as explained above.

Both NO_3^--N and NH_4^+-N probably occur as seasonally changing isotopic sources in soil. In addition, plant-available soil N is modified (to a yet unknown amount and variability in nature) by symbiotic relationships such as fungi and N_2-fixing bacteria or actinomycetes. The hypothesis above implies that instantaneous plant samples represent the integrated signal of δ^{15}N values of all plant N-sources and their temporally changing values in the past; they may not represent the soil δ^{15}N values at the time of sampling. As yet it is not known, in any quantitative detail how these pools change (see discussion below on analysing soil N pools and Chapter 5) and whether, for instance, they change with any systematic correlation among pools. Certainly, no one yet knows how these pools are related to plant δ^{15}N values in complex, multi-sourced soils.

6.2.3 The influence of mycorrhizas

Although it is generally agreed that mycorrhizas can have an effect on plant δ^{15}N, that effect has not been satisfactorily quantified, nor are the mechanisms known. In field studies, the influence of mycorrhizas is sometimes clear; in other instances their role is submerged in the variations of δ^{15}N. Högberg (1990) found a correlation between type of mycorrhizal association and foliar δ^{15}N in a semi-arid system in East Africa. In a later study (Högberg and Alexander, 1995) of a tropical wet forest in West Africa, no clear relationships between foliar δ^{15}N and fungal symbionts were found. In studies of semi-arid vegetation of South Africa and Australia (Pate et al., 1993 and Stock et al., 1993) there was no clear relationship between foliar δ^{15}N and type of root-zone symbiont of any kind, however, the most ^{15}N-enriched foliage was collected from non-mycorrhizal plants as confirmed in experiment by Handley et al. (1993).

Mycorrhizas may influence the δ^{15}N of whole plants by at least three means: (i) making N pools available which have different δ^{15}N values than the ones which the plant alone can access, (ii) discriminatory assimilation of the acquired N by the fungus so that the N passing to the plant host is different in δ^{15}N value from the original soil-sourced N pool, and (iii) altering the N metabolism of the host plant. The last mechanism may also alter the intra-plant distribution of δ^{15}N values, as explained above for NO_3^--N versus NH_4^+-N assimilation.

It is generally held that ecto-mycorrhizas chiefly assist plants in obtaining otherwise inaccessible organic N (e.g. Turnbull *et al.*, 1995) and that arbuscular mycorrhizas (AM) chiefly contribute to plant nutrition through providing otherwise inaccessible forms of phosphorus (Allen, 1991). While the first is undoubtedly true (Allen, 1991), it has recently been argued (e.g. Ibijbijen *et al.*, 1996; Tobar *et al.*, 1994) that arbuscular mycorrhizas (AM) allow plants to access more chemical forms of N than they would be able to do without the mycorrhiza-forming fungus.

Whether both or either type of fungal symbiont selectively assimilates ^{15}N-enriched N and passes ^{15}N-depleted N to the host plant is still an unanswered question. Data presented by Högberg *et al.* (1996) are consistent with such an hypothesis, but inconclusive. Handley *et al.*, (1993), however, showed that the δ^{15}N of ecto-mycorrhizal *Eucalyptus* was not affected when only mineral N was provided. Superficially, this appears to cast doubt on the hypothesis.

For AM, Handley *et al.* (1993) and Azcón, Handley and Scrimgeour (unpublished data), showed that substantial whole plant ^{15}N-enrichment (2‰ to 3‰) can occur relative to external N and was chiefly attributed to the interaction of external N concentration and species of AM-forming fungus in the presence of a single isotopic and chemical source of externally supplied N. This is consistent with the hypothesis that the fungus, itself, fractionates the source N or that plant metabolism is modified in the symbiotic association. Therefore, both types of mycorrhizas probably influence plant N metabolism via evoking or enhancing various enzyme activities (e.g. Cliquet and Stewart, 1993, Martin and Botton, 1993).

6.2.4 *Soil N pools*

Soil research has traditionally focused on amounts of N and categorised that N into mainly NO_3^--N, NH_4^+-N and organic N. Chemical analytical methods (e.g. various types of digestions and colorimetric techniques) were developed in order to quantify the amounts of these forms of N. There are, however, no universally applicable methods for isolating soil N pools for isotopic analysis at the natural abundance level. The problem becomes increasingly difficult as the amounts of these pools and their purity decreases, that is, as one moves from: (i) high N concentration hydroponic solutions containing a single N source, (ii) to soil solutions from agricultural soil which have relatively high N concentrations of the order of 1–2 mol m^{-3}, (iii) to natural soils in which the concentration of N may drop to < 1 mmol m^{-3}. For the first two types of samples, a modified microdiffusion technique (Sörensen and Jensen, 1991) will often suffice provided that vials are air-tight and that recovery of inorganic N is complete. For natural soils, where mineral N pools are often at the mmol m^{-3} level, there is frequently no method for separately analysing the δ^{15}N of NO_3^--N, NH_4^+-N, and organic N. In our experience, the problem of contamination becomes worse as soil organic matter content increases. In many soils which we have encountered, NH_4^+-N may be microdiffused successfully, but the remaining NO_3^--N cannot be separated from a relatively large contamination by other kinds of N (of unknown composition) even by pre-treatment with ion exchange resins. In some soils this contamination constitutes as much as 50% of the target N and occasionally ten times the colorimetrically measured amount of NO_3^--N.

We are aware of this contamination because modern methods of mass spectrometry make it possible to measure the amount, as well as the δ^{15}N, of the N in a sample.

Hence, the sample can be analysed for amounts of NO_3^--N and NH_4^+-N before purification and extraction. This expected amount can be compared, subsequently, with the results of mass spectrometry. N in amounts substantially in excess of that detected colorimetrically indicate contamination. Hence, the values obtained by older, non-automated methods (e.g., Farrell *et al.*, 1996; Feigin *et al.*, 1974; Kreitler, 1971; Turner *et al.*, 1987), which do not allow such comparisons of quantities, must now be treated with some caution. A more detailed discussion of soil N pools is given in Chapter 5, but in part the way forward may lie in developing new chemical methods specifically for the preparation of N samples for isotopic analysis.

6.3 Experimental systems

The central difficulty in designing experiments for the interpretation of $\delta^{15}N$ lies in the persistence of simultaneously occurring multiple isotopic N sources. Many of the processes which physically or chemically transform N also change the $\delta^{15}N$ of sources and sinks. Hence, research in the field into the mechanisms underlying plant $\delta^{15}N$ is usually done against a background of constantly changing and multiple isotopic sources. For $\delta^{15}N$ research, the challenge is to enhance the magnitude of changes in $\delta^{15}N$ against its own background levels so that the results can be interpreted. This can be done in three ways: (i) very lightly ^{15}N-enriched N (< 0.5 atom%) can be used; (ii) the external chemical and isotopic source of N can be controlled and limited to a single source; and (iii) the external concentration of N can be increased to enlarge the resulting effects on plant $\delta^{15}N$ (but potentially altering the discrimination expressed within the plant).

Whole plant experiments can be conducted in a number of growth media under varying amounts of experimental control and designed to answer different questions at different scales. Lacking ^{15}N-enrichment, each step which is more removed from strictly controlled hydroponics incurs costs in terms of data interpretation. Rorison and Robinson (1986) discussed various experimental systems for whole plant growth and cite specialist references on hydroponics. There is no one experimental system which answers all of the needs for exploring plant $\delta^{15}N$. However, each type of system can contribute something toward understanding the mechanisms underlying these isotopic signatures. The range of $\delta^{15}N$ to be found in nature in bulk samples, is, generally, no more than $\pm 20\permil$ (Handley and Scrimgeour, 1997). ^{15}N-enrichment at < 0.5 atom% can, in many systems, create a sufficiently clear signal to be traced while still allowing discriminations to be observed.

6.3.1 *Hydroponics*

If ^{15}N-enrichment is not used, treatment effects can be magnified by using a hydroponic system of growth in which the external concentration of N is large, relative to the plants' usual growth conditions. For agricultural plants, this might range from 1 to 15 mol m^{-3}. For wild plants a range of smaller concentrations (perhaps 25–100 mmol m^{-3}) are appropriate. Using high external concentrations (relative to demand by the plant) also helps to ensure that the isotopic composition of the solution N should remain constant and is not altered by plant assimilation between refreshments of the nutrient solution, although this should be routinely checked.

By using hydroponics, the source $\delta^{15}N$ can be controlled, but at the price of sacrificing normal root architecture and anatomy (Rorison and Robinson, 1986) as well as normal soil environment. However, lowering the external concentration in a well-mixed culture solution to that found in the field could cause problems for at least three reasons: (i) it risks inadequate plant nutrition; while (ii) perhaps losing the benefits of a magnified isotope effect; (iii) it also fails to simulate a normal soil environment in important ways, most of which are related to the fact that hydroponics are well-stirred and soil is not. Thus hydroponics would not simulate N-depleted zones near roots, the natural heterogeneity of soil N concentrations, the N-depletion-driven necessity for plants to grow into new areas of accessible N, constant mineralisation punctuated by large mineralisation events, diffusion in soil solution through a matrix of soil particles, ion exchange with soil particles, the changing predominance of chemical types of N over the growing season or finally competition from and associations with micro-organisms and other plants. Since it is labour intensive, hydroponic growth is best suited to experiments of short duration (hours to weeks).

Hydroponics are more easily conducted using NO_3^--N as the external N source. In water culture, vigorous aeration is necessary unless the experimental plants are tolerant of low oxygen conditions. This aeration, in the presence of micro-organisms in the culture solution, quickly converts NH_4^+-N to NO_3^--N . In an ideal world, sterile culture conditions should be possible (Evans et al., 1996). However, except for specialist laboratories, this is not usually practical in the course of growing whole plants in glasshouses or growth chambers. Alternatively, the culture solution can be changed frequently. Control of solution pH is especially important in NH_4^+-N nutrition (Rorison and Robinson, 1986).

Continuous-flow hydroponics, in which the plant roots are grown in rock wool or other solid media, could be used to overcome some of the effects of hydroponics on root architecture. But this method requires almost as much attention as hydroponics and does not simulate the soil environment in terms of N-depletion zones and the heterogeneity of N sources and concentrations in soil.

6.3.2 Solid media

Perlite and/or sand provide an inert, solid medium for root growth, which improves the normality of root architecture and anatomy without introducing the full complexities of soil. Vermiculite should be avoided because of the potential for fractionating ion-exchange. Use of an initially sterilised, inert solid medium does not eliminate ^{15}N-enrichment of the potting medium N over the course of the experiment. Bergersen et al. (1988) found evidence of source ^{15}N-enrichment (relative to the N supplied) in non-nodulated soybean. Handley et al. (1993) also documented ^{15}N-enrichment of initially sterilised sand used as a potting medium, as we have found (Azcón, Handley and Scrimgeour, unpublished data) for an initially sterilised mixture of N-poor soil mixed 1: 1 with quartz sand.

6.3.3 Soil

Soil-based experiments are usually best for long-term experiments (weeks to months duration). Because the soil buffers nutrient supplies to the plant, soil based experiments are less labour intensive than hydroponics or those using inert solid media. When soil is used in pot experiments, roots may grow normally, but little can be learned about the instantaneous source $\delta^{15}N$ of the plants. The act of excavating soil and putting it into pots increases $^{15}N/^{14}N$ fractionating mineralisation. Sub-sampling during the experiment causes the same disturbance/mineralisation problems as encountered when the soil was first potted-up. Additionally, there is seldom enough mineral N to analyse in a sub-sample. Serial harvests of a highly replicated experiment is another attractive design. However, there is only one way to ascertain whether the pots have stabilised at similar $\delta^{15}N$ values for whole soil and plant available N, thereby achieving the required replication. That way is to harvest or sub-sample the soils, causing disturbance or destruction and mineralisation or complete loss of the potted soil. These disturbance-related variations of $\delta^{15}N$ may not appear to be great on a whole-soil basis, while profoundly affecting plant-available, mineral N pools (Neilson, Wishart and Handley, unpublished data for Scottish soils of pasture, deciduous and coniferous woodlands).

6.3.4 *In the field*

In the field, plants are sampled against their normal growth environment. However, it is possible to know even less about the exact sources of plant N than in experiments using soil mixtures, because one is faced with the imponderables of the interactive effects of mycorrhizas, N_2-fixation, atmospheric deposition of N and increased heterogeneity of soil N sources. It is possible, however, to sample whole plant communities over a growing season and examine the resulting patterns of $\delta^{15}N$ for guidance in designing further mechanistic research. It is desirable to determine an appropriate sampling strategy statistically (e.g. Marriott *et al.*, 1997; Sutherland *et al.*, 1993) before designing the field sampling plan.

6.4 Needs for research

The considerations discussed above make it unlikely that in the near future plant $\delta^{15}N$ will be useful for revealing the exact or instantaneous source(s) of soil N for plants in the field. No one experimental system commonly in use can reveal, simultaneously, mechanisms of N isotope discrimination and emulate its behaviour in nature. Some logical interpolations among experimental designs are required. But this must be cautiously done, because the number of potentially interacting variables is large. As called for in Handley and Scrimgeour (1997), there is a real need for entirely new chemical methods for preparing samples for isotopic analysis. When appropriate methods are in hand for analysing the source N pools in most soils, progress can be made towards modelling the relationship between changing soil N isotopic sources and the resulting integration of this by plants and their symbionts.

References

Allen, M.F. (1991) The Ecology of Mycorrhizae. Cambridge University Press, New York, p. 1.

Bergersen, F.J., Peoples, M.B. and Turner, G.L. (1988) Isotopic discrimination during the accumulation of nitrogen by soybeans. *Aust. J. Plant Physiol.* **15**, 407–420

Chapin, F.S., Moilanen, L. and Kielland, K. (1993) Preferential use of organic nitrogen for growth by a non-mycorrhizal arctic sedge. *Nature* **361**, 150–153

Cleland, W.W. (1977) Determining the chemical mechanisms of enzyme-catalyzed reactions by kinetic studies. *Adv. in Enzymol.* **45**, 273–287

Cliquet, J.-B and Stewart, G.R. (1993) Ammonia assimilation in *Zea mays* L. infected with a vesicular-arbuscular mycorrhizal fungus *Glomus fasciculatum. Plant Physiol.* **101**, 865–871

Evans, R.D., Bloom, A.J, Sukrapanna, S.S. and Ehleringer, J.R. (1996) Nitrogen isotope composition of tomato (*Lycoperiscon esculentum* Mill. CV. T-5) grown under ammonium or nitrate nutrition. *Plant, Cell Environ.* **19**, 1317–1323.

Farquhar, G.D., O'Leary and M.H., Berry, J. (1982) On the relationship between carbon isotope discrimination and intercellular CO_2 concentration in leaves. *Aust. J. Plant Physiol.* **9**, 121–137.

Farquhar, G.D. and Richards, R.A. (1984) Isotopic composition of plant carbon correlates with water-use efficiency of wheat genotypes. *Aust. J. Plant Physiol.* **11**, 539–552

Farrell, R.E., Sandercock, P.J., Pennock, D.J. and van Kessel, C. (1996) Landscape-scale variations in leached nitrate – relationship to denitrification and natural N-15 abundance. *Soil Sci. Soc. Am. J.* **60**, 1410–1415

Feigin, A., Shearer, G., Kohl, D.H. and Commoner, B. (1974) The amount and nitrogen-15 content of nitrate in soil profiles from two Central Illinois fields in a corn–soybean rotation. *Soil Sci Soc Am. Proc.* **38**, 465–471

Handley, L.L., Brendel, O., Scrimgeour, C.M., Schmidt, S., Raven, J.A., Turnbull, M.H. and Stewart, G.R. (1996) The ^{15}N natural abundance patterns of field-collected fungi from three kinds of ecosystems. *Rapid Commun. Mass Spectrometry* **10**, 974–978

Handley, L.L., Daft, M.J., Wilson, J., Scrimgeour, C.M., Ingleby, K. and Sattar, M.A. (1993) Effects of the ecto- and VA-mycorrhizal fungi *Hydnagium carneum* and *Glomus clarum* on the $\delta^{15}N$ and $\delta^{13}C$ values of *Eucalyptus globulus* and *Ricinus communis. Plant, Cell Environ.* **16**, 375–382

Handley, L.L. and Raven, J.A. (1992) The use of natural abundance of nitrogen isotopes in plant physiology and ecology. *Plant,Cell Environ* **15**, 965–985

Handley, L.L, Robinson, D., Forster, B.P., Ellis, R.P., Scrimgeour, C.M., Gordon, D.C., Nevo, E., and Raven, J.A. (1997) Shoot $\delta^{15}N$ correlates with genotype and salt stress in barley. *Planta* **102**, 100–102

Handley, L.L. and Scrimgeour, C.M. (1997) Terrestrial plant ecology and ^{15}N natural abundance: the present limits to interpretation for uncultivated systems with original data from a scottish old field. *Adv. Ecol. Res.* **27**, 133–212

Högberg, P. (1990) ^{15}N natural abundance as a possible marker of the ectomycorrhizal habit of trees in mixed African woodlands. *New Phytol.* **115**, 483–486

Högberg, P. and Alexander, I.J. (1995) Roles of root symbioses in African woodland and forest: evidence from ^{15}N abundance and foliar analysis. *J. Ecol.* **83**, 217–224

Högberg, P., Högbom, L., Schinkel, H., Högberg, M., Johannisson, C. and Wallmark, H. (1996) ^{15}N abundance of surface soils, roots and mycorrhizas in profiles of European forest soils. *Oecologia* **108**, 207–214

Ibijbijen, J., Urquiaga, S., Ismaili, M., Alves, B.J.R. and Boddey, R.M. (1996) Effect of arbuscular mycorrhizas on uptake of nitrogen by *Brachiaria arrecta* and *Sorghum vulgare* from soils labelled for several years with [15]N. *New Phytol.* **134**, 353–360

Kielland, K. (1994) Amino acid absorption by Arctic plants: implications for plant nutrition and nitrogen cycling. *Ecol.* **75**, 2373–2383

Kielland, K. (1997) Role of free amino acids in the nitrogen economy of arctic cryptogams. *Ecosci.* **4**, 75–79

Kreitler, C.W. (1971) Determining the Source of Nitrate In Ground Water By Nitrogen Isotope Studies. Report No. 83 of the Bureau of Economic Geology, The University of Texas, Austin, Texas.

Marriott, C.A., Hudson, G., Hamilton, D., Neilson, R., Boag B., Handley, L.L., Wishart, J., Scrimgeour, C.M. and Robinson, D. (1997) Spatial variability of soil total C and N and their stable isotope signatures in an upland grassland system. *Plant Soil*, in press.

Marriotti, A, Marriotti, F, Champigny, M-L, Amanger, H, Moyse, A. (1980) Nitrogen isotope fractionation associated with nitrate reductase activity and uptake of nitrate by pearl millet. *Plant Physiol.* **69**, 880–884.

Martin, F. and Botton, B. (1993) Nitrogen metabolism of ectomycorrhizal fungi and ecto-mycorrhiza. *Adv. Plant Pathol.* **9**, 83–102.

Millard, P. (1988) The accumulation and storage of nitrogen by herbaceous plants. *Plant, Cell Environ.* **11**, 1–8.

O'Leary, M. (1988) Carbon isotopes in photosynthesis. *BioSci.* **38**, 328–336.

Pate, J.S., Stewart, G.R. and Unkovich, M. (1993) ¹⁵N natural abundance of plant and soil components of a *Banksia* woodland ecosystem in relation to nitrate utilisation, life form, mycorrhizal status and N_2-fixing abilities of component species. *Plant, Cell Environ.* **16**, 365–373.

Pennock, J.R. (1988) Isotope fractionation of nitrogen during the uptake of NH_4^+ and NO_3^- by *Skeletonema costatum*. *EOS* **69**, 1098.

Rorison, H. and Robinson, D. (1986) Mineral nutrition. In: *Methods in Plant Ecology*. P.D. Moore , S.B. Chapman (eds) Blackwell Scientific Publications, Oxford, pp. 145–213.

Shearer, G. and Kohl, D.H. (1986) N_2-fixation in field settings: estimations based on natural ¹⁵N abundance. *Aust. J. Plant Physiol.* **13**, 699–756.

Sørensen, P. and Jensen, E.S. (1991) Sequential diffusion of ammonium and nitrate from soil extracts to a polytetrafluoroethylene trap for ¹⁵N determination. *Anal. Chem. Acta* **252**, 201–203.

Steele, W. (1983) Quantitative measurements of nitrogen turnover in pasture systems with particular reference to the role of ¹⁵N. *Nuclear Techniques in Improving Pasture Management.* IAEA, Vienna, pp 17–35.

Stock, W.D., Wienand, K.T. and Baker, A.C. (1995) Impacts of invading N_2-fixing *Acacia* species on patterns of nutrient cycling in two Cape ecosystems: evidence from soil incubation studies and ¹⁵N natural abundance values. *Oecologia* **101**, 375–382.

Tobar, R., Azcón, R. and Barea, J.M. (1994) Improved nitrogen uptake and transport from ¹⁵N-labelled nitrate by external hyphae of arbuscular mycorrhiza under water-stressed conditions. *New Phytol.* **126**, 119–122.

Turnbull M., Goodall, R. and Stewart, G.R. (1995) The impact of mycorrhizal colonization upon nitrogen source utilization and metabolism in seedlings of *Eucalyptus grandis* Hill ex Maiden and *Eucalyptus maculata* Hook. *Plant, Cell Environ.* **18**, 1386–1394.

Turner, G.L., Gault, R.R., Morthorpe, L. and Chase, D.L. (1987) Differences in natural abundance of ¹⁵N in extractable mineral nitrogen of cropped and fallowed surface soils. *Austr. J. Agric. Res.* **38**: 15–25.

Wada, E. and Hattori, A. (1978) Nitrogen isotope effects in the assimilation of inorganic nitrogenous compounds by marine diatoms. *Geomicrobiol. J.* **1**, 85–101.

Yoneyama, T. (1995) Nitrogen metabolism and fractionation of nitrogen isotopes in plants. In: *Stable Isotopes in the Biosphere*. E. Wada, T. Yoneyama, M. Minagawa, T. Ando, and B.D. Fry (eds) Kyoto University Press, Kyoto, pp 92–102.

Yoneyama, T., Kamachi, K., Yamaya, T. and Mae, T. (1993) Fractionation of nitrogen isotopes by glutamine synthetase isolated from spinach leaves. *Plant Cell Physiol.* **34**, 489–491.

Yoneyama, T. and Kaneko, A. (1989) Variations in the natural abundance of ¹⁵N in nitrogenous fractions of komatsuna plants supplied with nitrate. *Plant Cell Physiol.* **30**, 957–962.

Yoneyama, T., Omata, T., Nakata, S. and Yazaki, J. (1991) Fractionation of nitrogen isotopes during the uptake and assimilation of ammonia by plants. *Plant Cell Physiol.* **32**, 1211–1217.

Zhang, J., Marshall, J.D., Jaquish, B.C. (1993) Genetic differentiation in carbon isotope discrimination and gas exchange in *Pseudotsuga menziesii*. *Oecologia* **93**, 80–87.

Zhen, R.-G., Koyro, H-W., Leigh, R.A., Tomos, A.D. and Miller, A.J. (1996) Compartmental nitrate concentrations in barley root cells measured with nitrate-selective microelectrodes and by single-cell sap sampling. *Planta* **185**, 356–361.

Variations in fractionation of carbon and nitrogen isotopes in higher plants : N metabolism and partitioning in phloem and xylem

T. Yoneyama, H. Fujiwara and J.M.Wilson

7.1 Introduction

The natural abundance of carbon and nitrogen isotopes in higher plants integrate both physiological and environmental effects (Deléens *et al.*, 1994). Carbon which comprises about 40% (dry weight) of plant tissues is acquired by photosynthetic CO_2 fixation, although significant anaplerotic CO_2 fixation is suggested in some tissues such as legume nodules (Vance and Heichel, 1991), and heterotrophic carbon gain is important in parasitic angiosperms (Richter *et al.*, 1995). The natural abundance of ^{13}C ($\delta^{13}C$) in plant carbon compounds reflects the signature of the sources of plant carbon as affected by the mechanisms of C metabolism (O'Leary, 1981, 1995; Raven and Farquhar, 1990; see Chapter 2) and CO_2 recycling (see Chapter 8). In addition, the effect of environmental factors such as temperature, water supply, and CO_2 concentration during CO_2 fixation are also reflected in plant $\delta^{13}C$ values (see Chapters 8 and 9). Nitrogen in plant tissues constitutes 1% to 6% (dry weight), and is chiefly acquired via root uptake of inorganic nitrogen and by nitrogen fixation in symbiotic systems such as the legume-rhizobia, *Azolla-Anabaena* and *Cycad-Nostoc*. Under certain conditions plants can also use organic nitrogen sources such as urea and amino acids. The natural abundance of ^{15}N ($\delta^{15}N$) in plant N compounds is also a reflection of the sources of plant nitrogen and the mechanisms of N metabolism (Handley and Raven, 1992; Yoneyama, 1995; Stewart *et al.*, 1995; see also Chapters 5 and 6).

Higher plants are composed of various tissues of different ages, and carbon and nitrogen are cycled amongst these tissues according to source–sink relationships. The green leaves are the major site of CO_2 fixation and the fixed C is converted to sugars

Stable Isotopes, edited by H. Griffiths.
© 1998 BIOS Scientific Publishers Ltd, Oxford.

for export. These sugars are transferred via the phloem to sink organs such as the fruits and roots. Nitrogen is absorbed from the soil solution via the roots and transferred to the shoot with partial transformations such as nitrate to amino acids in the roots. The root-derived nitrogen is transported through the xylem to sink organs. The remaining nitrate is reduced in the sink organs and assimilated to amino acids and many other N compounds, chiefly proteins. The N compounds in the sink organs are further metabolised through turnover of protein and amino acids, and the products transferred to newly forming sinks through the phloem, together with sugars. The cumulative uptake and cycling of carbon and nitrogen support the differentiation and growth of plant tissues, and the enzymatic reactions involved in these mechanisms may induce isotopic fractionations of carbon and nitrogen (Deléens et al., 1994, see Chapters 2 and 8).

Having firstly reviewed the extent and magnitude of fractionation during N uptake and transformation, we then go on to introduce new data on changes in $\delta^{13}C$ and $\delta^{15}N$ values in castor bean (Ricinus communis) plants grown in the field, based on analyses of xylem and phloem constituents, as compared to whole plant organic matter and soil nitrate.

7.2. Acquisition of nitrogen

The stable isotope composition of nitrogen may vary among the different forms available to the plant (NH_4^+, NO_3, organic N and atmospheric N_2) (Evans et al., 1996) and can change depending on soil depth (Gebauer et al., 1994; Mariotti et al., 1980a; Nadelhoffer and Fry, 1988; Shearer and Kohl, 1986). Whilst $\delta^{15}N$ values of soil N are largely positive, ranging between +2‰ and +10‰ (Yoneyama, 1996).

Isotopic fractionation during dinitrogen fixation by symbiotic systems has been found to be small (largely between -0.2‰ and -2.0‰) in legumes (Steele et al., 1983, Yoneyama et al., 1986), Frankia tree symbiosis (Domenach et al., 1989), and Azolla-Anabaena association (Yoneyama et al., 1987). The percentage of N derived from N_2 for field-grown N_2-fixing plants in homogeneous soils can be calculated (Amarger et al., 1980; Shearer and Kohl, 1986, Shearer et al., 1983, Wada et al., 1986; Yoneyama, 1987, 1990; Yoneyama et al., 1986, 1991b). However, the $\delta^{15}N$ values of soil N absorbed by plants are affected by nitrogen transformation in soils, and by the addition of fertilisers (Yoneyama et al., 1990; see Chapter 5), such that plant $\delta^{15}N$ values can approach zero even when no nitrogen fixation is occurring. This introduces difficulties in estimating the fractional contribution of N_2 fixation to the total N assimilated.

The discrimination of nitrogen isotopes during uptake of nitrate by hydroponically-grown higher plants was small (Kohl and Shearer, 1980, Mariotti et al., 1980b, Yoneyama and Kaneko, 1989), while that during uptake of ammonia was very large (Yoneyama et al., 1991a). When Steele (1983) grew Trifolium repens solely dependent on NH_4NO_3 ($\delta^{15}N$ values of NH_4 and NO_3 supplied were +10.6‰ and +10.5‰ respectively) as the N source, the $\delta^{15}N$ value of the vegetative N was +9.1‰, and that of the root N was +7.9‰, while the residual NH_4 and NO_3 in the sand culture had $\delta^{15}N$ values of +12.5‰ and +11.5‰, respectively. Interpreting this data using the observations of Yoneyama and Kaneko (1989) and Yoneyama et al. (1991a), the results of Steele (1983) suggest that NH_4 uptake was associated with lower plant $\delta^{15}N$ values and higher enrichment in residual N, while NO_3 uptake resulted in higher $\delta^{15}N$ values in the shoot as compared to the roots. This observation was also reported by Bergersen *et al.* (1988) and most recently, Evans *et al.* (1996) elegantly showed that patterns of root and shoot assimilation can alter the leaf $\delta^{15}N$ signal when grown at realistic (near-natural) NO_3- concentrations; an important consideration, given that in many species the mechansims responsible for uptake of nitrate and ammonium differ at relatively high (>1 mol m^{-3}) and low (<500 mmol m^{-3}) external concentrations (Evans *et al.* 1996, and references cited therein). However, they measured no such change when the nitrogen was supplied as ammonium.

7.3 Metabolism of nitrogen

Changes to the N signals of distinct soil and atmospheric N sources (see above), are brought about through isotopic fractionation which can occur at several steps in the sequence of assimilatory reactions responsible for converting nitrate and ammonia into amino acids and high-molecular N compounds in higher plants. An analysis of $\delta^{15}N$ values during nitrate assimilation and nitrogen fixation for *Brassica campestris* var. *rapa*, *Spinacia oleracea* and *Glycine max* (Yoneyama, 1995) showed that as compared to $\delta^{15}N$ value of whole tissues, those of nitrate were high in *Brassica* and *Spinacia* (+15‰ to +19‰). The amounts of ammonia, glutamine and asparagine were low, and their $\delta^{15}N$ values were lower than that of nitrate. For soybean solely dependant on nitrogen fixation, the $\delta^{15}N$ values of leaves and roots were negative, and the $\delta^{15}N$ values of glutamine (amide) were more negative than those of ammonia, while those of asparagine (amide) were less negative than those of glutamine (amide). These data indicate that the diversity of $\delta^{15}N$ values of individual nitrogenous compounds in different tissues is great and is influenced by different nitrogen nutrition.

Nitrate absorbed by plants undergoes reduction to nitrite and then ammonium by enzymatic reactions involving nitrate reductase and nitrite reductase. The nitrite reductase activity in higher plants is sufficiently great to reduce almost all the nitrite produced by nitrate reductase, whereas there may be unreduced nitrate which may accumulate in cells, mainly stored in vacuoles. The reaction of nitrate reductase is the site of isotopic fractionation (Ledgard *et al.*, 1985), and at high nitrate concentrations the nitrate accumulated in the cells is highly enriched in ^{15}N, while the reduced nitrogen (organic-N) is depleted in ^{15}N as compared to the $\delta^{15}N$ values of the nitrate supplied (Medina and Schmidt 1982; Yoneyama and Kaneko 1989).

Ammonium is assimilated into glutamine (amide) by glutamine synthetase (GS). The fact that the $\delta^{15}N$ values of tissue ammonia in rice leaves grown in ammonia medium are higher than those of assimilated N compounds suggests that fractionation of nitrogen isotopes occurs during the GS reactions (Yoneyama *et al.*, 1991a). If we consider that the direct substrate for GS is NH_3 which is in equilibrium with NH_4^+, and that the lighter nitrogen atom (^{14}N) favours NH_3, amide-N of glutamine may be depleted in ^{15}N. Recently, fractionation by GS (isolated from spinach leaves) showed a fractionation factor of about -17‰ during ammonia assimilation (Yoneyama *et al.*, 1993). Ammonia assimilation by GS, having a low K_m value for ammonia, is very efficient leaving only a small amount of ammonia unassimilated. Under such circumstances, the effect of overall isotopic fractionation during ammonia assimilation on the $\delta^{15}N$ values of tissues may be small (Evans *et al.*, 1996).

In nodules, fixed-N is converted to ammonium in bacteroids and assimilated into glutamine by GS. Shearer and Kohl (1989) showed that $\delta^{15}N$ values of free amides and amides bound to nodule fractions (plant cytosol, bacteroids, and cortex) were similar to the $\delta^{15}N$ values for total N in each fraction. However, the amide-N signal contains both the amide groups of glutamine and asparagine. The $\delta^{15}N$ values of ammonia and amide of asparagine were higher than those of amide of glutamine in nodules of *Glycine max* and *Sesbania cannabia* (Yoneyama, 1995).

Transfer of nitrogen among amino acids in plant cells occurs mainly through three reactions: (1) transfer of amide from glutamine by glutamate synthase (GOGAT) (Ito *et al.*, 1978); (2) transaminations; and (3) deamination and reassimilation of ammonia by GS (Ito and Kumazawa, 1978). Variations in $\delta^{15}N$ in different amino acids in plants have been documented by a number of workers (Gaebler *et al.*, 1966; Macko *et al.*, 1987; Minagawa *et al.*, 1992). When compared to the $\delta^{15}N$ value of glutamic acid, serine, leucine, lysine, and arginine gave low $\delta^{15}N$ values. These amino acids are principally produced during transamination by the reaction of amino acids (particularly glutamic acid) and keto acids, and occur as end-products of the reactions. The rates and directions of transamination reactions may depend on the organ (and organelle), age of tissues and also on the environment (nutrients, light, temperature etc.) (Yoneyama, 1995).

The molecules of glutamine, asparagine, lysine, arginine, histidine, tryptophan and many other nitrogenous compound contain more than 2 nitrogen atoms. Medina and Schmidt (1982) have shown that N in arginine from bacteria had $\delta^{15}N$ values of 0‰, +33‰ and +40‰ at N^2, N^5, and amidino-N positions, respectively. The heterogeneity of ^{15}N enrichment in a molecule is most likely to be a result of the different enzymatic reactions involved in forming molecules which may utilise different sources of nitrogen and different breakdown pathways.

In ureide-producing nodules, allantoin and allantoic acid, formed through metabolism of amide and purines are the major forms of fixed-N translocated, and showed low $\delta^{15}N$ values in soybean (Yoneyama, 1988). Ureide-N in root-sap exudate from N_2 grown soybean had a low $\delta^{15}N$ value (–1.4‰), while amino-N had a high $\delta^{15}N$ value (+6.7‰) (Bergersen *et al.*, 1988). Among plant organs, nodules have been found to be highly enriched in ^{15}N in some legumes and *Frankia*-infected trees, while other organs, fruits, leaves, stems and roots showed similar but lower $\delta^{15}N$ values (reviewed in Yoneyama, 1995).

7.4 Translocation of nitrogen

Internal cycling of N in plants results in a depletion of ^{15}N in sinks rather than in source organs (Goto *et al.*, 1989; Shearer *et al.*, 1983) and this may be due to the transport of amino acids depleted in ^{15}N, isotopic fractionation during nitrate reduction (Medina and Schmidt, 1982; Yoneyama and Kaneko, 1989), transamination (Macko *et al.*, 1986) and GS activity (Yoneyama *et al.*, 1993) and other enzymatic reactions in source organs. Cytosolic glutamine synthetase (GS1) in particular, plays an important role in the retranslocation of nitrogen from the leaves or stems during senescence, resulting in abundant glutamine in phloem fluids (Kamachi *et al.*, 1991; Ta, 1991). These enzymatic reactions lead to the production of ^{15}N-depleted amino acids which are transported to the sink organs. Roots are also important sinks of nitrogen (Pate, 1980; Yoneyama and Ishizuka 1982), and the $\delta^{15}N$ values of roots are sometimes lower than that of the shoot (Evans *et al.*, 1996; Shearer *et al.*, 1983; Steele *et al.*, 1983, Yoneyama and Kaneko 1989). However, in many cases of field-grown plants, the ^{15}N values of fruits do not differ from or indeed may sometimes be higher than those of the leaves. This may indicate that the fractionation of nitrogen isotopes during translocation is small and source effect is more important than kinetic effect in the overall ^{15}N concentration of plants (however, see Chapter 6). To date $\delta^{15}N$ values of phloem-transported N compounds have not been analysed.

7.5 C and N concentration and isotope composition in castor bean (*Ricinus communis*)

Data are presented from a field study undertaken at Tsukuba, Japan during the summer of 1995, and are compared with data for the previous year, when daily temperatures were higher and total rain fall levels much lower (320 mm in 1994 compared with 669 mm in 1995 during April and August). Changes in dry matter accumulation, C concentration, C content, C/N ratios, N concentration, and N content in the leaves, petioles, stems and fruits from the castor bean plants grown in 1995 are shown in Figure 7.1. The accumulation of dry matter in leaves increased to day 66 after sowing, and following fruit-set (day 80) the dry matter in the petioles and stems continuously increased. The N pool in the leaves was largest, followed by that in the stems. N accumulation by day 66 and day 80 was low in all of the vegetative plant parts. The C and

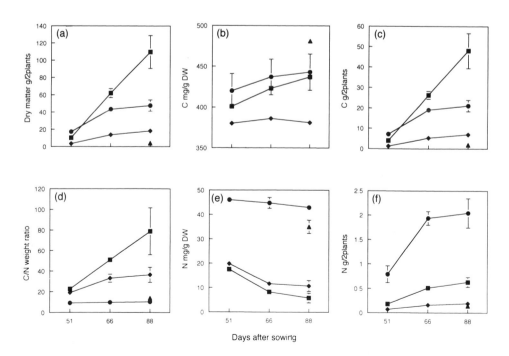

Figure 7.1. Dry matter content (A), C concentration (B), C content (C), C/N ratio (D), N concentration (E), and N content (F) for leaves (●), petioles (◆), stems (■) and fruits (▲) of field-grown castor bean plants in 1995. Experimental errors of three replicates are indicated by bars.

N concentrations in the fruits in 1995 were similar to those in the drier year, 1994, and N accumulation in the fruits smaller than in 1994, since dry matter accumulation in the fruits was small in 1995.

The $\delta^{13}C$ and $\delta^{15}N$ values of the leaves, petioles, stems and fruits are shown in Table 7.1. The $\delta^{13}C$ values in the leaves in 1995 were more negative than those in the drier year 1994, and exhibited no increase after fruit setting (day 80). The $\delta^{13}C$ values in the petioles and in the stems were similar on all three sampling occasions, and less negative by about 1‰ than those in the leaves; $\delta^{13}C$ values in the fruits was -27.3‰. The $\delta^{15}N$ values in the leaves in 1995 were similar to those in 1994, but there was a dramatic decrease of $\delta^{15}N$ values in the petioles and stems with age in 1995. The $\delta^{15}N$ value in the fruits was +2.0‰, which was higher than those in the vegetative plant parts on day 80.

Analysis of C and N concentrations, C/N ratios and $\delta^{13}C$ and $\delta^{15}N$ values in xylem and phloem fluids for 1995 are shown in Table 7.2. Phloem fluids were collected by applying capillary tubes to an incision made at mid-stem with a razor blade. The plants were later cut to obtain xylem fluids at the stem bases. Carbon concentrations in the xylem fluids in 1995 were similar to those in 1994, but N concentrations were much lower than in the previous dry year, resulting in high C/N ratios. The ratios of amino acid to total-N, in the xylem were 0.42 on day 51, 0.64 on day 66 and 1.00 on day 80 in 1995 (H.Fujiwara, unpublished data). Levels of carbon in the phloem fluids in 1995 were the same as in 1994, but N concentrations on days 66 and 80 of the 1995 investigation were low, resulting in high C/N ratios. The $\delta^{15}N$ values in the xylem fluids in 1995 were similar to whole plant material early in the season but then decreased by day 80. The $\delta^{13}C$ values in the phloem fluids were found to be slightly less negative than those in the leaves but more negative than those in the xylem fluids. The $\delta^{15}N$ values in the phloem fluids in 1995 were around 2‰ lower than in 1994, and decreased during the plant growth period.

In Table 7.3 the amount of water-extractable N and nitrate N in the soil near the castor bean plants and their $\delta^{15}N$ values are shown. Water-extractable N was largely composed of nitrate N, and relative quantities in 1995 were much less than in 1994, which may be due to the greater loss or plant uptake of nitrate. The $\delta^{15}N$ values of

Table 7.1. *$\delta^{13}C$ and $\delta^{15}N$ values in the leaves, petioles, stems, and fruits of field-grown castor bean plants in 1995*

Plant part	Days after sowing	$\delta^{13}C$ (‰)	$\delta^{15}N$ (‰)
Leaves	51	-29.9±0.1	+2.4±0.7
	66	-29.1±0.6	+2.0±0.2
	80	-29.1±0.4	+1.5±0.5
Petioles	51	-28.7±0.1	+3.7±0.5
	66	-28.0±0.7	+1.4±0.7
	80	-28.3±0.3	+0.7±1.0
Stems	51	-28.4±0.3	+3.0±0.0
	66	-27.8±0.6	+1.4±0.6
	80	-28.1±0.1	+0.8±1.3
Fruits	80	-27.3±0.9	+2.0±0.3

Table 7.2. C and N concentrations, C/N ratios, and $\delta^{13}C$ and $\delta^{15}N$ values in xylem and phloem fluids harvested in 1995

Fluids	Days after sowing	C mg ml^{-1}	N mg ml^{-1}	C/N	$\delta^{13}C$ (‰)	$\delta^{15}N$ (‰)
Xylem	51	0.80±0.53	0.11±0.05	6.7±2.0	-28.3±0.6	+2.9±0.1
	66	2.97±0.15	0.11±0.00	27.6±1.8	-28.0±0.3	+0.9±0.5
	80	1.30+0.42	0.14±0.02	9.7±4.1	-26.4±0.2	+1.7±0.5
Phloem	51	36.3±3.6	1.36±0.12	26.9±3.8	-28.8±0.6	+2.6±0.6
	66	39.9±1.1	0.86±0.02	46.6±0.0	-29.0±0.1	+0.1±1.0
	80	51.3±6.0	0.84±0.26	66.4±26.0	-28.6±0.7	+0.2±1.1

both types of N in 1995 were close, around ±1‰, but this was depleted by around 3‰ as compared to the $\delta^{15}N$ values for the drier soils in 1994 (data not shown).

In general, carbon isotope composition reflected water supply and environmental stress, with higher discrimination, that is, more negative $\delta^{13}C$ values in the wetter year, 1995 (Farquhar and Richards, 1984, see also Chapter 9). However, there were important shifts in carbon isotope composition both between plant parts and within xylem and phloem saps. There was a progressive decrease in $\delta^{13}C$ from leaves to petioles, stems and fruits, which was also seen in xylem samples over time. While in the drier year, phloem sap composition showed a similar shift (data not shown), this suggests that the relative rate of fruit formation may be associated with different partitioning patterns. The shift in xylem signature could either be associated with increased anaplerotic CO_2 fixation in the roots (Coker and Schubert, 1981), or may result from a shift in phloem export and increased allocation to the roots. The $\delta^{15}N$ values of water-extractable soil N (largely composed of nitrate) were 0‰ to 1‰, which were lower than $\delta^{15}N$ values of total soil N (+4.3‰), and the $\delta^{15}N$ values of the xylem fluids which were +1‰ to 3‰, again higher than those of the remaining soil nitrate. Nitrogen in the xylem fluids, was delivered to the above-ground parts through the season with increased percentages of amino acids being associated with a decrease in $\delta^{15}N$ values (Table 7.2), probably due to intensive cycling of previously stored amino acids through the roots (Cooper and Clarkson, 1989).

The vast majority of nitrogen in the phloem fluids was present as amino acids (more

Table 7.3. Water-extractable and nitrate-N and their $\delta^{15}N$ values from the castor bean-grown soils in 1995

Days after sowing	Water-extractable N		Nitrate-N	
	mg kg^{-1} soil (DW)	$\delta^{15}N$ (‰)	mg kg^{-1} soil (DW)	$\delta^{15}N$ (‰)
51	5.0±0.4	+1.0±0.5	4.9±0.2	+0.9±1.1
66	3.9±0.9	+0.4±1.0	3.5±1.1	+0.4±0.4
88	3.1±0.7	-0.2±0.4	3.0±1.0	-1.0±0.4

than 90%), with only trace levels of nitrate (Jeschke and Pate, 1991). The $\delta^{15}N$ values in the phloem fluids were lower than those in the corresponding xylem fluids. The $\delta^{15}N$ values of phloem on day 51 were between the $\delta^{15}N$ values of the leaves and stems plus petioles, but lower than the $\delta^{15}N$ values of all the above-ground parts on days 60 and 88, suggesting the active reduction of nitrate and export of the amino acid products, through the phloem.

The fruits were the sink for nitrogen as well as carbon transported by vascular systems (phloem and xylem). In 1994, the size of the fruits was large, while in 1995, they were small at the final harvest. The $\delta^{15}N$ values of fruits in 1994 was +1.8‰, which was very close to the values of the phloem fluids (+2.1‰) (data not shown). The $\delta^{15}N$ values of the fruits in 1995 was also +2.0‰, but in contrast to the results in 1994, this was close to the values of the xylem fluids (+1.7‰). These results suggest that the major transport of N to the fruits was by the phloem in 1994 but by the xylem in 1995. The transport of carbon to the fruits was mainly by the phloem in 1994 and by both phloem and xylem in 1995 as revealed by $\delta^{13}C$ values in phloem and xylem fluids.

7.6 Conclusions and future directions

Analysis of nitrogen isotopes alone, or when allied to $\delta^{13}C$, provide a powerful tool for investigating C and N trafficking in plants. In particular, data presented for *Ricinus communis* shows that the patterns of C and N utilisation and partitioning vary during the growth season, and are also highly dependent on annual growth conditions. While the use of combinations of stable isotopes often provide additional insight into biological transformations (e.g. Deléens *et al.*, 1994, see also Chapters 5 and 14), future work needs to resolve the major dichotomy between field- and lab-based studies: that of growth conditions and source nitrogen concentration. Care should be taken to provide a realistic rooting conditions for laboratory studies (see Evans *et al.*, 1996), or deal with heterogeneous natural soils where changes in N profile occur across spatial (laterally and with depth) and temporal scales (see Gebauer *et al.*, 1994). Within the plant, N cycling and transformations are regulated in part by inorganic N supply, as well as site of primary N assimilation and any subsequent re-mobilisation: as usual, more work is needed to investigate the interaction between N isotope composition and inorganic and organic N transformations.

References

Amarger, N., Mariotti, A., Mariotti, F., Durr, J.C., Bourguinon, C. and Agacherie, B. (1979) Estimate of symbiotically fixed nitrogen in field grown soybeans using variations in ^{15}N natural abundance. *Plant Soil* **52**, 269–280.

Bergersen, F.J., Peoples, M.B. and Turner, G.L. (1988) Isotopic discrimination during the accumulation of nitrogen by soybeans. *Aust. J. Plant Physiol.* **15**, 407–420.

Coker, G.T. III and Schubert, K.R. (1981) Carbon dioxide fixation in soybean roots and nodules I. Characterization of xylem exudate during early nodule development. *Plant Physiol.* **67**, 691–696.

Cooper, H.D. and Clarkson, D.T. (1989) Cycling of amino-nitrogen and other nutrients between shoots and roots in cereals-a possible mechanism integrating shoot and root in the regulation of nutrient uptake. *J.Exp. Bot.* **40**, 753–762.

Deléens, E., Cliquet, J.-B. and Prioul, J.-L. (1994) Use of ^{13}C and ^{15}N plant label near natural abundance for monitoring carbon and nitrogen partitioning. *Aust. J. Plant Physiol.* **21**, 133–146.

Domenach, A.M., Furdali, F. and Bardin, R. (1989) Estimation of symbiotic dinitrogen fixation in alder forest by the method based on natural ^{15}N abundance. *Plant Soil* **118**, 51–59.

Evans, R.D., Bloom, A.J., Sukrapanna, S.S. and Ehleringer, J.R. (1996) Nitrogen isotope composition of tomato (*Lycopersicon esculentum* Mill. cv. T-5) grown under ammonium or nitrate nutrition. *Plant, Cell and Environ.* **19**, 1317–1323.

Farquhar, G.D. and Richards, R.A. (1984) Isotopic composition of plant carbon correlates with water-use efficiency of wheat genotypes. *Aust. J. Plant Physiol.* **11**, 539–552.

Gaebler, O.H. Vitti, T.G. and Vukmirovich, R. (1966) Isotope effects in metabolism of ^{14}N and ^{15}N from unlabeled dietary proteins. *Can. J.Biochem.* **44**, 1249–1257.

Gebauer, G., Giesemann, A., Schulze, E.-D. and Jäger, H.-J. (1994) Isotope ratios and concentrations of sulfur and nitrogen in needles and soils of *Picea abies* stands as influenced by atmospheric deposition and sulfur and nitrogen compounds. *Plant Soil* **164**, 267–281.

Goto, S., Yoneyama, T., Matsui, E. and Kumazawa, K. (1989) Natural ^{15}N abundances in plant parts of groundnut (*Arachis hypogaea*). *Soil Sci. Plant Nutr.* **35**, 307–311.

Handley, L.L. and Raven, J.A. (1992) The use of natural abundance of nitrogen isotopes in plant physiology and ecology. *Plant, Cell and Environ.* **15**, 965–985.

Ito, O. and Kumazawa, K. (1978) Amino acid metabolism in plant leaf III. The effect of light on the exchange of ^{15}N-labelled nitrogen among several amino acids in sunflower discs. *Soil Sci. Plant Nutr.* **24**, 327–336.

Ito, O., Yoneyama, T. and Kumazawa, K. (1978) Amino acid metabolism in plant leaf IV. The effect of light on ammonium assimilation and glutamine metabolism in the cells isolated from spinach leaves. *Plant Cell Physiol.* **19**, 1109–1119.

Jeschke, W.D. and Pate, J.S. (1991) Modelling of the partitioning, assimilation and storage of nitrate within root and shoot organs of castor bean (*Ricinus communis* L.) *J. Exp. Bot.* **42**, 1091–1103.

Kamachi, K., Yamaya, T., Mae, T. and Ojima, K. (1991) A role for glutamine synthetase in the remobilization of leaf nitrogen during natural senescence in rice leaves. *Plant Physiol.* **96**, 411–417.

Kohl, D.H. and Shearer, G. (1980) Isotopic fractionation associated with symbiotic N$_2$ fixation and uptake of NO$_3^-$ by plants. *Plant Physiol.* **66**, 51–56.

Ledgard, S.F., Woo, K.C. and Bergersen, F.J. (1985) Isotopic fractionation during reduction of nitrate by extracts of spinach leaves. *Aust. J. Plant Physiol.* **12**, 631–640.

Macko, S.A., Estep, M.L.F., Engel, M.H. and Hare, P.E. (1986) Kinetic fractionation of stable nitrogen isotopes during amino acid transamination. *Geochim. Cosmochim. Acta* **50**, 2143–2146.

Macko, S.A., Fogel (Estep) M.L., Hare, P.E. and Hoering, T.C. (1987) Isotopic fractionation of nitrogen and carbon in the synthesis of amino acids by microorganisms. *Chemical Geology* **65**, 79–92.

Mariotti, A., Pierre, D., Vedy, J.C., Bruckert, S., Guillemot, J. (1980a) The abundance of natural nitrogen 15 in the organic matter of soils along an altitudinal gradient (Chablais, Haute Savoie, France). *Catena* **7**, 293–300.

Mariotti, A., Mariotti, F., Amargar, N., Pizelle, G., Ngambi, J.-M., Champigny, M.-L. and Moyse, A. (1980b) Fractionnements isotopiques de l-azote lors des processus d-absorption des nitrates et de fixation de l-azote atmosphérique par les plantes. *Physiol. Vég.* **18**, 163–181.

Medina, R. and Schmidt, H.-L. (1982) Nitrogen isotope ratio variations in biological material, indicator for metabolic correlations. In *Stable Isotopes* (eds H.-L. Schmidt, H. Förstel and K.Heinzinger), Springer, Amsterdam, pp. 465–473.

Minagawa, M. Egawa, S., Kabaya, Y. and Karasawa-Tsuru, K. (1992) Carbon and nitrogen isotope analysis for amino acids from biological sample. *Mass Spectrosc.* **40**, 47–56.

Nadelhoffer, K.J. and Fry, B. (1988) Controls on natural nitrogen-15 and carbon-13 abundances in forest soil organic matter. *Soil Sci. Soc. Am. J.* **52**, 1633–1640.

O'Leary, M.H. (1981) Carbon isotope fractionation in plants. *Phytochemistry* **20**, 553–567.

O'Leary, M.H. (1995) Environmental effects on carbon isotope fractionation in terrestrial plants. In *Stable Isotopes in the Biosphere* (eds E. Wada *et al.*), Kyoto University Press, Japan, pp. 78–91.

Pate, J.S. (1980) Transport and partitioning of nitrogen solutes. *Ann. Rev. Plant Physiol.* **31**, 313–340.

Raven, J.A., and Farquhar, G.D. (1990) The influence of N metabolism and organic acid synthesis on the natural abundance of isotopes of carbon in plants. *New Phytol.* **116**, 505–529.

Richter, A., Popp, M., Mensen, R., Stewart, G.R. and von Willert, D.J. (1995) Heterotrophic carbon gain of the parasitic angiosperm *Tapinanthus oleifolius*. *Aust. J. Plant Physiol.* **22**, 537–567.

Shearer, G., Kohl, D.H., Virgina, R.A., Bryan, B.A., Skeeters, J.L., Nilson, E.T., Sharifi, M.R. and Rundel, P.W. (1983) Estimates of N_2-fixation from variation in the natural abundance of ^{15}N in Sonoran desert ecosystems. *Oecologia* **36**, 365–373.

Shearer, G. and Kohl, D.H. (1986) N_2-fixation in field settings: estimations based on natural ^{15}N abundance. *Aust. J. Plant Physiol.* **13**, 699–756.

Shearer, G. and Kohl, D.H. (1989) Natural ^{15}N abundance of NH_4^+, amide N, and total N in various fractions of nodules of peas, soybeans and lupins. *Aust. J. Plant Physiol.* **16**, 305–313.

Steele, K.W. (1983) Quantitative measurements of nitrogen turnover in pasture systems with particular reference to the role of ^{15}N. In *Nuclear Techniques in Improving Pasture Management*. IAEA, Vienna, pp. 17–35.

Steele, K.W., Bonish, P.M., Daniel, R.M. and O'Hara, G.W. (1983) Effect of rhizobial strain and host plant on nitrogen isotopic fractionation in legumes. *Plant Physiol.* **72**, 1001–1004.

Stewart, G.R., Schmidt, S., Handley, L.L., Turnbull, M.H., Erskine, P.D., Joly, C.A. (1995) ^{15}N natural abundance of vascular rainforest epiphytes: implications for nitrogen source and acquisition. *Plant, Cell and Environ.* **18**, 85–90.

Ta, C.T. (1991) Nitrogen metabolism in the stalk tissue of maize. *Plant Physiol.* **97**, 1375–1380.

Vance, C.P. and Heichel, G.H. (1991) Carbon in N_2 fixation: limitation or exquisite adaptation. *Ann. Rev. Plant Physiol. Plant Mol. Biol.* **42**, 373–392.

Wada, E., Imaizumi, R., Kabaya, Y., Yasuda, T., Kanamori, T., Saito, G. and Nishimune, A. (1986) Estimation of symbiotically fixed nitrogen in field grown soybeans: An application of $^{15}N/^{14}N$ abundance and a low level of ^{15}N-tracer techniques. *Plant Soil* **93**, 269–286.

Yoneyama, T. and Ishizuka, J. (1982) ^{15}N study on the partitioning of the nitrogen taken by soybeans from atmospheric dinitrogen, medium nitrate or ammonium. *Soil. Sci. Plant Nutr.* **28**, 451–461.

Yoneyama, T., Nakano, H., Kuwahara, M., Takahashi, T., Kambayashi, I. and Ishizuka, J. (1986) Natural ^{15}N abundance of field grown soybean grains harvested in various locations in Japan. *Soil Sci. Plant Nutr.* **32**, 443–449.

Yoneyama, T. (1987) N_2-fixation and natural ^{15}N abundance of leguminous plants and *Azolla*. *Bull. Natl. Inst. Agrobiol. Resour.* **30**, 59–87.

Yoneyama, T., Ladha, J.K. and Watanabe, I. (1987) Nodule bacteroids and *Anabaena*: natural ^{15}N enrichment in the legume-*Rhizobium* and *Azolla-Anabaena* symbiotic systems. *J. Plant Physiol.* **127**, 251–259.

Yoneyama, T. (1988) Natural abundance of ^{15}N in root nodules of pea and broad bean. *J. Plant Physiol.* **132**, 59–62.

Yoneyama, T. and Kaneko, A. (1989) Variations in the natural abundance of ^{15}N in nitrogenous fractions of komatsuna plants supplied with nitrate. *Plant Cell Physiol.* **30**, 957–962.

Yoneyama, T. (1990) Use of natural ^{15}N abundance for evaluation of N_2 fixation by legumes and *Azolla* in remote fields. *Transactions of 14th ICSS*, Vol. IV, pp 126–130.

Yoneyama, T., Kouno, K. And Yazaki, J. (1990) Variation of natural ^{15}N abundance of crops and soils in Japan with special reference to the effect of soil conditions and fertilizer application. *Soil Sci. Plant Nutr.* **36**, 667–675.

Yoneyama, T., Omata, T., Nakata, S. and Yazaki, J. (1991a) Fractionation of nitrogen isotopes during the uptake and assimilation of ammonia by plants. *Plant Cell Physiol.* **32**, 1211–1217.

Yoneyama, T., Uchiyama, T., Sasakawa, H., Gamo, T., Ladha, J.K. and Watanabe, I. (1991b) Nitrogen accumulation and changes in natural ^{15}N abundance in the tissues of legumes with emphasis on N_2 fixation by stem-nodulating plants in upland and paddy fields. *Soil Sci. Plant Nutr.* **37**, 75–82.

Yoneyama, T., Kamachi, K. Yamaya, T. and Mae, T. (1993) Fractionation of nitrogen isotopes by glutamine synthetase isolated from spinach leaves. *Plant Cell Physiol.* **34**, 489–491.

Yoneyama, T. (1995) Nitrogen metabolism and fractionation of nitrogen isotopes in plants. In *Stable Isotopes in the Biosphere* (eds E.Wada *et al.*) Kyoto University Press, Japan, pp.92–102.

Yoneyama, T. (1996) Characterization of natural ^{15}N abundance of soils. In *Mass Spectrometry of Soils* (eds W. Boutton and S. Yamasaki), Marcel Dekker, New York, pp. 205–223.

Carbon isotope discrimination in terrestrial plants: carboxylations and decarboxylations

J.S. Gillon, A.M. Borland, K.G. Harwood, A. Roberts, M.S.J. Broadmeadow and H. Griffiths

8.1 Introduction

The process of CO_2 fixation by all terrestrial plants modifies the ratio of $^{13}C/^{12}C$ between plant material and atmospheric CO_2. Changes in the isotopic composition of plant material, $\delta^{13}C_p$, relative to source CO_2, $\delta^{13}C_a$, represent discrimination, or $\Delta^{13}C$, which is normally positive, reflecting organic material depleted in ^{13}C (Farquhar et al., 1982; O'Leary, 1981; Vogel, 1980). Primarily, the predominant photosynthetic pathway (C_3, C_4 or crassulacean acid metabolism (CAM)) governs large scale discrimination (+30 to –6‰) via the operation of varying proportions of the two carboxylating enzymes, ribulose-1, 5-bisphosphate carboxylase/oxygenase (RUBISCO) and phosphoenolpyruvate carboxylase (PEPC). Within these types, discrimination varies between the extremes of these enzymatic limits of fractionation and that for diffusion through air (+ 4.4‰), as regulated by diffusive supply of CO_2 to the sites of carboxylation. The contribution of CO_2 from respiration may represent a further source of discrimination (± 4‰), if the processes have discernible isotope effects. More fine-scale shifts in $\Delta^{13}C$ (± 2‰) may be the result of interspecific variation within the intrinsic fractionation expressed by both carboxylations and decarboxylations in leaves.

In this chapter we aim to provide the theoretical background to carbon isotope discrimination, with data to illustrate the variations observed. In addition, we aim to paramaterize some of the other mechanisms contributing to $\Delta^{13}C$ in plants, by manipulations of isotopic sources of carbon during photosynthesis and increasing carbon flux through respiratory pathways. Such extreme conditions in the laboratory may reflect actual situations in the field, and therefore aid interpretations of $\Delta^{13}C$ measured instantaneously in natural vegetation.

Stable Isotopes, edited by H. Griffiths.
© 1998 BIOS Scientific Publishers Ltd, Oxford.

8.2 Net carbon isotope discrimination

Fractionation during photosynthesis can be measured either by studying the change in isotopic composition of the substrate (gaseous CO_2) or the product (plant material). Initially, fractionation of ^{13}C resulting from net photosynthetic fixation of CO_2 was observed as a reduction in the $^{13}C/^{12}C$ ratio of leaf material (Craig, 1953; Nier and Gulbransen, 1939) relative to atmospheric CO_2 isotopic composition, calculated by

$$\Delta^{13}C = \frac{\delta^{13}C_a - \delta^{13}C_p}{1000 + \delta^{13}C_p}.1000 \tag{8.1}$$

where $\delta^{13}C_a$ and $\delta^{13}C_p$ are the isotopic compositions of air and plant material respectively, in units of permil (‰). This represents the integration of discrimination processes over the period in which the structural carbohydrates of the plant were synthesized. A more short-term measurement of discrimination is possible by analysis of extractable leaf carbohydrate pools (Brugnoli et al., 1988: Lauteri et al., 1993) which can bring the time resolution down to a few days or even hours (see Chapter 9). It is also possible to measure the corresponding enrichment of ^{13}C in the CO_2 passing over a leaf in a cuvette, in order to measure discrimination instantaneously, given as follows (Evans et al., 1986),

$$\Delta^{13}C_{obs} = \frac{\xi(\delta^{13}C_o - \delta^{13}C_e)}{1000 + \delta^{13}C_o - \xi(\delta^{13}C_o - \delta^{13}C_e)} \tag{8.2}$$

where $\xi = C_e/(C_e-C_o)$, C_e, C_o and $\delta^{13}C_e$, $\delta^{13}C_o$ referring to the CO_2 concentration (corrected to the same humidity) and isotopic composition of air, entering and leaving the cuvette, respectively. Discrimination measured either on a long-term basis or instantaneously is the net result of fractionation during both photosynthetic uptake (carboxylation) and respiratory release (decarboxylation) of CO_2. Where there is no fractionation during decarboxylation, then measured discrimination will only reflect that of photosynthetic uptake (Farquhar et al., 1982; O'Leary, 1988). However, the potential for fractionation during decarboxylation, albeit smaller than that for carboxylation, does exist (Ivlev, 1996; Rooney 1988; Troughton et al., 1974), thus net discrimination must include a proportion which is attributable to carbon loss via respiration.

8.3 Photosynthetic discrimination

During photosynthetic gas exchange, fractionation against $^{13}CO_2$ occurs during both diffusion from the ambient air surrounding the leaf to the sites of carboxylation and subsequently during fixation by RUBISCO. The overriding control of the extent of discrimination expressed by the plant will depend on the rate of supply of CO_2 to the sites of fixation. When CO_2 supply is not limiting, discrimination will reflect the fractionation of the respective carboxylation enzyme. However, when substrate supply is totally limiting, discrimination reflects the fractionation which occurs as the substrate diffuses to the site of carboxylation. These two extremes are rarely found in natural systems, and substrate supply is generally only partially limiting, hence the discrimination expressed usually lies between that for diffusive and enzymatic fractionation.

Early investigations demonstrated that, during higher plant photosynthesis in the gaseous environment, the supply of CO_2 was correlated to the ratio of intercellular to atmospheric CO_2 concentration, C_i/C_a (Farquhar et al., 1982; O'Leary, 1981; Vogel, 1980), and provided a breakthrough in the understanding of isotopic discrimination in plants. This, coupled with knowledge of both diffusional fractionation and fractionation due different carboxylation enzymes, has allowed successful development of models describing carbon isotope discrimination in plants with differing photosynthetic pathways.

8.3.1. C_3 carboxylation

The primary carboxylation enzyme in C_3 plants is RUBISCO. Initially, estimation of the isotope effect of RUBISCO, b_3, was achieved by combustion of the first product of the carboxylation reaction, namely 3-phosphoglycerate, 3-PGA (Christeller et al., 1976; Deleens et al., 1974; Estep et al., 1978; Park and Epstein, 1960; Wong et al., 1979), which provided estimates of fractionation, ranging from 20‰ to 40‰. Because the carboxylation reaction requires the 5 carbon compound ribulose-1,5-bisphosphate (RuBP), as a substrate, discrimination in the product is only reflected in 1 out of every six carbon atoms in PGA. Thus the fractionation measured must be multiplied by a factor of six to discover the isotope effect attributable to carboxylation alone (however, see Chapter 2). Unfortunately, any inaccuracies in analysis due to the presence of impurities or inhomogeneity in the isotopic composition of the residual carbon atoms will also be exaggerated during this calculation, and may have accounted for some of the variation in the determination of b_3 in earlier studies. This problem was circumvented by either direct analysis of CO_2 which had been decarboxylated from 3-PGA (Roeske and O'Leary, 1984) or by analysis of the remaining CO_2 substrate after exposure to RUBISCO (Guy et al., 1993), both of which agree on a value of $b_3 = 29$‰ for RUBISCO extracted from spinach. In gaseous systems, this figure was corrected to 30‰ to account for the dissolution of CO_2 into water, 1.1‰ (O'Leary, 1984; Vogel et al., 1970).

Theoretical models. By considering both the enzymatic and diffusive fractionation as a function of CO_2 supply (represented by C_i/C_a), discrimination can be calculated as $\Delta^{13}C_i$, under conditions minimising any input from respiratory sources (low oxygen partial pressure, high CO_2 assimilation rates) and was estimated using the following (Farquhar et al., 1982)

$$\Delta^{13}C_i = a + (b' - a)\frac{C_i}{C_a} \qquad (8.3)$$

where $\Delta^{13}C_i$ is the predicted theoretical discrimination, C_i and C_a refer to CO_2 concentration in the atmosphere and sub-stomatal cavity respectively, a is the fractionation during diffusion in air (4.4‰) and b' is the assumed net fractionation during carboxylations by RUBISCO and PEPC. This simple model appears to overestimate instantaneous $\Delta^{13}C$ (Figure 8.1A). Additional resistances to CO_2 diffusion occur within the leaf as CO_2 diffuses through the potentially tortuous intercellular air spaces and the aqueous cellular environment to the chloroplast, where carboxylation occurs. This is commonly expressed as a single value, the internal CO_2 transfer conductance, g_i, which not only reflects diffusional resistance from sub-stomatal cavity to chloroplast but also the area of exposed chloroplasts (von Caemmerer and Evans 1991;

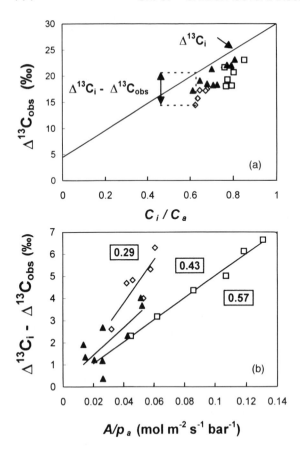

Figure 8.1. Measured instantaneous discrimination, $\Delta^{13}C$, for Quercus robur, Phaseolus vulgaris *and* Triticum aestivum, *measured under 2 %* O_2 *and varying light intensity. A) shown as the relationship with* C_i/C_a, *the solid line being predicted* $\Delta^{13}C$, *calculated from equation 3, where b' = 30‰. B) the difference between measured and predicted discrimination,* $\Delta^{13}C_i - \Delta^{13}C_{obs}$, *solid lines are linear regressions where* $r^2 = 0.585, 0.988, 0.705$ *and where b' is 29, 30 and 30.5 for* P. vulgaris, T. aestivum *and* Q. robur *respectively. Mesophyll conductance,* g_i *(estimated from the gradient of the relationship) for each species is shown inset* $(mol\ m^{-2}\ s^{-1}\ bar^{-1})$.

Evans *et al.* 1994). Thus, the CO_2 concentration within the chloroplast, C_c, will always be lower than C_i, and therefore any estimation of $\Delta^{13}C_i$ from C_i/C_a will overestimate instantaneous $\Delta^{13}C$. The predictive model can be adapted (Evans *et al.*, 1986), including the extra diffusional resistance to the chloroplast, as follows;

$$\Delta^{13}C_c = a\frac{C_a - C_i}{C_a} + a_i\frac{C_i - C_c}{C_a} + b'\frac{C_c}{C_a} \qquad (8.4)$$

where C_c refers to the partial pressure of CO_2 in the chloroplast, a_i is the combined fractionation during dissolution and diffusion through the liquid phase (1.8‰). It is not possible to measure C_c directly, however the drawdown between C_i and C_c is manifested in the difference between $\Delta^{13}C_i$ (from Equation 8.3) and $\Delta^{13}C_{obs}$ (from Equation 8.2), which increases with the quotient A/C_a (Figure 8.1B). By utilising the relationship $C_c = C_i - A/g_i$, where A is the net CO_2 assimilation rate, the following expression, adapted from Evans *et al.* (1986) to ignore the effects of respiratory inputs, combines measured and predicted discrimination to yield a relationship dependent on g_i;

$$\Delta^{13}C_i - \Delta^{13}C_{obs} = \frac{(b' - a_i)}{g_i} \cdot \frac{A}{C_a} \qquad (8.5)$$

Estimation of mesophyll conductance, g_i. Under non-respiratory conditions and assuming the correct estimation of b', the gradient of the relationship $\Delta^{13}C_i - \Delta^{13}C_{obs}$ versus A/C_a is inversely proportional to the internal CO_2 transfer conductance, as calculated from $g_i = (b' - a_i)/m$ (where m is the gradient of the response). This estimate of g_i from analysis of the complete $\Delta^{13}C_i - \Delta^{13}C_{obs}$ relationship is considered more 'robust' than an estimate based on a single point determination, achieved by simply solving equation 8.5 for individual Δ values. The latter depends too much on absolute values of instantaneous discrimination which may vary with uncertainty in fractionation values and/or photorespiratory fractionation (see Section 8.4.2), or cumulative errors during sample collection and analysis.

Internal conductance to CO_2 has been estimated between 0.05 and 0.6 mol m^{-2} s^{-1} bar^{-1} for a range of species (von Caemmerer and Evans, 1991; Epron *et al.*, 1995; Evans *et al.*, 1994; Lloyd *et al.*, 1992; Loreto *et al.*, 1992). Generally fast growing, herbaceous plants have the highest g_i in contrast to the low levels of internal conductance evident in woody perennials, with their thicker, sometimes sclerophyllous leaves. This is demonstrated in Figure 8.1B, where g_i for *Triticum aestivum*, *Phaseolus vulgaris* and *Quercus robur* is estimated as 0.57, 0.43 and 0.29 mol m^{-2} s^{-1} bar^{-1} respectively. This apparent interspecific variation in g_i has been implicated in causing the low A_{max} that occurs in woody perennials (Lloyd *et al.*, 1992; Syvertson *et al.*, 1995), and as such internal CO_2 conductance is being seen to being an increasingly significant constraint on photosynthetic production (for a full review on diffusion within leaves, see Evans and von Caemmerer, 1996). Furthermore, intraspecific variation of g_i has been observed under differing water regimes (see Chapter 9) and is thought to explain the elevational gradients observed in leaf $\delta^{13}C$ of *Metrosideros polymorpha* which correlated well with specific leaf mass (Vitousek *et al.*, 1990).

The above method for estimation of g_i, via analysis of the complete A/p_a relationship, facilitates *in vivo* determination of b' such that the relationship passes through the origin. For the three species shown in Figure 8.1B, b' varies between 29 and 30.5‰, confirming previous suggestions and observations that photosynthetic fractionation may vary interspecifically (von Caemmerer and Evans, 1991; O'Leary *et al.*, 1992), despite the agreement of independent *in vitro* determinations on RUBISCO alone (extracted from spinach, b_3 = 29‰; Roeske and O'Leary, 1984; Guy *et al.*, 1993). In C_3 plants, there is a small degree of CO_2 fixation by PEPC which may account for as much as 5‰ of total carboxylation (von Caemmerer and Evans, 1991; Holbrook *et al.*, 1984), thus b' (net fractionation of RUBISCO and PEPC) is likely to be less than b_3 (fractionation by RUBISCO alone). This would reduce the overall fractionation expressed during carboxylation (see Section 8.3.2), however the relative activity of PEPC may also vary among species, particularly with nitrogen nutrition. Even after accounting for PEPC activity, b_3 may still vary above and below the accepted value of 30‰.

8.3.2 C_4 *carboxylation*

Of the plants analysed for carbon isotopic composition by Craig (1953), one grass species was recorded as having a much more positive $\delta^{13}C$ with respect to the others, which was the first indication that an alternative photosynthetic pathway existed other than the conventional C_3 mechanism. It is now known that the C_4 pathway is characterised by the initial fixation of CO_2 into C_4 acids via PEPC in the mesophyll cells. These organic acids are then transported to the bundle sheath where CO_2 is concentrated by decarboxylation. Unlike RUBISCO, which utilises molecular CO_2 during carboxylation, PEPC utilises the bicarbonate ion, HCO_3- and so overall fractionation by PEPC includes dissolution and hydration of CO_2 as well as fractionation during carboxylation. Fractionation during dissolution and carboxylation are constant at 1.1‰ (O'Leary, 1984; Vogel et al., 1970) and 2‰ (Whelan et al., 1973) respectively. However, hydration of CO_2 fractionates by about –9‰ so that bicarbonate is enriched in ^{13}C relative to CO_2 and the overall fractionation, b_4, is –5.9‰. This means that C_4 plants have the potential to accumulate the heavier isotope. The equilibrium between CO_2 and bicarbonate is temperature dependent and therefore, the fractionation must be corrected according to temperature (Mook et al., 1974), for example, the overall fractionation expressed can vary between –6.3 and -5.2‰ as the temperature increases from 20 to 30 °C.

The discrimination due to PEPC alone can be investigated by studying night-time fixation of CO_2 (Phase I, Osmond, 1978) in plants with CAM. Direct measurements of instantaneous $\Delta^{13}C$ can be compared with estimates of $\Delta^{13}C$ obtained from measurements of C_i/C_a and by substituting b_4, the fractionation associated with C_4 carboxylation into Equation 8.3. In the CAM species *Clusia minor* and *C. fluminensis*, $\Delta^{13}C$ measured at night varied with vapour pressure deficit, VPD (Figure 8.2A, Roberts et al., 1997), both under natural conditions and artificially imposed VPDs. From comparison with C_i/C_a, it is evident that $\Delta^{13}C$ spans the extremes of diffusive and carboxylative fractionation (Figure 8.2C). Thus, according to the model (Farquhar et al., 1983), the limits of $\Delta^{13}C$ are first set by PEPC fractionation (ca. –6‰, stomata wide open, C_i/C_a at unity), and second, when products tend towards the diffusion limited isotope composition of CO_2 in the substomatal cavity (+4.4‰).

C4 photosynthesis. Similar measurements on the C_4 grass *Cynodon dactylon* during daytime photosynthesis still demonstrate a response of instantaneous $\Delta^{13}C$ to VPD (Figure 8.2B), although showing a much narrower range of $\Delta^{13}C$ compared to night-time CAM CO_2 uptake (Figure 8.2D). The dependence of $\Delta^{13}C$ on C_i/C_a in C_4 plants, typically leads to a narrow range of isotopic compositions, and may be related to the proportion of carbon which leaks from the bundle sheath cells (to be refixed by PEPC) thereby allowing discrimination by RUBISCO to be expressed (von Caemmerer and Hubick, 1989; Evans et al., 1986; Farquhar, 1983; Henderson et al., 1992). Initially this leakage was estimated to be ~10‰ of the inward CO_2 flux via C_4 acids (Hatch and Osmond, 1976), thereby introducing an inefficiency to the system because the energy invested in concentrating CO_2 internally is partially lost. A simple equation was developed by Farquhar (1983) which relates $\Delta^{13}C$ to C_i/C_a and bundle sheath leakiness:

$$\Delta^{13}C = a + (b_4 + \phi(b_3 - a_i) - a)\frac{C_i}{C_a} \qquad (8.6)$$

where b_4 is the net fractionation during carboxylation by PEPC, ϕ is the ratio rate of leakage from the bundle sheath relative to that of PEP carboxylation and a_i is the fractionation occurring during leakage (1.8‰). This theoretical consideration of discrimination during C_4 photosynthesis yielded much higher estimates of leakage ranging from 21‰ in *Zea mays* to 34‰ for *Amaranthis edulis* (Evans et al., 1986). Leakiness (ϕ) depends on the physical conductance of the bundle sheath cell wall and the balance between the biochemical capacities of the C_4 and C_3 cycles. From the theory of Δ in C_4 plants described above, a 40‰ reduction in RUBISCO content by antisense RNA in *Flaveria bidentis*, led to an increase in bundle sheath CO_2

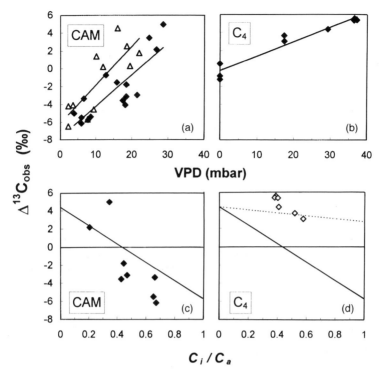

Figure 8.2. The response of measured instantaneous discrimination with vapour pressure deficit (VPD) for A) two CAM species, Clusia fluminensis (◆, glasshouse grown) and Clusia minor (Δ, field conditions), and B) the C_4 grass Cynodon dactylon, where solid lines are the linear regressions ($r^2 = 0.705, 0.654, 0.939$ respectively). Response of $\Delta^{13}C_{obs}$ versus C_i/C_a for C) C. fluminensis where the solid line is $\Delta^{13}C_i$ ($b_4 = -5.7$ ‰), and D) C. dactylon, the dashed line is $\Delta^{13}C$ predicted from Equation 5, assuming leakiness, ϕ, = 0.30. Redrawn from Roberts et al., (1997).

concentration and subsequent leakage from the bundle sheath (i.e. ϕ=24‰ in wild type, 37‰ in transgenic; von Caemmerer *et al.*, 1997). In contrast, the influence of mesophyll conductance on Δ in C_4 plants is almost insignificant. The reduced range in $\Delta^{13}C$ occurring due to the small fractionation associated with PEPC fixation combined with CO_2 leakage implies that drawdown in CO_2 partial pressure between the substomatal cavity and sites of carboxylation has little effect on $\Delta^{13}C$ (ca. 0.2‰ for a drop of 60 μmol mol^{-1}), unlike the large effects (up to 6‰, see Figure 8.1B) observed during C_3 fixation.

CAM photosynthesis. Whilst C_4 plants separate RUBISCO and PEPC carboxylation between different cell types, it is generally accepted that in CAM plants a temporal separation of C_3 and C_4 carboxylation confines PEPC activity to the dark (night) period (Phase I). Indeed, measurements of instantaneous $\Delta^{13}C$ in two obligate CAM species, *Tillandsia utriculata* and *Kalanchoë daigremontiana*, have demonstrated the rapid down-regulation of PEPC *in vivo* at the start of the photoperiod (Borland and Griffiths, 1997; Griffiths *et al.*, 1990). However, more recently, in species of the genus *Clusia*, PEPC activity has been shown to persist for ~4 hours into the light period, as illustrated by the continued increase in malate content (Borland *et al.*,

1993; Roberts *et al.*, 1996). Low values of instantaneous $\Delta^{13}C$ which have been measured for *C. fluminensis* (around -5‰) are indicative of extended C_4 carboxylation for 3–4 hours into the light (Roberts *et al.*, 1997) which, in turn, may reflect the delay in deactivation of PEPC by dephosphorylation (Borland and Griffiths, 1997). Whilst activation of RUBISCO during the first hours of daylight has yet to be determined in *C. fluminensis*, it is conceivable that the measured C_4-like $\Delta^{13}C$ is a result of high PEPC activity which masks RUBISCO activity as in C_4 plants. Figure 8.3 shows a comparison of measured and predicted $\Delta^{13}C$ for day-time C_3 photosynthesis for three species of *Clusia* which show different capacities for CAM under identical controlled conditions. The data suggest that the 'strength' of C_4 discrimination (magnitude of deviation between measured and predicted $\Delta^{13}C$) reflects the different *in vitro* biochemical capacities of PEPC and RUBISCO (Borland, A., unpublished data).

Values of instantaneous $\Delta^{13}C$ which are midway between those for pure C_3 and C_4 carboxylation suggest simultaneous fixation of CO_2 via both RUBISCO and PEPC. Such findings suggest a possible inefficiency in some CAM plants if futile cycling through malate synthesis/decarboxylation represents a significant proportion of day-time carbon flux (Borland and Griffiths, 1996). By calculating the extent to which measured discrimination deviates from discrimination predicted for C_3 and C_4

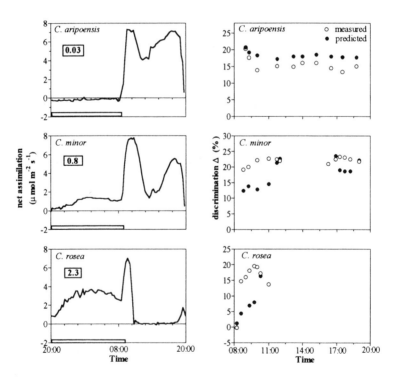

Figure 8.3. Rates of net CO_2 assimilation in leaves of three species of Clusia with different capacities for CAM under well-watered conditions. The boxed numbers indicate the ratio of maximum activities of PEPC: RUBISCO on a protein basis, measured in vitro. Also shown are the comparisons between discrimination ($\Delta^{13}C_{obs}$) measured directly on-line during daytime photosynthesis, with discrimination predicted from gas exchange measurements of C_i/C_a, assuming that RUBISCO is the dominant carboxylase (redrawn from Borland, A., unpublished data).

carboxylation, estimates of the relative rates *in vivo* of PEPC and RUBISCO may be obtained (Borland and Griffiths, 1997). Since the rate of PEPC activity calculated for *C. minor* compares closely with the amount of malate accumulated over the first 4 hours of the photoperiod (Phase II), it was concluded that futile cycling was negligible during Phase II although may occur later in the photoperiod during Phase IV when PEPC is re-activated (Borland, 1996; Osmond *et al.*, 1996, Borland and Griffiths, 1997). To date, evidence from measurements of instantaneous $\Delta^{13}C$ and PEPC activation *in vitro* suggests that processes of C_4 carboxylation/decarboxylation are tightly regulated to minimise any overlap in activity at the start of the day in CAM plants (Borland and Griffiths, 1997).

The CAM cycle also provides a period of exclusive C_3 carboxylation, using atmospheric CO_2, during the early part of phase IV, where only RUBISCO is active. From deviations in measured and predicted $\Delta^{13}C$ observed here, it is possible to estimate levels of mesophyll conductance for a range of species which express different degrees of CAM (Table 8.1). Compared to C_3 species (cf Figure 8.1B), the low estimates of g_i obtained from CAM species confirm the findings of Maxwell *et al.* (1997), which suggested that low mesophyll conductance, which would create significant diffusive limitations to C_3 photosynthesis, is tolerable in a CAM species where RUBISCO is presented with elevated internal [CO_2] from decarboxylation of organic acids. Furthermore, the range of CAM activity found in species of the genus *Clusia* illustrates the decrease in mesophyll conductance between obligate C_3 and CAM species. Coupled with this is an increase in leaf succulence (Table 8.1) and a reduction in volume of internal air space which are both prerequisite morphological characteristics for the vacuolar storage of C_4 acids in CAM species (Maxwell *et al.*, 1997).

8.4 Respiratory discrimination

Respiration is the release of CO_2 from plants via decarboxylase enzymes. Growth and/or maintenance respiration (also termed dark respiration) produce CO_2 via two decarboxylase enzymes resident in the tricarboxylic acid cycle, namely isocitrate dehydrogenase and 2-oxoglutarate dehydrogenase. Also, further CO_2 release in the light, photorespiration, occurs as a result of glycine decarboxylation in the glycollate

Table 8.1. Comparison of internal CO_2 conductance, estimated from instantaneous carbon isotope discrimination, and leaf succulence for species showing a range of crassulacean acid metabolism activity

Species	Photosynthetic type	g_i, internal CO_2 conductance ($\pm$ SE) (mol m^{-2} s^{-1} bar^{-1})	Leaf succulence (kg m^{-2})
Kalanchoe daigremontiana	Obligate CAM	0.06 (0.01)	3.4
Clusia fluminensis	Obligate CAM	0.082 (0.02)	1.15
Clusia minor	C_3 - CAM intermediate	0.100 (0.054)	0.77
Clusia aripoensis	Obligate C_3	0.192 (0.025)	0.62

pathway. This is present to recycle C_2 carbon lost from the C_3 Calvin cycle as a consequence of oxygenase activity by RUBISCO. Other decarboxylation reactions do occur within plants but the carbon fluxes via these are insignificant compared to the above. Most of the decarboxylase enzymes studied have measurable isotope effects (O'Leary et al., 1992). Indeed, fractionation occurring during photorespiration has been estimated in vivo to be substantial (Ivlev, 1996; Rooney, 1988; Troughton et al., 1974), albeit smaller than that for carboxylation by RUBISCO. After much evidence that dark respiratory fractionation was minimal, a recent study has shown there to be no discernible fractionation (Lin and Ehleringer, 1997). For these reasons, plant isotopic composition is governed primarily by discrimination during photosynthetic uptake, but small variations in isotopic composition (0 – 4‰) may occur due to loss of photorespiratory CO_2 or refixation of respiratory sources. Furthermore, since isotopic analysis is being more widely used as a tool for probing plant physiology, growth, ecology and palaeoclimatic reconstruction, the need for accurate interpretation of even the smallest changes in discrimination has increasingly important implications (see Chapters 13, 19 and 20).

8.4.1 Theory

In the theoretical situation where a leaf is only photosynthesizing, then net $\Delta^{13}C$ will only reflect that of photosynthetic uptake. According to present theory, even when respiration does occur, this will not affect the isotopic discrimination (both in the leaf material and instantaneously) if there is no fractionation between the leaf substrates and the CO_2 released (Farquhar et al., 1982; O'Leary, 1988). For example, non-discriminatory respiration from a leaf can be thought of as the complete loss of a portion of leaf carbon reserves, whilst leaving the residual leaf material neither ^{13}C-enriched nor depleted. In the case of instantaneous discrimination, where respired CO_2 is of the same isotopic composition as fixed carbon, the reduced enrichment observed in the CO_2 passing over the leaf is exactly matched by a reduced drawdown of CO_2 partial pressure in the cuvette, and measured $\Delta^{13}C_{obs}$ (as calculated from Equation 8.2) remains the same. This situation is shown by the filled symbols in Figure 8.4, where $\Delta^{13}C_i - \Delta^{13}C_{obs}$ shows the expected trend with A/p_a as dictated by measurements under low oxygen partial pressure. Slight variation does exist, which may be caused by short-term variations between $\delta^{13}C$ of carbon being assimilated and respired, and not respiratory fractionation per se. Furthermore, these effects may be exaggerated at low assimilation rates due to the increased relative contribution of dark respired CO_2.

However, where fractionation does occur during respiration, then the remaining leaf material will become either slightly enriched or depleted in ^{13}C, which is also evident in the instantaneous $\Delta^{13}C$. It is possible to induce artificially high levels of respiratory fractionation (> 20‰) through manipulating different substrate pools of carbon within the leaf. Since the turnover times for carbohydrate pools utilised during dark respiration can be as much as three hours (Parnik et al., 1972), it is possible to induce temporary isotopic differences between CO_2 being currently assimilated and that being respired. By growing P. vulgaris under an atmosphere containing a depleted isotopic composition of CO_2 ($\delta^{13}C_a$ = -40 ‰), leaf carbohydrate pools retain this depleted signal for a short period when the plant is subsequently exposed to air of ambient CO_2 isotopic composition ($\delta^{13}C_a$ = -8 ‰). If $\Delta^{13}C$ is measured during this period, net discrimination, $\Delta^{13}C_{obs}$ is much lower than that predicted for

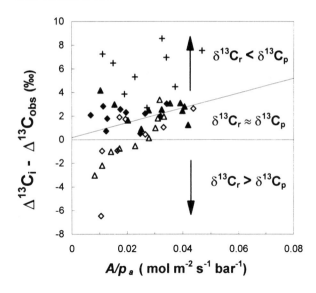

Figure 8.4. *The effect of respiratory CO_2 $\delta^{13}C$ on discrimination, $\Delta^{13}C_{obs}$, as related to that predicted, $\Delta^{13}C_p$, over a range of CO_2 assimilation in P. vulgaris, where filled symbols were measured and grown under ambient source gas $\delta^{13}C$ ($\delta^{13}C_r = \delta^{13}C_p$), open symbols measured under depleted source gas CO_2 ($\delta^{13}C_r < \delta^{13}C_p$), and crosses grown under depleted source gas CO_2 (to label leaf carbohydrates with depleted $\delta^{13}C$) but measured under ambient source gas $\delta^{13}C$. ($\delta^{13}C_r > \delta^{13}C_p$). The solid line represents the relationship of P. vulgaris under 2% O_2 (data shown in Figure 8.1B).*

gross photosynthesis alone, $\Delta^{13}C_i$ (positive $\Delta^{13}C_i - \Delta^{13}C_{obs}$, Figure 8.4, +), in accordance with the situation where respired CO_2 ($\delta^{13}C_R \sim$ -60 ‰) is depleted with respect to that being assimilated, ($\delta^{13}C_p \sim$ -30‰). The release of very depleted respired CO_2 in effect reduces the enrichment in ^{13}C observed downstream of the leaf, reducing $\Delta^{13}C_{obs}$.

The converse can also be replicated by exposing a leaf grown under ambient isotopic conditions to the depleted source CO_2. Before the labelling of leaf substrates occurs, $\Delta^{13}C$ can be measured instantaneously during the period where respired CO_2 ($\delta^{13}C_R \approx$ -30‰) is enriched relative to that being assimilated, ($\delta^{13}C_p \approx$ -60‰), and accordingly $\Delta^{13}C_{obs}$ is much higher relative to $\Delta^{13}C_i$ (negative $\Delta^{13}C_i - \Delta^{13}C_{obs}$, Figure 8.4, open symbols). In fact the latter situation is a common occurrence where the CO_2 source for air-handling systems is derived from commercially supplied CO_2 which is a by-product of industrial processes. Indeed, a study on discrimination in mosses, utilising such a depleted CO_2 source (Rice and Giles, 1996), recorded an overestimation of $\Delta^{13}C$ by up to 13‰, which was caused by release of respiratory CO_2 from carbohydrate pools not reflecting such depletion in ^{13}C.

8.4.2 *Photorespiratory fractionation*

Measurable isotope effects have been recorded during CO_2 release via glycine decarboxylase (*f*). Initially, experiments were conducted where photorespired CO_2 was collected in a stream of CO_2 free air, from which estimates of *f* ranged from -1.6 to -0.2‰ (Troughton *et al.*, 1974). However, CO_2 leaving the leaf may have been subject to subsequent discrimination during partial re-assimilation by RUBISCO. This would cause a ^{13}C enrichment in collected CO_2, resulting in an underestimation of *f*. Unlike dark respiratory fractionation, which can be measured in the dark without concurrent carbon fixation, in the light it is difficult to separate photorespiratory fractionation effects and photosynthesis when estimating *f*. Later experiments estimated photorespiratory fractionation under ambient photosynthetic conditions to yield an estimate of 7‰ (Rooney, 1988). Attempts at *in vitro* analysis have shown a wide range of fractionation (7.8 to -16.2‰) with significant interspecific variation (Ivlev, 1996). However, in this study, fractionation was also observed to vary according to the presence/absence of co-factors and the duration of the reaction. Thus, *in vitro* conditions may be variable and may not accurately present conditions *in vivo*.

As was mentioned above, measurement of net discrimination and the prediction of photosynthetic discrimination from simultaneous gas-exchange measurement allows inferences about the $\delta^{13}C$ of respired CO_2 to be made. Whereas substrates for dark respiration show low turnover rates (over 3 h), glycine decarboxylation substrate pools have rapid turnover times (about 10 min) (Parnik *et al*, 1972). Hence there should be no transient differences in the $\delta^{13}C$ of assimilated carbon and photorespiratory substrates; any difference between $\delta^{13}C_p$ and $\delta^{13}C_R$ must be attributable to enzymatic fractionation.

For *Triticum aestivum*, analysis of the $\Delta^{13}C_i - \Delta^{13}C_{obs}$ relationship with A/p_a revealed a reduction in $\Delta^{13}C_{obs}$ relative to $\Delta^{13}C_i$ as O_2 concentration increased, such that $\Delta^{13}C_i - \Delta^{13}C_{obs}$ was greater under photorespiratory conditions (Figure 8.5A). This is in accordance with the theory that respired CO_2 was depleted in ^{13}C with respect to that being assimilated, that is, f was positive. Due to the higher photosynthetic capacity of

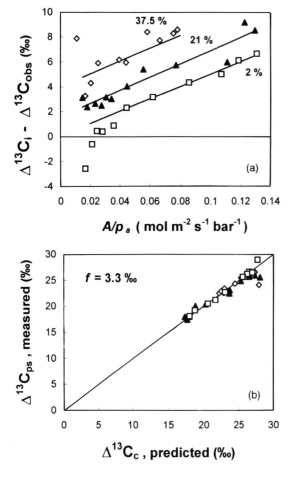

Figure 8.5. (A) The difference between measured and predicted discrimination, $\Delta^{13}C_i - \Delta^{13}C_{obs}$, under 2% (□), 21% (▲) and 37.5 % (◊) O_2 in T. aestivum, the lines represent linear regressions where r^2 = 0.988, 0.913, 0.492 respectively. For the 2 % O_2 treatment, points below 0.04 mol m^{-2} s^{-1} bar^{-1} are not considered in the regression, being caused by dark respiratory release under depleted source gas $\delta^{13}C$. (B) Comparison of measured photosynthetic discrimination, $\Delta^{13}C_{ps}$, assuming f = 3.3 ‰ and e = 0 ‰, with discrimination predicted from C_c/C_a, $\Delta^{13}C_c$ (see Equation 4). The solid line represents the 1: 1 line of unity. Reproduced from Gillon and Griffiths 1997 with permission from Blackwell Science.

T. aestivum as compared to *P. vulgaris*, higher photosynthetic rates were experienced and hence dark respiration was mostly negligible relative to assimilation rates. Only in the low oxygen treatment did dark respiration appear to affect $\Delta^{13}C$, which was the result of using source CO_2 depleted in ^{13}C (see Section 8.4.1). Unlike dark respiration, photorespiration increased linearly with oxygen partial pressure and so increased rates of glycine decarboxylation at the highest pO_2 values corresponded with the greatest differences between $\Delta^{13}C_i$ and $\Delta^{13}C_{obs}$ (open diamonds, Figure 8.5A). Also, because photorespiration occurs as a constant proportion of photosynthesis, the effect on $\Delta^{13}C$ remains constant over the full range of assimilation, and contrasts with dark respiration which only significantly affects $\Delta^{13}C$ at low assimilation rates where dark respired CO_2 becomes a significant proportion of net CO_2 exchange.

Estimation of photorespiratory fractionation, f. Because the $\delta^{13}C$ of photorespiratory substrates can be assumed to be equivalent to that of CO_2 recently assimilated (as predicted from C_c/C_a) and photorespiration rates can be calculated from estimates of the CO_2 compensation point in the absence of respiration in the day, Γ^* (Brooks and Farquhar, 1985), it is possible to quantify the effect of photorespiratory CO_2 release on $\Delta^{13}C$. Gillon and Griffiths (1997) adapted the initial mass balance assumptions describing the transfer of ^{13}C through a leaf cuvette, from which Evans *et al.* (1986) derived their expression defining instantaneous $\Delta^{13}C$ (Equation 8.2), to include the flux and isotopic composition of all respiratory CO_2,

$$J_e R_e + R_d R_r + F R_f = J_o R_o + V_c R_{ps} \tag{8.7}$$

where J_e, J_o, R_d, F and V_c are the molar fluxes of CO_2 into and out of the cuvette, through dark respiration, photorespiration and *gross* fixation respectively, and R_e, R_o, R_r, R_f and R_{ps} are their respective $^{13}C/^{12}C$ ratios. This can be solved for Δ_{ps}, where $\Delta_{ps} = (R_o/R_{ps}) - 1$, to obtain the following expression which describes the discrimination attributable to photosynthetic uptake alone, correcting for any respiratory CO_2 input and fractionation (actual or artificial due to depleted source gas $\delta^{13}C$)

$$(8.8)$$

$$\Delta^{13}C_{ps} = \cfrac{\dfrac{(uC_o+V_c)}{V_c}(\delta^{13}C_o-\delta^{13}C_e) - \dfrac{R_d}{V_c}(\delta^{13}C_n-e-\delta^{13}C_e) - \dfrac{\Gamma^*}{C_c}(\delta^{13}C_{fs}-f-\delta^{13}C_e)}{1000+\delta^{13}C_o - \dfrac{(uC_o+V_c)}{V_c}(\delta^{13}C_o-\delta^{13}C_e) + \dfrac{R_d}{V_c}(\delta^{13}C_n-e-\delta^{13}C_e) + \dfrac{\Gamma^*}{C_c}(\delta^{13}C_{fs}-f-\delta^{13}C_e)}$$

where u is the molar flux of air through the cuvette, C_o is the mole fraction of CO_2 in air leaving the cuvette (corrected to the humidity of air entering the cuvette), R_d is the day respiration rate, Γ is the CO_2 compensation point in the absence of day respiration, v_c is the carboxylation rate calculated from $V_c = (A + R_d)/(1 - \Gamma^*/p_c)$, $\delta^{13}C_{rs}$, e and $\delta^{13}C_{fs}$, f are the substrate $\delta^{13}C$ and fractionation for dark respiration and photorespiration respectively.

For *T. aestivum*, close agreement between Δ_{ps} and $\Delta^{13}C$ predicted from C_c/C_a, $\Delta^{13}C_c$, was obtained where f was estimated as 3.3 ± 1.6 ‰ (Figure 8.5B). This fully accounted for photorespiratory CO_2 release in all oxygen concentrations, observed by the elimination of treatment differences when plotting $\Delta^{13}C_i - \Delta^{13}C_{ps}$ (Gillon and Griffiths, 1997). The error associated with this estimate was determined by finding the range of f values for which the relationships of $\Delta_i - \Delta_{ps}$ and A/p_a were not significantly different ($p>0.2$). Unfortunately, the low range of assimilation rate evident in *P. vulgaris* prevented such empirical analysis of photorespiratory fractionation. However, at higher assimilation rates, the fact that all of the $\Delta^{13}C_i - \Delta^{13}C_{obs}$ relationships in *P. vulgaris* appeared to converge with that conducted in low oxygen partial pressure (Figure 8.4) indicates that photorespiratory fractionation is indeed close to zero in this species.

Previously, (photo)respiration effects on $\Delta^{13}C$ have been accounted for in the predictive model by addition of the two latter terms as follows (Farqhuar *et al.* 1982)

$$\Delta^{13}C = a\frac{C_a - C_i}{C_a} + a_i\frac{C_i - C_c}{C_a} + b'\frac{C_c}{C_a} + \frac{eR_d / k}{C_a} + \frac{f\Gamma *}{C_a} \qquad (8.9)$$

where k is the carboxylation efficiency. This aims to predict net discrimination, which is directly comparable to $\Delta^{13}C_{obs}$, in contrast to the comparison of gross photosynthetic discrimination ($\Delta^{13}C_{ps}$ vs. $\Delta^{13}C_c$) used above. Agreement between $\Delta^{13}C$ predicted by Equation 8.9 and $\Delta^{13}C_{obs}$ can also be used to estimate photorespiratory fractionation, yielding different values where $f = 12‰$ (*T. aestivum*) and 0 ‰ (*P. vulgaris*). Whilst both sets of estimate agree qualitatively, they do not agree quantitatively. Both the above approaches rely on initial assumptions which do not incorporate the potential refixation of (photo)respiratory CO_2, (Stulhfauth *et al.*, 1990) up to a third of that produced, which may be responsible for the discrepancies arising in estimates of f. These problems emphasise the need to account for the dynamic diffusive nature of CO_2 molecules in the leaf interior and concurrent discriminatory processes which are acting on the same molecular species when estimating carbon isotope effects *in vivo*. It is encouraging that isotopic analysis at present is adequate for a qualitative analysis of photorespiratory fractionation between species, demonstrated by *T. aestivum* and *P. vulgaris* here. This variation in photorespiratory fractionation between species is surprising in itself, considering the relatively conservative nature of fractionation in other enzymes (e.g. RUBISCO and PEPC). However, quantification of f appears beyond the scope of the present models at the current time.

8.5 Refixation of respiratory CO_2

Since the first isotopic analyses of plant material, it has been postulated that molecules of CO_2 released during respiration are just as likely to be assimilated by the leaf as are molecules diffusing in from the surrounding air. Initially, Wickman (1952) explained differences in the $\delta^{13}C$ of plants from locations with varying windspeed by suggesting that depleted CO_2 from respiration could accumulate to a greater extent around plants where conditions were less windy. Photosynthetic uptake of this depleted source of CO_2 would therefore be reflected in the $\delta^{13}C$ of the leaf material. Subsequently, this 'cyclic enrichment' theory has been rejected by Craig (1953) and since then the debate has continued.

8.5.1 *Canopy photosynthesis*

A similar situation occurs in forest canopies, where depleted CO_2 from soil and plant respiration accumulates at the forest floor due to poor coupling with the atmosphere (Medina et al.,1986). Attempts have been made to quantify the extent of respiratory recycling in forest canopies (Lloyd et al., 1996; Sternberg, 1989) yet most studies suggest that the proportion of respiratory CO_2 which is re-assimilated by the forest is small, between 1 and 5‰ (Buchmann et al., 1997; Lloyd et al.,1996). Gross fluxes of

CO_2 in and out of the canopy, are generally up to three orders of magnitude greater than the photosynthetic flux (3000 μmol m^{-2} s^{-1} as compared to 1–10 μmol m^{-2} s^{-1}, Lloyd et al., 1996), hence molecules of respired CO_2 are far more likely to diffuse out of the canopy than to be re-assimilated.

8.5.2 Leaf photosynthesis

Re-assimilation of respired CO_2 in plants has been recognised as a significant portion of the total carbon flux through plants (Griffiths, 1988, 1996; Ludwig and Canvin 1971; Osmond et al., 1980). Certain environmental and physiological situations increase this refixation where external CO_2 uptake is restricted by high diffusive resistance and/or high rates of internal respiration, notably during nocturnal CO_2 fixation in Crassulacean acid metabolism plants, and aquatic plant photosynthesis (Griffiths, 1988). Similarly, under water-stressed conditions in C_3 plants, refixation may be increased by low stomatal conductances (Stuhlfauth et al., 1990), which is supported by conductance-based flux estimates where gross fluxes of CO_2 in and out of leaves (30–50 μmol m^{-2} s^{-1}) may be much closer to the CO_2 assimilation flux (10–20 μmol m^{-2} s^{-1}), in contrast to forest canopies which are well-coupled to the atmosphere. Moreover CO_2 refixation, especially of photorespired CO_2, may be essential to maintain an adequate electron sink under high light intensities and drought conditions (Osmond et al., 1980).

The effect on carbon isotope composition is less well understood, but in theory refixation should represent the assimilation of a slightly depleted source CO_2 which reflects the combination of internal (respiratory) and external (atmospheric) CO_2 (Griffiths et al., 1990; Griffiths, 1996). Such source modification has been observed during progressive drought experiments (J. Ghashghaie, personal communication) and in cereal crops (Bort et al., 1996), where high respiration and/or low conductances lead to greater refixation and more negative $\delta^{13}C$ of assimilates. However, the effects of refixation per se need to be distinguished from changes in $\delta^{13}C$ due to physiological responses of C_i/C_a to experimental conditions.

8.6 Discrimination under natural conditions

Discrimination processes and the models which describe them have been successfully studied in both the laboratory and also in common garden environments. However, to interpret the isotopic composition of plants growing under natural conditions which experience constantly fluctuating environmental variables, instantaneous $\Delta^{13}C$ needs to be studied in the field. Typically, experimental species have tended to be crop plants, or plants with high photosynthetic capacities, but these do not represent the full spectrum of species and conditions in the natural environment. Therefore, the diurnal variation in gas exchange and $\Delta^{13}C$ was studied in a pioneering C_3 rain forest shrub, *Piper aduncum*, during the onset of the dry season in both 1992 and 1995, Trinidad, West Indies (M.S.J. Broadmeadow, unpublished data; Harwood et al., 1998). Conditions were characterised by high light intensity and limited water availability, and consequently temperatures were high and CO_2 assimilation rates were low, never exceeding 8 mmol m^{-2} s^{-1}. Measured $\Delta^{13}C$ showed enormous variation from 5 to 35‰ (Figure 8.6A) which did not appear to be correlated with C_i/C_a, as indicated by the relatively constant $\Delta^{13}C_i$ over the day (Figure 8.6B). The variation in $\Delta^{13}C_{obs}$ may be a function of leaf-to-leaf variation, since individual

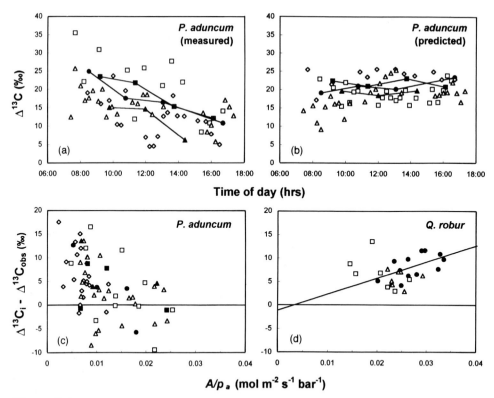

Figure 8.6. *Instantaneous discrimination in field conditions. (A) Diurnal variation in $\Delta^{13}C_{obs}$ measured in* Piper aduncum *(Trinidad, West Indies), for random leaves (open symbols) in exposed leaves, Feb 1992 (◊), shaded leaves, Feb 1992 (Δ) and exposed leaves, Feb 1995 (□), and for three individual leaves followed over a second day (closed symbols, solids lines joining points of the same leaf) in Feb. 1995; (B) Diurnal variation in $\Delta^{13}C_i$ (predicted from Equation 3, where b' = 29‰) for* P. aduncum, *symbols as for A. The difference between $\Delta^{13}C_i$ and $\Delta^{13}C_{obs}$ for (C)* P. aduncum, *symbols as for (A), and (D)* Quercus petraea *(Northumberland, UK) in July 1995 for leaves at different heights in the canopy, 1m (□), 6.5m (Δ) and 10m (●), under two light intensities (400 or 800 mmol m^{-2} s^{-1}). The solid line represents the linear regression for leaves at 10 m, from which g$_i$ was estimated (r^2 = 0.317). A, B and C are redrawn from Harwood* et al. *(1998) with data added from M.S.J. Broadmeadow (unpublished). D is redrawn from Harwood (1997).*

leaves appear to follow a regular downward trend through the day (joined closed symbols, Figure 8.6A. By analysis of the difference between both predicted and measured discrimination, $\Delta^{13}C_i - \Delta^{13}C_{obs}$, the greatest deviation occurred at the lowest assimilation rates, where the range $\Delta^{13}C_{obs}$ extended above and below that predicted (Figure 8.6C). This was a similar response to that seen in *P. vulgaris* (Figure 8.4), indicating an increasing effect of respiration on net $\Delta^{13}C$ at low assimilation rates. However, leaf respiration rates in the tropics can be as much as 2 μmol m^{-2} s^{-1}, 10 fold more than that experienced for *P. vulgaris* in the laboratory (Gillon and Griffiths, 1997), which may explain some of the excessive deviations in $\Delta^{13}C_{obs}$ observed in *Piper aduncum* in Trinidad. Again, fractionation is likely to be due to refixation or transient differences between the isotopic signal of CO_2 assimilated and respiratory substrates due to slow turnover rates of the latter, and not dark respiratory fractionation (Lin and Ehleringer, 1997).

Whilst such large deviations from $\Delta^{13}C_i$ are both intriguing and difficult to explain with conventional theory, they have been consistently observed when measuring on-line $\Delta^{13}C$ under field conditions in the tropics (Harwood et al., 1998; J.S. Gillon, unpublished data). Interpretation of the data must be cautious since errors may be as high as 4‰, caused by on-line analysis at low assimilation rates (high ξ in Equation 8.2) amplifying mass spectrometer precision limits (0.1‰). Nevertheless, large deviations in $\Delta^{13}C_{obs}$ occur at both extremes of ξ (total range is 16–35; Harwood, 1997), suggesting that they may in part be a function of actual leaf processes. This may, in fact, represent the observable effect of respiratory re-fixation on $\Delta^{13}C$; stomatal conductance to CO_2 (generally <0.1 mol m^{-2} s^{-1}) was low due to high leaf–air vapour pressure deficits (25–50 mbar), possibly increasing the proportion of respired CO_2 being re-assimilated (see Section 8.4).

In a temperate woodland, $\Delta^{13}C$ measured at different heights under high and low light within a *Quercus petraea* canopy (Harwood, 1997) appeared to more closely represent the expected response; indeed, the leaves at 10 m appear to show a reasonable linear increase in $\Delta^{13}C_i - \Delta^{13}C_{obs}$, from which g_i was estimated to be 0.08 mol m^{-2} s^{-1} bar^{-1}. Yet despite the slight increase in assimilation and lower rates of respiration in the temperate climate, the data still illustrate the changeover point between photosynthetic and respiratory contribution to net discrimination. And here, ξ was also low, never exceeding 16, thus reducing the potential error. However, more importantly, it is possible to see the large leaf-to-leaf variation present between leaves at different heights even when measured under comparable light conditions. So not only does temporal variation exist in discrimination as environmental factors fluctuate, but also large spatial variation is evident within the same canopy, possibly as a function of leaf morphological and physiological adaptations.

8.7 Conclusions

As the applications of carbon isotope analysis appear to encompass an ever increasing scope of subjects, examining discrimination in a wider variety of natural and artificial environments is important to probe all of processes governing carbon isotope composition in plants. Whereas the qualitative relationships of $\Delta^{13}C$ have been known for some time, current research, including palaeoclimatic reconstruction and analysis of mesophyll limitations to photosynthesis, require quantitative analysis of carbon isotope discrimination, which depends more upon absolute fractionation effects than causal relationships between physiology and $\Delta^{13}C$ (see Chapters 19 and 20). In any area that utilises carbon isotopes as a tool, it is of the utmost importance that the enormous potential for spatial and temporal variation in $\Delta^{13}C$ is acknowledged, not only between different leaves and plants, but also within the same leaves and plants, before making further interpretations of $\Delta^{13}C$.

Acknowledgements

Much of the work presented has resulted from research supported by the Natural Environmental Research Council (NERC, UK), both as Ph.D. studentships (Nos. GT4/92/27 to KGH, GT4/93/35 to AR and GT4/94/379 to JSG), full research grants (GR3/09805 to AMB) and small grants for the work in Trinidad (Nos. GR9/541 in

1992 and GR9/1676 in 1995). Their financial support is gratefully acknowledged, as well as the continued support from all members of the Potter Lab at the University of Newcastle-upon-Tyne.

References

Borland A.M., Griffiths H., Broadmeadow M.S.J., Fordham M.C. and Maxwell C. (1993) Short-term changes in carbon isotope discrimination in the C_3-CAM intermediate *Clusia minor* L. growing in Trinidad. *Oecologia* **95**, 444–453.

Borland A.M. (1996) A model for the partitioning of photosynthetically fixed carbon during the C_3-CAM transition in *Sedum teliphium*. *New Phytologist* **134**, 433–444.

Borland A.M. and Griffiths H. (1996) Variations in the phases of CAM and regulation of carboxylation patterns determined by carbon isotope discrimination techniques. In: *Crassulacean Acid Metabolism. Biochemistry, Ecophysiology and Evolution* (eds. K. Winter and J.A.C. Smith) Springer-Verlag, Berlin, pp. 230–246.

Borland A.M. and Griffiths H. (1997) A comparative study on the regulation of C_3 and C_4 carboxylation processes in the constitutive CAM plant *Kalanchoe daigremontiana* and the C_4-CAM intermediate *Clusia minor*. *Planta* (in press).

Bort J., Brown R.H. and Araus J.L. (1996) Refixation of respiratory CO_2 in the ears of C_3 cereals. *J. Exp. Bot.* **47**, 1567–1575.

Brooks A. and Farquhar G.D. (1985) Effect of temperature on the carbon dioxide/oxygen specificity of ribulose-1,5-bisphosphate carboxylase/oxygenase and the rate of respiration in the light. *Planta* **165**, 397–406.

Brugnoli E., Hubick K.T., Von C.S., Wong S.C. and Farquhar G.D. (1988) Correlation between the carbon isotope discrimination in leaf starch and sugars of C_3 plants and the ratio of inter-cellular and atmospheric partial pressures of carbon dioxide. *Plant Physiology* **88**, 1418–1424.

Buchmann N. Guehl, J.M., Barigah, T.S. and Ehleringer, J.R. (1997) Interseasonal comparison of CO_2 concentrations, isotopic composition and carbon dynamics in an Amazonian rainforest (French Guiana). *Oecologia* (in press).

Christeller J.T., Laing W.A. and Troughton H.T. (1976) Isotope discrimination by ribulose-1–5-diphosphate carboxylase. *Plant Physiology* **57**, 580–582.

Craig H. (1953) The geochemistry of the stable carbon isotopes. *Geochim. Cosmochim. Act* **3**, 53–92.

Deleens E., Lerman J.C., Nato A. and Moyse A. (1974) Carbon isotope discrimination by the carboxylating reactions in C_3, C_4 and CAM plants. In: *3rd Int. Conference Photosynthesis* (ed. M Avron) Elsevier, Amsterdam, pp. 1267–1276.

Epron D., Doddard D., Cornic G. and Genty B. (1995) Limitation of net CO_2 assimilation by internal resistances to CO_2 transfer in the leaves of two tree species (*Fagus sylvatica* L. and *Castanea sativa* Mill.). *Plant Cell Environ.* **18**, 43–51.

Estep M.F., Tabita R., Parker P.L. and Van Baalen C. (1978) Carbon isotope fractionation by ribulose-1–5-bisphosphate caboxylase from various organisms. *Plant Physiology* **61**, 680–687.

Evans J.R., Sharkey T.D., Berry J.A. and Farquhar G.D. (1986) Carbon isotope discrimination measured concurrently with gas exchange to investigate carbon dioxide diffusion in leaves of higher plants. *Aust. J. Plant Physiology* **13**, 281–292.

Evans J.R., Von Caemmerer S., Setchell B.A. and Hudson G.S. (1994) The relationship between CO_2 transfer conductance and leaf anatomy in transgenic tobacco with a reduced content of Rubisco. *Aust. J. Plant Physiology* **21**, 475–495.

Evans J.R. and Von Caemmerer S. (1996) Carbon dioxide diffusion within leaves. *Plant Physiology* **110**, 339–346.

Farquhar G.D., O'Leary M.H. and Berry J.A. (1982) On the relationship between carbon isotope discrimination and the intercellular carbon dioxide concentration in leaves. *Aust. J. Plant Physiology* **9**, 121–137.

Farquhar G.D. (1983) On the nature of carbon isotope discrimination in C$_4$ species. *Aust. J. Plant Physiology* 10, 205–226.

Gillon J.S. and Griffiths H. (1997) The influence of (photo)respiration on carbon isotope discrimination in plants. *Plant Cell Environ.* 20, 1217–1230.

Griffiths H., Broadmeadow M.S.J., Borland A.M. and Hetherington C.S. (1990) Short-term changes in carbon isotope discrimination identify transitions between C$_3$ and C$_4$ carboxylation during Crassulacean acid metabolism. *Planta* 181, 604–610.

Griffiths H. (1988) Crassulacean acid metabolism: a re-appraisal of physiological plasticity in form and function. *Advances Bot. Res.* 15, 43–92.

Griffiths H. (1996) Evaluation and integration of environmental stress using stable isotopes. In: *Photosynthesis and the Environment* (ed N.R. Baker), Advances in Photosynthesis Vol 5, Kluwer, Dordrecht, pp. 451–468.

Guy R.D., Fogel M.L. and Berry J.A. (1993) Photosynthetic fractionation of the stable isotopes of oxygen and carbon. *Plant Physiology* 101, 37–47.

Harwood K.G., Gillon J.S., Broadmeadow M.S.J. and Griffiths H. (1998) Diurnal variation of $\Delta^{13}CO_2$, $\Delta C^{18}O^{16}O$ and evaporative site enrichment of $\delta H_2^{18}O$ in *Piper aduncum* under field conditions in Trinidad. *Plant Cell Environ.* (in press).

Harwood K.G. (1997) Variation in the stable isotopes of oxygen and carbon within forest canopies. Ph.D. Thesis, Newcastle-upon-Tyne.

Hatch, M.D. and Osmond, C.B. (1976) Compartmentation and transport in C$_4$ photosynthesis. In: *Transport in Plants III. Intracellular Interactions and Transport Processes* (eds C.R. Stocking, and U. Heber). Encyclopedia of Plant Physiology New Series, Vol. 3, Sringer-Verlag, Berlin, pp. 144–184.

Henderson S.A., Von C.S. and Farquhar G.D. (1992) Short-term measurements of carbon isotope discrimination in several C$_4$ species. *Aust. J. Plant Physiology* 19, 263–285.

Holbrook G.P., Keys A.J. and Leech R.M. (1984) Biochemistry of photosynthesis in species of *Triticum* of different ploidy. *Plant Physiology* 74, 12–15.

Ivlev A.A. (1996) Fractionation of carbon ($^{13}C/^{12}C$) isotopes in glycine decarboxylase reaction. *FEBS Lett.* 386, 174–176.

Lauteri M., Brugnoli E. and Spaccino L. (1993) Carbon isotope discrimination in leaf soluble sugars and in whole-plant dry matter in *Helianthus annuus* L. grown under different water conditions. In: *Stable isotopes and plant carbon–water relations* (eds J.R. Ehleringer, A.E. Hal and G.D. Farquhar) Academic Press, San Diego, CA, pp. 93–108.

Lin G. and Ehleringer J.R. (1997) Carbon isotopic fractionation does not occur during dark respiration in C$_3$ and C$_4$ plants. *Plant Physiology* 114, 391–394.

Lloyd J., Syvertsen J.P., Kriedemann P.E. and Farquhar G.D. (1992) Low conductances for carbon dioxide diffusion from stomata to the sites of carboxylation in leaves of woody species. *Plant Cell Environ.* 15, 873–899.

Lloyd J., Kruijt B., Hollinger D.Y., Grace J., Francey R.J., Wong S.C., Kelliher F.M., Miranda A.C., Farquhar G.D., Gasj J.H.C., Vygodskaya N.N., Wright I.R., Miranda H.S. and Schulze E.D. (1996) Vegetation effects on the isotopic compostion of atmospheric CO$_2$ at Local and regional scales: Theoretical aspects and a comparison between rain forest in Amazonia and a Boreal forest in Siberia. *Aust. J. Plant Physiology* 23, 371–399.

Maxwell K., von Caemmerer S. and Evans (1997) Is a low *conductance* to CO$_2$ diffusion a consequence of Crassulacean acid metabolism? *Aust. J. Plant Physiology.* (in press).

Medina E., Montes G., Cuevas E. and Roksandic Z. (1986) Profiles of CO$_2$ and $\delta^{13}C$ values of the upper Rio Negro Basin, Venezuela. *Journal Tropical Ecology* 2, 207–217.

Mook W.G., Bommerson J.C. and Staverman W.H. (1974) Carbon isotope fractionation between dissolved bicarbonate and gaseous carbon dioxide. *Earth Planetary Sci. Lett.* 22, 169–176.

Nier A.O. and Gulbransen E.A. (1939) Variations in the relative abundance of the carbon isotopes. *J. Am. Chem. Soc.* 61, 697–698.

O'Leary M.H. (1981) Carbon isotope fractionation in plants. *Phytochemistry* 20, 553–567.

O'Leary M.H. (1984) Measurement of the isotopic fractionation associated with diffusion of carbon dioxide in aqueous solution. *J. Phys. Chem.* **88**, 823–825.

O'Leary M.H. (1988) Carbon isotopes in photosynthesis. *Bioscience* **38**, 328–336.

O'Leary M.H., Madhavan S. and Paneth P. (1992) Physical and chemical basis of carbon isotope fractionation in plants. *Plant Cell Environ.* **15**, 1099–1104.

Osmond C.B. (1978) Crassulacean acid metabolism: a curiosity in context. *Ann. Rev. Plant Physiology* **29**, 379–414.

Osmond C.B., Winter K. and Powles S.B. (1980) Adaptive significance of carbon dioxide cycling during photosynthesis in water-stressed plants. In: *Adaptions of Plants to Water and High Temperatures and Stress* (eds N.C. Turner and P.J. Kramer) Wiley, New York, pp. 139–154.

Osmond C.B., Popp M. and Robinson S.A. (1996) Stoichiometric nightmares: studies of photosynthetic O_2 and CO_2 exchanges in CAM plants. In: *Crassulacean Acid Metabolism. Biochemistry, Ecophysiology and Evolution* (eds K. Winter and J.A.C. Smith) Springer-Verlag, Berlin, pp. 230–246.

Park R. and Epstein S. (1960) Carbon isotope fractionation during photosynthesis. *Geochim. Cosmochim. Acta* **21**, 110–126.

Parnik T., Keerberg O. and Viil J. (1972) $^{14}CO_2$ evolution from bean leaves at the expense of fast-labelled intermediates of photosynthesis. *Photosynthetica* **6**, 66–74.

Rice S.K. and Giles L. (1996) The influence of water content and leaf anatomy on carbon isotope discrimination and photosynthesis. *Plant Cell Environ.* **19**, 118–124.

Roberts A., Griffiths H., Borland A.M. and Reinert F. (1996) Is Crassulacean acid metabolism activity in sympatric species of hemi-epiphytic stranglers such as *Clusia* related to carbon cycling as a photoprotective process? *Oecologia* **106**, 28–38.

Roberts A., Borland A.M. and Griffiths H. (1997) Discrimination processes and shifts in carboxylation during the phases of Crassulacean Acid Metabolism. *Plant Physiology* **113**, 1283–1292.

Roeske C.A. and O'leary M.H. (1984) Carbon isotope effects on the enzyme-catalyzed carboxylation of ribulose bisphosphate. *Biochemistry* **23**, 6275–6284.

Rooney M.A. (1988) Short-term carbon isotope fractionation by plants. Ph.D. Thesis, Wisconsin.

Sternberg L. (1989) A model to estimate carbon dioxide recycling in forests using $^{13}C/^{12}C$ ratios and concentrations of ambient carbon dioxide. *Agricultural Forest Meteorology* **48**, 163–173.

Stuhlfauth T., Scheurrmann R. and Fock H.P. (1990) Light energy dissipation under water stress conditions: contribution of reassimilation and evidence for additional processes. *Plant Physiol.* **92**, 1053–1061.

Syvertsen J.P., Lloyd J., McConchie C., Kriedemann P.E. and Farquhar G.D. (1995) On the relationship between leaf anatomy and CO_2 diffusion through the mesophyll of hypostomatous leaves. *Plant Cell Environ.* **18**, 149–157.

Troughton J.H., Card K.A. and Hendy C.H. (1974) Photosynthetic pathways and carbon isotope discrimination by plants. *Carnegie Institute Washington Yearbook* **73**, 768–780.

Vitousek P.M., Field C.B. and Matson P.A. (1990) Variation in foliar $\delta^{13}C$ in Hawaiian *Metrosideros polymorpha*: a case for internal resistance? *Oecologia* **84**, 362–370.

Vogel J.C., Grootes P.M. and Mook W.G. (1970) Isotopic fractionation between gaseous and dissolved carbon dioxide. *Zeitschrift fur Physik,* **230**, 225–228.

Vogel J.C. (1980) Fractionation of the carbon isotopes during photosynthesis. In: *Sitzungsberichte der Heidelberger Akadamie der Wissenschaften*, Springer-Verlag, Berlin, pp. 111–135.

von Caemmerer S. and Hubick K.T. (1989) Short-term carbon isotope discrimination in C_3-C_4 intermediate species. *Planta* **178**, 475–481.

von Caemmerer S. and Evans J.R. (1991) Determination of the average partial pressure of carbon dioxide in chloroplasts from leaves of several C_3 plants. *Aust. J. Plant Physiology* **18**, 287–306.

von Caemmerer S., Millgate A., Farquhar G.D. and Furbank R.T. (1997) Reduction of ribulose-1,5-bisphosphate carboxylase/oxygenase by antisense RNA in the C_4 plant *Flaveria bidentis* leads to reduced assimilation rates and increased carbon isotope discrimination. *Plant Physiology* **113**, 469–477.

Whelan T., Sackett W.M. and Benedict C.R. (1973) Enzymatic fractionation of carbon isotopes by phosphoenolpyruvate carboxylase in C_4 plants. *Plant Physiology* **51**, 1051–1054.

Wickman F.E. (1952) Variations in the relative abundance of the carbon isotopes in plants. *Geochim. Cosmochim. Acta* **2**, 243–254.

Wong W.W., Benedict C.R. and Kohel R.J. (1979) Enzymatic fractionation of the stable carbon isotopes of carbon dioxide by ribulose-1–5-bisphosphate carboxylase. *Plant Physiology* **63**, 852–856.

Carbon isotope discrimination in structural and non-structural carbohydrates in relation to productivity and adaptation to unfavourable conditions

E. Brugnoli, A. Scartazza, M. Lauteri, M.C. Monteverdi and C. Máguas

9.1 Introduction

During photosynthetic CO_2 assimilation by C_3 plants, discrimination occurs against the heavier stable isotope ^{13}C. Therefore, the isotope ratio $^{13}C/^{12}C$ in photosynthetic products is always lower than that of atmospheric CO_2. Carbon isotope composition ($\delta^{13}C$) of plant material is largely dependent on the photosynthetic pathway, with large changes in the isotope abundance ratio between C_3, C_4 and crassulacean acid metabolism (CAM) species (for a review see Farquhar et al., 1989; see Chapter 8). In addition, within each of these plant groups there is considerable variation in carbon isotope composition, and C_3 species show large changes in carbon isotope composition depending on plant species, morpho-anatomical characteristics and environmental variability (Brugnoli and Farquhar, 1998; Farquhar et al., 1989).

It has been shown that carbon isotope discrimination ($\Delta^{13}C$) in C_3 leaves correlates with the ratio of intercellular and atmospheric concentration of CO_2 (C_i/C_a). This relationship has been demonstrated for different time scales of integration: from the entire life of the plant using $\Delta^{13}C$ in plant dry matter (Farquhar et al., 1982), to instantaneous estimates using the isotope composition of the air measured on-line in a gas exchange system (Evans et al., 1986; see Chapter 8). It has also been shown that carbon isotope

Stable Isotopes, edited by H. Griffiths.
© 1998 BIOS Scientific Publishers Ltd, Oxford.

:ion in leaf sucrose and starch gives a time scales of integration of C_i/C_a of ~ut 1–2 days (Brugnoli *et al.*, 1988) and, therefore, intermediate between the two above methods. The analysis of $\Delta^{13}C$ in carbohydrates has been used for studying short-term changes in C_i/C_a and to quantify diffusional and biochemical limitation to photosynthesis (Borland *et al.*, 1994; Lauteri *et al.*, 1993)   .

Plant water-use efficiency (WUE), the ratio of biomass production to water loss, is also dependent on C_i/C_a. Plant WUE is considered a relevant trait for conferring tolerance to environmental stresses, such as drought and salinity. Therefore, carbon isotope discrimination has been proposed as a tool for detecting differences in C_i/C_a and water-use efficiency in ecophysiological studies and in breeding programmes for improved drought tolerance and/or productivity.

Ten years ago, in the analysis of perspectives for genetic improvements for resistance to environmental stresses, Passioura (1986) wrote: 'apart from substantial and deliberate improvements in the timing of flowering, almost all of the useful genetic improvements in resistance to drought or salinity have been quite empirical, as in traditional breeding programs, or marvelously fortuitous'. Hence, he indicated carbon isotope discrimination as one of the most promising, physiologically based, techniques for screening genotypes with improved water-use efficiency. In the last decade, an increasing number of investigators have been studying the relationships between carbon isotope discrimination, water-use efficiency and productivity, either in the ecological and physiological contexts, or in the attempt to obtain genotypes with improved productivity or water-use efficiency in crop species. The application of stable isotopes has brought about significant improvements in the knowledge of the physiological ecology of photosynthesis. However, the application of stable isotopes in breeding has not been as successful as expected. Partly, in contrast to the initial enthusiasm, conflicting results have been reported. Various relationships have been demonstrated between biomass productivity and carbon isotope discrimination, depending on species and experimental conditions. Positive correlations have been found in wheat (e.g. Condon *et al.*, 1987), barley (Acevedo, 1993; Craufurd *et al.*, 1991) and crested wheatgrass (Read *et al.*, 1991); negative correlations have been described in sunflower (Lauteri *et al.*, 1993; Virgona *et al.*, 1990; Virgona and Farquhar, 1996) and peanut (Hubick *et al.*, 1986), and no relationships were found in common bean (White *et al.*, 1990).

In this chapter, recent results obtained in the study of carbon isotope discrimination, especially in carbohydrate $\Delta^{13}C$, in relation to photosynthesis, productivity and WUE in C_3 species are discussed. We first discuss the relevance of the analysis of carbon isotope discrimination in plant carbohydrates for studying variation in plant productivity and WUE during the ontogeny. This analysis may be particularly relevant for increasing tolerance or yield stability under environmental stress. Secondly, it is shown that the analysis of $\Delta^{13}C$ in leaf carbohydrates may be used to estimate relevant physiological features such as mesophyll conductance (i.e. the conductance to CO_2 diffusion from the substomatal cavities to the sites of carboxylation). Hence, we show that carbohydrate $\Delta^{13}C$ may be used to study limitations of photosynthesis.

9.2 Theoretical background to carbon isotope discrimination

The fractionation of carbon isotopes during photosynthesis depends on physical and chemical processes. Extensive reviews have covered the area of carbon isotope discrimination and its physical and chemical basis (Farquhar *et al.*, 1989; O'Leary,

1981; O'Leary *et al.*, 1992; see Chapter 8). Hence, here the theoretical basis of carbon isotope discrimination in plants will be limited to few fundamental definitions.

During photosynthesis, CO_2 must diffuse from the atmosphere to the chloroplast stroma. Since $^{12}CO_2$ diffuses faster than $^{13}CO_2$, several fractionation processes occur along this diffusion path, so that the CO_2 available at the sites of carboxylation is always significantly depleted in ^{13}C compared to the atmospheric stable isotope ratio (Farquhar *et al.*, 1989). The extent of fractionation during diffusion from the surrounding atmosphere, through the leaf intercellular air spaces, to the sites of carboxylation depends on the different conductances to CO_2 transfer (e.g. conductances in the gas phase, during dissolution, in the liquid phase) and on the resulting CO_2 gradients. Additional isotope fractionation processes are determined by carboxylating enzymes. In species with the C_3 photosynthetic pathway, ribulose-1,5-bisphosphate carboxylase-oxygenase (RUBISCO) discriminates against ^{13}C by about 30‰, compared to gaseous CO_2. The variations associated with C_4 and CAM pathways are described in detail in Chapter 8 of this volume. In C_3 photosynthesis, the combination of different isotope effects leads to a predicted linear relationship between $\Delta^{13}C$ and the ratio of chloroplastic (C_c) and atmospheric (C_a) concentrations of CO_2. This model, described by equation 8.9 in Chapter 8, has been widely used in a simplified form where $\Delta^{13}C_i$ is related to the ratio of intercellular and atmospheric partial pressures of CO_2 (C_i/C_a), according to the equation (Farquhar *et al.*, 1982)

$$\Delta^{13}C_i = a + (b-a)C_i/C_a \qquad (9.1)$$

where, a is the fractionation occurring during diffusion in air (4.4‰), b is the discrimination due to carboxylation reactions (about 28‰ taking into account that a proportion, thought to be near to 5%, of the carbon assimilated in C_3 plants is fixed by β-carboxylation by PEP carboxylase).

The underlying hypothesis of this equation is that the discrimination processes during photorespiratory and respiratory CO_2 release are small (however, see Chapter 8). Furthermore, the internal conductance to CO_2 diffusion from the intercellular air spaces to the chloroplast, the so-called mesophyll conductance, is thought to be high in most leaves; hence C_i would be equal, or very close to, the chloroplastic CO_2 partial pressure C_c. When these assumptions are true, then $\Delta^{13}C$ in plant material can be used as a convenient estimate for the mean C_i/C_a, also being equivalent to C_c/C_a. This ratio C_i/C_a is considered an important physiological parameter because it reflects the balance between the degree of stomatal aperture and photosynthetic capacity. Hence, a change in C_i/C_a, reflected in a change in $\Delta^{13}C$, may be caused either by variations in photosynthetic capacity or in stomatal conductance, or both. Since the flux of water transpired shares, at least partly, the same path with CO_2 diffusion through stomatal pores (although in the opposite direction), variations in stomatal conductance will strongly affect the WUE of plants. In fact, instantaneous WUE_i, derived from gas exchange and also called the transpiration efficiency, defined as the instantaneous ratio of carbon gain to water loss, is defined as

$$WUE_i = \frac{A}{E} = \frac{1}{1.6}\frac{C_a(1-C_i/C_a)}{(e_i-e_a)} \qquad (9.2)$$

where A is the CO_2 assimilation rate, E is the transpiration rate and e_i and e_a indicate the vapour pressures inside the leaf and in the atmosphere respectively. The factor 1.6

derives from the ratio of binary diffusivity of CO_2 and water vapour in air. Seasonal water use, WUE_s is linked to WUE_i, when the proportion of carbon fixed and respired at night or by non-photosynthetic organs at any time, and the proportion of water transpired at night or by organs other than leaves during night and day are taken into account (Farquhar *et al.*, 1989).

From Equation 9.2 it is evident that WUE_i is linearly inversely dependent on C_i/C_a and, hence, on $\Delta^{13}C$. Therefore, C_i/C_a and $\Delta^{13}C$ are seen as indicators of the set point for gas exchange in C_3 plants (Ehleringer, 1993). The relationship between $\Delta^{13}C$ and C_i/C_a has been demonstrated in several species and in different conditions (for a review see Brugnoli and Farquhar, 1998; Farquhar *et al.*, 1989).

9.3 WUE and plant productivity

On the basis of the relationships between $\Delta^{13}C$ and C_i/C_a and WUE, one would expect that, at least under water-limiting conditions, genotypes with low $\Delta^{13}C$ should have higher WUE and higher productivity compared to genotypes with high $\Delta^{13}C$. Accordingly, we have found a negative correlation between $\Delta^{13}C$ and WUE in several experiments with sunflower genotypes grown in different experimental conditions. Figure 9.1 shows the typical inverse relationship between $\Delta^{13}C$ measured in leaf dry matter and WUE determined gravimetrically in different sunflower genotypes, grown either under well-watered conditions or under drought. These results are in good agreement with previous reports for sunflower (Lauteri *et al.*, 1993; Virgona *et al.*, 1990; Virgona and Farquhar, 1996) and for numerous other species (e.g., Farquhar and Richards, 1984; Hall *et al.*, 1993; Hubick and Farquhar, 1989; Menéndez and Hall, 1995; Wright *et al.*, 1988). However, several studies have indicated that $\Delta^{13}C$ is sometimes directly related to plant biomass or grain yield (e.g. Acevedo, 1993; Condon *et al.*, 1987; Craufurd *et al.*, 1991). A possible explanation for these conflicting results is

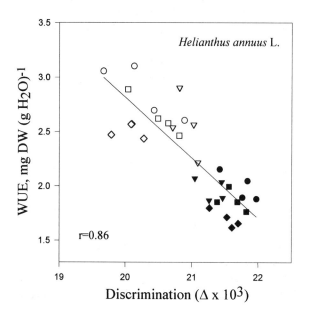

Helianthus annuus L.

Figure 9.1. *Relationship between plant water-use efficiency (WUE) and carbon isotope discrimination ($\Delta^{13}C$) in leaf dry matter in four sunflower genotypes exposed to drought or fully irrigated conditions (E. Brugnoli, M. Lauteri, M.C. Guido and L. Spaccino, unpublished results). Plants were grown in pots inside a glasshouse under the natural photoperiod. Closed symbols: fully irrigated controls; open symbols; plants subjected to drought. Different symbols correspond to different genotypes.*

that in the latter experiments, water was not limiting. Under well-watered conditions, genotypes with high $\Delta^{13}C$ would show high stomatal conductance and therefore, high photosynthesis rates and high yield.

However, a positive correlation between $\Delta^{13}C$ and yield does not necessarily mean that $\Delta^{13}C$ and WUE are positively correlated, as speculated by Condon *et al.* (1987). Under conditions of optimal water availability, WUE may not be the critical feature for determining plant productivity. In the same experiment with sunflower, where an inverse correlation was found between $\Delta^{13}C$ and WUE (Figure 9.1), different correlations were observed between biomass accumulation and $\Delta^{13}C$. While plant biomass was negatively correlated ($r=0.66$) with $\Delta^{13}C$ among genotypes grown under drought conditions, a positive correlation, although weak ($r=0.33$), was found when those genotypes were grown under well-watered conditions. Similar correlations were observed between harvest index and $\Delta^{13}C$ (E. Brugnoli and M. Lauteri, unpublished results). These results indicate that, under well-watered conditions, genotypes with higher WUE may be not the most productive. Hence, selection for low $\Delta^{13}C$ may be promising under drought, but not necessarily where water is not limiting, possibly leading to the exclusion of high-yielding genotypes. Of course, genotypes with increased WUE mostly attributable to increased photosynthetic capacity (cf. Hubick *et al.*, 1986) would be highly desirable both under drought and well-watered conditions. However, in most of the cases studied, variations in WUE and $\Delta^{13}C$ appear to be mostly related to changes in stomatal conductance. Therefore, it is concluded that selection for both high and low $\Delta^{13}C$ may be the most appropriate strategy for increasing productivity under drought-prone environments.

Environmental parameters may also affect the relationship between $\Delta^{13}C$ and WUE. From Equation 9.2 it is evident that this relationship is strongly affected by the vapour pressure difference (VPD) between the leaf and the surrounding atmosphere, and changes in VPD may lead to dramatic changes in the correlation between $\Delta^{13}C$ and WUE. In glasshouse experiments, with rice grown under different environmental conditions, we obtained no correlation between $\Delta^{13}C$ and WUE. This result was entirely attributable to differences in VPD. In fact, when WUE was multiplied by VPD to obtain the intrinsic WUE (the term introduced by Tanner and Sinclair, 1983, and referred to as κ by Hubick and Farquhar, 1989) a strong negative correlation was obtained between this derived term and $\Delta^{13}C$ ($r=0.84$, A. Scartazza, M. Lauteri and E. Brugnoli, unpublished results). These studies introduce further difficulties in comparing genotypes grown in different environments. The study of oxygen isotopes in plant organic material may be extremely useful for investigating the effects of variation in environmental parameters on WUE (see Chapter 3).

9.4 Carbon isotope discrimination in soluble carbohydrates and productivity

In most studies, WUE and yield have been compared with $\Delta^{13}C$ measured in whole-plant dry matter, which gives an integration of WUE and C_i/C_a over the entire period of growth. However, in many crop species, productivity and yield may be strongly affected by conditions experienced during special developmental stages, such as flowering and grain filling. In these stages drought is known to be especially detrimental for grain yield because of increased flower sterility and reduced seed weight. Most of the carbon contributing to grain yield is assimilated during flowering and grain filling stages, although significant and variable amounts (about 20–30‰ under unstressed conditions, but up to 70‰ under drought) of the grain dry weight can derive from relocation of carbon assimilated prior to anthesis (Austin *et al.*, 1980; Craufurd *et al.*, 1991). Therefore, the study of C_i/C_a and WUE during flowering and grain filling may be especially relevant to improve drought tolerance and productivity in crop species in water-limited environments.

It is now well documented that the analysis of $\Delta^{13}C$ in soluble sugars and/or starch may be useful to assess short-term variation in C_i/C_a and WUE in C_3 species (Brugnoli *et al.*, 1988; Brugnoli *et al.*, 1994; Lauteri *et al.*, 1993). Particularly, carbon isotope discrimination measured in leaf soluble extracts, mostly represented by sucrose, or in starch has been shown to be highly correlated with the assimilation weighted average of C_i/C_a over a short-term period of about 1–2 days (Brugnoli *et al.*, 1988). This relationship has been validated in a range of species (e.g. poplar, bean, sunflower, cotton, chestnut, wheat and rice) and under different environmental conditions (drought, salinity, high VPD, increased CO_2), both under glasshouse or field conditions (Brugnoli and Björkman, 1992; Brugnoli *et al.*, 1994; E. Brugnoli, A. Scartazza, M. Lauteri and A. Augusti, unpublished results; E. Brugnoli and O. Björkman, unpublished results; C. Picon, A. Ferhi, J.-M. Guehl, personal communication). It has also been demonstrated that when leaves show rapid changes in C_i/C_a, $\Delta^{13}C$ in soluble carbohydrates also changes quite rapidly. However, $\Delta^{13}C$ measured in structural carbon is not affected significantly by short-term changes in C_i/C_a and fluctuations in other environmental parameters (Brugnoli *et al.*, 1994). For example, only after a prolonged period of drought is a significant change in the isotopic signature of whole-plant dry matter recorded, while leaf sugars, with a much faster turn-over, are affected even by variations in C_i/C_a during a short-term period of hours (Brugnoli *et al.*, 1988). Therefore, the analysis of $\Delta^{13}C$ in soluble carbohydrates may be extremely useful for studying short-term variation in WUE.

The analysis of $\Delta^{13}C$ in soluble carbohydrates has been used to study WUE and drought tolerance in several species. In recent experiments, carbon isotope discrimination measured in leaf and peduncle carbohydrates has been found to represent a reliable indicator of drought tolerance in upland rice. In field and greenhouse experiments with several upland rice genotypes, a negative correlation between $\Delta^{13}C$ measured in sugars extracted from peduncles and grain yield has been found (B. Da Silveira Pinheiro, E. Brugnoli, R.B. Austin, M.P. do Carmo and A. Scartazza, unpublished results). This relationship was significant both when plants were fully-irrigated, and when the same genotypes were exposed to a 15 day period of drought starting from panicle emergence-flowering. In contrast, in the same experiment, no clear relationships were observed between grain Δ and grain yield.

Further glasshouse experiments by the authors have shown that drought, imposed during panicle emergence, strongly reduced $\Delta^{13}C$ in soluble sugars extracted from leaves and stems. In contrast, the effects on $\Delta^{13}C$ of grain and of whole-plant dry matter were much less pronounced. Carbon isotope discrimination measured in grain, leaves and other structural carbon pools was positively correlated with grain yield and, generally with biomass production. However, the sign of the correlation between $\Delta^{13}C$ measured in soluble sugars extracted from rice leaves and peduncles and yield was dependent on the time when sugars were sampled in relation to imposition of stress. Soluble carbohydrates extracted from peduncles after flowering include a combination of the pool of recently fixed carbon (mostly by flag leaves) and that of carbohydrate assimilated before anthesis and relocated toward the growing grains. Our experiments have demonstrated that $\Delta^{13}C$ measured in peduncle sugars is highly correlated with an assimilation weighted average of C_i/C_a integrated over a period of about 10–15 days. Figure 9.2 shows the relationship between peduncle sugar $\Delta^{13}C$ and the average of C_i/C_a, weighted by assimilation rate, over a period of 15 days after flowering. It is evident that the relationship across different genotypes is rather strong, both under irrigated and drought conditions. The correlation coefficient for the pooled data (drought and control) was even higher ($r=0.88$). Hence, $\Delta^{13}C$ of this carbon pool gives an intermediate integration time of C_i/C_a between that of leaf sugars and that of cellulose. We have also found that carbon isotope discrimination in peduncle carbohydrates was correlated with flower fertility, which is influenced by drought, plant water potential and abscissic acid (unpublished results). Therefore, the analysis of $\Delta^{13}C$ in peduncle sugars may be particularly relevant for studying photosynthesis and water relations in plants during the period of anthesis, which is crucial for flower fertility and seedset.

9.5 Soluble carbohydrate $\Delta^{13}C$ and CO_2 mesophyll conductance

Carbon isotope discrimination in C_3 leaves depends on several fractionation processes, and Equation 9.2 represents a useful simplification of the model described in detail by

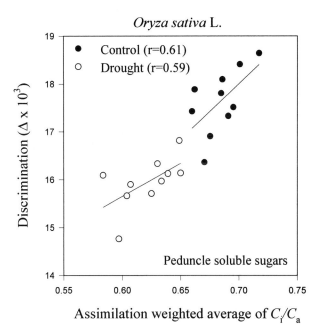

Figure 9.2. Relationship between carbon isotope discrimination measured in peduncle ethanol-soluble sugars and C_i/C_a in 10 upland rice (Oryza sativa L.) genotypes exposed to control condition or to a period of drought starting from panicle emergence-flowering. Potted plants were grown inside a glasshouse. The values of C_i/C_a were calculated as assimilation weighted averages over a period of about 15 days from the start of drought imposition. Each data point represents a different genotype. Closed symbols: fully irrigated controls; open symbols: genotypes exposed to drought. The fitted linear regressions to the different treatments are statistically significant. The significance level for the regression fitted to the pooled data (control and drought) was higher ($r=0.88$).

Farquhar *et al.* (1989). However, substantial differences between carbon isotope discrimination measured on-line and that expected on the basis of Equation 9.1 were observed in several species, and specially in woody plants (von Caemmerer and Evans, 1991; Evans and von Caemmerer, 1996; Evans *et al.*, 1986; Lloyd *et al.*, 1992). The deviation between observed and expected $\Delta^{13}C$ has been attributed mostly to a substantial drop in the CO_2 concentration between the substomatal cavities and the sites of carboxylation, due to low internal conductance (g_i) to CO_2 diffusion (also called mesophyll conductance). A method comparing predicted and observed on-line $\Delta^{13}C$ has been developed to estimate g_i and used in a range of species (e.g. von Caemmerer and Evans, 1991; Evans *et al.*, 1986; Lloyd *et al.*, 1992; see Chapter 8). As discussed above, $\Delta^{13}C$ in soluble carbohydrates ($\Delta^{13}C_{sug}$) can be used to detect temporal variation in photosynthetic capacity and in the conductance to CO_2 diffusion. Deviations between the expected discrimination ($\Delta^{13}C_i$) and that measured in soluble carbohydrates were also observed in some species, including cotton, sunflower and chestnut (E. Brugnoli and M. Lauteri, unpublished data). These deviations were comparable with those observed on the same species using on-line discrimination (Lauteri *et al.*, 1997). Hence, the equation described by Lloyd *et al.* (1992) and previously used with on-line discrimination ($\Delta^{13}C_{obs}$) was adapted to calculate the internal conductance to CO_2 diffusion using measurements of $\Delta^{13}C_{sug}$ in leaf soluble carbohydrates (E. Brugnoli and M. Lauteri, unpublished results). Hence, g_i was calculated for each leaf sample as

$$g_i = \frac{(b - e_s - a_1)\dfrac{A}{C_a}}{(\Delta^{13}C_i - \Delta^{13}C_{sug}) - \dfrac{f\Gamma^*}{C_a}} \tag{9.3}$$

where e_s (approximately 1.1‰ at 25°C) is the fractionation occurring when CO_2 enters in solution and a_1 (=0.7‰) is the fractionation during diffusion in the liquid phase, $\Delta^{13}C_i$ and $\Delta^{13}C_{sug}$ are discrimination predicted from Equation 9.2 and that observed in leaf soluble sugars respectively, f is the fractionation associated with photorespiration and Γ^* is the compensation point in the absence of dark respiration. This equation assumes that there is no significant fractionation during dark respiration, as recently shown in protoplasts and intact leaves (Lin and Ehleringer, 1997; Gillon and Griffiths, 1997). Further uncertainties in the calculation of g_i using Equation 9.3 are represented by the values of b and f. The estimation of internal conductance is rather sensitive to changes in the values of these parameters, which may be estimated from a series of on-line measurements on the same leaf (see Chapter 8), but not using single destructive measurement of soluble carbohydrate $\Delta^{13}C$. The sensitivity of g_i to changes in b and f would be maximum at high g_i, decreasing at low conductances. As stated above, the actual *in vivo* value of b in C_3 plants is thought to be close to 28‰, although higher values have also been reported (Gillon and Griffiths, 1997). However, f has been often assumed to be close to zero (von Caemmerer and Evans, 1991; Evans *et al.*, 1986), while a value of 7‰ has also been reported (Rooney, 1988). More recently, Gillon and Griffiths (1997) have shown that f can vary among species, with values of about 0.5‰ in bean and 3.3‰ in wheat (see Chapter 8). According to the latter report, also based on other concurrent on-line measurements, values of $b=28‰$ and $f=3.5‰$ were used for the determination of g_i. While changes in these parameters would affect the absolute value of g_i, the comparison within species and growth conditions will be still valid. Hence, soluble sugar $\Delta^{13}C$ can be used to estimate internal conductance to CO_2 diffusion.

Because of the purification procedure used to remove amino acid and organic acids after soluble-sugar extraction (Brugnoli et al., 1988), the leaf carbohydrates obtained may be considered to be representative of the initial C_3 products of photosynthesis, not including the products of anaplerotic CO_2 fixation by PEP carboxylase in C_3 plants. Therefore, sucrose should show a discrimination ($\Delta^{13}C_3$) which is greater than that of total CO_2 uptake as measured on-line ($\Delta^{13}C_{obs}$). Brugnoli et al. (1988) demonstrated that this offset in the discrimination of sucrose compared to the overall discrimination during photosynthesis is expressed by the equation

$$\Delta^{13}C_3 = \Delta^{13}C_{obs} + \beta \, (b_3 - b_4) \tag{9.4}$$

where β is the proportion of carbon which is fixed by phosphoenolpyruvate carboxylase (PEPC) in C_3 species, and b_3 and b_4 are the fractionations associated with ribulose-1,5-bisphosphate and PEPC carboxylations, respectively. Figure 9.3 shows the relationships between C_i/C_a and $\Delta^{13}C_{obs}$ and $\Delta^{13}C_3$ and the relative offset between these two parameters. We calculated the internal conductance using $\Delta^{13}C_i$ predicted from Equation 9.1 or using the value of $\Delta^{13}C$ according to Equation 9.4. Then we compared

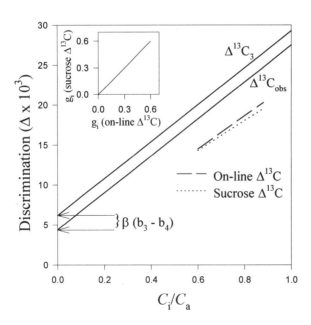

Figure 9.3. Relationship between C_i/C_a and carbon isotope discrimination in C_3 leaves in the initial C_3 products of photosynthesis and that associated to the overall CO_2 uptake, including the CO_2 fixed by PEPC. The solid lines represent the discrimination expected on the basis of the theory in the initial C_3 products of photosynthesis ($\Delta^{13}C_3$) and in the total CO_2 uptake ($\Delta^{13}C_{obs}$), respectively. The shift between these parameters is equal to $\beta(b_3-b_4)$. The broken line is the on line $\Delta^{13}C$ and the dotted line is the discrimination in sucrose, both measured in leaves showing relatively low internal conductance to CO_2 diffusion (g_i). The inset shows the relationship between g_i (mol m^{-2} s^{-1}) calculated from on- line $\Delta^{13}C$ and g_i calculated from sucrose $\Delta^{13}C$ according to Equation 9.3 with b=28‰ and f=3.5‰.

these two methods with the results obtained by conventional on-line $\Delta^{13}C$ measurements. Our results showed that the values of g_i calculated from soluble sugar $\Delta^{13}C$ using Equation 9.1 were almost perfectly matched by those calculated from on-line measurements (inset of Figure 9.3). Figure 9.3 also shows typical relationships between C_i/C_a and $\Delta^{13}C$ measured on-line and in sugars: these results are quite similar, with no significant offset. Therefore, we conclude that $\Delta^{13}C_{sug}$ is not significantly greater than $\Delta^{13}C_{obs}$ and Equation 9.4 is not necessary for calculating g_i. There is no obvious explanation for this finding, however our recent results are in good agreement with those reported previously for other species (Brugnoli *et al.*, 1988). Each newly fixed carbon is effectively diluted 1 in 3 by existing carbon skeletons, and it is possible that the carbon of initial C_3 products of photosynthesis is rapidly exchanged with that of organic acid and amino acids. Another possibility is that the isotopic signature by PEP carboxylase in organic acids is lost or reduced because of subsequent decarboxylations.

The analysis of mesophyll conductance described above gives a longer integration time compared with the estimation obtained from on-line $\Delta^{13}C$, which may be regarded as an instantaneous measurement. However, the analysis of soluble sugar $\Delta^{13}C$ can be used to detect seasonal variation in g_i in C_3 leaves. Furthermore, using gas exchange measurements and samples for $\Delta^{13}C$ analysis in sugars, several measurements can be taken in the same day either in the controlled environment or in the field, whereas on-line $\Delta^{13}C$ measurements are difficult to perform. This method was tested on several species, with different photosynthetic capacity and leaf anatomy, exposed to different environmental conditions. Substantial variations in g_i have been demonstrated in several plant species, according to previous reports (von Caemmerer and Evans, 1991; Evans *et al.*, 1986, Lloyd *et al.*, 1992; Loreto *et al.*, 1992). Particularly low g_i values were found in woody species. Furthermore, the rate of CO_2 assimilation was linearly dependent on g_i in all species, indicating that the draw-down between C_i and C_c was more or less constant for these particular species. Figure 9.4 shows the relationship between assimilation rate and g_i in three

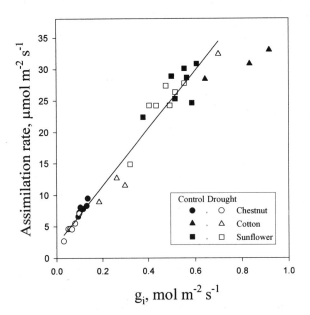

Figure 9.4. Relationship between CO_2 net assimilation rate and internal conductance to CO_2 diffusion (g_i) in sunflower, cotton and chestnut leaves under irrigated conditions or exposed to long term-drought. Internal conductance was calculated according to Equation 9.3 with b = 28‰ and f = 3.5‰. Open symbols: drought; closed symbols: control. The regression equation ($y=2.3+45.9x$, $r^2=0.96$) is fitted by least square method, excluding the two data points for cotton with g_i higher than 0.8 mol m^{-2} s^{-1}.

different species exposed to either well-watered conditions or drought. The relationship between A and g_i was linear and showed a departure from linearity at very high g_i. It is evident that chestnut showed lowest values of g_i compared with sunflower and cotton. The values of g_i found for chestnut are in good agreement with previous reports using a different method based on chlorophyll fluorescence for estimating g_i (Epron *et al.*, 1995).

Figure 9.4 also shows that drought stress always caused a significant decline in g_i, in all species analysed. Therefore, under stress conditions, a reduction in mesophyll conductance might be regarded as an acclimatory response of photosynthetic apparatus. Also changes in nitrogen content and possibly in RUBISCO might contribute to reductions in g_i and in photosynthetic capacity.

Variations in mesophyll conductance were observed during different developmental stages. We have observed a marked decline in g_i during ontogeny in rice (unpublished results). These changes may also be associated to leaf aging and related variations in morphological and anatomical characteristics and in leaf nitrogen content, according to previous reports (Loreto *et al.*, 1994). The implications of variations in mesophyll conductance on photosynthesis have been covered by recent reviews (Evans and von Caemmerer, 1996; Parkhurst, 1994); however, the ecological and functional significance of mesophyll conductance in terms of water economy and adaptation to harsh environments deserves more attention. In particular, the possible role of low mesophyll conductance on water conservation at the level of mesophyll cells needs further investigation, although the problems of quantifying g_i at low rates of assimilation under stress have already been discussed in Chapter 8.

Certainly different ecological significance should also be ascribed to the CO_2 transfer conductance in lichens. These poikilohydric organisms, without the ability to actively regulate water loss, have thallus water content largely dependent on atmospheric humidity. Lichens with different photobiont associations exhibit different water requirements for the activation of photosynthesis. However, if water is necessary to take up CO_2, it can represent a significant barrier to gas diffusion. The resistance to CO_2 diffusion would be also dependent on the thallus structure, and on the relative proportion of mycobiont and algal partner, particularly in those lichens containing green algae but lacking a CO_2-concentrating mechanism (CCM), which is present in all cyanobiont lichens and some other green algal lichens. Recently, it has been demonstrated that spatial variation in carbon isotope discrimination across the thallus of several lichen species reflected differences in the CO_2 transfer resistance, induced by age-related variations in thallus structure and anatomy and in the concentration of photobiont partner (Máguas and Brugnoli, 1996). In particular, the marginal younger part of the thallus of *Lobaria pulmonaria* (without a CCM) showed higher $\Delta^{13}C$, associated with higher chlorophyll content and reduced thickness of the upper and lower cortex. Therefore, variations in $\Delta^{13}C$ were attributed to changes in diffusion conductance, which is thought to be inversely related to the cortex thickness and porosity and to its hydration state (Máguas and Brugnoli, 1996). Therefore, these layers of mycobiont biomass may play an important role in photosynthetic efficiency and water retention in these organisms.

9.6 Concluding remarks

Carbon isotope discrimination, when measured in soluble carbohydrates, is associated with measurements of $\Delta^{13}C$ in structural carbon pools, and can be useful for estimating variation in photosynthetic performances and WUE integrated over different time- scales. In particular, the analysis of $\Delta^{13}C$ in different plant compounds and pools allows the estimation of the average C_i/C_a and WUE over integration-times ranging from one day (leaf sugars), to weeks (stem sugars) and to the plant life (structural carbon). The estimation of the conductance to CO_2 diffusion in the leaf mesophyll using the analysis of $\Delta^{13}C$ in leaf carbohydrates appears particularly promising for understanding the significance of g_i in different plant types and life forms, and the importance in the ecological context of adaptation to the environment. Further studies are needed for investigating possible variations in the fractionation during photorespiration and their effects on the estimation of internal conductance (see Chapter 8).

It can be concluded that the analysis of $\Delta^{13}C$ in soluble carbohydrates is a reliable indicator of C_i/C_a changes and in drought tolerance. This analysis can be relevant for integrating short-term changes in C_i/C_a, which would otherwise be masked by the long-term $\Delta^{13}C$ in structural carbohydrates. Therefore, carbon isotope discrimination in soluble sugars can be used to predict WUE and, possibly, crop productivity, particularly when the timing of sampling is targetted towards drought-susceptible developmental sequences during flowering and seedset. Analysis of carbon isotope discrimination in plant carbohydrates may then be extremely relevant in ecological studies and in breeding programmes for improved tolerance to drought and other environmental stresses.

Acknowledgements

The authors thank Luciano Spaccino for skillful assistance in carbon isotope analysis and in sample preparation. This work was partially supported by CNR-Special Project RAISA, Sub-project 2, paper No. 2861. The financial support from the Commission of European Union, DG XII, Contract TS3-CT93*0200, is also acknowledged.

References

Acevedo, E. (1993) Potential of carbon isotope discrimination as a selection criterion in barley breeding. In: *Stable Isotopes and Plant Carbon–Water Relations* (eds J.R. Ehleringer, A.E. Hall and G.D. Farquhar). Academic Press, San Diego, pp. 155–172.

Austin, R.B., Morgan, C.L., Ford, M.A. and Blackwell, R.D. (1980) Contributions to grain yield from pre-anthesis assimilation in tall and dwarf barley phenotypes in two contrasting seasons. *Ann. Bot.* **45**, 309–319.

Borland, A.M, Griffiths, H., Broadmeadow, M.S.J., Fordham, M.C., Maxwell, C. (1994) Carbon isotope composition of biochemical fractions and the regulation of carbon balance in leaves of the C_3-crassulacean acid metabolism intermediate *Clusia minor* L. growing in Trinidad. *Plant Physiol.* **106**, 493–501.

Brugnoli, E. and Björkman, O. (1992) Growth of cotton under continuous salinity stress: influence on allocation pattern, stomatal and non-stomatal components of photosynthesis and dissipation of excess light energy. *Planta* **187**, 335–347.

Brugnoli, E. and Farquhar, G.D. (1998) Photosynthetic fractionation of carbon isotopes. In: *Photosynthesis: physiology and metabolism, Advances in Photosynthesis.* (eds R.C. Leegood, T.D. Sharkey and S. von Caemmerer). Kluwer Academic Publishers, in press.

Brugnoli, E., Lauteri, M. and Guido, M.C. (1994) Carbon isotope discrimination and photosynthesis: response and adaptation to environmental stress. In: *Plant Sciences* (eds Y. De Kouchkovsky and F. Larher). SFVP Universite de Rennes, pp. 269–272.

Brugnoli, E., Hubick, K.T., von Caemmerer, S., Wong, S.C. and Farquhar, G.D. (1988) Correlation between the carbon isotope discrimination in leaf starch and sugars of C_3 plants and the ratio of intercellular and atmospheric partial pressures of carbon dioxide. *Plant Physiol.* **88**, 1418–1424.

von Caemmerer, S. and Evans, J.R. (1991) Determination of the average partial pressure of CO_2 in chloroplasts from leaves of several C_3 plants. *Aust. J. Plant Physiol.* **18**, 287–305.

Condon, A.G., Richards, R.A. and Farquhar, G.D. (1987) Carbon isotope discrimination is positively correlated with grain yield and dry matter production in field-grown wheat. *Crop Sci.* **27**, 996–1001.

Craufurd, P.Q., Austin, R.B., Acevedo, E. and Hall, M.A. (1991) Carbon isotope discrimination and grain-yield in barley. *Field Crops Res.* **27**, 301–313.

Ehleringer, J.R. (1993) Carbon and water relations in desert plants: an isotopic perspective. In: *Stable Isotopes and Plant Carbon–Water Relations* (eds J.R. Ehleringer, A.E. Hall, G.D. Farquhar). Academic press, San Diego, pp. 155–172.

Epron, D., Godard, D., Cornic, G. and Genty, B. (1995) Limitation of net CO_2 assimilation rate by internal resistances to CO_2 transfer in the leaves of two tree species (*Fagus sylvatica* L. and *Castanea sativa* Mill.). *Plant, Cell Environ.* **18**, 43–51.

Evans, J.R., Sharkey, T.D., Berry, J.A. and Farquhar, G.D. (1986) Carbon isotope discrimination measured concurrently with gas exchange to investigate CO_2 diffusion in higher plants. *Aust. J. Plant Physiol.* **13**, 281–292

Evans, J.R. and Caemmerer, S. von. (1996) Carbon dioxide diffusion inside leaves. *Plant Physiol.* **110**, 339–346.

Farquhar, G.D. and Richards, R,A. (1984) Isotopic composition of plant carbon correlates with WUE of wheat genotypes. *Aust. J. Plant Physiol.*, **11**, 359–552

Farquhar, G.D, Ehleringer, J.R. and Hubick, K.T. (1989) Carbon isotope discrimination and photosynthesis. *Ann. Rev. Plant Physiol. Plant Mol. Biol.* **40**, 503–537.

Farquhar, G.D., O'Leary, M.H. and Berry, J.A. (1982) On the relationship between carbon isotope discrimination and the intercellular carbon dioxide concentration in leaves. *Aust. J. Plant Physiol.* **9**, 121–137.

Gillon, J.S. and Griffiths, H. (1997) The influence of (photo)respiration on carbon isotope discrimination in plants. *Plant Cell Environ.* **20**, 1217–1230.

Hall, A.E., Ismail, A.M. and Menendez, C.M. (1993) Implications for plant breeding of genotypic and drought-induced differences in water-use efficiency, carbon isotope discrimination, and gas exchange. In: *Stable Isotopes and Plant Carbon–Water Relations* (eds J.R. Ehleringer, A.E. Hall and G.D. Farquhar). Academic Press, San Diego, pp. 349–369.

Hubick, K.T. and Farquhar, G.D. (1989) Carbon isotope discrimination and the ratio of carbon gained to water lost in barley cultivars. *Plant Cell Environ.* **12**, 795–804.

Hubick, K.T., Farquhar, G.D. and Shorter, R. (1986) Correlation between water-use efficiency and carbon isotope discrimination in diverse peanut (*Arachis*) germoplasm. *Aust. J. Plant Physiol.* **13**, 803–816.

Lauteri, M., Scartazza, A., Guido, M.C. and Brugnoli, E. (1997) Comparison of photosynthetic capacity, carbon isotope discrimination and mesophyll conductance among provenances of *Castanea sativa* adapted to different environments. Submitted to *Functional Ecology*, in press.

Lauteri, M., Brugnoli, E. and Spaccino, L. (1993) Carbon isotope discrimination in leaf soluble sugars and in whole-plant dry matter in *Helianthus annuus* L. grown under different water conditions. In: *Stable Isotopes and Plant Carbon–Water Relations* (eds J.R. Ehleringer, A.E. Hall and G.D. Farquhar). Academic Press, San Diego, pp. 93–108.

Lin, G. and Ehleringer, J.R. (1997) Carbon isotope fractionation does not occur during dark respiration in C_3 and C_4 plants. *Plant Physiol.* **114**, 391–394.

Lloyd, J., Syvertsen, J.P., Kriedemann, P.E. and Farquhar, G.D. (1992) Low conductances for CO_2 diffusion from stomata to sites of carboxylation in leaves of woody species. *Plant, Cell Environ.* **15**, 873–899.

Loreto, F., Harley, P.C., Di Marco, G. and Sharkey, T.D. (1992) Estimation of mesophyll conductance to CO_2 flux by three different methods. *Plant Physiol.* **98**, 1437–1443.

Loreto, F., Harley, P.C., Di Marco, G. and Sharkey T.D. (1994) Measurements of mesophyll conductance, photosynthetic electron transport and alternative electron sinks of field grown wheat leaves. *Photosynth. Res.* **41**, 397–403.

Máguas, C. and Brugnoli, E. (1996) Spatial variation in carbon isotope discrimination across the thalli of several lichen species. *Plant Cell Environ.* **19**, 437–446.

Menéndez, C.M. and Hall, A.E. (1995) Heritibility of carbon isotope discrimination and correlations with earliness in cowpea. *Crop Sci.* **35**, 673–678.

O'Leary, M.H. (1981) Carbon isotope fractionation in plants. *Phytochemistry* **20**, 553–567.

O'Leary, M.H., Madhavan, S. and Paneth, P. (1992) Physical and chemical basis of carbon isotope fractionation in plants. *Plant, Cell Environ.* **15**, 1099–1104.

Parkhurst, D.F. (1994) Tansley Review No. 65. Diffusion of CO_2 and other gases inside leaves. *New Phytol.* **126**, 449–479.

Passioura, J.B. (1986) Resistance to drought and salinity: avenues for improvement. *Aust. J. Plant Physiol.* **13**, 191–201.

Read, J.J., Johnson, D.A., Asay, K.H. and Tieszen L.T. (1991) Carbon isotope discrimination, gas exchange, and WUE in crested wheatgrass clones. *Crop Sci.* **31**, 1203–1208.

Rooney, M.A. (1988) Short-term carbon isotope fractionation by plants. *Ph.D. Thesis,* University of Wisconsin, Madison, WI.

Tanner, C.B. and Sinclair, T.R. (1983) Efficient water use in crop production: research or re-search. In: *Limitations to Efficient Water Use in Crop Production* (ed H. Taylor). ASA-CSSA-SSSA, Madison, WI, USA, pp. 1–28.

Virgona, J.M. and Farquhar, G.D. (1996) Genotypic variation in relative growth rate and carbon isotope discrimination in sunflower is related to photosynthetic capacity. *Aust. J. Plant Physiol.* **23**, 227–236.

Virgona, J.M., Hubick, K.T., Rawson, H.M., Farquhar, G.D. and Downes, R.W. (1990) Genotypic variation in transpiration efficiency, carbon-isotope discrimination, and carbon allocation during early growth in sunflower. *Aust. J. Plant Physiol.* **17**, 207–214.

White, J.W., Castillo, J.A. and Ehleringer, J.R. (1990) Associations between productivity, root growth and carbon isotope discrimination in *Phaseolus vulgaris* under water deficit. *Aust. J. Plant Physiol.* **17**, 189–198.

Wright, G.C., Hubick, K.T. and Farquhar, G.D. (1988) Discrimination in carbon isotopes of leaves correlates with water-use efficiency of field grown peanut cultivars. *Aust. J. Plant Physiol.* **15**, 815–825.

Oxygen-18 of leaf water: a crossroad for plant-associated isotopic signals

Dan Yakir

10.1 Introduction

The oxygen isotopic composition of plant organic matter and the CO_2, O_2 and water vapour exchanged between the plants and the atmosphere are imprinted with the ^{18}O signature of leaf water. This is a useful signal because the ^{18}O content of leaf water is unique and is influenced by environmental conditions. The hydrological and meteorological systems greatly influence the regional and local scale ^{18}O signature of the water that feed plants (Gat, 1996; see Chapter 23). Superimposed on this climatic signal is a marked ^{18}O enrichment associated with evapotranspiration in the leaves (Farris and Strain, 1978, Gonfiantini *et al.*, 1965; also see Yakir, 1992 for a review). The extent of this enrichment is influenced by temperature, humidity and the original isotopic signature, and by characteristics of the evaporating surface (Craig and Gordon, 1965).

The ^{18}O signature of leaf water is transduced to CO_2 during photosynthetic CO_2 exchange due to hydration with leaf water. This hydration/dehydration process, catalysed in the chloroplasts by carbonic anhydrase (CA), allows for oxygen isotopic exchange between CO_2 and water (DeNiro and Epstein, 1979). Only a certain fraction of the CO_2 that diffuses into leaves is fixed into carbohydrates while the rest escapes back to the atmosphere carrying the ^{18}O signature of leaf water. The quantitative use of this isotopic signal in atmospheric CO_2 for assessing plant–atmosphere interactions has already been demonstrated (Ciais *et al.*, 1997a,b; Farquhar *et al.*, 1993; Flanagan and Varney, 1995; Francey and Tans, 1987; Friedli *et al.*, 1987; Yakir and Wang, 1996; see Chapters 12 and 24), but is still critically dependent on estimation of the source isotopic signature in leaf water.

Water in leaf chloroplasts is also the substrate for the water-splitting reaction in photosynthesis and controls the ^{18}O signature of the photosynthetically produced O_2 (Guy *et al.*, 1987; 1993). On regional and global scales this signal represents the terrestrial contribution to the so-called Dole effect (the displacement of 23.5‰ in the $\delta^{18}O$ value of

Stable Isotopes, edited by H. Griffiths.
© 1998 BIOS Scientific Publishers Ltd, Oxford.

atmospheric O_2 with respect to that of mean ocean water; Bender *et al.*, 1985; 1994). The Dole effect can be a useful indicator of changes in the balance between terrestrial and oceanic productivity because on land the ^{18}O signature of O_2 is influenced by leaf water enrichment, resulting in proportional ^{18}O enrichment of atmospheric O_2. Here too, reliable estimates of the ^{18}O signature of O_2 produced on land are still restricted by uncertainties in estimating the source ^{18}O signal within leaves.

The ^{18}O signature of plant organic matter is imprinted with that of chloroplast water because of the aforementioned isotopic exchange between CO_2 and water, and subsequent exchange between leaf water and organic matter (Sternberg and DeNiro, 1983; Sternberg *et al.*, 1986; Yakir and De Niro, 1990; Chapter 3). Variations in the isotopic signature of plant organic matter are therefore useful indicators of changes in environmental conditions in which the plant grew. If this signal is dominated by changes in the ^{18}O signature of the source water feeding the plant, it can be related to changes in temperature (Edwards and Fritz, 1986; Epstein *et al.*, 1977; see Chapter 18). If, however, the variations in the ^{18}O signature of the organic matter are dominated by changes in the extent of leaf water enrichment, it can be related to relative humidity (Burk and Stuiver, 1981; Edwards and Fritz, 1986; Lipp *et al.*, 1996; Yakir *et al.*, 1994b). One of the advantages of the approaches using isotopes lies in the fact that plant organics, such as cellulose, are accumulated over the lifetime of the leaf or the plant, providing a well preserved, time-integrated record (e.g. Edwards and Fritz, 1986, 1988; Feng and Epstein, 1994; see Chapter 18).

The ^{18}O signature of leaf water is therefore an important crossroads for both utilisation and interpretation of this powerful tracer in a wide range of environmental studies. This chapter aims to briefly review the various factors influencing the ^{18}O content of leaf water, and attempts to point out the major sources of uncertainty still involved in estimating this isotopic signature.

10.2 Evaporative enrichment

Water is taken up by the roots and transported with relatively little isotopic modification to the leaves (Figure 10.1). In leaves, water is lost to the atmosphere in the process of evapotranspiration, coupled with the leaf CO_2 and O_2 exchange. During evaporation, water molecules containing the lighter isotopes, ^{1}H and ^{16}O, evaporate slightly faster than water molecules having the heavy isotopes, $^{2}H(D)$ and ^{18}O. Moreover, provided the air into which the vapour is transported is not fully turbulent, the water molecules containing the lighter isotopes will also diffuse faster away from the liquid–vapour interface. These effects are quantitatively described for well mixed, constant volume systems by the model developed by Craig and Gordon (1965):

$$\delta_E = \frac{\alpha^* \cdot \delta_L - h^* \cdot \delta_a - \epsilon_{eq} - (1 - h^*) \cdot \epsilon_k}{(1 - h^*) + (1 - h^*) \cdot \epsilon_k / 1000} \cong \frac{\delta_L - h^* \cdot \delta_a - \epsilon_{eq} - (1 - h^*) \cdot \epsilon_k}{1 - h^*} \quad (10.1)$$

where, $\delta = [(R_{sample}/R_{standard}) - 1] \times 10^3 ‰$ (R is the ratio $^{18}O/^{16}O$, or $^{2}H/^{1}H$ and the conventional standard is SMOW), and subscripts E, L and a stand for evaporating water, liquid water and ambient air moisture respectively; α^* is the fractionation factor for the liquid–vapour equilibrium and, for ^{18}O is related by $\epsilon_{eq} = (\alpha^* - 1) \times 10^3$ (Majoube 1971); ϵ_k is the kinetic fractionation (which depends on the molecular diffusion of the

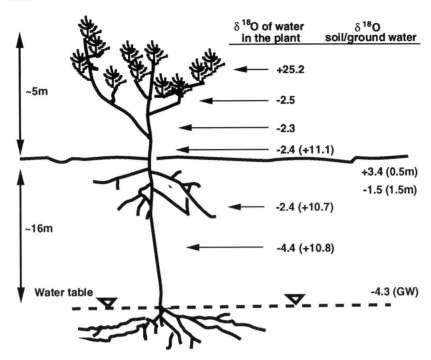

Figure 10.1. Variations in the $\delta^{18}O$ values of water in a Tamarix jordanis tree and its relationship to the $\delta^{18}O$ values of soil and ground water. Water from the stem and roots was extracted from cores that were separated to inner xylem sap and outer phloem layers (values in parentheses). The tree with a dual root system, was growing on a partially collapsed sand dune in the central Negev, Israel and showed, at the time of sampling, utilisation of both ground and soil water. Large enrichment of water was observed in tissue associated with the phloem sap at the outer part of the stem and root (adapted from Adar et al., 1995).

isotopic species, and is influenced by the aerodynamic nature of the boundary condition; Merlivat, 1978); and h^* was traditionally used for relative humidity of the ambient air at the surface temperature (and can be substituted for leaves by e_i/e_a, the ratio of the vapour pressures inside the leaf and in the atmosphere).

The evaporation of depleted vapour (δ_E) from the water surface results in ^{18}O enrichment of the liquid water left behind, δ_L. All else being constant, the increase in δ_L will obviously be followed in time with an increase in δ_E. This process will continue until isotopic steady state conditions are attained when δ_E is identical to that of the input water, δ_i, and δ_L is constant. The steady state value, δ_{ss}, can then be estimated by solving equation 10.1 for δ_L, substituting δ_i for δ_E, as approximated by:

$$\delta_{ss} = \delta_s + \epsilon_{eq} + \epsilon_k + h^* \times (\delta_a - \delta_s - \epsilon_k)$$ (10.2)

Equation 10.2 is widely applied to leaves assuming they represent a small, constant volume water body with a rapid turnover allowing it to be near isotopic steady state with respect to environmental conditions at any time. There are, however, uncertainties involved in almost all the parameters used in the above model, the importance of which depends on the scale of observation. Some of the uncertainty is environmentally related. Such is the case with respect to the $\delta^{18}O$ values of source water and of atmospheric vapour. In other cases, the uncertainty is more specifically related to plant characteristics. Such is the case with respect to the kinetic fractionation, and the assumptions of steady state, constant volume and homogeneity of the system.

10.3 Source water

Any isotopic enrichment occuring in leaf water is superimposed on the background isotopic signature of the input water. It is important to consider possible variations in the $\delta^{18}O$ value of the water that feeds plants (Dawson, 1993; Ehleringer and Dawson, 1992; White, 1989). Variations in the ^{18}O signature of environmental water utilised by plants are driven primarily by the isotope effects in the hydrological cycle (Craig, 1961; Epstein and Mayeda, 1953; Gat, 1996; see Chapter 23). It is useful to note that at any given location, the annual mean ^{18}O signature of precipitation has a relatively conservative value which is largely dependent on geographic location (Craig, 1961; Dansgard, 1964). Data on the isotopic composition of precipitation are generally available from global sampling networks (e.g. IAEA, 1990) or model simulations (Jouzel et al., 1987). Such annual mean values, conservative as they may be, can still integrate substantial temporal and spatial variations that can be useful for a variety of environmental studies (e.g. Dawson, 1993; Ehleringer and Dawson, 1992; Emerman and Dawson, 1996; Matsuo and Freidman, 1967; Sternberg and Swart, 1987; White, 1989; Yakir and Yechieli, 1995).

In the soil, isotopic enrichment associated with evaporation from the soil surface can introduce additional variations to δ_i. Soil evaporation often results with a characteristic and predictable isotopic gradient in the top ~ 1 m of the soil column (Allison et al., 1983; Barnes and Allison, 1983; Mathieu and Bariac, 1996a,b; Zimmerman et al., 1967, see Chapter 24). This trend influences the variations observed in the $\delta^{18}O$ values of plant water because different plants use water from different depths, providing useful information on water usage patterns in plants (Dawson, 1993; Dawson and Ehleringer, 1991; Midwood et al., 1993). Interestingly the evaporative enrichment near the soil surface can be very large (10‰ or more in the top 5–10 cm of the soil), but it is also associated with a sharp decrease in moisture content to levels which are not conducive to plant uptake. As a result, much of this extreme enrichment is ignored by the plants (e.g. Figure 10.1). The extent of the variations in the isotopic composition of the water actually utilised by plants can be assessed by measuring the isotopic composition of water in woody stems or in the xylem sap flow of plants (Dawson and Ehleringer, 1993). This is a useful approach because there is, in most cases, no change in isotopic composition of the water during uptake (Dawson, 1993; Ehleringer and Dawson, 1991, 1992; but see also Lin and Sternberg, 1993). Notably, in a recent survey of 90 plant species grown within the Jerusalem Botanical Garden (Wang, unpublished data) and fed only by local precipitation, a range as large as 8‰ was observed in the $\delta^{18}O$ values of water extracted from woody stems.

It is generally assumed that there is no modification to the isotopic composition of water within the plant conductive system before it reaches the leaf. Several exceptions to this assumption, however, have been reported. In woody plants, it was shown that relatively small changes occur only near the leaves in the green part of the stems (Dawson and Ehleringer, 1993). In Tamarix trees, an enrichment of more than 2‰ in $\delta^{18}O$ values was observed between the main branch and the twigs in the summer, but not in the winter (Adar et al., 1995). Isotopic enrichment in the stems could occur as a result of effects of evaporation, particularly if water transport is slow and small in non-woody plants. For example, variations of about 3‰ were observed in the $\delta^{18}O$ values of water along the stem of the desert shrub *Zygophyllum domsoum* (J.R.Gat, Lipp and Yakir, unpublished data) Isotopic enrichment of stem water can also be due to exchange with highly enriched leaf water flowing down in the phloem system. For example, highly enriched water was observed in parts of the stem associated with the phloem sap in Tamarix trees, resulting with a sharp gradient of more than 10‰ across the stem to deeper xylem sap layers that maintained the $\delta^{18}O$ value of the source water (Figure 10.1).

10.4 Atmospheric moisture and boundary layers

During evaporation from the leaves, there is an isotopic interaction between the liquid and the vapour phases (Craig and Gordon, 1965). As a consequence, the $\delta^{18}O$ value of the air water vapour influences that of the leaf water (cf. Equations 10.1 and 10.2). On local or regional scales, the isotopic composition of the atmospheric water vapour is influenced, in turn, by the source of the moisture and interactions with the surface during its transport (Gat, 1996; see Chapter 23). Monitoring variations in atmospheric moisture is difficult and has not been done as extensively as for precipitation water (IAEA, 1990). In the most simple case, air moisture is near equilibrium with the local precipitation water (Dansgaard, 1964; Gat, 1996; Gat and Matsui, 1991; see Chapter 23). This is often the case in continental regions, but can be markedly different in coastal areas (Tzur, 1971). In the latter case, marked variations in the $\delta^{18}O$ values of atmospheric moisture can result from changes in atmosphere sea-surface interactions and rain-out history before an air mass reaches the site of interest (e.g. Gat and Rindsberger, 1985; Leaguy *et al.*, 1983). At the field or local scales, significant variations can occur in the $\delta^{18}O$ value of atmospheric water vapour during the day, which can be associated with changes in wind direction. For example, we observed in a recent field study in Israel a pseudo-cyclic change in $\delta^{18}O$ (vapour) from about -11‰ at 7 am to about -14‰ at mid-day and back to about -11‰ at 7 PM (Figure 10.2). Such changes could not be correlated with local evapotranspiration. As indicated in Equation 10.2, the effects of such variations in δ_a values on estimates of the $\delta^{18}O$ values of leaf water depend on relative humidity, and would, for example, result with 1.5‰ variation in estimates of δ_{ss} for $h^* = 0.5$.

The large flux of evapotranspiration (ET) from vegetation can export water into the canopy and the atmospheric boundary layers with distinct isotopic signatures, modifying the local $\delta^{18}O$ values of the air moisture (Gat and Matsui, 1991). Since it is generally assumed that vegetation is near isotopic steady state with respect to evapotranspiration (see below), the $\delta^{18}O$ values of the evaporation flux must be near that of the source (soil or ground) water. Its effect on the mixed atmospheric value can, therefore, be estimated from background isotopic measurements and the increase in specific humidity. Such an approach has been used both in the planetary boundary layer (e.g. White and Gedzelman, 1984) and at the field canopy scales (e.g. Yakir *et al.*, 1990). Co-variations in the specific humidity and its $\delta^{18}O$ values within the canopy boundary layer can be useful for estimating the magnitude of the ET flux (Bariac *et al.*, 1989; Brunel *et al.*, 1992; Yakir and Wang, 1996). Moreover, such measurements have the potential to allow the partitioning of the ET flux into its leaf-transpiration and soil-evaporation components, if the two fluxes are isotopically distinct.

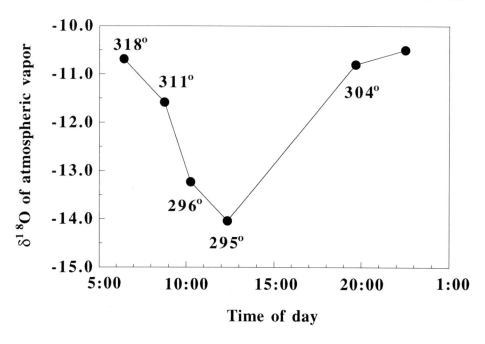

Figure 10.2. *Daily variations in the δ¹⁸O values of atmospheric moisture during August 16/17 1995 in the central Negev, Israel. Sampling was carried out from a mast extending above the canopy boundary layer of a small* Acacia saligna *plantation (8.5 m above ground and 3 m above the canopy of a 50 m long plot; wind direction is indicated for each sampling time, with N=360° ; Wang and Yakir, unpublished data).*

Although concomitant variations of $\delta^{18}O$ and humidity in the canopy boundary layer can be useful, such effects are also a source of uncertainty in estimating the $\delta^{18}O$ values of leaf water. In large scale studies it is often the background, regional data that are used to estimate the $\delta^{18}O$ value of the atmospheric moisture in conjunction with Equation 10.2. A consideration of the canopy boundary layer effects would be likely to improve such estimates, but further research is still needed to provide a better understanding of such effects (see Chapter 24).

In addition to the planetary and canopy boundary layers, all leaves have a character-istic, species-dependent, boundary layer separating the evaporating leaf surface from the turbulent atmosphere around it. Such intrinsic boundary layers also significantly influence leaf water enrichment and $\delta^{18}O$ value (Buhay *et al.*, 1996; Cooper and DeNiro, 1989). A maximum diffusional fractionation of about 28‰ (or effectively $28(1-h^*)$‰) for ^{18}O in water (Equation 10.2; Merlivat, 1978) would be expected in the fully stagnant boundary layer such as within the leaf sub-stomatal air spaces where transport of water vapour is dominated by molecular diffusion. Although no discrim-ination is expected in fully turbulent air where mass is transported solely by eddies, an intermediate effect can be expected near the leaf surface where laminar flow is more likely. In this case the ^{18}O fractionation is often estimated by the maximum diffusion effect to the 2/3 power, or about 19‰. The overall fractionation factor in the water vapour pathway from the evaporating sites in the leaves to the mixed atmosphere should reflect the weighted effects of the different boundary layers (Farquhar and Lloyd, 1993):

$$\epsilon_{k} = \frac{28r_{s} + 19r_{bl}}{r_{s} + r_{bl}} \qquad (10.3)$$

where r_{s} and r_{bl} are the leaf stomatal and boundary layer resistances to diffusion respectively.

Recently, Buhay *et al.* (1996) have argued that variations in the kinetic fractionation of leaves can be predicted from the aerodynamic properties of a leaf, and accounting for these effects can help explain many of the discrepancies between observed $\delta^{18}O$ values of leaf water and those predicted by equation 10.2. Wang (unpublished data) has recently estimated the range of kinetic fractionation among 90 plant species growing under similar conditions at a botanical garden to be between 20‰ and 26‰. The effect of such a range on the variations in the $\delta^{18}O$ value of leaf water is proportional to $(1 - h^{*})$, or about 3‰ for $h^{*} = 0.5$ (Equation 10.2).

10.5 Bulk leaf water

Equation 10.2 as presented above, indicates that for a given location and time where δ_{i} and δ_{a} are fairly constant, the dominant factor influencing the $\delta^{18}O$ value of leaf water, δ_{LW}, is relative humidity. This is clearly seen in the typical diel cycle in δ_{LW} (e.g. Allison *et al.*, 1985; Dongmann *et al.*, 1974; Förstel, 1978; Zundel *et al.*, 1978). Leaf water becomes enriched during the morning when humidity decreases, and more depleted in the afternoon when humidity increases. Although equation 10.2 is a good description of the natural diel cycle observed in δ_{LW}, relatively large discrepancies are observed between the predicted values and measured δ_{LW} values of bulk leaf water at any given time of the day (e.g. Allinson *et al.*, 1985; Bariac *et al.*, 1991; Farris and Strain, 1978; Flanagan *et al.*, 1991; Yakir *et al.*, 1990). It is now realised that such discrepancies may be associated with some of the underlying assumptions in the model depicted by Equation 10.2. These include the assumptions that leaves are at isotopic steady state, that leaf water has a constant volume and that leaf water is isotopically well mixed and homogeneous.

10.5.1 *Isotopic steady state*

The assumption of isotopic steady state is often made *a priori* for leaves because the ratio between the evaporative flux through the leaves and their volume is normally large (see Chapter 3). Yet, this ratio can greatly vary among plant species and environmental conditions. Since isotopic steady state for leaf water requires that the $\delta^{18}O$ values of the input and output water are the same (see Section 10.2), the attainment of steady state can be checked by directly comparing these two $\delta^{18}O$ values (Flanagan *et al.*, 1991; Wang and Yakir, 1996; Yakir, 1992; Yakir *et al.*, 1994a). Such measurements showed that although leaf water is often near isotopic steady state, the assumption may not always be strictly valid. Following changes in ambient humidity, there will be a slow approach to a new isotopic steady state in leaves with slow leaf water turn-over time, which can vary from a few minutes to several hours between plant species (Wang and Yakir, 1996). Under natural conditions, the deviations from steady state will also be influenced by the rate of change in relative humidity during the day.

Notably, the capacity of leaves to accumulate the heavy isotopes in water is of course very limited. Integrated on an annual, seasonal and perhaps even daily time-scales, leaves must therefore be near isotopic steady state. Short-term deviations from this state, as noted above, are likely to be compensated for by opposing shifts at other times. For example, in a simulated daily cycle, δ_{LW} was lower than predicted by equation 10.2 in the morning and higher in the afternoon, (Wang and Yakir, 1996). Short-term deviations from steady state may, however, be important when coupled with, for example, CO_2 gas exchange which carries the isotopic signature of leaf water. If the diurnal cycle in δ_{LW} lags behind that in the rate of CO_2 exchange, estimating the $\delta^{18}O$ values of the CO_2 based on Equation 10.2 may introduce a bias toward higher $\delta^{18}O(CO_2)$ values (see Chapters 12 and 24).

While checking for isotopic steady state at the single leaf scale under laboratory conditions is now relatively simple, carrying out such tests on a canopy scale is more demanding. It is, however, possible based on measurements of the co-variations in specific water composition and its $\delta^{18}O$ value in the canopy boundary layer (see previous section). When such measurements are obtained, a simple mixing model such as that proposed by Keeling (1961) can be applied to estimate the $\delta^{18}O$ signature of the vapour flux from the canopy and compare it with measurements of mean stem water $\delta^{18}O$ values of the underlying vegetation. It is encouraging to note that at least in canopies of highly active crop fields, such measurements confirmed that the canopies were near isotopic steady state (Yakir and Wang, 1996).

10.5.2 Constancy of leaf volume

The assumption of constant volume often assumed for leaves, is also not strictly valid. Figure 10.3 shows some examples of the changes in leaf volume of an arbitrary selection

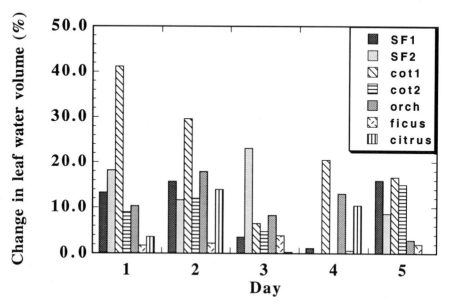

Figure 10.3. Relative changes in leaf volume (weight per unit area) during five consecutive days. Plants (unidentified varieties) of sunflower (SF), cotton (cot), orchid (orch.), ficus and citrus were sampled on campus in Rehovot Israel. SF2 and cot2 were water stressed. (Wang and Yakir, unpublished data).

of plant species. Changes in volume (as weight per cm² leaf area) ranged from almost zero up to about 40%. As expected, although day-to-day variation in the extent of leaf dehydration was observed, the sensitivity to this effect seems to be species dependent. Dehydration would tend to increase the δ_{LW} and may explain some of the observations where measured δ_{LW} values were higher than predicted by Equation 10.2 (e.g. Farris and Strain, 1978; Flanagan and Varney, 1995; Walker and Lance, 1991). When leaf water volume, V, is not constant with time it reflects the imbalance between the evaporation flux (E) and the input flux (F_{in}) of the leaf:

$$\frac{dV}{dt} = F_{in} - F_{out} \tag{10.4}$$

and the isotopic composition of leaf water R_L will change accordingly:

$$\frac{dVR_L}{dt} = F_{in}R_{in} - F_{out}R_{out} \tag{10.5}$$

where R_{in} the isotopic ratio (e.g. $^{18}O/^{16}O$) of the input water and R_{out} of the evaporation flux (Equation 10.1). Equations 10.4 and 10.5 can be combined and rearranged to yield:

$$\frac{dR_L}{dt} = \frac{1}{V}\left[F_{in}(R_{in} - R_L) + E(R_L - R_{out})\right] \tag{10.6}$$

indicating the inverse dependency between leaf water volume and its isotope ratio, as will be further discussed below. Note that a leaf dehydrating during the day, will recharge with depleted soil water during the night, returning to its original volume. Since this process occurs at night when the leaf stomata are closed, it is likely to reflect simple mixing between ^{18}O enriched leaf water and depleted soil water (Yakir et al., 1990; Yakir et al., 1994c).

10.5.3 *Isotopic homogeneity*

The assumption of isotopic homogeneity in the evaporating water body, made in the original Craig and Gordon (1965) model, has been shown to be incompatible with leaves in a number of studies. By pressurising a leaf in a pressure chamber, leaf water can be forced out through the petiole in a reversal of the normal water-flow. Analysing the $\delta^{18}O$ values of the water fractions successively extracted from a leaf in this way showed isotopic heterogeneity to be as large as 10‰ in $\delta^{18}O$ values (Yakir et al., 1989). Spatial heterogeneity was subsequently observed in several studies in which small leaf water samples were obtained from different parts of leaves (Bariac et al., 1994; Luo and Sternberg, 1992; Tissue et al., 1991; Wang and Yakir, 1996; Yakir, 1992).

The details of the isotopic heterogeneity in leaf water are not well understood at present and different approaches have been proposed to describe it. The most simple approach has been to assume that leaf water is composed of two distinct fractions, the first interacting with the atmosphere having a $\delta^{18}O$ value as predicted by Equation 10.2, and a second of unfractionated source water. By adjusting the proportions of these fractions, it is possible to reconcile Equation 10.2 with measured values of bulk-leaf water (Bariac et al., 1994; Walker and Lance, 1991; Walker et al., 1989). This approach was also expanded to include an intermediate mesophyll water fraction that only slowly exchanges with the two other pools (Yakir, 1992; Yakir et al., 1990). A one-dimensional advection–diffusion model was recently proposed to describe continuous isotopic gradients in leaves due to the contrasting effects of the convective flux of source water into the leaf, and the diffusion of ^{18}O enriched water from the evaporating surfaces into the leaf tissue (Farquhar and Lloyd, 1993; see Chapter 3). The exponential decay of the ^{18}O enrichment signal at the surface (described by Equation 10.2) to the source isotopic signature of the water entering the leaf, is integrated to give the $\delta^{18}O$ value of bulk leaf water, δ_{LW}:

$$\delta_{LW} = \delta_i + \left(\delta_{ss} - \delta_{in}\right) \times \frac{1 - e^p}{\mathscr{P}} \qquad (10.7)$$

where δ_{ss}, is given by equation 10.2, a leaf péclet number, $\mathscr{P}$, is used where $\mathscr{P} = EL/C_wD$ and E is the rate of transpiration, L the effective mixing pathlength, C_w the molar concentration of water and D the diffusivity of $H_2^{18}O$. This model links the $\delta^{18}O$ values of bulk leaf water with the rate of transpiration by a specific peclet number, which describes the ratio between the advection and diffusion fluxes in the leaf. It also considers an anatomical dimension, L, that is not well characterised at present, but can be estimated empirically by comparing measured and calculated values of δ_{LW}.

The transport of water within leaves is complex and the above models obviously provide simplifications. Water movement in leaves and the resulting isotopic gradients are likely to have 3-dimensional aspects. For example, Equation 10.7 considers the isotopic enrichment between the vein (source water) and the evaporating surface, but spatial isotopic gradients can occur both in the source water and in the evaporating surfaces within the leaves. This may be the case in a corn leaf which shows a progressive isotopic enrichment, for which the integration which may be represented only by Equations 10.2 or 10.7. A quantitative description of such progressive enrichment has been proposed by Gat and Bowser (1991) for a system consisting of a string of pools:

$$\delta^l_{ss} = \frac{\delta^l_{ss}{}^{-1} \times \theta \cdot \left(1-h\right)/h + \left(\delta_a + \epsilon/h\right)}{1 + \theta \times \left(1-h\right)/h} \qquad (10.8)$$

where l represents an element in a string, θ is the ratio of the input flux over the evaporation flux in an element of the string (F_{in}/E) and $\epsilon = \epsilon_{eq} + (1-h)\ \epsilon_K$. As indicated above, such a string will show a progressive ^{18}O enrichment in its elements, but the integrated average of the string should agree with Equation 10.2. Note that in leaves, this would occur only in cases where all the elements of the string are within the sampled tissue. The spatial heterogeneity described by Equation 10.8 may have implications for the use of ^{18}O in, for example, investigating CO_2 exchanged by leaves. The rate of such gas exchange may not be uniformly distributed across the leaf, resulting with a bias in the ^{18}O signature transduced from leaf water to atmospheric CO_2. In the case of the corn leaf, higher rates of CO_2 exchange occur at the base or centre of the leaf, as compared with the leaf tip, and this can lead to an overestimation of the $\delta^{18}O$ values in CO_2 by using Equation 10.2 directly.

10.6 2-D simulation of ^{18}O in leaf water

As indicated above, the complex pathway of water in leaves is difficult to fully describe with two-compartment or unidimensional models. A somewhat more realistic picture can perhaps be obtained with a two or three dimensional model of leaf water. Analytical solutions to such a systems are difficult to obtain, making a numerical approach more practical. We used a simple 2-D advection diffusion software for a passive tracer (PDE2D, G. Sewell, El Paso, Texas) to model the time-dependent evolution of the isotopic tracer concentration in a leaf exposed to atmospheric conditions that vary with the diurnal cycle. The leaf was replaced in the model by a square domain the default thickness and length of which were set to 0.1 and 10 cm respectively (but could be modified). Rates of diffusion within the leaf domain were that for water in the horizontal direction, and an order of magnitude larger in the vertical direction. The water carrying the isotopic tracer enters the leaf from the side (having a constant isotopic composition) and exits by evaporation from the top (with isotopic composition governed by Equation 10.1), and with no flux at the other boundaries. The input, F_{in}, and evaporation flux E, was dependent on atmospheric humidity which was specified (humidity, temperature and $\delta^{18}O$ values of stem water and atmospheric humidity for a typical case scenario were taken from Förstel, 1978; the evapotraspiration flux was adjusted to reach a mid-day maximum of 2 mmol m^{-2}sec^{-1}).

Although this model still greatly simplifies the situation in a leaf, it offers several features which may be more realistic than obtained in an analytical solution. For example, it is a dynamic, non-equilibrium model which is not restricted to steady-state situations, and estimation of the isotopic composition of leaf water can be made at any point in the leaf domain. It can be used to simulate the behaviour of any leaf by adjusting the leaf characteristics, such as rates of fluxes through and within the leaf, its shape and dimensions. This may have important practical implications, first in estimation of the $\delta^{18}O$ values of specific water fractions within the leaf (such as in the chloroplasts), and second, used together with vegetation maps and specific plant characteristics, to obtain a canopy or regional scale estimates of the $\delta^{18}O$ values of leaf water. Figure 10.4 shows some results (a more detailed report will appear separately) of the leaf model simulation at mid-day (maximum leaf water enrichment). As would be expected, the $\delta^{18}O$ value of bulk leaf water decreased as E increased (cf. Equation 10.7). As is normally observed in the field, the $\delta^{18}O$ values of bulk leaf water were significantly lower than the value predicted by Equation 10.2. An intermediate value was obtained for the leaf layer comprising the top 10% of the leaf thickness (i.e. top 100 μm, equivalent to 1–2 cell layers, in the case depicted in Figure 10.4). This water layer became more enriched with increasing flux rates, approaching the predicted steady state value (Figure 10.4a). Interestingly, the timing of the diurnal maximum in $\delta^{18}O$ value of leaf water was also sensitive to the evaporative flux rate. This is due to the non-equilibrium conditions in the leaf and the relatively slow mixing process of the enriched water at the surface acting against the convective flux of depleted water flowing through the leaf (Figure 10.4b). In our typical case scenario ($E = 2$ mmol m^{-2} sec^{-1}), the diurnal maximum in $\delta^{18}O$ values of bulk leaf water was delayed by more than three hours with respect to changes in humidity and the predicted steady state value (Equation 10.2). Such delays are consistent with observations (e.g. Förstel, 1978), and may have implications for isotopic measurements in gases exchanged by leaves. For example, as already mentioned above, if the lag in leaf water $\delta^{18}O$ involves water in the chloroplasts (e.g. top 100 μm) where CO_2–H_2O isotopic exchange occur, the productivity weighted ^{18}O signature of CO_2 may not directly reflect the humidity weighted leaf ^{18}O signature, as would be predicted from Equation 10.2.

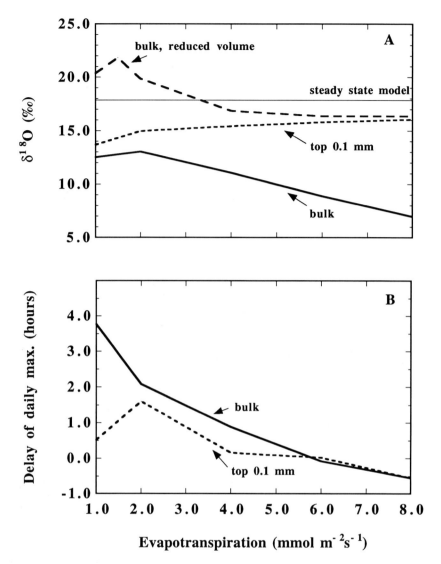

Figure 10.4. Effects of the rate of evapotranspiration on (A) the simulated δ¹⁸O values of bulk leaf water, and of the water at the top 0.1 mm of the leaf; and (B) on the time displacement of the daily maximum in δ¹⁸O values of bulk leaf water. Simulations were carried out using a 2-D advection diffusion model described in the text (for bulk and top 0.1 mm data), or with a simple mixed box model (for reduced volume data).

As expected, the mean values presented in Figure 10.4 integrate large isotopic gradients within the leaf (Figure 10.5). Notably, in spite of the small leaf thickness, these gradients have clear 2-D characteristics (which greatly diminish during the night when evaporation stops; Figure 10.5).

To check for the effects of decreasing leaf water volume during the day, a separate box model was employed. Well mixed isotopic composition and a constant input flux were assumed, but the output flux, E, depended on atmospheric humidity as before. A

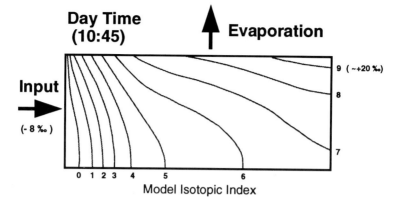

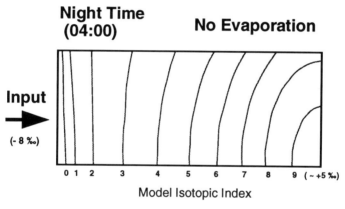

Figure 10.5. Simulation of the two-dimensional ^{18}O gradients in leaf water formed during the daily cycle in evapotranspiration (see text for details). Contours describe the distribution of ^{18}O along the simulated leaf from source water of -8‰ to the maximum value indicated on the right.

limit to the change in the leaf volume, typically up to 50%, was imposed in this case. Other boundary conditions were as for the 2-D model, and a NAG library routine was used in conjunction with the two first order differential equations discussed above (Equations 10.4 and 10.6) to simulate the development of the isotopic composition of the leaf water over the daily cycle. The results for mid-day maximum values (Figure 10.4a) showed that leaves which decrease in volume during the day had significantly higher $\delta^{18}O$ values as compared to constant volume leaves, with this value often higher than predicted by Equation 10.2. This effect diminished with increasing flux, because in these cases the limit to volume change was likely to be reached early in the day, after which the leaf maintained a constant volume (e.g. by stomatal closure) and then behaved like a constant volume leaf.

10.7 ^{18}O of water in chloroplasts

As noted at the outset, it is the $\delta^{18}O$ value of the water in leaf chloroplasts, δ_c, which is transduced to photosynthetically produced organic matter and to CO_2 and O_2 exchanged by leaves. The challenge is therefore to estimate this specific isotopic signature in spite of the difficulties associated with the large spatial and temporal variations in the isotopic composition of leaf water. Fortunately, the problem may be less important since for efficient gas exchange the chloroplasts are required to be near the liquid-gas interfaces (e.g. the evaporating surfaces) in the leaves. This, in turn, indicates that, at least *a priori*, the $\delta^{18}O$ value of the water in the chloroplast should be close to that estimated by Equation 10.2. Several attempts have been made to test this assumption and provide data for improving estimates of δ_c.

10.7.1 *Exchange of* O_2

Perhaps the most direct way to estimate $\delta^{18}O_2$ has used isotopic analysis of photosynthetically produced O_2. Since there is no isotopic fractionation in the water splitting reaction during photosynthesis (Guy *et al.*, 1987; 1993; Yakir *et al.*, 1994a), the $\delta^{18}O$ value of the photosynthetically produced oxygen directly reports that of the water in chloroplasts, from which it was produced. Laboratory experiments with sunflower plants, showed the potential of this approach and indicated that, under the experimental conditions, δ_c was indistinguishable from the $\delta^{18}O$ value of bulk leaf water, but lower than predicted by Equation 10.2 (Yakir *et al.*, 1994a).

10.7.2 *Exchange of* CO_2 *and associated diffusional fractionation*

Several other experimental protocols have relied on the $\delta^{18}O$ value of CO_2 exchanged with leaves. In this approach it is assumed that rapid oxygen isotope exchange occurs between CO_2 and H_2O predominantly in the chloroplasts (where it is catalyzed by CA) and therefore any CO_2 diffusing out of the leaf will carry the ^{18}O signature of the chloroplast water (see Chapters 12 and 24). The isotopic fractionations associated first with the equilibrium between CO_2 and water, and second with the diffusion of CO_2 between the chloroplasts and the atmosphere must be considered in this case (Farquhar and Lloyd, 1993; Flanagan *et al.*, 1994; Yakir *et al.*, 1994a; Williams *et al.*, 1996).

While the CO_2–H_2O equilibrium isotopic fractionation is well characterized (Friedman and O'Neil, 1977; Brenninkmeijer *et al.*, 1983), significant uncertainty is still associated with the effective diffusional isotope effects. In the pathway between the chloroplast and the atmosphere, CO_2 has to diffuse in solution through the chloroplast stroma and limiting membrane, the cytoplasm, the cellular membrane and the cell walls; and in the gas phase through the intercellular spaces, the stomates and a leaf boundary layer. There is a large difference between the isotopic fractionation in the air phase (8.8‰ in stagnant air, ~5.8‰ in a laminar boundary layer) and in the aqueous phase (~0.8‰). The relative importance of each diffusion step must be estimated from the respective resistances to CO_2 diffusion, or the CO_2 concentration gradient across the specific resistance.

Traditionally, the resistance to diffusion in the air phase is estimated from measurements of transpiration and is used to estimate the CO_2 concentrations in the intercellular spaces C_i. The additional resistance between the cell walls surfaces and the site of CO_2 fixation in the chloroplasts is often estimated from the difference between this estimate of C_i and that based on ^{13}C discrimination (von Caemmerer and Evans, 1991). This is because the ^{13}C discrimination must reflect the total diffusion pathway between the atmosphere and the actual site of carboxylation, and produces therefore an estimate of the CO_2 concentration inside the chloroplasts, C_c.

Note, however, that the chloroplast stroma can represent 20–50% of the internal resistance to diffusion between the cell wall surfaces and the site of carboxylation (Nobel, 1991; Evans and von Caemmerer, 1996). This is important because such resistance would influence the ^{13}C gradient and the effective ^{13}C discrimination with respect to the atmosphere, but not the ^{18}O discrimination. Catalyzed-exchange of oxygen between CO_2 and H_2O would impose the ^{18}O signature of chloroplast water at least up to the chloroplast limiting membrane. Existance of any membrane-bound or external CA activity would eliminate other components in the resistance pathway leading to the liquid–gas interface.

Thus, while for ^{13}C measurements C_c is the appropriate quantity for estimating discrimination, an intermediate value between C_c and C_i, that would apply to the CO_2 concentration at the chloroplast surface (say, $C_{cs} = 0.5 \ (C_a + C_i)$), would be more appropriate for ^{18}O measurements. Marked differences between C_i and C_c (and therefore between C_c and C_{cs}) can occur particularly under conditions of high photosynthetic activity. In these cases, differences between C_c and C_{cs} can lead to variations of 0.5–1‰ in estimates of the $\delta^{18}O$ values of chloroplast water (with lower values obtained using C_{cs}). An estimate of the total diffusional fractionation, $\bar{a}$ can be obtained from (see Farquhar and Lloyd, 1993):

$$\bar{a} = \frac{(C_a - C_s)a_b + (C_s - C_w)a + (C_w - C_{cs})a_w}{C_a - C_{cs}} \tag{10.9}$$

where C denotes CO_2 concentration and subscripts a, s, w and cs denote atmosphere, leaf surface, cell walls surfaces and chloroplast surface respectively, and a denotes the kinetic fractionation associated with boundary layer (subscript b, ~ 5.8‰), stagnant air inside the leaf (8.8‰) and in water (subscript w, ~ 0.8‰ Craig and Gordon, 1965).

An experimental protocol that avoids the uncertainties associated with the diffusional fractionation uses re-circulated air around a leaf in a closed system. In time, the leaf reaches its CO_2 compensation point, where there are no concentration gradients between the leaf and the atmosphere (i.e. any diffusional fractionations are cancelled out), and the CO_2 reaches isotopic equilibrium with chloroplast water. Analysing an aliquot of the atmospheric CO_2 in the system, and accounting for the equilibrium isotope effect between CO_2 and water, has permitted the reconstruction of δ_c, yielding results similar to those obtained with O_2 (Yakir et al., 1994a).

10.7.3 Respired CO_2 signal

In other protocols using the CO_2 approach, both the equilibrium and diffusional isotopic fractionations have been considered. In one such protocol a leaf was allowed to respire CO_2 in the light into a flow of CO_2-free air. The CO_2 is then removed from the air flow for isotopic analysis. In this case there is a net flux of CO_2 from the leaf with a predictable diffusional effect, but no fractionation during the trapping of the CO_2 sample. The expected ^{18}O fractionation in CO_2 can be estimated from:

$$\Delta^{12}C^{18}O^{16}O = \frac{(C_{cs} - C_a)}{C_{cs}}\bar{a} \tag{10.10}$$

where $\overline{a}$ is the effective diffusional fractionation, C_{cs} and C_a are the CO_2 concentrations at the chloroplast surface and in the atmosphere respectively. This calculation is similar to that developed for ^{13}C discrimination in photosynthesis (Farquhar et al., 1978) but with the flux in the reverse direction (respiration) and with no biochemical fractionation (quantitative trapping of the CO_2). Estimating the CO_2 concentration gradient between the atmosphere and the chloroplast can be made by substituting measured conductances, g, for the CO_2 concentrations:

$$\Delta = \frac{g_f(g_f + g_i)}{\overline{a}} \tag{10.11}$$

where subscripts f and i stand for air flow through the chamber and leaf, respectively. As mentioned above, if measured leaf conductance is limited to that from the intercellular spaces to the atmosphere an additional component to the site of the CO_2-water equilibration needs to be considered (Evans and von Caemmerer, 1996). Alternatively, $\delta^{13}C$ values of the respired CO_2 and the leaf organics can be used to estimate the effective diffusional fractionation and C_{cs}/C_a ratios. The $\delta^{18}O$ value of chloroplast water can then be evaluated from the isotopic analysis of the respired CO_2 $\delta^{18}O_R$ and $\Delta^{12}C^{18}O^{16}O$ according to $\delta^{18}O_c = \delta^{18}O_R + \Delta - \epsilon_{eq}$, where ϵ_{eq} is the equilibrium fractionation between water and CO_2 in permils (Friedman and O'Neill, 1977; Brenninkmeijer et al., 1983). Applying this protocol in measurements of sunflower plants under laboratory conditions, yielded results consistent with those obtained by the direct O_2 approach (Yakir et al., 1994a).

10.7.4 Instantaneous discrimination

Perhaps the simplest protocol employs the on-line discrimination method proposed by Evans et al. (1986) for estimating the instantaneous ^{13}C discrimination of leaves during photosynthesis (see Chapter 8 for a detailed discussion of this method.) In this case, the leaf enclosed in the chamber can be maintained at close to ambient conditions and $\delta^{18}O_c$ can be calculated according to (Farquhar and Lloyd, 1993):

$$\delta^{18}O_c = \frac{\Delta - \overline{a}}{f} + \delta^{18}O_a - \epsilon^{\circ} \tag{10.12}$$

where δ_a is the $\delta^{18}O$ values of atmospheric CO_2, $\overline{a}$ is the effective diffusional fractionation for ^{18}O in CO_2 (weighted by conductances), and $f = C_{cs}/(C_a - C_{cs})$. The isotopic discriminations (Δ) against $C^{18}O^{16}O$ during photosynthetic gas exchange can be calculated based on measurements, as for $^{13}CO_2$ (Evans et al., 1986), according to:

$$\Delta = \frac{\zeta(\delta_o - \delta_e)}{1000 + \delta_o - \zeta(\delta_o - \delta_e)} \cdot 1000 \tag{10.13}$$

where δ_c and δ_o are the $\delta^{13}C$ or $\delta^{18}O$ values of the CO_2 leaving and entering a leaf chamber in which the measured leaf is enclosed; $\xi = C_e/(C_e - C_o)$; C_e and C_o are CO_2 concentrations of the air entering and leaving the leaf chamber, respectively. Estimates of C_{cs} can be obtained as discussed above (see Section 10.7.2) by assuming that C_{cs} represents an intermediate value between C_i and C_c. Estimates of C_i are obtained from gas exchange measurements and of C_c from concurrent measurements of Δ

based on ^{13}C values (Δ_{obs}, Evans *et al.*, 1986; von Caenmerer and Evans, 1991; see Chapter 8).

$$\frac{C_c}{C_a} = \frac{C_i}{C_a} - \frac{\Delta_i - \Delta_{obs} - \left(eR_d / k + f\Gamma^*\right) / C_a}{b - a_w} \qquad (10.14)$$

where Δ_i is the expected discrimination, calculated as:

$$\Delta_i = \bar{a}\frac{-C_a - C_i}{C_a} + b\frac{C_i}{C_a} \qquad (10.15)$$

where $\bar{a}$ is the overall ^{13}C fractionation during diffusion from the atmosphere to the intercellular spaces; b is the effective fractionation of RUBISCO and any PEP-carboxylation ($27 - 29‰$); e is the ^{13}C fractionation associated with dark respiration R_d; f is the fractionation associated with photorespiration; k is the carboxylation efficiency of RUBISCO and Γ^* is the CO_2 compensation point in the abscence of dark respiration during the day (Brooks and Farquhar, 1985; see Chapter 8). Results obtained by the on-line discrimination approach were interpreted as indicating that $\delta^{18}O_c$ values are similar to those predicted by equation 10.2 (Farquhar *et al.*, 1993).

Notably, uncertainties are still associated with all available methods for estimating $\delta^{18}O_c$. For the first three protocols described above, conditions during the measurements are different from ambient. In essentially all protocols uncertainty is still associated with the assumption of a full isotopic equilibrium between CO_2 and water in the chloroplast. In some cases, reconstructed $\delta^{18}O_c$ values lower than expected based on equation 10.2, were interpreted as indicating incomplete equilibrium between CO_2 and water in the chloroplast (Farquhar *et al.*, 1993; Flanagan *et al.*, 1994; Williams *et al.*, 1996). This required that the ratio of rates of CO_2 carboxylation to that of CO_2 hydration in the chloroplasts (τ) is about 0.02 (Farquhar and Lloyd, 1993). Estimates of $\delta^{18}O_c$ are also sensitive to estimates of the effective kinetic (diffusion) fractionation and to the assumed C_{cs}/C_a and C_c/C_a ratio. The limited constraint we have on the kinetic fractionation, a, has been noted above and its importance is apparent from equations 10.10 and 10.12. Estimates of C_c/C_a according to equation 10.13 are, in turn, sensitive to the value used for the biochemical fractionation, b. This value was shown to be 29‰ for RUBISCO, including green tissue (e.g. Guy *et al.*, 1993), but it was also argued that effects of other carboxylation processes in the leaves can reduce the effective fractionation to about 27‰ (Farquhar and Lloyd, 1993). This uncertainty can introduce variations of a few permils in estimated $\delta^{18}C$ values. Little information is available on the fractionation associated with photorespiration. A value of $f = 8‰$, is often adopted (Rooney, 1988), but recent evidence (Gillon and Griffiths 1997; see Chapter 8) suggest that this value can vary among species and at least in some cases can be much smaller than 8‰, (which would increase estimated δ_c). In contrast to the factors discussed above, f has a relatively minor effect on estimating $\delta^{18}O_c$ values.

10.8 Conclusions

Standing at the crossroads between source water and gaseous exchange of O_2, water vapour and CO_2, the importance of estimating the $\delta^{18}O$ signature of leaf and chloroplast water has been highlighted. The complexity involved in making such estimates has been discussed, and various factors that can contribute to the observed variations in the $\delta^{18}O$ values of leaf water have been described. The major sources of uncertainty that require further research have been identified in order to improve estimates and predictions of the $\delta^{18}O$ values of leaf water. It is important to emphasise, however, that in spite of the intricacy associated with measurement of the ^{18}O signature of leaf water, and the various isotopic signatures associated with it, ^{18}O in plant material and in atmospheric CO_2 and O_2 already yields new insights into plant-environment inter-actions: as shown in the final Chapter in this volume, reducing the uncertainties noted here will only improve these applications and the interpretations based on them.

Acknowledgement

The author is grateful to Professor E. Tziperman and Dr. H. Jarosh for help with the 2D model. This work was supported by grants from the Bi-National US-Israel Sci. Foundation (BSF) and from the Israel–Germany Binational, BMBF.

References

Adar, E. M., Gev, I., Lipp, J., Yakir, D. and Gat, J. R. (1995) Utilization of oxygen-18 and deu-terium in stem flow for the identification of transpiration source: soil water versus ground-water in sand dune terrain. In *Application of Tracers in Arid Zone Hdrology*, E. Adar, C. Leibundgut eds. 1st. Assoc. Sci. Hydrol. Publ. **232**, 329–338.

Allison, G.B., Barnes, C.J. and Hughes, M.W. (1983) The distribution of deuterium and ^{18}O in dry soils, 2: Experimental. *J. Hydrol.* **64**, 377–397.

Allison, G. B., Gat, J. R. and Leaney, F. W. J. (1985) The relationship between deuterium and oxygen-18 delta values in leaf water. *Chemical Geol.* **58**, 145–156.

Bariac, T., Deleens, E., Gerband, A., Andre, M. and Marioti A. (1991) La composition iso-topique (^{18}O, ^2H) de al vapeur d'eau transpiree: etude en conditions assrevies. *Geochim. Cosmochim. Acta* **55**, 3391–3402.

Bariac, T., Gonzalez-Dunia, J., Katerji, N., Bethenod, O., Bertolini, J. M. and Mariotti, A. (1994) Spatial variation of the isotopic composition of water (^{18}O, ^2H) in the soil–plant–atmosphere system, 2. Assessment under field conditions. *Chemical Geol.* **115**, 317–333.

Bariac, T., Rambal, S., Jussrand, C. J. and Berger, A. (1989) Evaluating water fluxes of field-grown alfalfa from diurnal observations of natural isotope concentrations, energy budget and ecophysiological parameters. *Agric. For. Meteorol.* **48**, 263–283.

Barnes, C. J. and Allison, G. B. (1983) The distribution of deuterium and ^{18}O in dry soils 1. the-ory. *J. Hydrol.* **60**, 141–156.

Bender, M., Sowers, T. and Labeyrie, L. (1994) The Dole effect and its variations during the last 130,000 years as measured in the Vostok ice core. *Global Biogeochem. Cycles* **8**, 363–376.

Bender, M. L., Labeyrie, L., Raynaud, D. and Loris, C. (1985) Isotopic composition of atmos-pheric O_2 in ice linked with deglaciation and global primary productivity. *Nature* **318**, 349–352.

Brenninkmeijer, C. A. M., Kraft, P. and Mook, W. G. (1983) Oxygen isotope fractionation between CO_2 and H_2O. *Isotope Geosci.* **1**, 181–190.

Brooks, A. and Farquhar, G. D. (1985) Effect of temperature on the CO_2/O_2 specificity of ribu-lose-1,5-bisphosphate carboxylase/oxygenase and the rate of respiration in the light. *Planta* **165**, 397–406.

Brunel, J. P., Simpson, H. J., Herczeg, A. L., Whitehead R. and Walker G. R. (1992) Stable iso-tope composition of water vapour as an indicator of transpiration fluxes from rice crops. *Water Resources Res.* **28**, 1407–1416.

Buhay, W. M., Edwards, T. W. D. and Aravena, R. (1996) Evaluating kinetic fractionation factors used for ecologic and paleoclimatic reconstructions from oxygen and hydrogen isotope ratios in plant water and cellulose. *Geochim. Cosmochim. Acta* **60**, 2209–2218.

Burk, R. L. and Stuiver, M. (1981) Oxygen isotope ratios in trees reflect mean annual temperature and humidity. *Science* **211**, 1417–1419.

Caemmerer, S. v. and Evans, J. R. (1991) Determination of the average partial pressure of CO_2 in chloroplasts from leaves of several C3 plants. *Aust. J. Plant Physiol.* **18**, 287–305.

Ciais, P. et al. (1997a) A three-dimentional synthesis study of $\delta^{18}O$ in atmospheric CO_2 1. Surface fluxes. *J. Geophys. Res.* **102**, 5873–5883.

Ciais, P. et al. (1977b) A three-dimensional synthesis study of $\delta^{18}O$ in atmospheric CO_2 2. Simulations with the TM2 transport model. *J. Geophys. Res.* **102**, 5857–5872.

Cooper, L. W. and DeNiro, M. J. (1989) Covariance of oxygen and hydrogen isotopic composition in plant water: species effect. *Ecology* **70**, 1619–1628.

Craig, H. (1961) Istopic variations in meteoric waters. *Science* **133**, 1702–1703.

Craig, H. and Gordon, L. I. (1965) Deuterium and oxygen-18 variations in the ocean and the marine atmosphere. In: *Proc. Conf. on stable isotopes in oceanographic studies and paleotemperatures Pisa* (ed. E. Tongiorgi) Laboratory of Geology and Nuclear Science, pp. 9–130.

Dansgaard, W. (1964) Stable isotopes in precipitation. *Tellus* **XVI**, 436–468.

Dawson, T. E. (1993) Water sources of plants as determined from xylem-water isotopic composition: pespectives on plant competition, distribution, and water relations. In *Stable Isotopes and Plant Carbon/Water Relations.* J. R. Ehleringer, A. E. Hall and G. D. Farquhar (eds.), Academic Press, New York, NY. pp. 465–496.

Dawson, T.E. and Ehleringer, J.R. (1991) Streamside trees that do not use stream water: Evidence from hydrogen isotope ratios. *Nature* **350**, 335–337.

Dawson, T. E. and Ehleringer, J. R. (1993) Isotopic enrichment of water in the 'woody' tissues of plants: Implications for plant water source, water uptake, and other studies which use the stable isotopic composition of cellulose. *Geochim. Cosmochim. Acta* **57**, 3487–3492.

DeNiro, M. J. and Epstein, S. (1979) Relationship between the oxygen isotope ratios of terrestial plant cellulose, carbon dioxide, and water. *Science* **204**, 51–53.

Dongmann, G., Nurnberg, H. W., Förstel, H. and Wagener, K. (1974) On the enrichment of $H_2^{18}O$ in the leaves of transpiring plants. *Radiat. Environ. Biophys.* **11**, 41–52.

Edwards, T. W. D. and Fritz, P. (1986) Assessing meteoric water composition and relative humidity from ^{18}O and ^2H in wood cellulose: paleoclimatic implications from southern Ontario, Canada. *App. Geochem.* **1**, 715–723.

Edwards, T. W. D. and Fritz, P. (1988) Stable-isotope paleoclimate records for southern Ontario. Canada: comparison of results from marl and wood. *Can. J. Earth Sci.* **25**, 1397–1406.

Ehleringer, J. R. and Dawson, T. E. (1992) Water uptake by plants: perspectives from stable isotope composition. *Plant Cell Environ.* **15**, 1073–1082.

Emerman, S. H. and Dawson, T. E. (1996) The role of macropores in the cultivation of Bell pepper in salinized soil. *Plant Soil* **181**(2), 241–249.

Epstein, S. and Mayeda, T. (1953) Variation of ^{18}O content of water from natural sources. *Geochim. Cosmochim. Acta* **42**, 213–224.

Epstein, S., Thompson, P. and Yap, C. J. (1977) Oxygen and hydrogen isotopic ratios in plant cellulose. *Science* **198**, 1209–1215.

Evans, J.R. and von Caemmerer, S. (1996) Carbon dioxide diffusion inside leaves. *Plant Physiol.* **110**, 339–346.

Evans, J. R., Sharkey, T. D., Berry, J. A. and Farquhar, G. D. (1986) Carbon isotope discrimination measured concurrently with gas exchange to investigate CO_2 diffusion in leaves of higher plants. *Aust. J. Plant Physiol.* **13**, 281–292.

Farquhar, G. D. and Lloyd, J. (1993) Carbon and oxygen isotope effects in the exchange of carbon dioxide between plants and the atmosphere. In: *Stable Isotopes and Plant Carbon/Water Relations.* J. R. Ehleringer, A. E. Hall and G. D. Farquhar (eds.), Academic Press, New York, NY, pp. 47–70.

Farquhar, G. D., Lloyd, J., Taylor, J. A., Flanagan, L. B., Syvertsen, J. P., Hubick, K. T., Wong, S. C. and Ehleringer J. R. (1993) Vegetation effects on the isotope composition of oxygen in the atmospheric CO_2. *Nature* **363**, 439–443.

Farquhar, G. D., O'Leary, M. H. and Berry, J. A. (1982) On the relationship between carbon isotope discrimination and the intercellular carbon dioxide concentration in leaves. *Aust. J. Plant Physiol.* **9**, 121–137.

Farris, F. and Strain, B. R. (1978) The effects of water stress on leaf $H_2^{18}O$ enrichment. *Radiat. Environ. Biophys.* **15**, 167–202.

Feng, X. and Epstein, S. (1994) Climatic implications of an 8000-year hydrogen isotope time series from Bristlecone pine trees. *Science* **265**, 1079–1081.

Flanagan, L. B., Comstock, J. P. and Ehleringer, J. R. (1991) Comparison of modeled and observed environmental influences on the stable oxygen and hydrogen isotope composition of leaf water in *Phaseolas vulgaris* L. *Plant Physiol.* **96**, 588–596.

Flanagan, L. B. and Ehleringer, J. R. (1991) Effects of mild water stress and diurnal changes in temperature and humidity on the stable oxygen and hydrogen isotopic composition of leaf water in *Cornus stolonifera* L. *Plant Physiol.* **97**, 298–305.

Flanagan, L. B., Phillips, S. L., Ehleringer, J. R., Lloyd, J. and Farquhar, G. D. (1994) Effect of changes in leaf water oxygen isotopic composition on discrimination against $C^{18}O^{16}O$ during photosynthetic gas exchange. *Aust. J. Plant Physiol.* **21**, 221–234.

Flanagan, L.B. and Varney, G.T. (1995) Influence of vegetation and soil CO_2 exchange on the concentration and stable isotopic ratio of atmospheric CO_2 within a *Pinus rsinosa* canopy. *Oecologia* **101**, 37–44.

Förstel, H. (1978) The enrichment of ^{18}O in leaf water under natural conditions. *Radiat. Environ. Biophys.* **15**, 323–344.

Francey, R. J. and Tans, P. P. (1987) Latitudinal variation in oxygen-18 of atmospheric CO_2. *Nature* **327**, 495–497.

Friedli, H., Siegenthaler, U., Rauber, D. and Oeschger, H. (1987) Measurements of concentration, $^{13}C/^{12}C$ and $^{18}O/^{16}O$ ratios of tropospheric carbon dioxide over Switzerland. *Tellus* **39B**, 80–88.

Gat, J. R. (1996) Oxygen and Hydrogen Isotopes in the Hydrologic-Cycle. *Ann. Rev. Earth Planet Sci.* **24**, 225–262.

Gat, J. R. and Bowser, C. (1991) The heavy isotope enrichment of water in coupled evaporative systems. In: *Stable Isotope Geochemistry: A Tribute to Samuel Epstein* H. P. Taylor, J. R. O'Neil and I. R. Kaplan (eds.), Lancaster, pp. 159–168.

Gat, J.R. and Matsui, E. (1991) Atmospheric water balance in the Amazon basin: an isotopic evapotranspiration model. *J. Geophys. Res.* **96**, 13179–13188.

Gat, J.R. and Rindsberger, M. (1985) The isotope signature of precipitation originaing in the Mediterranean sea area: A possible monitoring of climate modification? *Israel J. Earth Sci.* **34**, 80–85.

Gillon, J. S. and Griffiths, H. (1997) The influence of (photo) respiration on carbon isotope discrimination in plants. *Plant Cell Environ.*, in press.

Gonfiantini, R., Gratziu, S. and Tongiorgi, E. (1965). Oxygen isotope composition of water in leaves. In *Isotopes and Radiation in Soil Plant Nutrition Studies*, Tech. Rep. Ser. No 206, IAEA, Vienna, pp. 405–410.

Guy, R. D., Fogel, M. F., Berry, J. A. and Hoering, T. C. (1987) Isotope fractionation during oxygen production and consumption by plants. In: *Progress in Photosynthetic Research III*. J. Biggins (ed.), Kluwer, Dordrecht, pp. 597–600.

Guy, R. D., Fogel, M. L. and Berry, J. A. (1993) Photosynthetic fractionation of stable isotopes. *Plant Physiol.* **101**, 37–47.

IAEA (1990). *Environmental Isotope Data No.9: World Survey of Isotope Concentration in Precipitation (1984–1987)*. International Atomic Energy Agency (IAEA), Vienna.

Jouzel, J., Russell, G. L., Suozzo, R. J., Koster, R. D., White, J.W.C. and Broecker, W.S. (1987) Simulations of the HDO and $H_2^{18}O$ atmospheric cycles using the NASA/GISS general circulation model: The seasonal cycle for present-day conditions. *J. Geophys. Res.* **92**, 14739–14760.

Keeling, C.D. (1968) The concentration and isotopic abundances of carbon dioxide in rural and marine air. *Geochim. Cosmochim. Acta* **24**, 277–298.

Leaguy, C., Rindesberger, R., Zangvil, A., Issar, A. and Gat, J.R. (1983) The relation between the ^{18}O and deuterium contents of rain water in the Negev desert and air-mass trajectories. *Isotope Geosci.* **2**, 929–948.

Lin, G. and Sternberg, L.S.L. (1993) Hydrogen isotopic fractionation by plant roots during water uptake in coastal wetland plants. In J. R. Ehleringer, A. E. Hall and G. D. Farquhar (eds.), *Stable Isotopes and Plant Carbon/Water Relations*. Academic Press, New York, pp. 497–510.

Lipp, J., Trimborn, P., Edwards, T., Waisel, Y. and Yakir, D. (1996) Climatic Effects on the Delta-O-18 and Delta-C-13 of Cellulose in the Desert Tree Tamarix Jordanis. *Geochim. Cosmochim. Acta* **60**(17), 3305–3309.

Luo, Y. H. and Sternberg, L. (1992) Spatial D/H heterogeneity of leaf water. *Plant Physiology* **99**, 348–350.

Matsuo, S. and Friedman, E. (1967) Deuterium content of fractionally collected rainwater. *J. Geohpys. Res.* **72**, 6374–6376.

Majoube, M. (1971) Fractionnement en oxygene-18 et en deuterium entre l'eau et sa vapeur. *J. Chim. Phys.* **68**, 1423–1436.

Mathieu, R. and Bariac, T. (1996a) An isotopic study (^2H and ^{18}O) of water movements in clayed soils under a semiarid climate. *Water Resources Res.* **32**(4), 779–789.

Mathieu, R. and Bariac, T. (1996b) A numerical model for the simulation of stable isotope profiles in drying soils. *J. Geophys. Res.* **101**, 12,685–12,696.

Merlivat, L. (1978) Molecular diffusivities of $H_2{}^{18}O$ in gases. *J. Chem. Phys.* **69**, 2864–2871.

Midwood, A. J., Boutton, T. W., Watts, S. E. and Archer, S. R. (1993) Natural abundance of ^2H and ^{18}O in rainfall, soil moisture and plants in a subtropical thorn woodland ecosystem: Implications for plant water use. In: *Proceedings Isotope Techniques in the study of past and current environmental changes in the hydrosphere and the atmosphere*. IAEA, Austria, pp. 419–431.

Nobel, P. S. (1992) *Physiochemical and Environmental Plant Physiology*, Academic Press, San Diego, CA.

Rooney, M. A. (1988) *Short-term carbon isotope fractionation by plants*. Ph.D. thesis, University of Wisconsin, Madison.

Sternberg, L. S. L. and DeNiro, M. J. (1983) Biogeochemical implications of the isotopic equlibrium fractionation factor between the oxygen atmos of acetone and water. *Geochim. Cosmochim. Acta* **47**, 2271–2274.

Sternberg, L. S. L. and Swart, P.K. (1987) Utilization of freshwater and ocean water by coastal plants of southern Florida. *Ecology* **68**, 1898–1905.

Sternberg, L. S. L., DeNiro, M. J. and Savidge, R. A. (1986) Oxygen isotope exchange between metabolites and water during biochemical reactions leading to cellulose synthesis. *Plant Physiol.* **82**, 423–427.

Tissue, D. T., Yakir, D. and Nobel, P. S. (1991) Diel water movement between parenchyma and chlorenchyma of two desert CAM plants under dry and wet conditions. *Plant Cell Environ.* **14**, 407–413.

Tzur, Y. (1971) *Isotope separation in the evaporation of water*. Ph.D. thesis, Weizmann Institute of Science, Rehovot, Israel.

Walker, C. D. and Lance, R. C. M. (1991) The fractionation of ^2H and ^{18}O in leaf water of Barley. *Aust. J. Plant Physiol.* **18**, 411–425.

Walker, C. D., Leaney, F. W., Dighton, J. C. and Allison, G. B. (1989) The influence of transpiration on the equilibration of leaf water with atmospheric water vapour. *Plant, Cell Environ.* **12**, 221–234.

Wang, X. F. and Yakir, D. (1995) Temporal and Spatial Variations in the O-18 content of leaf water in different plant-species. *Plant, Cell Environ.* **18**(12), 1377–1385.

White, J. W. C. (1989) Stable hydrogen isotope ratios in plants. A review of current theory and some potential applications. In: *Stable Isotopes in Ecological Research* P. W. Rundel, J. R. Ehleringer and K. A. Nagy (eds.), Springer-Verlag, Berlin, pp. 142–162.

White, J. W. C. and Gedzelman S. D. (1984) The isotopic composition of atmospheric water vapour and the concurrent meteorological conditions. *J. Geophys. Res.* **89**, 4937–4939.

Williams, T. G., Flanagan, L. B. and Coleman, J. R. (1996) Photosynthetic gas exchange and discrimination against $^{13}CO_2$ and $C^{18}O^{16}O$ in tobacco plants modified by an antisense construct to have low chloroplastic carbonic anhydrase. *Plant Physiol.* **112**, 319–326.

Yakir, D. (1991) Water compartmentation in plant tissue: Isotopic evidence. In: *Water and Life* G. N. Somero, C. B. Osmond and C. L. Bolis (eds.), Springer-Verlag, Berlin, pp. 205–222.

Yakir, D. (1992) Variations in the natural abundance of oxygen-18 and deuterium in plant carbohydrates. *Plant, Cell Environ.* **15**, 1005–1020.

Yakir, D. and DeNiro, M. J. (1990) Oxygen and hydrogen isotope fractionation during cellulose metabolism in *Lemna gibba* L. *Plant Physiol.* **93**, 325–332.

Yakir, D., DeNiro, M. J. and Gat J. R. (1990) Natural deuterium and oxygen-18 enrichment in leaf water of cotton plants grown under wet and dry conditions: evidence for water compartmentation and its dynamics. *Plant, Cell Environ.* **13**, 49–56.

Yakir, D., DeNiro, M. J. and Rundel, P. W. (1989) Isotopic inhomogeneity of leaf water: evidence and implications for the use of isotopic signals transduced by plants. *Geochim. Cosmochim. Acta* **53**, 2769–2773.

Yakir, D., Berry, J. A., Giles, L. and Osmond, C. B. (1994a) Isotopic heterogeneity of water in transpiring leaves: identification of the component that controls the $\delta^{18}O$ of atmospheric O_2 and CO_2. *Plant, Cell Environ.* **17**, 73–80.

Yakir, D., Issar, A., Gat, J. R., Adar, E., Trimborn, P. and Lipp, J. (1994b) ^{13}C and ^{18}O of wood from the Roman siege rampart in Masada, Israel (AD 70–73): Evidence for a less arid climate for the region. *Geochim. Cosmochim. Acta* **58**, 3535–3539.

Yakir, D., Ting, I. and DeNiro, M.J. (1994c) Natural abundance $^2H/^1H$ ratios of water storage in leaves of *Peperomia Congesta* HBK during water stress. *J. Plant Physiol.* **144**, 607–612.

Yakir, D. and Wang, X. F. (1996) Fluxes of CO_2 and Water Between Terrestrial Vegetation and the Atmosphere Estimated from Isotope Measurements. *Nature* **380**, 515–517.

Yakir, D. and Yechieli, Y. (1995) Plant Invasion of Newly Exposed Hypersaline Dead-Sea Shores. *Nature* **374**, 803–805.

Zimmermann, U., Ehhalt, D. and Munnich, K. O. (1967) Soil-water movement and evaporation: changes in isotopic composition of the water. In: *Proceedings of the Symposium on Isotopes in Hydrology* M. Knippner (ed.), IAEA, Vienna, pp. 567–584.

Zundel, G., Miekeley, W., Breno, M. G. and Förstel, H. (1978) The $H_2^{18}O$ enrichment in the leaf water of tropic trees: comparison of species from the tropical rain forest and the semi-arid region in Brazil. *Radiat. Environ. Biophys.* **15**, 203–212.

The role of hydrogen and oxygen stable isotopes in understanding water movement along the soil–plant–atmospheric continuum

Todd E. Dawson, Roman C. Pausch and Hester M. Parker

11.1 Introduction

Plants cover over 70% of the Earth's land surfaces (Costanza *et al.*, 1997; Whittaker, 1975) and can regulate the speed and magnitude by which water within these surfaces re-enters the terrestrial compartment of the hydrological cycle (Chahine, 1992; Covich, 1993; Dawson and Ehleringer, 1997 see Chapter 2, 3; Gleick, 1993; Newson, 1994; Rind *et al.*, 1992). The movement of water from land surfaces, particularly forested ones, back into the atmosphere should therefore be largely through a soil–plant–atmospheric continuum (SPAC) (Hunt *et al.*, 1991; Whitehead and Hinckley, 1991). Water movement along this continuum is determined by both the physical properties of the soil and atmospheric environment, and the biological properties of plants themselves. Below ground, soil properties such as specific hydraulic conductivity as well as moisture retention and chemical characteristics will determine what water is available for plants to take up and use. Above ground, atmospheric conditions that are known to influence plant transpiration can have a significant impact on the rates of water loss and water movement through plants (Hollinger *et al.*, 1994; Jarvis and McNaughton, 1986; McNaughton and Jarvis, 1983; Meinzer, 1993). At the interface between the soil and the atmosphere are the plants. It is aspects of plant form (such as rooting depth and distribution and canopy architecture) and plant function (such as stomatal behaviour and xylem hydraulic properties) which interact with the

Stable Isotopes, edited by H. Griffiths.
© 1998 BIOS Scientific Publishers Ltd, Oxford.

below- and above-ground environments to influence the rates and magnitudes of water movement through the SPAC and thus through the terrestrial compartment of the hydrological cycle.

Historically, a variety of methods have been employed to provide information on water movement through the SPAC (Kramer and Boyer, 1995; Rutter, 1968). Despite this, what regulates water movement though each individual part of the SPAC is not always well understood. This may be in part due to a lack of methods which provide a clear picture of what is going on in places like the soil where water movement is very difficult to observe. Studies utilizing natural variation in H and O stable isotopes of water in the environment and in plant tissues are providing new and important information for understanding the water relations of plants and the environments they inhabit. This information touches on a wide array of topics from the uptake and use of water by plants, to how differential utilization of water sources by different plant species in the same community may permit co-occurrence, to how rooting depth and pattern for different plant taxa relates to water availability and how water sources and water use patterns effect the hydrology of vegetated lands. Reviews by Dawson (1993a), Dawson and Ehleringer (1997) Ehleringer and Dawson (1992), Ehleringer *et al.* (1993), Lajtha and Michner (1994); Griffiths (1991) and White (1988) discuss much of this research. Only recently have stable isotope methods been applied towards understanding the SPAC (see for example Walker and Brunel, 1990). However, no single study has fully explored each compartment of the SPAC in an explicit way so as to provide a complete picture of water movement along this continuum.

In order to apply isotope methods towards enhancing our understanding of water movement through the SPAC, we must first determine what controls water movement through each separate compartment. Water movement in soils and into plant root systems will be a complex function of soil properties, root anatomy and morphology and the stability of water sources which plants may use at different times or from different depths (space). Root water uptake will also be determined by 'demand' by the plant canopy and patterns by which stomata regulate water loss from leaves (Whitehead and Hinckley, 1991). Water movement from leaves and canopies into the atmosphere will be determined by the stomatal and hydraulic properties of the plants and the strength to which leaves are coupled to the aerial environment which surrounds them (Hollinger *et al.*, 1994; Jarvis and McNaughton, 1986; Meinzer, 1993; Spittlehouse and Black, 1981). The analysis of stable H and O stable isotope ratios (δD and $\delta^{18}O$ respectively) of water in soils and soil profiles, plant roots and stems, leaves, and in the water vapour surrounding the leaves can link the soil, plant and atmospheric compartments of the SPAC (Dawson and Ehleringer, 1997, and below). In the sections which follow we review some of what we know about how stable isotopes are providing information about each compartment of the SPAC. We then attempt to draw together information which may provide an isotopic perspective on the SPAC and in so doing point to where further research is needed.

11.2 Stable isotopes and our understanding of water movement along the soil–plant–atmospheric continuum

11.2.1 *Plant water sources*

The use of stable hydrogen and oxygen isotopes in studies of plant water relations has dramatically improved our ability to link plants to their source water through non-destructive means (Ehleringer and Dawson, 1992). H and O stable isotopes can be used as tracers to compliment the information on spatial patterns of water uptake gained from studying rooting patterns. A growing number of studies have successfully used stable isotopes to understand, and indirectly, to measure the depth of water uptake by plants in a wide variety of environments (Dawson and Ehleringer, 1991; Flanagan et al., 1992; Sternberg et al., 1991; Thorburn et al., 1993 a, b, 1994; White et al., 1985; for additional references see reviews by White, 1988; Walker and Richardson, 1991; Dawson, 1993 a; Dawson and Ehleringer, 1997; see Chapter 14). This work has recently been extended to document spatial patterns of water uptake in an applied, agroforestry context. Smith et al. (1997) for example have demonstrated that crops and adjacent windbreak trees in Niger competed for the same water source in areas where the trees were shallow-rooted and could not gain access to deep water in the aquifer at 35 m. Interestingly, in areas where the water table was shallow (6–8 m), these same trees and crop plants did not compete for water because they used moisture from different soil depths.

Isotopes have also allowed researchers to examine temporal shifts in water use that can not be determined from documenting the location of roots in the soil profile. Examples of temporal variation in water source utilization are the differential use of winter versus summer precipitation by desert plants (Ehleringer et al., 1991 see Chapter 14) and Pinyon-Juniper woodland species (Flanagan et al., 1991, 1992) living in the arid southwestern regions of North America. More recently, Dawson and Pate (1996) documented a temporal and spatial shift in which water sources were used by the woody tree, Banksia prionotes. These shifts were correlated to a change in the water content of the upper and lower soil layers. As the upper soils became dry with the onset of the long, hot, dry season in southwestern Australia, shallow lateral roots within the soil profile stopped taking up water and water uptake shifted to deeper, moist layers (Figure 11.1). The root systems of these phreatophytic plants are dimorphic, not only with respect to their location within the soil profile, but also with respect to root water uptake activity and the availability of water during the annual wet and dry seasons which characterize Mediterranean environments.

In environments which become flooded, stable isotope analyses are also providing new insights. For example, a study by Yakir and Yechieli (1995) has shown that perennial species inhabiting the hypersaline soils adjacent to the Dead Sea preferentially used occasional floodwater events as their main water source over the saline soil water in their immediate environment. This strategy was believed to account for the ability of these perennials to colonize and persist in these saline and otherwise physiologically stressful environments. Jolly and Walker (1996) also compared temporal shifts in the source of water taken up by flooded and non-flooded Eucalyptus largiflorens before and after flooding. In addition to comparing the H and O stable isotope signatures of the soil and ground water with those of the trees, the authors measured osmotic and soil matric potential to construct soil water potential profiles. This allowed them to determine with more accuracy the location of water uptake than by using the isotope composition alone. Knowing the minimum xylem water potential recorded for that species (-3.5 MPa), they located the zones of available water in the soil profile at those depths where total soil water potential did not go below -3.5 MPa. Thus, Jolly and Walker as well as Mensforth et al. (1994) could identify a more precise location of water uptake by eliminating those soil depths for which there was either a mismatch of isotope signature or of soil water availability.

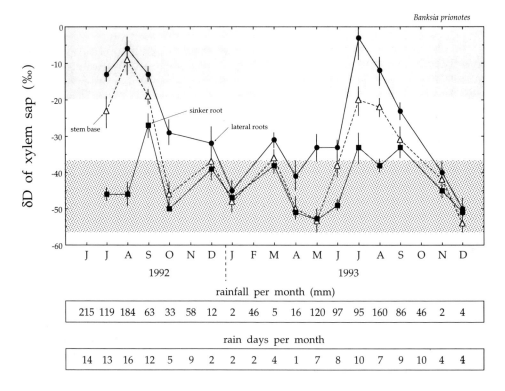

Figure 11.1. The mean (bar = s.d) δD values of xylem water (sap) from the lateral roots (●), the base of the stems (△) and the tap (sinker) roots (■) from a 6-year group of Banksia prionotes trees. The samples were obtained for 15 months during an 18-month period at Yanchep, western Australia. The amount of rainfall and the number of days per month when it rained were also recorded and are shown in boxes in the lower part of the figure. Rainfall δD values ranged from near zero to -20‰ during the study; shown as the light grey bar across the top of the figure. Ground water δD values varied from -38 to -56‰ for this same period; shown as the stippled bar across the bottom of the figure. The data show a greater dependence by plants on water from lateral roots (look at the stem base data) in the winter (July and August). In contrast, plants appear to be obtaining the majority of their water from ground water sources in the summer (December/ January) keeping their lateral roots and above ground stems well-suplied with water (from Dawson and Pate, 1996).

Jolly and Walker's effort to examine the proximate (soil layer) as well as the ultimate (ground versus rain water) sources of water emphasizes the important role that the soil compartment of the SPAC plays in the movement of water from each source into the plant. Soil properties have long been known to influence the uptake of water by plants, yet, this fact has been largely ignored in previous studies that use isotopes to link plants to their source water.

11.2.2 *Water in soils*

As mentioned above, many studies have reported a correlative relationship between the isotopic value of plant xylem water and their source water. However, few studies have attempted to address and then quantify the importance of edaphic factors for understanding the role that soil water plays during movement through the SPAC. Soil water can be a mixture of sources depending upon the soil type, texture, structure, rainfall patterns and hydrological properties of the regions in which a particular soil type is found (Allison and Hughes, 1983; Allision et al., 1983, 1984; Barnes and Allison 1988) These factors, coupled with isotope fractionation within and off of the soil profile, will determine the isotopic values of waters available for plant uptake (Barnes and Allison, 1983). Under certain circumstances, what determines the isotope ratio of soil water will be complex and multifaceted, and therefore must be carefully characterized if the goal is to trace water from soils into and through the plant and atmospheric compartments of the SPAC (Walker and Brunel, 1990).

It is known that variation in the isotope ratio of water within the rhizosphere (the edaphic zone inhabited by plant roots) of plants does exist. What leads to this variation is often unknown or poorly described for most soil-plant systems. For example, Figure 11.2 shows that the isotopic 'signature' of rhizosphere soil water can vary dramatically in both space and time (Liu et al., 1995). Although the observed variation will depend on the mixing of different precipitation sources (i.e. rain, fog, and/or snow melt) with existing ground water, the δD and $\delta^{18}O$ composition of the soil water within the rhizosphere will also be dramatically influenced by the soil's physical characteristics as well as the ambient environmental factors which influence water movement into, through and from the soil profile (Barnes and Allison, 1989). It is very important therefore to understand how soil structure and ambient environmental factors interact to determine the mixing proportions as well as the evaporation gradients within a soil profile. Because soil structure and the ambient environmental conditions can influence the δD and $\delta^{18}O$ of soil water, simplistic soil characterization is likely to be misleading when interpreting plant-water-source data. To avoid this, robust soil characterization must be performed.

A handful of studies have characterized isotope profiles in soils: the investigations by Allison et al. (1983, 1983), Barnes and Allison (1983, 1988, 1989), and Barnes et al.

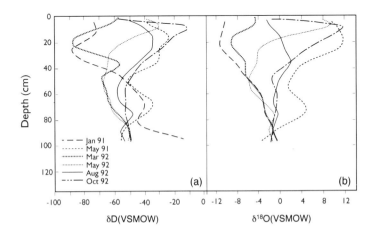

Figure 11.2. The seasonal variation in the soil water δD (a) and $\delta^{18}O$ (b) isotope profile of a desert soil in Arizona, USA. Note the nearly 90‰ variation in the δD profile in January 1991 alone as well as the 50‰ variation in δD between winter (January 91) and summer (August 92) at only one depth (-25 cm) (from Liu et al., 1995).

(1989) are particularly notable. All have demonstrated how evaporative conditions can influence the isotopic gradients of soils and that soil moisture content, the thermal conditions within the profile, and soil texture, all affect the shape of the isotopic profile (see Figure 11.2). Most often these data are derived from soil columns under known and controlled conditions and show that soil structure can have an important influence on the fractionation process and thus the isotopic profile. In the field, Walker and Brunel (1990) also showed strong gradients in δD and $\delta^{18}O$ within the soil profile. They showed that soil temperature had a clear influence on the shape of the isotope profile; temperature varied considerably between shaded and sunlit surfaces and under these surfaces, they discovered different δD and $\delta^{18}O$ of the soil water. The interaction of thermal and evaporative gradients operating in different soils must therefore be considered if we want to understand what leads to the observed variation in soil water isotopic composition and in the water available for uptake by plant roots.

Soil porosity and texture may also affect the isotopic signatures of soil water (Emerman and Dawson, 1997). With the exception of a small set of laboratory data, however, this is poorly documented. In one study, Leaney et al. (1993) investigated subsurface flow in a podzolic soil, and using $\delta^{18}O$ these authors were able to partition the contribution of different source waters to the subsurface soil water $\delta^{18}O$. Interestingly, some of the variation they observed was not what one might predict based on simple evaporative enrichment or percolation. The authors suggested that the unusual variation they observed in the isotopic signatures of water within the soil profile could be attributed to the flow patterns of water through a network of soil macropores. Bengtsson et al. (1991), also discussed the influence of soil macropores on the $\delta^{18}O$ values of water within a forest soil in Finland. Water within these macropores was derived from a mixture of snow melt and shallow ground water. During the spring the melt water mixed with ground water in poorly drained areas and in the zones of low water permeability. This water 'mixture' then flowed into the shallow ground water table through macropores. If plant xylem-water had been sampled during these subsurface flow and mixing events something very different would have been concluded about sources of water used by the plants compared to times when mixing and subsurface flow was at a minimum or non-existent. Another study conducted in Alaska by Cooper et al. (1993) found marked variation in the $\delta^{18}O$ of soil and stream water during snow melt, and attributed this to isotopic fractionation during the phase-change from solid to liquid since no mixing of the soil water and ground water could occur because of the impermeable permafrost layer in the subsoil. Most recently, Liu et al. (1995) reported variation in the $\delta^{18}O$ of soil water at different soil depths in an undisturbed desert soil which was correlated with soil development. These soils varied in age, and the authors suggested that an increase in the proportion of finer soil particles (and therefore a microporous structure) in the older soils probably caused greater evaporation and hence a heavier isotopic signature.

What we know about the mechanisms which lead to isotopic variation within the rhizosphere of the SPAC is limited. Some information about what leads to isotopic variation in soil water does exist and has been well documented in a few cases cited above. However, the physical and biological processes which combine to influence the isotopic composition of water near and at the root surface are poorly understood or simply unknown. Information about this soil–root interface and the possible isotopic variation which exists there could be critical in understanding how the rhizosphere has an impact on water movement through the SPAC.

11.2.3 Linking water source to plant performance

Hydrogen and oxygen isotopes have enabled researchers to link the source of water uptake to whole plant water status, water use efficiency, and leaf level gas exchange in different species (reviewed in Ehleringer *et al.*, 1993). Several comparative studies have demonstrated greater water stress and water-use efficiency (WUE) in species that depend on shallow or less predictable water sources than in co-occuring species with access to permanent stream or ground water (Flanagan *et al.*, 1992; Thorburn *et al.*, 1994; Valentini *et al.*, 1992). Likewise, Dawson (1993b) linked the water status of forest herbs with their access to water that was hydraulically lifted at night into shallow soil layers by *Acer saccharum*. But again, these studies were concerned with linking one end member of the SPAC, the source water, to plant water status without examining the role of the proximal source of water for plant uptake in the soil. A study by Jackson *et al.* (1995) compared the isotope signature of soil water at different depths with that of tree species that varied in rooting depth and leaf phenology in a semi-deciduous lowland tropical forest on Barro Colorado Island, Panama (Figure 11.3). The authors determined that an evaporative gradient in the soil profile influenced water availability for plants and was responsible for the different isotope signatures of tree species with their roots at different depths in this profile. They also showed that there was a negative relationship between the δD of xylem water and both leaf water potential and transpiration rate (Figure 11.3). Shallow-rooted species had more negative leaf-water potentials (greater leaf-water stress) and lower transpiration rates than species with access to deeper soil-water. In contrast to previous studies, Jackson and her colleagues found no relationship between rooting depth and instantaneous water use efficiency WUE_i. This study demonstrated the important role of evaporative processes in the soil to water status and uptake patterns of plants. In addition, their study showed that stable isotope analyses could be used to investigate water use in

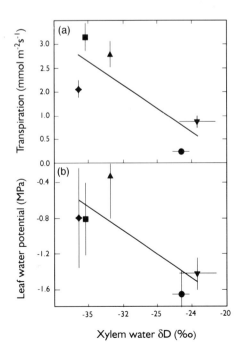

Figure 11.3. The relationship between xylem water δD and (a) plant transpiration (upper panel) and (b) leaf water potential at the end of the dry season for five plant species inhabiting gas in a lowland tropical forest, Barro Colorado Island, Panama. Each symbol is a different species (mean ± se; n = 2–6): Cecropia obtusiflora (▲), Miconia argentea (■), Piper cordulatum (◆), Psychotria limonensis (▼), and Palicourea guianensis (●). Note that for this study a more negative leaf water potential indicates greater plant water stress while a more negative δD value indicates deep soil water (from Jackson et al., 1995).

tropical systems where seasonal differences in the isotope composition of source waters are less-pronounced or non-existent, unlike studies conducted in temperate systems (Dawson, 1993a).

The use of hydrogen and oxygen isotopes in these studies of plant performance in relation to water in the environment demonstrates that isotopic studies of this type might also be used to understand water movement through the SPAC at the entire community and/or ecosystem scale.

11.2.4 *Linking water source to communities and ecosystems*

There have been several studies that use isotopes to link the source of water uptake with ecosystem-level water flux by scaling up from the individual plant to make inferences about stand-level hydrological processes (Brunel *et al.*, 1992; Busch *et al.*, 1992; Dawson and Ehleringer, 1997; Thorburn *et al.*, 1993).

Dawson (1996a) for example has recently demonstrated that older individuals of *Acer saccharum* transpire more than younger trees. Older (larger) trees have access to both the ground water (a perennial water source) and shallow soil water reservoirs that the trees themselves create through hydraulic lift. However, younger trees only had access to shallow soil water and therefore transpired less and were more sensitive to soil water deficit and increased evaporative demand. This study showed the utility of linking isotope information on water sources available to different aged trees with other information (e.g. sap flow and Bowen ratio methods) which permits one to estimate transpiration. The outcome permitted scaling up to the stand-level and estimating how water flux and water source are interdependent (Dawson, 1996a; also see Waring *et al.*, 1980 using radioisotopes). Based on his findings, Dawson suggested that ground water discharge from older stands would be greater than from younger stands, but that mixed stands could result in even greater annual water flux due to the exploitation of ground water and soil water by older trees, soil water by younger trees, and enhanced transpiration overall due to the utilization of hydraulically lifted water by all trees.

A more recent use of stable isotopes for extrapolating to the ecosystem-level water balance has been the estimation of hydrologic inputs due to fog drip. We (1996b; T.E. Dawson, unpublished data) have now demonstrated that fog interception by Coastal Redwoods (*Sequoia sempervirens*) may account for 22–46% of yearly moisture input to the Coastal Redwood forests of northern California. This input comprises approximately 19% of all the water used each year by the Redwood trees themselves. In addition, understory species appeared to benefit directly from this increased input, obtaining an average of 66% of their water from fog that had been intercepted by the tree foliage and had dripped into the soil, during the otherwise dry summer. The study showed the importance of trees to the hydrological balance of fog-inundated ecosystems. Moreover, the isotopic information permitted an understanding of how this water source was important for the entire community from both an ecological and hydrological perspective (see also Bruijnzeel, 1991). Without being able to 'trace' plant xylem water back to its source, understanding the link between water input and the source of that input for community and ecosystem function would not have been possible.

11.2.5 *Linking soils and plants to the atmosphere*

Understanding fully what regulates water movement through the SPAC requires that we understand the fate of water as it re-enters the atmosphere from plant leaves. During the re-entry process the isotopic composition of both leaf water and the water vapour will be changed by both equilibrium and kinetic fractionation processes (Craig and Gordon, 1965; Flanagan and Ehleringer, 1991; see Chapters 10, 12 and 24). The extent of this fractionation has the potential to tell us something about the atmospheric conditions at the time of water flux (e.g. vapour pressure difference between the leaf and the air; Flanagan and Ehleringer, 1991; Flanagan et al., 1991) as well as leaf water status (Dongmann et al., 1974; Wang and Yakir, 1995; Yakir et al., 1989, 1994; Zundel et al., 1978), the rates of evapotranspiration (Bariac et al., 1983, 1987, 1989; Brunel et al., 1992; Gat and Bowser, 1991; Gat and Matusi, 1991, personal communication) and leaf transpiration (Flanagan et al., 1991; Walker and Brunel, 1990). Reviewing these studies it becomes clear that our current models for linking leaf water and water vapour isotope composition are incomplete and require further refinement and data (see Chapters 10 and 23). However, the potential for using isotopes to eventually link soil and plant water status to the atmospheric compartment of the SPAC holds great promise. For example, ongoing research by several groups is now investigating the isotopic signatures of leaf water, leaf water vapour, atmospheric water vapour and the water sources taken up by plant roots (see also Chapters 10 and 12). Such information may then be used eventually to determine the proportion of water above a particular ecosystem that comes directly from the plant canopy or from evaporation and which water sources (soil or ground water) are being transpired by different species of plants in the community or ecosystem, and feed into the models described in Chapter 24.

Efforts to use the leaf water isotopic signal to determine the atmospheric conditions during plant transpiration or the transpiration rate itself have met with mixed success. This may, in part, be due to that fact that the current model (CG) first developed by H. Craig and L. Gordon (1965) used to interpret the data or to predict what we might expect is based on 'open' evaporating systems, like lakes, where equilibrium fractionation between the liquid and vapour phases of water is assumed to be at steady state (Flanagan et al., 1991). However, a number of observations point out that this assumption may be invalid or only valid under certain circumstances (Wang and Yakir, 1995) because plants possess a series of 'pools' (stem, vein and mesophyll water). Each of these pools may have their own unique isotopic values (Yakir et al., 1989) and flux through them may never be at isotopic steady state because transpiration rates vary over time (e.g. daily) and space (e.g. within a canopy). Moreover, the water volume within the series of pools may be quite large relative to the flux of water moving through them (Wang and Yakir, 1995). All of these conditions would prevent leaves from ever reaching a steady state equilibrium, especially in large plant canopies exposed to a highly variable and non-linear set of environmental conditions, as has been discussed in detail in Harwood et al. (1988; see also Chapters 3 and 10).

Under laboratory conditions using single leaves, Flanagan and his colleagues (1991) performed an elegant set of experiments on isotopic enrichment of leaf water aimed at predicting leaf transpiration rate (E). They showed that if the isotope composition of the atmospheric water vapour and the conductance of water vapour through the leaf boundary layer are known and if E and the leaf-to-air vapour pressure difference (VPD) are constant, then the leaf-water isotopic enrichment and that predicted from their model (a modified version of the CG model) agreed fairly well. However, their model always predicted a greater degree of H and ^{18}O enrichment than was observed.

This discrepancy was likely to be caused by at least some of the factors stated above (e.g. non-steady state conditions) and therefore the authors recommended that these be quantified if the model was to be applied (Flanagan *et al.*, 1991). They also pointed out that using leaf water enrichment may be problematic for obtaining quantitative information about VPD. Some of these same conclusions are echoed by Wang and Yakir (1995) in a recent survey they performed on isotopic variation in the leaf water of several plant species. They caution against making too many assumptions about isotopic steady state in leaves and show that isotopic heterogeneity of leaf water can add further complications to the interpretation of data about leaf water enrichment (see Chapter 10). All of this information points out to us that the current models are in need of refinement and modification before they can be used to make robust predictions about leaf E and VPD based solely on isotopic leaf water enrichment.

Although it is clear that the CG model, or some variant, may not yet provide the precise link between leaf water enrichment and water loss from leaves in a predictive fashion, under field conditions some data point to the promise of using isotopes to evaluate fluxes of water vapor from plant canopies. For example, in a series of papers, Bariac and his colleagues show a strong relationship between the δD and $\delta^{18}O$ of leaf water and evapotranspirational flux (Figure 11.4) over a relatively uniform alfalfa canopy. Although the causes underlying the daily cycle of leaf water enrichment could

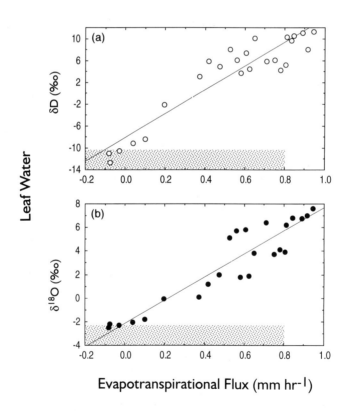

Figure 11.4. *The relationship between evapotranspirational flux determined with aerodynamic-energy balance methods and (a) the δD and (b) the $\delta^{18}O$ of leaf water. The data were obtained from field-grown alfalfa plants (redrawn from Bariac et al., 1989 in Dawson and Ehleringer, 1997).*

not be fully explained, the data showed that it is possible to predict water fluxes based on the isotopic composition of leaf water (see Bariac *et al.*, 1983, 1989). Walker and Brunel (1990) were also able to establish a relationship between the degree of isotopic leaf water enrichment and evapotranspiration working in a relatively heterogeneous *Eucalyptus* 'mallee' community. Moreover, they extended these approaches by measuring the isotopic composition of water vapour. Using an atmospheric transport model provided a promising method, partially supported by data, for predicting water flux from complex landscapes based on the isotopic composition of water in the soils, the plants, and the water vapour. While we believe that there are parts of their study which need further verification and clarification, the approach may provide one of the best attempts to link the different compartments of the SPAC using isotopic and modeling methods in concert. Lastly, the idea of measuring the isotope composition of water sources along the entire SPAC can be extended by combining it with eddy correlation methods (Brunel *et al.*, 1992). These methods allow one empirically to determine both the gaseous fluxes from entire plant stands (e.g. Hollinger *et al.*, 1994) and the isotopic 'fingerprint' of these gases. Known as relaxed eddy-accumulation (Pattey *et al.*, 1993), this combined method has the potential to close the SPAC and identify the sources and sinks of water and other gases moving into and out of any ecosystem.

11.3 Future research on isotopes and the SPAC

Plants act as important 'regulators' of water movement through the SPAC (Dawson and Ehleringer, 1997) particularly in forested ecosystems which cover nearly a third of all land surfaces (Costanza *et al.*, 1997) and where water use by plants is high and rooting patterns are diverse (Canadell *et al.*, 1996). Either at a single species or ecosystem level, few studies have linked plant water sources, soil properties, plant behaviour, and atmospheric conditions to the rates and patterns of water flux through the SPAC. This lack of an integrated picture may be due, in part, to the difficulty of working on each compartment of the SPAC simultaneously. However, stable H and O isotope analyses of water in plants and the environment can help to understand each compartment and then to link them into a more robust perspective on the SPAC. Isotope analyses provide information which may not be obtainable with other methods (e.g. tracing water through the SPAC). They can also provide a long-term picture which is lacking when making measurements on an instantaneous basis at a particular point in space or time (Ehleringer and Dawson, 1992).

One particularly important 'compartment' of the SPAC which has been poorly researched by plant biologists is the soil. Isotope analyses can be especially powerful here and are providing more detailed information about how soil structure and plant roots interact to influence uptake and movement of water (Emerman and Dawson, 1997). This is important because isotope tracing in some sense allow us to 'see' into the soil environment without disrupting soil structure. This is essential if we hope to understand the ways in which water moves naturally through the SPAC. At the other end of the SPAC, the leaf–atmosphere interface, there is still a considerable amount of research to be done. Recent advances in modelling and measuring water movement at this interface using isotopes have been made (see Section 11.2.5 above and Chapters 10, 12 and 24). However, the models need refinement, the measurements need verification across a range of plants and conditions, and these data need explicit integration into work being performed on soils and root systems. To date such an integration has not been fully accomplished. The research of Bariac and his co-workers and of Walker and Brunel (1990) comes closest but the regulatory role of the plants via stomatal behaviour and rhizosphere interactions were not well quantified. Our ongoing work in sugar maple and Coastal Redwood forests is now incorporating the dynamic nature of whole-plant water use and root water uptake (Dawson, 1993, 1996 a,b) with soil characterization, microclimatic analyses and detailed stable isotope information from soil, plant, leaf, and atmospheric water vapour samples. These different, yet complimentary, sets of data and analyses we believe will provide some of most comprehensive data on the SPAC and perhaps point the way to how hydrogen and oxygen stable isotopes can enhance our understanding of water movement along the soil–plant–atmospheric continuum.

Acknowledgments

We thank Howard Griffiths for his invitation to attend the meeting in Newcastle upon Tyne which prompted this review. Conversations with Howard, Jim Ehleringer, Larry Flanagan, Joel Gat, Dave Hollinger, and Dan Yakir were particularly helpful. Financial support for some of our own research was provided by the National Science Foundation and the A.W. Mellon Foundation.

References

Allison, G.B. and Hughes, M.W. (1983) The use of natural tracers as indicators of soil-water movement in a temperate semi-arid region. *J. Hydrol.* **60**, 157–173.

Allison, G.B., Barnes, C.J. and Hughes, M.W. (1983) The distribution of deuterium and ^{18}O in dry soil. *Exp. J. Hydrol.* **64**, 377–397.

Allison, G.B., Barnes, C.J., Hughes, M.W. and Leaney, F.W. (1984) Effect of climate and vegetation on oxygen-18 and deuterium profiles in soils. In: *Isotope Hydrology*, IAEA, Vienna. pp. 105–122.

Bariac, T., Ferhi, A., Jusserand, C. and Létolle R. (1983) Sol–plante–atmosphère: contribution à l'étude dé la composition isotopique de l'eau des differentes composantes de ce systeme. In: *Proc. Symp. Isotope Rad. Tech. Soil Phys. Irrig. Stud.* IAEA, Vienna, pp. 561–576.

Bariac, T., Klamecki, S., Jusserand, C. and Létolle, R. (1987) Evolution de la composition isotopique de l'eau (^{18}O) dans le continuum sol-plante-atmosphère (example d'une parcelle cultivée en blé, Versailles, France, Juin, 1984). *Catena*, **14**, 55–72.

Bariac, T., Rambal, S., Jusserand, C. and Berger, A. (1989) Evaluating water fluxes of field-grown alfalfa from diurnal observations of natural isotope concentrations, energy budget and ecophysiological parameters. *Agric. For. Meteorol.* **48**, 263–283.

Barnes, C.J. and Allison, G.B. (1983) The distribution of deuterium and ^{18}O in dry soils. *J. Hydrol.* **60**, 141–156.

Barnes, C.J. and Allison, G.B. (1988) Tracing of water movement in the unsaturated zone using stable isotopes of hydrogen and oxygen. *J. Hydrol.* **100**, 143–176.

Barnes, C.J. and Allison, G.B. (1989) Temperature gradient effects on stable isotope and chloride profiles in dry soils. *J. Hydrol.* **112**, 69–87.

Bengtsson, L., Lepistö, A., Saxena, R.K. and Seuna, P. (1991) Mixing of meltwater and groundwater in a forested basin. *Aqua Fennica*, **21**, 3–12.

Bruijnzeel, L.A. (1991) Hydrologic impacts of tropical forest conversion. *Nature and Resources* **27**, 85–95.

Brunel, J-P., Simpson, H.J., Herczeg, A.L., Whitehead, R. and Walker, G.R. (1992) Stable isotope composition of water vapor as an indicator of transpiration fluxes from rice crops. *Water Res.* **28**, 1407–1416.

Busch, D.E., Ingraham, N.L. and Smith, S.D. (1992) Water uptake in woody riparian phreatophytes of the southwestern United States: a stable isotope study. *Ecological Applications* **2**, 450–459.

Canadell, J., Jackson, R.B., Ehleringer, J.R., Mooney, H.A., Sala, O.E. and Schulze, E-D. (1996) Maximum rooting depth of vegetation types at the global scale. *Oecologia* **108**, 583–595.

Chahine, M.T. (1992) The hydrological cycle and its influence on climate. *Nature* 359: 373–380.

Cooper, L.W., Solis, C., Kane, D.L. and Hinzman, L.D. (1993) Application of oxygen-18 tracer techniques to Arctic hydrological processes. *Arctic and Alpine Research* **25**, 247–255.

Costanza, R. *et al.* (1997) The value of the world's ecosystem services and natural capital. *Nature* **387**, 253–260.

Covich, A.P. (1993) Water and ecosystems. In: *Water In Crisis. A guide to the world's fresh water resources* (ed P.H. Gleick). Oxford University Press, New York, pp. 40–55.

Craig, H., Gordon, L.I. (1965) Deuterium and oxygen-18 variations in the ocean and the marine atmosphere. In: *Proceedings of the Conference on Stable Isotopes in Oceanographic Studies and Paleotemperatures* (ed E. Tongiorgi). Lab. Geol. Nucl., Pisa, Italy, pp. 9–130.

Dawson, T.E. (1993a) Water sources of plants as determined from xlyem-water isotopic composition: perspectives on plant competition, distribution, and water relations. In: *Stable Isotopes and Plant Carbon–Water Relations* (eds J.R. Ehleringer, A.E. Hall, G.D. Farquhar). Academic Press, San Diego, pp. 465–496.

Dawson, T.E. (1993b) Hydraulic lift and water use in plants: implications for performance, water balance and plant–plant interactions. *Oecologia* **95**, 565–574.

Dawson, T.E. (1996a) Determining water use by trees and forests from isotopic, energy balance and transpiration analyses: the roles of tree size and hydraulic lift. *Tree Physiology* **16**, 263–272.

Dawson, T.E. (1996b) The use of fog precipitation by plants in coastal redwood forests. In: *Proceedings of the Conference on Coast Redwood Forest Ecology and Management* (ed J. LeBlanc). Humbolt State University, pp. 90–93.

Dawson, T.E. and Ehleringer, J.R. (1991) Streamside trees that do not use stream water. *Nature*, **350**, 335–337.

Dawson, T.E. and Ehleringer, J.R. (1993) Isotopic enrichment of water in the woody tissues of plants: implications for plant water source, water uptake, and other studies which use stable isotopes. *Geochim Cosmochim Acta*, **57**, 3487–3492.

Dawson, T.E. and Ehleringer, J.R. (1997) Plants, isotopes, and water use: a catchment-level perspective. In: *Isotope Tracers in Catchment Hydrology* (eds J.J. McDonnell and C. Kendall). Elsevier Science Publishers, Amsterdam.

Dawson, T.E. and Pate, J.S. (1996) Seasonal water uptake and movement in root systems of Australian phreatophytic plants of dimorphic root morphology: a stable isotope investigation. *Oecologia* **107**, 13–20.

Dongmann, G., Nurnberg, H.W., Förstel, H. and Wagner, K. (1974) On the enrichment of $^2H^{18}O$ in the leaves of transpiring plants. *Radiat. Environ. Biophys.* **11**, 41–52.

Ehleringer, J.R. and Dawson TE. (1992) Water uptake by plants: perspectives from stable isotopes. *Plant Cell Environ.* **15**, 1073–1082.

Ehleringer, J.R., Phillips, S.L., Schuster, W.S.F. and Sandquist, D.R. (1991) Differential utilization of summer rains by desert plants. *Oecologia* **88**, 430–434.

Ehleringer, J.R., Hall, A.E. and Farquhar, G.D. (eds) (1993) *Stable Isotopes and Plant Carbon–Water Relations*. Academic Press, Inc., San Diego.

Emerman, S.H. and Dawson, T.E. (1997) Experiments using split-root chambers on water uptake from soils macropores by sunflower. *Plant and Soil* **189**, 57–63.

Flanagan, L.B. and Ehleringer, J.R. (1991) Stable isotopic composition of stem and leaf water: applications to the study of plant water use. *Functional Ecology,* **5**, 270–277.

Flanagan, L.B., Comstock, J.P. and Ehleringer, J.R. (1991) Comparison of modeled and observed environmental influences in stable oxygen and hydrogen isotope composition of leaf water in *Phaseolus vulgaris*. *Plant Physiol.* **96**, 588–596.

Flanagan, L.B., Ehleringer, J.R. and Marshall, J.D. (1992) Differential uptake of summer precipitation among co-occurring trees and shrubs in a pinyon-juniper woodland. *Plant Cell Environ.* **15**, 831–836.

Gat, J.R. and Browser, C. (1991) The heavy isotope enrichment of water in coupled evaporative systems. In: *Stable isotope geochemistry: a tribute to Samuel Epstein* (eds H.P. Taylor Jr., R. O'Neil and I.R. Kaplan). The Geochemical Society, Special Publication No. 3. pp. 159–168.

Gat, J.R. and Matusi, E. (1991) Atmospheric water balance in the Amazon Basin: an isotopic evapotranspiration model. *J. Geophys. Res.* **96(D7)**, 13179–13188.

Gleick, P.H. (ed) (1993) *Water In Crisis. A guide to the world's fresh water resources.* Oxford University Press, New York. 473 pp.

Griffiths, H. (1991) Applications of stable isotope technology in physiological ecology. *Functional Ecology* **5**, 254–269.

Harwood, K.G., Gillon, J.S., Griffiths, H. and Broadmeadow, M.S.J. (1997) Diurnal variation of $\Delta^{13} CO_2$, $\Delta C^{18}O^{16}O$ and evaporative site enrichment of δH^2 ^{18}O) in *Piper aduncum* under field conditions in Trinidad. *Plant Cell Environ.* In press.

Hollinger, D.Y., Kelliher, F.M., Schulze, E-D and Köstner, B.M.M. (1994) Coupling of tree transpiration to atmospheric turbulence. *Nature* **371**, 60–62.

Hunt, E.R. Jr, Running, S.W. and Federer, C.A. (1991) Resistances and capacitances in liquid water pathway for soil–plant–atmosphere models at various temporal and spataila scales. *Agric. Forest Meteorol.* **54**, 169–195.

Jackson, P.C., Cavelier, J., Goldstein, G., Meinzer, F.C. and Holbrook, N.M. (1995) Partitioning of water resources among plants of a lowland tropical forest. *Oecologia,* **101**, 197–203.

Jarvis, P.G. and McNaughton, K.G. (1986) Stomatal control of transpiration: scaling up from leaf to region. *Adv. Ecological Res.* **15**, 1–49.

Jolly, I.D. and Walker, G.R. (1996) Is the field water use of *Eucalyptus largiflorens* F. Muell. affected by short-term flooding? *Aust. J. Ecology* **21**, 173–183.

Kramer, P. and Boyer, J. (1995) *Plant and soil water relations.* Academic Press, San Diego.

Lajtha, K. and Michener, R.H (eds). (1994) *Stable Isotopes in Ecology and Environmental Science.* Blackwell Scientific Publications, Oxford.

Leaney, F.W. Jr., Smettem, J. and Chittleborough, D.J. (1993) Estimating the contribution of preferential flow to subsurface runoff from a hillslope using deuterium and chloride. *J. Hydrol.* **147**, 83–103.

Liu, B., Phillips, F., Hoines, A., Campbell, A.R. and Sharma, P. (1995) Water movement in desert soil traced by hydrogen and oxygen isotopes, chloride, and chlorine-36, southern Arizona. *J. Hydrol.* **168**, 91–110.

McNaughton, K.G. and Jarvis, P.G. (1983) Predicting the effects of vegetation changes on transpiration and evaporation. In: *Water deficits and plant growth.* Vol. VII (ed. T.T. Kozlowski). Academic Press, New York. pp. 1–47.

Meinzer, F.C. (1993) Stomatal control of transpiration. *Tr. Ecology Evolution,* **8**, 289–294.

Mensforth, L.J., Thorburn, P.J., Tyerman, S.D. and Walker, G.R. (1994) Sources of water used by riparian *Eucalptus camaldulensis* overlying highly saline groundwater. *Oecologia* **100**, 21–28.

Newson, M. (1994) *Hydrology and the River Environment.* Clarendon Press, Oxford, 221 pp.

Pattey, E., Desjardins, R.L. and Rochette, P. (1993) Accuracy of the relaxed eddy-accumulation technique, evaluated using CO_2 flux measurements. *Boundary-Layer Meteorol.* **66**, 341–355.

Rind, D., Rosenzweig, C. and Goldberg, R. (1992) Modelling the hydrological cycle in assessments of climate change. *Nature* **358**, 119–122.

Rutter, A.J. (1968) Water consumption by forests. *In:* *Water Deficits and Plant Growth.* Vol. II (ed T.T. Kozlowski). Academic Press, New York. pp. 23–84.

Smith, M.D., Jarvis, P.G. and Odongo, J.C.W. (1997) Sources of water used by trees and millet in Sahelian Wwndbreak systems. *J. Hydrol.* **198**, 140–153.

Spittlehouse, D.L. and Black, T.A. (1981) Measuring and Modelling forest evapotranspiration. *Can. J. Chem. Eng.* **59**, 173–200.

Sternberg, L.d.S.L., Ish-Shalom-Gordon, N., Ross, M. and O'Brien, J. (1991) Water relations of coastal plant communities near the ocean/freshwater boundary. *Oecologia* **88**, 305–310.

Thorburn, P.J., Walker, G.R. and Brunel, J-P. (1993a) Extraction of water from *Eucalyptus* trees for analysis of deuterium and oxygen-18: Laboratory and field techniques. *Plant Cell Environ.* **16**, 269–277.

Thorburn, P.J., Hatton, T.J. and Walker, G.R. (1993b) Combining measurements of transpiration and stable isotopes of water to detemine groundwater discharge from forests. *J. Hydrol.* **150**, 563–587.

Thorburn, P.J., Mensforth, L.J. and Walker, G.R. (1994) Reliance of creek-side river red gums on creek water. *Aust. J. Marine and Freshwater Res.* **45**, 1439–1443.

Valentini, R., Scarascia Mugnozza, G.E. and Ehleringer, J.R. (1992) Hydrogen and carbon isotope ratios of selected species of a mediterranean macchia ecosystem. *Functional Ecology* **6**, 627–631.

Walker, C.D., Leaney, F.W., Dighton, J.C. and Allison, G.B. (1989) The influence of transpiration on the equibration of leaf water with atmospheric water vapour. *Plant Cell Environ.* **12**, 221–234.

Walker, C.D., Brunel, J-P. (1990) Examining evapotranspiration in a semi-arid region using stable isotopes of hydrogen and oxygen. *J. Hydrol.* **118**, 55–75.

Walker, C.D., Richardson, S.B. (1991) The use of stable isotopes of water in characterizing the sources of water in vegetation. *Chemical Geology (Isotope Geoscience Section)*, **94**, 145–158.

Wang, X-F. and Yakir, D. (1995) Temporal and spatial variations in oxygen-18 content of leaf water in different plant species. *Plant Cell Environ.* **18**, 1377–1385.

Waring, R.H., Whitehead, D. and Jarvis, P.G. (1980) Comparison of an isotopic method and the Penman-Monteith equation for estimating transpiration from Scots pine. *Can. J. Forest Res.* **10**, 555–569.

White, J.W.C. (1988) Stable hydrogen isotope ratios in plants: a review of current theory and some potential applications. In: *Stable Isotopes in Ecological Research* (eds P.W. Rundel, J.R. Ehleringer, and K.A. Nagy). Ecological Studies, Vol. 68. Springer-Verlag, Heidelberg. pp. 142–162.

White, J.W.C., Cook, E.R., Lawrence, J.R. and Broecker, W.S. (1985) The D/H ratios of sap in trees: implications for water sources and tree ring D/H ratios. *Geochim. Cosmochim. Acta*, **49**, 237–246.

Whitehead, D. and Hinckley, T.M. (1991) Models of water flux through forests stands: critical leaf and stand parameters. *Tree Physiol.* **9**, 35–57.

Whittaker, R.H. (ed) (1975) *Communities and Ecosystems*. Macmillan, New York. 385 pp.

Yakir, D., DeNiro, M.J and Rundel, P.W. (1989) Isotopic inhomogeneity of leaf water: evidence and implications for use of isotopic signals transduced by plants. *Geochim. Cosmochimic. Acta* **53**, 2769–2773.

Yakir, D., Ting, I., DeNiro, M.J. (1994) Natural abundance $^2H/^1H$ ratios of water storage in leaves of *Peperomia congesta* HBK during water stress. *J. Plant Physiol.* **144**, 607–612.

Yakir, D. and Yechieli, Y. (1995) Plant invasion of newly exposed hypersaline Dead Sea shores. *Nature* **374**, 803–805.

Zundel, G., Miekeley, W., Grisi, B.M. and Förstel, H. (1978) The $H_2^{18}O$ enrichment of leaf water of tropic trees: comparison of species from the tropical rain forest and the semi-arid region in Brazil. *Radiat. Environ. Biophys.* **15**, 203–212.

Oxygen isotope effects during CO$_2$ exchange: from leaf to ecosystem processes

Lawrence B. Flanagan

12.1 Introduction

Terrestrial ecosystems play a major role in the global carbon cycle (Sundquist, 1993). There is a great need, therefore, to understand how global changes, such as increasing atmospheric CO$_2$ and enhanced levels of nitrogen deposition, will influence physiological responses of terrestrial vegetation, and how the vegetation can feedback to influence the global carbon cycle and the earth's climate system (Sellers *et al.*, 1996). In order to understand large scale vegetation–atmosphere interactions we need to develop techniques to scale up or extend physiological measurements beyond the leaf or individual plant level. Despite significant progress in using micrometerological and remote sensing techniques (Ehleringer and Field, 1993), attempts to extend physiological measurements to large areas are often difficult for a variety of technical reasons. However, stable isotope analyses of atmospheric carbon dioxide are one promising tool which can scale processes from the soil and individual plant level through to the troposphere (see Chapter 24).

Berry (1992) has suggested that if we assume that the total volume of our atmosphere is constant, we can treat our global atmosphere as a closed system and take approaches to studying it as we would a physiological system in a closed chamber. Recent work has illustrated that measurements of atmospheric CO$_2$ concentration can be used to study aspects of large scale ecosystem gas exchange processes (Keeling *et al.*, 1995, 1996; Myneni *et al.*, 1997). The slow mixing of the atmosphere and the fact that uptake (photosynthesis) and release (respiration) of CO$_2$ are not uniformly distributed in space and time, cause local changes in atmospheric CO$_2$ concentration from the global average value (Conway *et al.*, 1988, 1994; see Chapter 13). For example, superimposed on the annual increase of atmospheric CO$_2$ are large seasonal fluctuations, which are most apparent in the high latitudes of the northern hemisphere. These seasonal fluctuations occur as a result of differences in the timing and

Stable Isotopes, edited by H. Griffiths.
© 1998 BIOS Scientific Publishers Ltd, Oxford.

magnitude of photosynthesis and respiration in terrestrial ecosystems (Keeling *et al.*, 1996; Myneni *et al.*, 1997).

Associated with changes in atmospheric CO_2 concentrations are changes in the carbon and oxygen stable isotope ratio of atmospheric CO_2 (Ciais *et al.*, 1995; Ciais *et al.*, 1997b; Keeling *et al.*, 1995; Mook *et al.*, 1983; Trolier *et al.*, 1996). For example, Mook *et al.* (1983) and Trolier *et al.* (1996) have noted a seasonal change in the oxygen isotope ratio of CO_2 that correlated with the decline in atmospheric CO_2 concentration. There is also a latitudinal gradient, with atmospheric CO_2 having lower $\delta^{18}O$ values in more northerly regions (Ciais *et al.*, 1997b; Francey and Tans, 1987; Trolier *et al.*, 1996). The global-scale changes in the stable isotope ratio of atmospheric CO_2 result from isotope effects that occur during ecosystem gas exchange processes and, therefore provide a useful tracer of interactions that occur between ecosystems and the atmosphere. The purpose of this chapter is to review studies of factors that influence the oxygen isotope ratio of atmospheric CO_2 within terrestrial plant canopies. Inclusion of this mechanistic process information in global models of atmospheric transport may allow measurements of the stable isotope ratio of atmospheric CO_2 to be used as a tool in studies of large scale vegetation–atmosphere CO_2 exchange (Ciais *et al.*, 1997a,b; Farquhar *et al.*, 1993; see Chapter 24).

The stable isotope ratio of atmospheric CO_2 within a plant canopy is primarily influenced by three factors (Lloyd *et al.*, 1996): (i) isotope discrimination that occurs during photosynthetic gas exchange; (ii) the isotope ratio of CO_2 input into the canopy by plant and soil respiration; and (iii) turbulent exchange with the atmosphere above the canopy. A description of these processes and the factors that control them is presented below.

12.2 Discrimination against $C^{18}O^{16}O$ during photosynthetic gas exchange

12.2.1 *Model of $C^{18}O^{16}O$ discrimination during leaf gas exchange*

There are two major processes that influence discrimination against $C^{18}O^{16}O$ during photosynthetic gas exchange (Farquhar and Lloyd, 1993). The first process is fractionation that occurs during diffusion. The CO_2 molecules containing ^{18}O are heavier and, therefore, diffuse at a slower rate than molecules containing only ^{16}O. The second process is an oxygen isotope exchange reaction that occurs between CO_2 and H_2O molecules in the chloroplast. During photosynthetic gas exchange, a portion of the CO_2 that enters the leaf and equilibrates with chloroplast water is not fixed and diffuses back out of the leaf with an altered oxygen isotope ratio. The amount of CO_2 that escapes from the leaf depends on the partial pressure of CO_2 in the chloroplast and conductance to diffusion within and outside the leaf. The oxygen isotope exchange reaction occurs via a mechanism that involves the hydration of dissolved CO_2. The CO_2 molecule must be resident in water for approximately 30 seconds to allow for hydration and oxygen isotopic exchange to occur (Francey and Tans, 1987). This equilibration time is long relative to the average residence time of CO_2 in leaves (less than 1 second) during photosynthetic gas exchange. However, the presence of the enzyme, carbonic anhydrase (CA), in plant tissue facilitates the equilibration reaction by rapidly catalysing the hydration of CO_2 and dehydration of HCO_3^-. The oxygen isotope ratio of CO_2 leaving the leaf will depend on: (i) the oxygen isotope ratio of chloroplast water; (ii) leaf temperature, which influences the equilibrium fraction-ation factor for the CO_2–H_2O exchange reaction; and (iii) fractionation during diffu-sion of CO_2 out of the leaf. The Farquhar and Lloyd (1993) model quantitatively describes the influence of discrimination during photosynthetic gas exchange. In a simplified form the model is shown below:

$$\Delta C^{18}O^{16}O = \frac{a + \dfrac{C_c}{C_a - C_c}(\delta^{18}O_c - \delta^{18}O_a)}{1 - \dfrac{C_c}{C_a - C_c}(\delta^{18}O_c - \delta^{18}O_a)} \qquad (12.1)$$

where a is the average fractionation during diffusion of CO_2 into and out of a leaf; C_c and C_a are the partial pressures of CO_2 in the chloroplast and ambient air, respectively; $\delta^{18}O_c$ is the oxygen isotope ratio of CO_2 in the chloroplast (VPDB–CO_2 scale); and $\delta^{18}O_a$ is the oxygen isotope ratio of CO_2 in the well mixed atmosphere outside of a leaf.

In this model (Equation 12.1) it is assumed that CO_2 in the chloroplast ($\delta^{18}O_c$) is at complete isotopic equilibrium with leaf (chloroplast) water. Depending on the relative activities of CA and ribulose bisphosphate carboxylase (RUBISCO), isotopic equilib-rium may not be complete, and Equation 12.1 can be modified as follows (Farquhar and Lloyd, 1993)

$$\Delta C^{18}O^{16}O = \frac{a(1 + 3\tau) + \dfrac{C_c}{C_a - C_c}((\delta^{18}O_c - \delta^{18}O_a) + 3\tau b)}{1 - \dfrac{C_c}{C_a - C_c}(\delta^{18}O_c - \delta^{18}O_a) + 3\tau \dfrac{C_c}{C_a - C_c}} \qquad (12.2)$$

where τ is the ratio of the rate of carboxylation by RUBISCO to the rate of hydration by CA, and b represents discrimination against $C^{18}O^{16}O$ during carboxylation (taken as 0‰).

Shown graphically, the Farquhar and Lloyd (1993) model predicts that $\Delta C^{18}O^{16}O$ is a strong function of the ratio of chloroplast and ambient CO_2 concentrations (C_c/C_a) (Figure 12.1). At a constant C_c/C_a, $\Delta C^{18}O^{16}O$ values should increase when chloroplast water has a higher ^{18}O content (Figure 12.1).

12.2.2 *Comparison of modelled and measured leaf $C^{18}O^{16}O$ discrimination*

In the initial experimental tests of the model of $C^{18}O^{16}O$ discrimination, Farquhar *et al.* (1993) made comparisons between observed and predicted $\Delta C^{18}O^{16}O$ values when environmental conditions were altered to induce variation in the ratio of photosyn-thetic capacity and stomatal conductance and thereby cause large changes in C_c/C_a. A large amount of scatter was apparent in the observed $\Delta C^{18}O^{16}O$ values for a given C_c/C_a value, probably as a result of variation in the leaf–air vapour pressure difference (VPD) and associated changes in leaf water oxygen isotope ratio. In a second set of experiments conducted by Flanagan *et al.* (1994), plant leaves were maintained at dif-ferent VPDs in a controlled environment chamber in order to generate a range of leaf water $^{18}O/^{16}O$ ratios. Measurements of 'on-line' discrimination showed that $\Delta C^{18}O^{16}O$ values increased in association with the change in leaf water ^{18}O content,

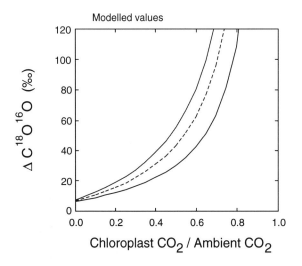

Figure 12.1. *Modelled influence of changes in the ratio of chloroplast CO_2 concentration and ambient CO_2 concentration on discrimination against $C^{18}O^{16}O$ during photosynthetic gas exchange. The different lines represent calculations done with increasing ^{18}O content of chloroplast water.*

and the observed $\Delta C^{18}O^{16}O$ values were strongly correlated to values predicted by the mechanistic model (Flanagan *et al.*, 1994). However, the extent to which $\Delta C^{18}O^{16}O$ values respond to changes in VPD, depends on the contrasting effects of leaf water isotopic enrichment and changes in stomatal conductance associated with alterations in VPD. A plant with strong stomatal closure in response to decreases in humidity, may not show any change in $\Delta C^{18}O^{16}O$ values, despite the fact leaf water ^{18}O content increased in association with the decrease in humidity (Williams *et al.* 1996). This is because stomatal closure results in a lower C_c/C_a ratio and therefore lower discrimination, which can counteract any increase in $\Delta C^{18}O^{16}O$ values expected because of increased leaf water ^{18}O content associated with high evaporative enrichment of leaf water at low humidity (Williams *et al.*, 1996).

Recently we have conducted two additional types of experiments to test the $\Delta C^{18}O^{16}O$ model. In the first experiment we made use of the fact that in *Sphagnum* moss, large changes occur in C_c/C_a in association with changes in moss water content (Figure 12.2), while VPD remains quite stable (Williams and Flanagan, 1996). Measurements of 'on-line' discrimination, taken during the time required for the moss to dry out from full hydration, showed large changes in $\Delta C^{18}O^{16}O$ values that were strongly correlated to values predicted by the model (Figures 12.2 and 12.3). In the second type of experiment, transgenic tobacco (*Nicotiana tabacum*) plants, modified by antisense suppression of chloroplast CA, were compared to wild type plants during 'on-line' $\Delta C^{18}O^{16}O$ measurements under a range of environmental conditions (Williams *et al.*, 1996). On average the low CA plants had only 8% of the CA activity of the wild type plants. As expected, there was a clear difference in $\Delta C^{18}O^{16}O$ between low CA and wild type plants (Figure 12.4), a result similar to that seen by Price *et al.* (1994). The Farquhar and Lloyd (1993) model assumes that $\Delta C^{18}O^{16}O$ will be strongly influenced by the extent to which isotopic equilibrium between CO_2 and chloroplast water is achieved. In plants with low CA activity, there will be incomplete equilibration between CO_2 and chloroplast water and, therefore, low $\Delta C^{18}O^{16}O$ values are expected. The degree to which isotopic equilibration is achieved is reflected in the ratio of the number of fixations of CO_2 to hydrations of CO_2 (τ in equation 12.2). A value for τ of 0.019 established a good fit between the observed and predicted values (Figure 12.4) for control

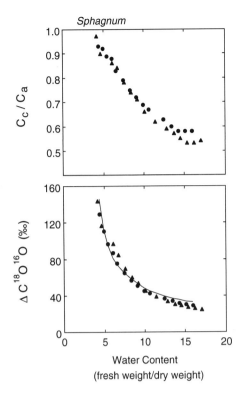

Figure 12.2. *Effect of variation in water content on changes in the ratio of chloroplast CO_2 concentration and ambient CO_2 concentration (C_c/C_a), and discrimination against $C^{18}O^{16}O$ during photosynthetic gas exchange in* Sphagnum. *Redrawn from data in Williams and Flanagan (1996).*

plants. If chloroplast water and CO_2 were in perfect isotopic equilibrium in the control plants, as would be described by a τ of 0 in Equation 12.2., this would result in an increase in $\Delta C^{18}O^{16}O$ values of approximately 24‰ ± 6.2, above the observed values for the control plants. The average τ value of 0.019 ± .003 (S.E. $n=32$) for tobacco is close to the value calculated for *Phaseolus* by Flanagan *et al.* (1994) using the same methodology. In contrast, a τ of approximately 0.5 was calculated for tobacco plants with low carbonic anhydrase activity, reflecting the lower level of hydrations.

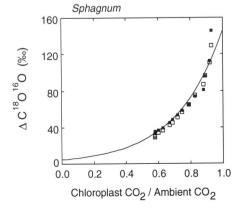

Figure 12.3. *Comparison between modelled and observed effect of changes in the ratio of chloroplast CO_2 concentration and ambient CO_2 concentration on discrimination against $C^{18}O^{16}O$ during photosynthetic gas exchange in* Sphagnum. *The open symbols represent observed data and the dark symbols represent modelled values. Based on data in Williams and Flanagan (1996).*

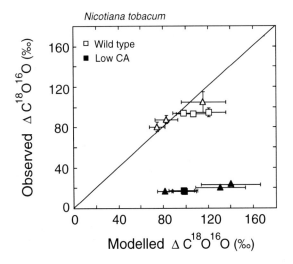

Figure 12.4. Comparison between observed and modelled discrimination against C^{18}O^{16}O during photosynthetic gas exchange in transgenic tobacco plants. The low CA tobacco plants were transformed using an antisense construct, and had approximately 8% the level of carbonic anhydrase activity as the control plants. Reproduced from Williams et al. (1996) with permission from the American Society of Plant Physiologists.

12.2.3 *Oxygen isotope ratio of leaf and chloroplast water*

The application of Equations 12.1 and 12.2 requires information about the isotope ratio of chloroplast water. In most applications it is assumed that chloroplast water has the same oxygen isotope ratio as water at the evaporative sites within leaves (hereafter simply referred to as leaf water). The isotope composition of leaf water can be estimated based on a model of isotopic fractionation during transpiration (Craig and Gordon, 1965; Flanagan, 1993). There is some controversy about the assumption that chloroplast water has the same isotopic composition as water at the sites of evaporation within leaves (see Chapter 10). Yakir *et al.* (1994) have suggested that the isotopic signature of water in chloroplasts is closer to that of total leaf water, than to water at the sites of evaporation. The δ^{18}O of total leaf water can be up to 6‰ lower than that predicted by the evaporative enrichment model (Flanagan *et al.*, 1991, Yakir *et al.*, 1994; see Chapter 10). The discrepancy between predicted and measured total leaf water is a function of the transpiration rate, probably as a result of the shifting balance between the bulk flow of unfractionated source water into the leaf, and the back diffusion of water enriched in ^{18}O away from the sites of evaporation (Flanagan *et al.*, 1991, 1994; Farquhar and Lloyd, 1993; see also Chapters 3 and 10). A discrepancy of 6‰ between the isotopic signature of water in the chloroplast and water at the sites of evaporation could generate a difference of approximately –20‰ in the ΔC^{18}O^{16}O values predicted using Equations 12.1 and 12.2. However, a number of 'on-line' measurements of ΔC^{18}O^{16}O values under a variety of environmental conditions are consistent with the assumption that chloroplast water has an isotopic signature close to that of water at the evaporative sites within leaves (Farquhar *et al.*, 1993; Flanagan *et al.*, 1994; Williams and Flanagan, 1996; Williams *et al.*, 1996). We feel justified, therefore, in using the Craig–Gordon (CG) evaporative enrichment model to estimate the ^{18}O content of chloroplast water in studies of factors influencing the isotope ratio of atmospheric CO$_2$ (but see Appendix 2 in Chapter 3).

12.3 Oxygen isotope effects during respiratory CO$_2$ exchange

The major processes that influence C^{18}O^{16}O discrimination during photosynthetic gas exchange, also have important effects on the oxygen isotope ratio of respired CO_2. Equilibration of respired CO_2 with plant and soil water, and fractionation during diffusion are the dominant processes. Temperature and the oxygen isotope ratio of the water pool associated with the source of the respiratory CO_2 exert primary control on the oxygen isotope ratio of respired CO_2. The magnitude of the discrimination factor during diffusion through the soil is approximately 8.8‰, but its precise value depends on the relative rates of; (i) CO_2 production, (ii) equilibration between CO_2 and H_2O, (iii) diffusion of CO_2 out of the soil, and (iv) invasion of atmospheric CO_2 into the soil (Hesterberg and Siegenthaler, 1991; Tans, 1998; see Chapter 24). Within a plant canopy, the oxygen isotope ratio of respired CO_2 is likely to be much less uniform than the carbon isotope ratio of respired CO_2 (Farquhar et al., 1993; Flanagan and Varney, 1995; Friedli et al., 1987), because of the diverse nature of the isotope content of water in various plant and soil components. A description of these components is provided below.

The source of water taken up by plants is precipitation and/or ground water, and since no fractionation occurs during water uptake by most plants, the isotope ratio of water in plant roots is the same as that of the water available in the soil (Ehleringer and Dawson, 1992; Flanagan et al., 1992; White et al., 1985; see Chapters 11 and 23). As noted earlier, fractionation does occur during transpiration, so water in above-ground plant tissues can have an oxygen isotope ratio substantially different from that of source water, depending on the rate of evaporation relative to the input of unfractionated source water in that tissue (Farquhar and Lloyd, 1993; Flanagan, 1993; Flanagan et al., 1994). Most non-green stem tissue shows no evaporative enrichment of [18]O (Dawson and Ehleringer, 1992). In contrast green, unsuberized stem tissue can have water with a substantially different [18]O content than source water (Dawson and Ehleringer, 1992). However, usually it is only the leaves that have water very enriched in [18]O. In addition there can be a strong diurnal cycle observed for leaf water oxygen isotope ratios. At night when respiration can cause a significant increase in the concentration of atmospheric CO_2 within a plant canopy, leaf water should have a lower [18]O content than that observed during the day, with values closer to that of stem water. During the day, leaf and stem components may have respired CO_2 with very different oxygen isotope ratios because of the potential difference in the oxygen isotope content of their respective water pools. However, the quantitative importance of the oxygen isotope ratio of respired CO_2 contributed by above-ground plant components during the day is likely to be very small because it forms such a low proportion of the total CO_2 exchange between the plant canopy and the atmosphere.

The isotopic ratio of water in soils, and that of soil respired CO_2, is dependent on precipitation inputs, ground water transport, and the amount of evaporation that occurs from the soil (Allison and Hughes, 1983; Barnes and Allison, 1983, 1984; Forstel et al., 1990; see Chapter 11). The oxygen isotope ratio of water in a soil covered by a grass meadow has been shown to vary seasonally in association with the seasonally changing isotope ratio of precipitation inputs (Hesterberg and Siegenthaler, 1991). Water at different depths (30 cm and 80 cm) had different isotopic compositions because of the time lag for water movement down through the soil and because of soil water evaporation (Hesterberg and Siegenthaler, 1991). In soils covered with a dense plant canopy there should be less influence of evaporation on soil water isotopic ratios (Allison and Hughes, 1983; Barnes and Allison, 1983, 1984), but throughfall precipitation can have oxygen isotope ratios that are enriched above that of total precipitation because of interception and evaporation of water from the canopy (Dewalle and Swistock, 1994). As a consequence soil water values are slightly enriched in ^{18}O relative to what would be predicted by total precipitation inputs. Soil water oxygen isotope ratio is expected to change temporally and spatially because of differences in precipitation inputs and above-ground vegetation, and as a consequence the oxygen isotope ratio of soil-respired CO_2 should be quite variable in different ecosystems (see Chapter 24).

For modelling purposes, it is convenient to assume that respired CO_2 has equilibrated with the water pool associated with the respiratory source (i.e. leaf, stem or soil water). The isotopic equilibrium reaction between oxygen in CO_2 and oxygen in water takes a long time to complete when it is uncatalysed (Francey and Tans, 1987). However, for CO_2 released from plant tissues the equilibrium assumption is strongly supported by the rather ubiquitous presence of CA in plant tissues (Graham et al., 1984). Carbonic anhydrase may also be present in soils because it is produced by common soil algae such as *Chlamydomonas*, and the enzyme may be released from decaying leaf litter. The equilibration between CO_2 and water in soils will also be enhanced by inorganic catalytic sites and the long pathlengths for diffusion of CO_2 through soil air spaces (Francey and Tans, 1987). Ciais et al. (1997a) calculated that it can take approximately 6 hours for a CO_2 molecule to diffuse 30 cm through the soil to reach the atmosphere. Such a long residence time should allow for equilibration, and studies by Allison et al. (1987) and Hesterberg and Siegenthaler (1991) have provided support for the idea that soil CO_2 is in isotopic equilibrium with soil water (see Chapter 24).

12.3.1 *Oxygen isotope ratio of respired CO₂ in boreal forest ecosystems*

We have used a simple mixing model developed by Keeling (1958, 1961) to calculate the isotope ratio of CO_2 respired in a forest canopy, which can also be used for ^{13}C: see Chapter 13. For $\delta^{18}O$:

$$\delta^{18}O_f = \frac{[CO_2]_o}{[CO_2]_f}(\delta^{18}O_o - \delta^{18}O_R) + \delta^{18}O_R \qquad (12.3)$$

where $[CO_2]$ is concentration of CO_2 and $\delta^{18}O$ is the stable oxygen isotope ratio of atmospheric CO_2, and the subscripts 'f' and 'o' represent the atmosphere at a point within the forest canopy and above (outside) the forest canopy respectively. It can be seen from Equation 12.3 that a plot of $1/[CO_2]_f$ and $\delta^{18}O_f$ gives a straight line relationship with slope, $[CO_2]_o (\delta^{18}O_o - \delta^{18}O_R)$, and an intercept $\delta^{18}O_R$. Estimates for $\delta^{18}O_R$, the isotopic composition of respired CO_2, were obtained from the y-intercept of a geometric mean linear regression between $\delta^{18}O_f$ and $1/[CO_2]_f$ values measured on air samples collected within a forest canopy at night. Measurements were made in Black Spruce and Jack Pine ecosystems in the boreal forest of Canada. A discussion of some our results is provided below in order to illustrate the influence of factors affecting the oxygen isotope ratio of respired CO_2 in terrestrial ecosystems.

For samples collected in the relatively open, arid Pine site, strong linear relationships were observed between $1/CO_2$ concentration and the oxygen isotope ratio of CO_2 in air samples collected at night (Figure 12.5). The $\delta^{18}O_R$ values calculated from the linear regression analysis were similar to values predicted from estimates of soil water oxygen isotope ratio (based on plant stem water isotope analysis) (Figure 12.6). However, the calculated $\delta^{18}O_R$ values were slightly enriched (by 2–4‰) in ^{18}O relative

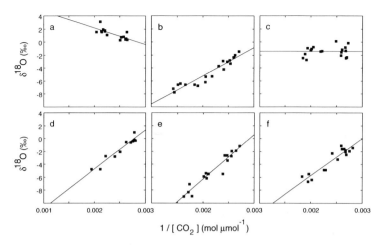

Figure 12.5. *Relationship between the oxygen isotope ratio and $1/CO_2$ concentration of air samples collected at night in Black Spruce and Jack Pine ecosystems in northern Canada. The black spruce site during (a) spring, (b) summer, and (c) fall sampling periods. The jack pine site during (d) spring, (e) summer, and (f) fall sampling periods. Redrawn from data in Flanagan et al. (1997) with permission.*

to predicted values, likely to be a result of the upper few centimeters of soil containing water enriched in ^{18}O, relative to the water taken up by the Pine trees, which comes predominantly from a water source deeper in the soil. The contribution of above ground plant respiration, particularly leaf respiration, could also result in the input of CO_2 with a higher $\delta^{18}O$ value, as discussed above. Michael G. Ryan (USDA Forest Service, Rocky Mountain Experiment Station, Fort Collins, CO, 80526, USA; unpublished data) has determined that foliage, stem and soil respiration contribute to total night time respiration in the Jack Pine site in the following proportions: foliage 0.315, stem 0.079 and soil 0.606. Given the large contribution of leaf respiration to total system respiration, leaf water at night must have had a similar oxygen isotope ratio as soil water in order to obtain such strong linear regressions between changes in $1/CO_2$ and $\delta^{18}O$ values of atmospheric CO_2 (Figure 12.5).

In contrast to the patterns observed in the Pine site, $\delta^{18}O_R$ values calculated from the linear regression analysis were substantially enriched in ^{18}O relative to predictions from stem water isotope ratios in the Black Spruce site (Figure 12.6). The forest floor of the Black Spruce site has continuous carpet of feather moss and *Sphagnum* moss. Evaporation of water from the moss can result in significant changes to the oxygen isotope ratio of moss water. Any CO_2 diffusing out of the moss/soil surface has the potential to equilibrate isotopically with this moss water pool, which can be substantially enriched in ^{18}O, and very different in composition from the rest of the water in the soil profile. For example, during the mid-summer (July) sampling period at the Black Spruce site, the moss had been recently wetted by precipitation and had a relatively negative $\delta^{18}O$ (standard mean ocean water) value of -7.3 ±1.3‰ (mean ± SD, $n=5$), which was enriched in ^{18}O relative to Black Spruce stem water (-18.0 ±0.2‰), but still significantly lower than the Black Spruce leaf water $\delta^{18}O$ (SMOW) value of 5.5 ±2.2‰, measured on leaf samples collected at midday. The $\delta^{18}O_R$ value calculated using the linear regression technique (-14.4‰, Figure 12.5) was very similar to that predicted for respired CO_2 in

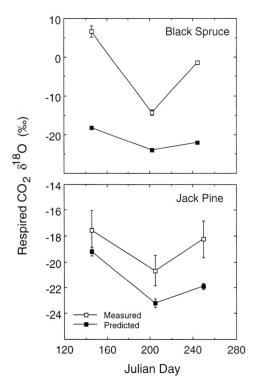

Figure 12.6. *Comparison of measured and predicted values of the oxygen isotopic composition of carbon dioxide released during plant and soil respiration in boreal forest ecosystems. Calculated values were based on CO$_2$ in isotopic equilibrium with soil water, which was estimated based on tree stem water isotopic composition. Measured values were determined from the y-intercept of a linear regression between 1/[CO$_2$] and oxygen isotope ratio of atmospheric CO$_2$ collected at night within a forest canopy. Redrawn from data in Flanagan* et al. *(1997) with permission.*

isotopic equilibrium with moss water (–13.4‰), but enriched in ^{18}O relative to that of respired CO$_2$ in equilibrium with bulk soil water (as estimated by tree stem water, Figure 12.6). In contrast, there was no significant linear relationship between changes in 1/CO$_2$ concentration and the δ^{18}O value for atmospheric CO$_2$ during the fall (September) sampling period (Figure 12.5). Prior to that sampling, the moss had not received any precipitation for some time, and moss water must have been relatively enriched in ^{18}O with a similar δ^{18}O (SMOW) value to that of tree leaf water (1.3±1.3‰, measured at midday). The increase in CO$_2$ concentration within the canopy at night, resulting from plant and soil respiration, did not cause a significant change in isotope ratio, because CO$_2$ released from the soil/moss surface had equilibrated with moss water, and so was similar in isotopic composition to that of atmospheric CO$_2$ above the forest (Figure 12.5). If moss water had a δ^{18}O (SMOW) value of approximately 4‰, the expected value of respired CO$_2$ in equilibrium with moss water (–1.4‰: VPDB–CO$_2$ scale), would be in close agreement with the average of δ^{18}O measurements made on CO$_2$ samples collected at night (Figure 12.5). During the spring sampling period in the Black Spruce site (May), we observed a significant increase in the δ^{18}O value of atmospheric CO$_2$ at night in association with the increase in CO$_2$ concentration (Figure 12.5). This indicated that respired CO$_2$ had equilibrated with water that was very enriched in ^{18}O, with the linear regression analysis indicating a δ^{18}O$_R$ value of 6.6‰ (Figure 12.6). Black Spruce leaf water had δ^{18}O (SMOW) values at midday of 10.6‰ during the spring sampling. Assuming that moss water had a similar oxygen isotope ratio as tree leaf water, the expected δ^{18}O value of respired CO$_2$ in equilibrium with moss water was 6.6‰ (VPDB–CO$_2$ scale), in complete agreement with the value calculated using the linear regression technique.

Our data indicate that the isotope ratio of CO_2 respired from the moss/soil surface in the Black Spruce sites was significantly influenced by short-term changes in precipitation and the oxygen isotope ratio of moss water. Since Black Spruce ecosystems and other vegetation-types with an abundance of moss dominate northern regions, moss water enriched in ^{18}O could have a substantial influence on the oxygen isotope ratio of atmospheric CO_2 in these areas. The ^{18}O-enriched water signal in the moss can be passed on to respired CO_2 diffusing through the moss layer to the forest floor surface, and could also influence the atmosphere during moss photosynthetic gas exchange. In addition, both trees and moss in the Black Spruce ecosystems are strongly influenced by short-term changes in the isotope ratio of summer precipitation, which in general is more enriched in ^{18}O than ground water. This would tend to cause terrestrial biosphere–atmosphere CO_2 exchange to have less of a depletion effect on the ^{18}O content of the atmosphere than global-scale model calculations using average annual precipitation inputs as an estimate for soil water isotopic composition (see below; see also Chapter 24).

12.4 Relative influence of photosynthesis, respiration and turbulent fluxes on the isotope ratio of atmospheric CO_2 in plant canopies

A mass balance approach can be used to illustrate the relative importance of discrimination during photosynthetic gas exchange, the isotope ratio of respired CO_2, and turbulent exchange, on the oxygen isotope ratio of atmospheric CO_2 within plant canopies. The influence of the three major fluxes is weighted by their associated isotope effects, as shown in the equation below (Lloyd et al. 1996):

$$M_i[CO_2]_i \frac{d\delta^{18}O_i}{dt} = A\Delta C^{18}O^{16}O + R(\delta^{18}O_R - \delta^{18}O_i) + F_{oi}(\delta^{18}O_o - \delta^{18}O_i) \quad (12.4)$$

where M_i is the molar density of air within the forest canopy space (mol m^{-3}); $[CO_2]_i$ is the average CO_2 concentration within a forest canopy space (μmol mol^{-1}); $\delta^{18}O_i$ is the average isotope ratio of CO_2 within the forest canopy space (‰); A is the net CO_2 assimilation rate (μmol m^{-2} s^{-1}); $\Delta C^{18}O^{16}O$ is discrimination against $C^{18}O^{16}O$ during photosynthetic gas exchange (‰); R is the respiration rate of plants and soil (μmol m^{-2} s^{-1}); $\delta^{18}O_R$ is the oxygen isotopic composition of plant and soil respired CO_2 (‰); F_{oi} is the one-way turbulent flux of CO_2 (mmol m^{-2} s^{-1}) into the canopy from the atmosphere above (outside) the canopy; and $\delta^{18}O_o$ is the average isotope ratio of atmospheric CO_2 above (outside) the forest canopy (‰).

During our collection of air samples for isotope analysis in the boreal forest ecosystems, simultaneous measurements were made, by other research groups, of ecosystem CO_2 exchange using the eddy covariance technique. Data from these combined studies have indicated that during the day photosynthetic CO_2 exchange can have the dominant effect on the isotope ratio of atmospheric CO_2, acting to enrich canopy CO_2 in ^{18}O, followed closely by turbulent exchange with the atmosphere above the canopy, which acts in an opposite direction (Figure 12.7). At night, input of respiratory CO_2 can cause substantial changes (usually depletions) in the ^{18}O content of atmospheric CO_2 within plant canopies as shown in Section 12.3.1. The influence of the isotope effects are weighted by the magnitude of the associated CO_2 flux. For

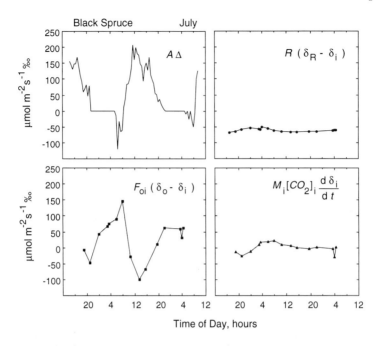

Figure 12.7. *Comparison of the relative influence of photosynthesis, respiration and turbulent exchange on changes in the oxygen isotope ratio of atmospheric CO$_2$ within a Black Spruce canopy during July 1994 in northern Canada (see Equation 4). Reproduced from Flanagan et al. Global Biogeochemical Cycles, pp. 83–98 (1997) copyright by the American Geophysical Union*

example, despite similar $\Delta C^{18}O^{16}O$ values during the July and September sampling periods in the southern Jack Pine site, the influence of discrimination during photosynthetic gas exchange was reduced during September because of a low CO$_2$ assimilation rate (Figure 12.8), likely to be caused by water stress effects on the vegetation. In contrast, discrimination during photosynthetic gas exchange in the Black Spruce site had similar effects during July and September, because no reduction in CO$_2$ assimilation was apparent during the fall sampling period. The data in Figures 12.7 and 12.8 clearly indicate the important influence of biological activity in terrestrial ecosystems on the isotope ratio of atmospheric CO$_2$.

12.5 Regional and global variation in the influence of terrestrial ecosystems on the oxygen isotope ratio of atmospheric CO$_2$

Terrestrial ecosystems are expected to influence the oxygen isotope ratio of atmospheric CO$_2$ to a different extent primarily because of variation in: (i) factors that influence, photosynthetic capacity, stomatal conductance and C_c/C_a; (ii) factors that influence ecosystem respiration; (iii) the oxygen isotope ratio of precipitation inputs to the ecosystem; and (iv) factors that influence the leaf and chloroplast water oxygen isotope ratio. We now have a good mechanistic understanding of all the major processes described above.

It is possible to use the mass balance equations developed by Lloyd *et al.* (1996) to

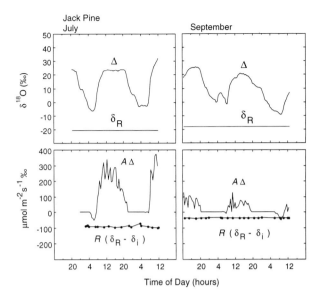

Figure 12.8. *The influence of photosynthesis and respiration on the oxygen isotope ratio of atmospheric CO_2 within a jack pine forest canopy during July and September 1994 in northern Canada (see Equation 4). In the upper panels the isotope effects during photosynthesis (Δ) and respiration (δ_R) are shown alone. The lower panels show the isotope effects multiplied by the associated flux term (solid line indicates the AΔ values, and the solid line with dots shows the R (δ_R-δ_i) values). Redrawn from data in Flanagan et al. (1997).*

study the regional influence of isotope effects during photosynthesis and respiration on atmospheric CO_2. The analyses require atmospheric CO_2 samples to be collected by aircraft in vertical profiles through the convective boundary layer (CBL), and are based on the CBL budgeting technique developed by Raupach *et al.* (1992). The CBL budgeting approach allows more direct links to be made between the study of ecosystem processes and global modelling approaches. Recent measurements of the oxygen isotope composition of atmospheric CO_2 made on samples collected by aircraft over the boreal forest in Russia are consistent with our understanding of oxygen isotope effects during terrestrial ecosystem gas exchange (Nakazawa *et al.*, 1997).

Farquhar *et al.* (1993) used the simple models of the fractionation processes described above to calculate how oxygen isotope effects during photosynthetic gas exchange and respiration should vary among ecosystems across the globe. The Farquhar *et al.* (1993) model was tested by showing that it: (i) correctly calculated the overall $\delta^{18}O$ value of atmospheric CO_2, and (ii) was qualitatively consistent with the latitudinal gradient in $\delta^{18}O$ observed by Francey and Tans (1987). The latitudinal gradient in the oxygen isotope ratio of atmospheric CO_2 is primarily influenced by the very strong latitudinal gradient in the $\delta^{18}O$ value of precipitation (Figure 12.9). In northern latitudes, precipitation is very depleted in ^{18}O because of fractional distillation of water during atmospheric transport away from tropical regions (Dansgaard, 1964).

Ciais *et al.* (1997a, b) used a three-dimensional atmospheric transport model in conjunction with estimates of surface fluxes to calculate the influence of terrestrial photosynthesis and respiration, ocean CO_2 exchange, and anthropogenic emissions on the

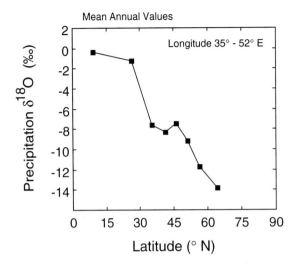

Figure 12.9. Latitudinal variation in the oxygen isotope ratio of precipitation in the northern hemisphere. Based on data in IAEA (1986).

$\delta^{18}O$ of atmospheric CO$_2$. Their results illustrated the important contrasting effects of terrestrial photosynthesis and respiration and are explored in detail in Chapter 24. In particular, the strong influence of terrestrial respiration was shown because of its effect on the shape of the latitudinal change in the $\delta^{18}O$ of atmospheric CO$_2$, and on the lag between seasonal changes in CO$_2$ concentration and $\delta^{18}O$ observed at northern locations. Ciais *et al.* (1997a,b) concluded that measurements of the $\delta^{18}O$ of atmospheric CO$_2$ could be used to provide information on the large scale distribution of gross CO$_2$ fluxes, although they noted that their model could not yet be used for quantitative flux estimates because they did not examine the sensitivity of their calculations to model parameters (see Chapter 24). In order to adequately interpret global scale differences in ecosystem net CO$_2$ exchange, it is necessary to have knowledge of the source of variation in the gross CO$_2$ fluxes. This is because photosynthesis and respiration respond differently to important environmental changes such as temperature and CO$_2$ concentration.

Carbon dioxide exchange between the terrestrial biosphere and the atmosphere has a different effect on the oxygen isotope ratio of atmospheric CO$_2$ than does CO$_2$ exchange with the oceans. This is because the ^{18}O content of ocean water is relatively constant throughout the globe, varying only slightly in association with ocean salinity (Craig and Gordon, 1965). In contrast, the ^{18}O content of fresh water (precipitation, soil water, plant water) is very different from that of ocean water and varies substantially among different biomes (IAEA 1986; see Chapter 23). It may be possible, therefore, to use measurements of change in the oxygen isotope ratio of atmospheric CO$_2$ in an analogous manner to ^{13}C in studies of partitioning of anthropogenic CO$_2$ emissions between terrestrial and ocean sinks (Ciais *et al.*, 1997a,b; Farquhar *et al.*, 1993). Because terrestrial biomes across the globe have soil water and plant water with very different $\delta^{18}O$ values, it may also be possible to identify the influence of specific biomes by monitoring changes in the oxygen isotope ratio of atmospheric CO$_2$ (Farquhar *et al.*, 1993). Since processes controlling the ^{18}O composition of atmospheric CO$_2$ are completely independent from those affecting ^{13}C, the ^{18}O data could provide important checks on the results of ^{13}C studies.

References

Allison, G.B. and Hughes, M.W. (1983) The use of natural tracers as indicators of soil-water movement in a temperate semi-arid region. *J. Hydrol.* 60, 157–173.

Allison, G.B., Colin-Kaczala, C., Filly, A. and Fontes, J.Ch. (1987) Measurement of isotopic equilibrium between water, water vapour and soil CO_2 in arid zone soils. *J. Hydrol.* 95, 131–141.

Barnes, C.J. and Allison, G.B. (1983) The distribution of deuterium and ^{18}O in dry soils. 1. Theory. *J. Hydrol.* 60, 141–156.

Barnes, C.J. and Allison, G.B. (1984) The distribution of deuterium and ^{18}O in dry soils. 3. Theory for non-isothermal water movement. *J. Hydrol.* 74, 119–135.

Berry, J.A. (1992) Biosphere, atmosphere, ocean interactions: a plant physiologist's perspective. In: *Primary Productivity and Biogeochemical Cycles in the Sea* (eds P.G. Falkowski and A.D. Woodhead). Plenum Press, New York, pp. 441–454.

Ciais, P., Tans, P.P., Trolier, M., White. J.W.C. and Francey, R.J. (1995) A large northern hemisphere terrestrial CO_2 sink indicated by the $^{13}C/^{12}C$ ratio of atmospheric CO_2. *Science* 269, 1098–1102.

Ciais, P., Denning, A.S, Tans, P.P., Berry, J.A., Randall, D.A., Collatz, G.J., Sellers, P.J., White, J.W.C., Trolier, M., Meijer, H.J., Francey, R.J., Monfray, P. and Heimann, M. 1997a. A three dimensional synthesis study of ^{18}O in atmospheric CO_2 Part 1: Surface fluxes. *J. Geophys. Res.* 102, 5857–5872.

Ciais, P., Tans, P.P., Denning, A.S., Francey, R.J., Trolier, M., Meijer, H.J., White, J.W.C., Berry, J.A., Randall, D.A., Collatz, G.J., Sellers, P.J., Monfray, P. and Heimann, M. (1997b) A three dimensional synthesis study of ^{18}O in atmospheric CO_2 Part 2: Simulations with the TM2 transport model. *J. Geophys. Res.* 102, 5873–5883.

Conway, T.J., Tans, P.P., Waterman, L.S., Thoning, K.W., Kitzis, D.R., Masarie, K.A. and Zhang, N. (1994) Evidence for interannual variability of the carbon cycle from the NOAA Climate Monitoring and Diagnostics Laboratory global air sampling network. *J. Geophys. Res.* 99, 22831–22855.

Conway, T.J., Tans, P.P., Waterman, L.S., Thoning, K.W., Masarie, K.A. and Gammon, R.H. (1988) Atmospheric carbon dioxide measurements in the remote global troposphere, 1981–1984. *Tellus* 40B, 81–115.

Craig, H. and Gordon, L.I. (1965) Deuterium and oxygen-18 variations in the ocean and the marine atmosphere. In: *Proceedings of a Conference on Stable Isotopes in Oceanographic Studies and Paleotemperatures, Spoleto, Italy* (ed E. Tongiorgi). Lischi and Figli, Pisa, pp. 9–130.

Dansgaard, W. (1964) Stable isotopes in precipitation. *Tellus* 16, 436–468.

Dewalle, D.R. and Swistock, B.R. (1994) Differences in oxygen-18 content of throughfall and rainfall in hardwood and coniferous forests. *Hydrol. Proc.* 8, 75–82.

Dawson, T.E. and Ehleringer, J.R. (1993) Isotopic enrichment of water in the 'woody' tissues of plants: implications for plant water source, water uptake, and other studies which use the stable isotopic composition of cellulose. *Geochim. Cosmoschim. Acta* 57, 3487–3492.

Ehleringer, J.R. and Dawson, T.E. (1992) Water uptake by plants: perspectives from stable isotope composition. *Plant Cell Environ.* 15, 1073–1082.

Ehleringer, J.R. and Field, C.B. (1993) *Scaling physiological processes: leaf to globe.* Academic Press, San Diego.

Farquhar, G.D., Lloyd, J., Taylor, J.A., Flanagan, L.B., Syvertsen, J.P., Hubick, K.T., Wong, S.C. and Ehleringer, J.R. (1993) Vegetation effects on the isotope composition of oxygen in atmospheric CO_2. *Nature* 363, 439–443.

Farquhar, G.D. and Lloyd, J. (1993) Carbon and oxygen isotope effects in the exchange of carbon dioxide between terrestrial plants and the atmosphere. In: *Stable Isotopes and Plant Carbon – Water Relations* (eds J.R, Ehleringer, A.E. Hall and G.D. Farquhar). Academic Press, San Diego, pp. 47–70.

Flanagan, L.B. (1993) Environmental and biological influences on the stable oxygen and hydrogen isotopic composition of leaf water. In: *Stable Isotopes and Plant Carbon – Water Relations* (eds J.R. Ehleringer, A.E. Hall and G.D. Farquhar). Academic Press, San Diego, pp. 71–90.

Flanagan, L.B., Brooks, J.R., Varney, G.T. and Ehleringer, J.R. (1997) Discrimination against C^{18}O^{16}O during photosynthesis and the oxygen isotope ratio of respired CO$_2$ in boreal forest ecosystems. *Global Biogeochem. Cycles* 11, 83–98.

Flanagan, L.B. and Varney, G.T. (1995) Influence of vegetation and soil CO$_2$ exchange on the concentration and stable oxygen isotope ratio of atmospheric CO$_2$ within a *Pinus resinosa* canopy. *Oecologia* 101, 37–44.

Flanagan, L.B., Phillips, S.L., Ehleringer, J.R., Lloyd, J. and Farquhar, G.D. (1994) Effect of changes in leaf water oxygen isotopic composition on discrimination against C^{18}O^{16}O during photosynthetic gas exchange. *Aust. J. Plant Physiol.* 21, 221–234.

Flanagan, L.B., Ehleringer, J.R. and Marshall, J.R. (1992) Differential uptake of summer precipitation among co-occurring trees and shrubs in a pinyon-juniper woodland. *Plant Cell Environ.* 15, 831–836.

Flanagan, L.B., Comstock, J.P. and Ehleringer, J.R. (1991) Comparison of modelled and observed environmental influences on the stable oxygen and hydrogen isotope composition of leaf water in *Phaseolus vulgaris* L. *Plant Physiol.* 96, 588–596.

Forstel, H., Frinken, J., Hutzen, H., Lembrich, D., Putz, T. (1990) Application of H$_2$^{18}O as a tracer of the water flow in the soil. In: *International Symposium on the Use of Stable Isotopes in Plant Nutrition, Soil Fertility, and Environmental Studies*. International Atomic Energy Agency, Vienna.

Francey, R.J. and Tans, P.P. (1987) Latitudinal variation in oxygen-18 of atmospheric CO$_2$. *Nature* 327, 495–497.

Friedli, H., Siegenthaler, U., Rauber, D. and Oeschger, H. (1987) Measurements of concentration, ^{13}C/^{12}C and ^{18}O/^{16}O ratios of tropospheric carbon dioxide over Switzerland. *Tellus* 39B, 80–88.

Graham, D., Reed, M.L., Patterson, B.D., Hockley, D.G. and Dwyer, M.R. (1984) Chemical properties, distribution and physiology of plant and algal carbonic anhydrases. *Ann. NY Acad. Sci.* 429, 222–237.

Hesterberg, R. and Siegenthaler, U. (1991) Production and stable isotopic composition of CO$_2$ in a soil near Bern, Switzerland. *Tellus* 43B, 197–205.

IAEA (1986) Environmental Isotope Data Number 8: World Survey of Isotope Concentration in Precipitation (1980–1983). International Atomic Energy Agency, Technical Report Series 264, Vienna.

Keeling, C.D. (1958) The concentration and isotopic abundances of atmospheric carbon dioxide in rural areas. *Geochim. Cosmochim. Acta* 13, 322–334.

Keeling, C.D. (1961) The concentration and isotopic abundances of carbon dioxide in rural and marine air. *Geochim. Cosmochim. Acta* 24, 277–298.

Keeling, C.D., Whorf, T.P., Wahlen, M., van der Plicht, J. (1995) Interannual extremes in the rate of rise of atmospheric carbon dioxide since 1980. *Nature* 375, 666–670.

Keeling, C.D., Chin, J.F.S. and Whorf, T.P. (1996) Increased activity of northern vegetation inferred from atmospheric CO$_2$ measurements. *Nature* 382, 146–149.

Lloyd, J., Kruijt, B., Hollinger, D.Y., Grace, J., Francey, R.J., Wong, S.C., Kelliher, F.M., Miranda, A.C., Farquhar, G.D., Gash, J.H.C., Vygodskaya, N.N., Wright, I.R., Miranda, H.S. and Schulze, E.D. (1996) Vegetation effects on the isotopic composition of atmospheric CO$_2$ at local and regional scales: theoretical aspects and a comparison between rain forest in Amazonia and boreal forest in Siberia. *Aust. J. Plant Physiol.* 23, 371–399.

Mook, W.G., Koopmans, M., Carter, A.F. and Keeling, C.D. (1983) Seasonal, latitudinal, and secular variations in the abundance and isotopic ratios of atmospheric carbon dioxide. 1. Results from land stations. *J. Geophys. Res.* 88, 10915–10933.

Myneni, R.B., Keeling, C.D., Tucker, C.J., Asrar, G. and Nemani. R.R. (1997) Increased plant growth in the northern high latitudes from 1981 to 1991. *Nature* 386, 698–702.

Nakazawa, T., Sugawara, S., Inoue, G., Machida, T., Makshyutov, S. and Mukai, H. (1997) Aircraft measurements of the concentrations of CO_2, CH_4, N_2O, and CO and the carbon and oxygen isotopic ratios of CO_2 in the troposphere over Russia. *J. Geophys. Res.* 102, 3843–3859.

Price, G.D., Caemmerer, S. von, Evans, J.R., Yu, J., Lloyd, J., Oja, V., Kell, P., Harrison, K., Gallagher and A., Badger, M. (1994) Specific reduction of chloroplast carbonic anhydrase activity by antisense RNA in transgenic tobacco plants has a minor effect on photosynthetic CO_2 assimilation. *Planta* 193, 331–340.

Raupach, M.R., Denmead, O.T. and Dunin, F.X. (1992) Challenges in linking atmospheric CO_2 concentrations to fluxes at local and regional scales. *Aust. J. Bot.* 40, 697–716.

Sellers, P.J., Bounoua, L., Collatz, G.J., Randall, D.A., Dazlich, D.A., Los, S.O., Berry, J.A., Fung, I., Tucker, C.J., Field, C.B., Jensen, T.G. (1996) Comparison of radiative and physiological effects of doubled atmospheric CO_2 on climate. *Science* 271, 1402–1406.

Sundquist, E.T. (1993) The global carbon dioxide budget. *Science* 259, 934–941.

Tans, P.P. (1997) Oxygen isotopic equilibration between carbon dioxide and water in soils. *Tellus* 50B, in press.

Trolier, M., White, J.W.C., Tans, P.P., Masarie, K.A. and Gemery, P.A. (1996) Monitoring the isotopic composition of atmospheric CO_2: measurements from NOAA global air sampling network. *J. Geophys. Res.* 101, 25 897–25 916.

White, J.W.C., Cook, E.R., Lawrence, J.R. and Broecker, W.S. (1985) The D/H ratios of sap in trees: implications for water sources and tree ring D/H ratios. *Geochim. Cosmochim. Acta* 49, 237–246.

Williams, T.G. and Flanagan, L.B. (1996) Effect of changes in water content on photosynthesis, transpiration and discrimination against $^{13}CO_2$ and $C^{18}O^{16}O$ in *Pleurozium* and *Sphagnum*. *Oecologia* 108, 38–46.

Williams, T.G., Flanagan, L.B. and Coleman, J.R. (1996) Photosynthetic gas exchange and discrimination against $^{13}CO_2$ and $C^{18}O^{16}O$ in tobacco plants modified by an antisense construct to have low chloroplastic carbonic anhydrase. *Plant Physiol.* 112, 319–326.

Yakir, D., Berry, J.A., Giles, L. and Osmond, C.B. (1994) Isotopic heterogeneity of water in transpiring leaves: identification of the component that controls the $\delta^{18}O$ of atmospheric O_2 and CO_2. *Plant Cell Environ.* 17, 73–80.

Carbon isotope discrimination of terrestrial ecosystems

N. Buchmann, J.R. Brooks, L.B. Flanagan and J.R. Ehleringer

13.1 Introduction

13.1.1 *Use of ^{13}C at the ecosystem level*

Detailed knowledge of the interactions between the atmosphere, biosphere and pedosphere is essential for an understanding of global carbon dynamics in terrestrial ecosystems. Understanding these linkages has become even more crucial as atmospheric carbon dioxide concentrations ($[CO_2]$) continue to increase (Conway *et al.*, 1988; Conway *et al.*, 1994; Komhyr *et al.*, 1985). In this context, stable isotopes have proved useful as indicators of the constraints on global carbon budgets (Tans *et al.*, 1996). Since tropospheric and respired CO_2 from terrestrial ecosystems have very different carbon isotope ratios (approximately -8‰ and -27‰, respectively), these CO_2 sources can be differentiated, and the coupling of terrestrial and atmospheric carbon fluxes can be addressed (Keeling, 1958; Lloyd *et al.*, 1996; Sternberg, 1989; Tans *et al.* 1990). Global circulation models have suggested a large carbon sink (either oceanic or terrestrial) in the Northern Hemisphere (Bender *et al.*, 1996; Ciais *et al.*, 1995; Denning *et al.*, 1995; Francey *et al.*, 1995; Keeling *et al.* 1996; Tans *et al.* 1990; see Chapter 24). Carbon isotope ratios ($\delta^{13}C$) of atmospheric carbon dioxide have been used in inverse global models (i.e. using a top-down approach) not only to determine the global distribution of carbon sinks, but also to quantify the relative contribution of oceans and terrestrial plants to the carbon removal from the atmosphere (Ciais *et al.*, 1995; Francey *et al.*, 1995; Keeling *et al.*, 1984; Mook *et al.*, 1983). However, lacking sufficient data from terrestrial ecosystems, important physiologically-based input parameters such as the ratio of internal to atmospheric $[CO_2]$ (C_i/C_a), and the carbon isotopic composition of respired CO_2 have been estimated using models for the dominant plant species within different biomes (but see Chapter 8). Further uncertainties

Stable Isotopes, edited by H. Griffiths.
© 1998 BIOS Scientific Publishers Ltd, Oxford.

arise because of the regional and temporal variability of these estimates, and due to disequilibrium effects. Better quantification of ecophysiological parameters and an understanding of seasonal variation in these parameters could prove useful in identifying potential mechanisms constraining atmosphere/terrestrial ecosystem models.

13.1.2 *Carbon isotope discrimination within ecosystems*

Lloyd and Farquhar (1994) modelled the carbon isotope discrimination ($\Delta^{13}C^{16}O^{16}O$) of entire canopies (Δ_A), with the aim of integrating across plants present within an ecosystem. Separating biomes, they estimated Δ_A globally, using data sets for ecophysiological parameters from the dominant plant species (stomatal responses to leaf-to-air vapour pressure differences (VPD) mol fraction differences), and for climate (temperature, precipitation, and relative humidity), which was then scaled for the global distribution of vegetation. Lloyd and Farquhar (1994) then validated their model by calculating the difference between the $\delta^{13}C$ of atmospheric CO_2 within the nocturnal boundary layer and estimates of $\delta^{13}C$ of respired CO_2: differences from the 1:1 line were less than 2.5 ‰. This modeling effort provided encouraging support for such a bottom-up approach, but some initial caution is necessary since the two terms compared integrate over very different temporal as well as spatial scales: Δ_A estimates integrate over periods of as long as a single growing season and represent values of a few dominant canopy forming (overstory) species only.

However, the estimates for $\delta^{13}C$ of respired CO_2 (Keeling, 1958; see Section 13.2.3) represent the carbon respired by all species (over- and understory vegetation), and also that respired by roots and microorganisms in the soil. In addition, the time scale represented by the respiration estimates is longer than a single growing season: turnover rates vary with mean annual temperature (Trumbore *et al.*, 1996), with relatively fast turnover in tropical forests (~5 years), but much slower turnover in temporal and boreal forests (as long as 100 years; Bird *et al.*, 1996). Thus, $\delta^{13}C$ of respired CO_2 will integrate carbon isotope composition of foliage and litter over much longer time spans than the Δ_A estimates. Additionally, land use changes account for the vast majority of the current global vegetation change (Houghton, 1995), and therefore $\delta^{13}C$ of soil organic carbon and soil respired CO_2 is more likely to be representative of the actual carbon released than is the current plant cover. Hence, in order to describe the impact that ecosystems have on the $\delta^{13}C$ of the atmosphere, one should consider not only the carbon isotope discrimination during current photosynthesis, but also the large historic signal recorded in the soil carbon stocks (Bird *et al.*, 1996; Trumbore *et al.*, 1996). Therefore, it seems appropriate to use a temporally and spatially integrated measure of ecosystem carbon isotope discrimination that includes soil carbon.

As a next step from the canopy discrimination Δ_A (Lloyd and Farquhar, 1994), we introduce Δ_e, a term that describes the carbon isotope discrimination of an entire ecosystem, including the soil compartment (Equation 13.1).

$$\Delta_e = \frac{\delta^{13}C_T - \delta^{13}C_R}{1 + \delta^{13}C_R} \tag{13.1}$$

where $\delta^{13}C_T$ is the carbon isotope ratio of the troposphere and $\delta^{13}C_R$ the carbon isotope ratio of CO_2 respired by soil microorganisms and vegetation (below- and aboveground). We use the $\delta^{13}C$ of respired CO_2 as an integrated value for the carbon isotopic composition of all organic matter in an ecosystem, representing carbon both in soils and vegetation. Since there is no fractionation during mitochondrial respiration (Lin and Ehleringer, 1997), the $\delta^{13}C$ of ecosystem respiration best represents all organic carbon in this ecosystem. Thus, the familiar concept of carbon isotope discrimination at the leaf level is transferred to the ecosystem level (Figure 13.1). The individual terms of the equation for leaf level carbon isotope discrimination (Δ_l; see Chapter 8) are substituted by the appropriate terms for the ecosystem level: $\delta^{13}C_a$ with $\delta^{13}C_T$ and $\delta^{13}C_l$ with $\delta^{13}C_R$. Δ_e is based on $\delta^{13}C$ of tropospheric CO_2 and on field measurements of canopy air to estimate $\delta^{13}C$ of ecosystem respired CO_2. Thus, Δ_e integrates not only over Δ_l of all photosynthesising leaves but also includes information about the $\delta^{13}C$ of soil organic matter (for detailed information about how to calculate Δ_e see Section 13.3.3).

Recent studies have built on the pioneering work of Keeling (1958) to describe relationships between CO_2 concentration ($[CO_2]$) and $\delta^{13}C$ within and above canopies. The results of these studies can be used to estimate $\delta^{13}C$ of respired CO_2 for a wide range of different ecosystems, including tropical forests (Broadmeadow et al., 1992; Buchmann et al., 1997a; Francey et al., 1985; Lancaster, 1990; Lloyd et al., 1996; Quay et al., 1989; Sternberg et al., 1989), temperate forests (Buchmann et al., 1997b,c; Keeling 1961a,b; Lancaster, 1990), boreal forests (Flanagan et al., 1996), and agricultural ecosystems (Buchmann and Ehleringer, 1997; Yakir and Wang 1996). Models of turbulent mixing of canopy air with the convective boundary layer (CBL) or the free troposphere provided detailed insight into how canopy gradients of $[CO_2]$ and $\delta^{13}C$ develop (Grace et al., 1995; Kruijt et al., 1996; Lloyd et al., 1996; Quay et al., 1989; Yakir and Wang 1996). Thus, combining these datasets with an understanding of

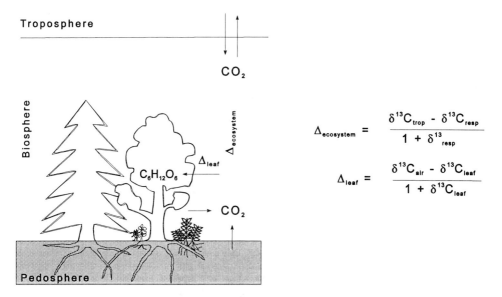

Figure 13.1. *Conceptual model of carbon isotope discrimination within a terrestrial ecosystem.*

canopy processes and their interactions with the atmosphere could be a valuable and promising approach to describe ecosystem functioning, especially with regard to climate change (Houghton *et al.*, 1996).

In this chapter, we will address ecosystem level variation of Δ_e, possible impacts of seasonality, of characteristic features of the dominant plant species within an ecosystem (life form, canopy structure, age) as well as the influence of climate and soil nutrient availability.

13.2 Methodology

This section introduces the methods needed to calculate ecosystem carbon isotope discrimination as described in Equation 13.1. We first focus on where to obtain tropospheric baseline data, and then discuss how to determine $\delta^{13}C$ of respired CO_2. Potential sources of errors associated with Δ_e estimates will be addressed before we will give an overview about recent research about Δ_e in Section 13.3.

13.2.1 *Isotope ratios*

Carbon isotope ratios ($\delta^{13}C$) are calculated as

$$\delta^{13}C = 1000 \cdot \left(\frac{R_{sample}}{R_{standard}} - 1 \right) \tag{13.2}$$

where R_{sample} and $R_{standard}$ are the $^{13}C/^{12}C$ ratios of the sample and the standard (VPDB), respectively (Farquhar *et al.*, 1989). The overall precision of carbon isotope measurements with modern dual-inlet isotope ratio mass spectrometers is generally better than 0.03‰ for carbon in gas samples (with 100 µL CO_2; 10 µL CO_2 with a coldfinger). If a continuous flow mass spectrometer with a preconcentrator is available (conflow/precon), then precision for carbon isotope analysis is about 0.3‰ (with 0.2 µL CO_2).

13.2.2 *Tropospheric baseline data for [CO₂] and δ¹³C*

Tropospheric baseline $[CO_2]_T$ have been measured on a regular basis since the International Geophysical Year (1958), and since then a steadily increasing number of land and oceanic stations have been established to monitor background levels of $[CO_2]$ globally. Keeling started $[CO_2]_T$ and $\delta^{13}C_T$ measurements at the South Pole in 1957 (Keeling *et al.*, 1976), and at Mauna Loa, Hawaii in 1958 (Pales and Keeling, 1965), which later became the SIO network (Scripps Institution of Oceanography). A second CO_2 flask sampling program, measuring first at Niwot Ridge, Colorado, and at Ocean Station 'Charlie' in the North Atlantic, followed in 1968. It was continuously expanded to become the present NOAA/CMDL Cooperative Flask Sampling Network (National Oceanic and Atmospheric Administration/Climate Monitoring and Diagnostics Laboratory), which today includes 43 stations and two cruise tracks in the South China Sea and the Pacific Ocean (Ciais *et al.*, 1995; Conway *et al.*, 1988; Conway *et al.*, 1994; Komhyr *et al.*, 1985). Since 1990, the INSTAAR (Stable Isotope Laboratory at the Institute of Arctic and Alpine Research) has been measuring tropospheric $\delta^{13}C_T$ ratios, complementing the concentration measurements of the

NOAA/CMDL network (Ciais *et al.*, 1995). Since 1982, $[CO_2]_T$ and $\delta^{13}C_T$ data have also been available from the CSIRO network (Commonwealth Scientific and Industrial Research Organization), including sites at high southern latitudes such as Cape Grim (Francey *et al.*, 1995). Recent efforts have expanded the existing sampling networks by creating a Global Carbon Cycle Observing System, which has the advantage of increasing spatial integration and achieving a better separation between natural variability and expected signals in response to increased atmospheric $[CO_2]$ (Tans *et al.*, 1996; see Chapter 24). All $[CO_2]$ data (unfortunately not $\delta^{13}C$) have been archived and summarized through the Carbon Dioxide Information Analysis Center at the Oak Ridge National Laboratory (e.g. Boden *et al.*, 1994). Data compilations are updated regularly, and also provide data from other national networks.

Within these networks, air samples are collected mainly in remote areas (oceanic or coastal sites, deserts, maintaintops) where $[CO_2]$ is not or least affected by local CO_2 sources and sinks. Thus, they represent almost free tropospheric air, although flasks are collected at 1 m height above the ground (SIO, at 7 and 27 m; NOAA/CMDL, some tall towers up to 500 m). Generally, replicate flasks are sampled weekly or biweekly during daytime hours (SIO, hourly), and analysed at one central laboratory for each network. We recommend using tropospheric $[CO_2]_T$ and $\delta^{13}C_T$, measured at a station as close as possible to the site under study (for discussion of errors, see Section 13.2.5).

13.2.3 *Estimates of $\delta^{13}C$ of respired CO_2*

For estimating ecosystem carbon isotope discrimination, the $\delta^{13}C$ of respired CO_2 should represent a weighted average of all respiration processes within the ecosystem. Because soil respiration chambers cover only a relatively small area and do not include foliage and branch/stem respiration, we discourage the use of such chambers to determine $\delta^{13}C_R$. Instead, to incorporate all respiratory processes within an ecosystem, we recommend using the so-called 'Keeling plot' method: canopy air samples are collected at times providing a wide range of $[CO_2]$, and their $[CO_2]$ and $\delta^{13}C$ ratios are analysed. Using a regression approach (see below), $\delta^{13}C_R$ of ecosystem respiration can be determined. Thus, all respiration fluxes are covered and weighted by their respective flux rates. In addition, this technique will average respiration over a larger ground area than is typically possible by using chamber or enclosure techniques.

Keeling (1958) observed that when measuring atmospheric air, the $[CO_2]_a$ and its $\delta^{13}C_a$ changed in concert in a predictable manner. If $1/[CO_2]_a$ is plotted against the corresponding $\delta^{13}C_a$ values (so-called 'Keeling plot'), a linear relationship is obtained (Figure 13.2). This relationship reflects the mixing of tropospheric CO_2 with an additional CO_2 source that is depleted in ^{13}C compared to the troposphere. The following linear equation describes the mixing model adequately (Keeling, 1961a,b):

$$\delta^{13}C_a = \frac{[CO_2]_T}{[CO_2]_a} \cdot (\delta^{13}C_T - \delta^{13}C_R) + \delta^{13}C_R \qquad (13.3)$$

The intercept of this equation has been used to identify the carbon isotope ratio of the additional CO_2 source, for example, in forest canopies, of respired CO_2 ($\delta^{13}C_R$). The linear regression equation describes day and nighttime data very well, and r^2 are generally greater than 0.9.

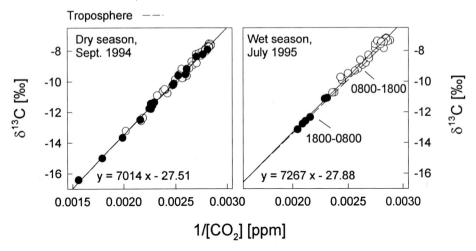

Figure 13.2. *Relationship between 1/[CO₂] and δ¹³C of canopy air in a primary tropical rainforest in French Guiana (after Buchmann* et al., *1997a).*

13.2.4 *Collection and analysis of canopy air*

To determine the $\delta^{13}C$ of respired CO_2 accurately by using the 'Keeling plot' method, it is crucial to obtain very reliable estimates of the regression coefficients. Thus, canopy air must be sampled over a wide range of $[CO_2]$ (> 50 ppm), and both $[CO_2]$ and $\delta^{13}C$ need to be measured with high precision. How does one achieve a wide $[CO_2]$ range? $[CO_2]$ and $\delta^{13}C$ values change vertically and daily. Vertical CO_2 profiles are typically rather uniform during the day within the main canopy, except near to the soil surface, where $[CO_2]$ and $\delta^{13}C$ change, sometimes drastically (Brooks et al., 1997a; Buchmann and Ehleringer, 1997; Buchmann *et al.*, 1996, 1997a-c; Flanagan *et al.*, 1996; Lloyd *et al.*, 1996). If understory vegetation is present, profiles can be strongly influenced by understory gas exchange (Buchmann *et al.*, 1997b). Six canopy air samples per height profile proved to be sufficient to characterise a canopy, with a sampling emphasis on lower heights where changes in $[CO_2]$ and $\delta^{13}C$ are more pronounced. If air is collected above the actual plant canopy, fetch size should be considered to identify potential additional sources of CO_2 (Gash 1986; Schuepp *et al.*, 1990). For example, fetch size for a 1.5 m tall crop canopy increases from 15 m to 125 m if the instrumental plane is moved from 1.5 m up to 2.5 m height; for a 10 m tall forest canopy, fetch size increases from 100 m to 613 m if the instrumental plane is moved from 10 m up to 15 m height (90% fetch, Gash 1986).

$[CO_2]$ and $\delta^{13}C$ ratios exhibit a distinct daily course, driven by turbulent mixing with the troposphere as well as by photosynthesis and respiration. $[CO_2]$ is typically high during times with high respiration and low turbulent mixing (at night). This nocturnal buildup dissipates in the morning when turbulent mixing (due to increasing temperatures) and active photosynthesis set in. $[CO_2]$ normally stays low during the day until the evening when temperatures, and therefore turbulent mixing, decrease and respiration again prevails. We recommend collecting flasks during the nocturnal build-up, but also include some daytime samples to increase the coverage of a wide range of $[CO_2]$, and maybe even include flasks collected during a pronounced photosynthetic depletion in canopy $[CO_2]$ below tropospheric background concentrations. We have measured nighttime $[CO_2]$ gradients between 75 ppm and 140 ppm and daytime gradients between 6 ppm and 100 ppm in forest canopies, but for adequate accuracy to estimate $\delta^{13}C_R$, we recommend a range of 50 ppm or greater. Our experiences from several dozen forest stands showed that day and night data lie on the same regression line. Thus, increasing the range of the independent variable $(1/[CO_2])$ for the regression analysis by daytime samples will increase the precision of the regression technique and yield a more realistic estimate of the regression coefficients.

Canopy air should be collected dry, and stored in inert flasks that do not leak or exchange CO_2 (e.g. made of glass), with seals that do not exchange with CO_2. We collected air for isotopic analyses by pumping dry air through pre-evacuated glass sampling flasks with two high-vacuum stopcocks. Care should be taken to flush the flask several times with canopy air, before the flask is closed, and returned to the lab for isotopic analysis. The size of the flask will depend on the method of CO_2 purification and carbon isotope analysis. For example, if the flasks are extracted manually using a vacuum extraction line (see below), and a modern mass spectrometer is used for isotope analysis, 1.7 to 2 l flasks filled with ambient $[CO_2]$ are sufficient for high precision carbon isotope determinations. Flask volumes might decrease if a cold finger or a pre-concentrator is used. We generally collected 15–30 flasks to estimate Δ_e for a given ecosystem.

The concentration of CO_2 in the flask should be determined as precisely as possible, because the accuracy of this measurement will strongly affect the error associated with the intercept (see Section 13.2.5). We generally used an infra-red gas analyser (IRGA) to measure the $[CO_2]$ of the air being pumped out of the sampling flask while we were collecting samples in the field. These analysers are accurate to 0.1 ppm.

Before isotope ratios of CO_2 can be analysed, CO_2 is separated from other gases present in the canopy air sample. This can be done either on-line (e.g. using a gas chromatograph on-line with the mass spectrometer) or off-line. In the case of off-line preparation, CO_2 is extracted cryogenically using a three-trap vacuum line (each trap with a double loop), and transferred into a sampling tube (for more details, see Ehleringer, 1991). The first trap (ethanol-dry ice slurry at -86 °C) is used to freeze water, the other two traps (liquid nitrogen) to collect CO_2. This technique will also trap nitrous oxide (N_2O) which condenses with CO_2 at liquid nitrogen temperatures. CO_2 should be separated from N_2O prior to isotope analysis, either by combustion with copper oxide wire (not applicable if $\delta^{18}O$ is of interest as well) or by using a gas chromatograph. If $\delta^{18}O$ is of interest as well, but a gas chromatograph is unavailable, a correction factor can be applied after carbon and oxygen analyses of the CO_2/N_2O gas mixture (Friedli and Siegenthaler, 1988; Mook and van der Hoek, 1983; Mook and Jongsma, 1987). A detailed discussion of the limitations to sampling and analysis of $\delta^{18}O$ in CO_2 is given in Chapter 24.

13.2.5 *Expected errors associated with* Δ_e *estimates*

Two major sources of error should to be considered for Δ_e estimates, these are errors associated with the $\delta^{13}C$ of tropospheric CO_2 and the $\delta^{13}C$ of respired CO_2, The precision of the tropospheric background data, for example collected by NOAA/CMDL is < 0.5 ppm for $[CO_2]$, and $\pm$ 0.03‰ for $\delta^{13}C$. Natural temporal variability for baseline data, collected at the same station over a one month period is less than 0.2‰ (M. Trolier personal communication). Further isotopic shifts may occur as the air mass moves above continental areas before it reaches the study area. However, variability between stations at a similar latitude is between 0.5 and 1 ppm and around 0.25‰ (see Conway *et al.*, 1994; Ciais *et al.*, 1995 respectively; see Chapter 24).

The largest error for the Δ_e estimates is associated with the estimates of $\delta^{13}C_R$. Owing to the nature of regression analyses, the accuracy of the regression equation increases with sample number, and the spread of the data over the entire range of the independent variable (in our case, $1/[CO_2]$). This is of special importance, because the 'Keeling plot' method requires extrapolating an intercept that is far beyond the range of measured values. Furthermore, slopes and intercepts should be calculated by geometric mean regressions, because both x and y variables are associated with an error ($1/[CO_2]$ as well as $\delta^{13}C$) (Sokal and Rohlf, 1981). After using this approach for 49 different stands in ten different ecosystems, the standard error for $\delta^{13}C$ of respired CO_2 averaged $\pm$0.98 (SE $\pm$0.10) ‰, with an absolute range from 0.2‰ to 4.0‰.

13.3 Variation in Δ_e estimates

13.3.1 *Evergreen versus deciduous forests*

Natural landscapes comprise mosaics of evergreen and deciduous ecosystems, most pronounced in temperate and boreal regions. Information about the influence of different life forms on ecophysiological processes is therefore critical for our understanding what controls carbon dynamics in evergreen and deciduous forest ecosystems, and how to integrate over diverse regions.

Higher intrinsic water-use efficiencies (WUEs), lower stomatal conductance and lower photosynthetic rates have often been observed for evergreen compared with deciduous trees (e.g. Chabot and Hicks, 1982; Körner, 1994; Schulze, 1982), but this is not always the case (Sobrado and Ehleringer, 1997). Although evergreen forests may tend to support a larger leaf area index (LAI), a comparison of $[CO_2]$ and $\delta^{13}C$ profiles between deciduous and evergreen canopies has shown that $[CO_2]$ and $\delta^{13}C$ gradients are larger in deciduous than in evergreen forests, despite a similar overall shape of these profiles (Brooks *et al.*, 1997a; Buchmann *et al.*, 1997b,c). This observation was made for forest ecosystems in very different environments (for the boreal forest and for dry temperate mountain sites) as well as for non-average years with regard to precipitation and temperature.

However, Δ_e estimates for deciduous and evergreen forests growing within the same region tend to be similar, independent of biome (Table 13.1). Differences between deciduous and evergreen stands were smaller than the natural variability within a life form (< 0.9‰). These results illustrate a potential difficulty associated with scaling leaf-level patterns to the ecosystem level. At the leaf level, ecophysiological differences between deciduous and evergreen shrubs and trees might be reflected in different Δ_l values (Brooks *et al.*, 1997b; Marshall and Zhang, 1994). However, at the higher organisational level of the ecosystem, carbon isotope discrimination seemed to be dominated by different mechanisms and differences between deciduous and evergreen forests were smaller. This overall similarity of Δ_e might be a potential advantage to model carbon dynamics in highly diverse landscapes such as a mosaic of evergreen and deciduous forests.

Table 13.1. *Comparison of Δ_e among deciduous and evergreen forest ecosystems, averaged over one growing season (mean ± SE, $3 \leq n \leq 8$). From Flanagan et al., (1996); Buchmann et al., (1997b).*

Life form	Species	Δ_e
Boreal (Man., Canada)		
Deciduous	*Populus tremuloides*	18.8 ± 0.5‰
Evergreen	*Picea mariana*	19.0 ± 0.2‰
	Pinus banksiana	18.9 ± 0.2‰
Boreal (Sask., Canada)		
Deciduous	*Populus tremuloides*	19.3 ± 0.6‰
Evergreen	*Picea mariana*	18.5 ± 0.1‰
	Pinus banksiana	19.0 ± 0.1‰
Temperate (UT, USA)		
Deciduous	*Populus tremuloides*	18.0 ± 0.7‰
Evergreen	*Pinus contorta*	18.3 ± 0.9‰

13.3.2. *Stand structure*

Forest architecture can be described by an array of different parameters such as the height of the dominant vegetation, the stand density, the LAI, the foliage distribution as well as the presence or absence of understory vegetation. All of these parameters can be highly variable, spatially as well as temporally. We tested the effect of stand LAI on Δ_e by comparing stands ranging in LAI from 1.4 to 9.2, including evergreen and deciduous forests from boreal and temperate ecosystems (Figure 13.3). In boreal ecosystems, coniferous forests (either *Picea mariana* or *Pinus banksiana*) showed similar Δ_e estimates to deciduous forests (*Populus tremuloides*) (Figure 13.3), even though the LAI of the *Picea mariana* stands were about four times larger than those of the *Pinus banksiana* stands (Flanagan *et al.*, 1996). Furthermore, these estimates were relatively stable at the regional scale. Stands at both southern and northern borders of the Canadian boreal forest showed similar Δ_e values: 18.9‰ (*Pinus banksiana*) and 19.0‰ (*Picea mariana*) at the northern border (Manitoba) versus 19.0‰ (*Pinus banksiana*) and 18.5‰ (*Picea mariana*) at the southern border (Saskatchewan). Differences between the two locations averaged about 0.4‰ throughout the growing season (recalculated from Flanagan *et al.*, 1996).

Three *Abies amabilis* stands from the Pacific Northwest, differing in age (between 40 and >220 years old), height (between 8 and 42 m tall) as well as in stand and foliage density (D. G. Sprugel, unpublished data) were also included in this analysis (Buchmann *et al.*, 1997c). [CO_2] and $\delta^{13}C$ profiles of canopy air were least pronounced in the tall old-growth stand that allowed for relatively free turbulent mixing with the air above the canopy (stand density: 488 trees per ha; LAI 6.4). However, the two younger stands showed lower daytime [CO_2] and more enriched canopy $\delta^{13}C$ values than the old-growth stand, clearly demonstrating less air turbulence and a large photosynthetic effect in these denser canopies (stand densities between 2241 and 70000 trees per ha, LAI around 9; Buchmann *et al.*, 1997c). However, Δ_e estimates for these wet temperate ecosystems differed at maximum by 0.9‰ (Figure 13.3), although LAI were almost 3 m^2/m^2 larger for the younger stands than for the old-growth stand. Only a weak trend

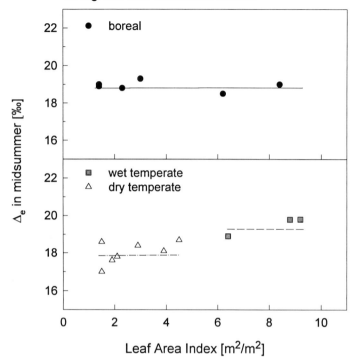

Figure 13.3. *Midsummer Δ_e estimates for evergreen and deciduous forest ecosystems in relation to stand leaf area index. Data for boreal forests from Flanagan et al., (1996; Populus tremuloides, Picea mariana, Pinus banksiana), for wet temperate forests from Buchmann et al., (1997c; Abies amabilis), and for dry temperate forests from Buchmann et al., (1997b; Acer ssp., Populus tremuloides, Pinus contorta).*

was observed as Δ_e increased slightly with increasing foliage density: the old-growth *Abies amabilis* stand (LAI 6.4, foliage density <1 m³/m² throughout the canopy) had a Δ_e value of 18.9‰. The densest stand (LAI 8.8, peak foliage density of about 4 m³/m²) showed a Δ_e of 19.8‰.

Deciduous stands of *Acer* spp. and *Populus tremuloides*, growing in the very different environment of dry temperate Utah, also exhibited rather constant Δ_e values as well (Buchmann *et al.*, 1997b). Doubling the overstory LAI in relatively open canopies (LAI <4.5) did not affect upper canopy [CO_2] or $\delta^{13}C$ nor Δ_e estimates (Figure 13.3). This integrative measure for the entire ecosystem was very similar for open (LAI 2.1: 17.4‰) as well as for dense riparian *Acer* spp. stands (LAI 4.5: 16.9‰), independent of the presence of a vigorous understory vegetation in the open stand. Thus, differences in forest architecture between stands growing in a similar abiotic environment did not have an effect on Δ_e estimates.

13.3.3. *Seasonality*

Seasonal effects in tropical forests might be expected to be minor because day length varies little throughout the year, and temperatures also remain relatively constant. The Δ_e values estimated for a primary tropical rainforest in French Guiana (5° North) were very similar for the wet and the dry seasons (Buchmann *et al.*, 1997a). The seasonal difference between the dry season estimate (20.3‰) and the wet season estimate (20.5‰) was within the experimental error (see Section 13.2.5).

However, temperate and boreal regions are characterised by pronounced seasonal differences of incoming radiation, and temperature. Thus, differences in Δ_e values throughout the growing season were expected to be larger than in tropical forests. However, the seasonal course of Δ_e for three *Abies amabilis* forests in the Pacific Northwest (wet temperate) revealed only very weak patterns throughout the growing season (Figure 13.4;

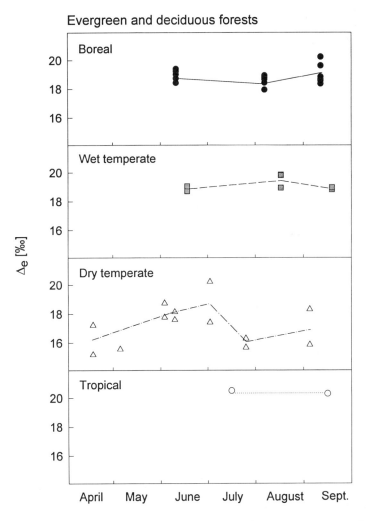

Figure 13.4. *Seasonality of Δ_e for evergreen and deciduous forest ecosystems. Data for boreal forests from Flanagan* et al., *(1996; Populus tremuloides, Picea mariana, Pinus banksiana), for wet temperate forests from Buchmann* et al., *(1997c; Abies amabilis), for dry temperate forests from Buchmann et al., (1997b; Acer ssp.), and for tropical forests from Buchmann* et al., *(1997a).*

Buchmann *et al.*, 1997c). Maximum differences between the stands were less than 1‰, well within the natural variability of $\delta^{13}C$ values of respired CO_2. Although seasonal changes in Δ_e were small, they were consistent for both younger stands: August Δ_e values tended to be slightly higher than those of June or September. A similar trend with higher summertime Δ_e estimates was observed in a very different environment, in dry temperate Utah (Buchmann *et al.*, 1997b). We measured $[CO_2]$ and $\delta^{13}C$ of canopy air during two growing seasons in two riparian *Acer* spp. stands, differing in overstory LAI (LAI 2.1 and 4.5). Although the seasonal pattern of Δ_e within these deciduous multi-layered canopies as larger than in the tropical and the wet temperate forests, seasonal effects were not significant. A similar weak seasonal trend was seen in boreal forests. This common pattern for evergreen as well as deciduous forest ecosystems suggests either a strong regulation of canopy carbon and water fluxes, keeping Δ_e and therefore ecosystem intrinsic WUE constant, or the effects of changing leaf level carbon discrimination are swamped out by more constant soil biological processes.

13.3.4. *Site history*

Estimates of Δ_e are dependent on the $\delta^{13}C$ of ecosystem respiration (see Equation 13.1). Generally, one assumes that $\delta^{13}C$ of respired CO_2 is close to the carbon isotopic composition of the dominant vegetation, even if it represents a pooled value integrating over several decades. However, under certain circumstances, this might not be the case. Two examples illustrate this point: the agricultural practice of crop rotation, and land use changes such as the conversion of tropical forests to C_4 pastures.

Schönwitz *et al.*, (1986) measured the $\delta^{13}C$ values of CO_2 evolved during soil respiration on sites with C_3 and C_4 crops. A C_3 soil cropped with corn for about one year showed intermediate $\delta^{13}C$ values of soil respired CO_2, indicating that about 30% of the respired carbon originated from C_4 residues. Other studies supported these results and have shown that soil organic carbon was a mixture of different crop generations (Table 13.2). Our study with C_4 and C_3 crop canopies also clearly demonstrated the importance of site history (Buchmann and Ehleringer, 1997). Although the recent plant cover had very distinct leaf $\delta^{13}C$ values (Table 13.2), the Δ_e estimates were not very different from each other, and averaged 13.2‰ for corn and 13.8‰ for alfalfa canopies. However, this implies that in spite of differences in the photosynthetic pathway of both crop species (C_4 vs. C_3), no differences existed in the $\delta^{13}C$ value of respired CO_2. Why? The agricultural practice at these particular sites included crop rotation each 5 years. Thus, soil organic carbon (SOC) and CO_2 efflux were mixtures of litter from previous years of cropping and the current vegetation cover. Calculating the origin of respired carbon revealed that for both crop sites, about 35–40 % originated from C_4 material and about 60% from C_3 material. These observations indicated a fast turnover rate for soil carbon and were within the range given in other studies (Table 13.2). The clearing of tropical forests and the subsequent conversion into C_4 pastures has a similar effect as crop rotation. Neill *et al.* (1996) measured $\delta^{13}C$ of soil organic matter and soil respired CO_2 on a sequence from C_3 forest to 81-year-old C_4 pastures. They found steadily increasing $\delta^{13}C$ ratios with pasture age, with values changing faster in the soil-respired CO_2 than those in SOC. However, how differences in decomposition rates, fractionation during decomposition and/or preferential decomposition of certain soil organic matter compounds might affect the $\delta^{13}C$ value of SOC and respired CO_2 is still unclear.

Table 13.2. Origin of C in soil organic matter (SOC) and of CO_2 released by soil respiration (resp)

History	$\delta^{13}C_{leaf}$ C3	C4	$\delta^{13}C_{SOC}$	$\delta^{13}C_{resp}$	% C4 origin	Reference
C3 → 1 yr C4	−26.0	−13.1		−19.1	30	Schönwitz et al. (1986)
C3 → 13 yr C4	−26.0	−12.0	−23.0		22	Balesdent et al. (1987)
C3 → 17 yr C4	−26.0	−12.5	−20.9		38	Balesdent et al. (1990)
C3 → 17 yr C4			−22.6		25	
C3 → 5 yr C4	−27.0	−12.0	−26.0		7	Arrouays et al. (1995)
C3 → 20 yr C4			−24.0		20	
C3 → 32 yr C4			−23.5		23	
C3 → 25 yr C4	−27.6	−13.4	−22.9		33	Gregorich et al. (1995)
C3 → 4 yr C4	−26.4	−12.0	−21.8		32	Wedin et al. (1995)
C3 → 4 yr C4			−22.3		28	
C4 → 5 yr C3	−28.2 ± 0.2	−12.8 ± 0.1	−23.6 ± 1.00	−22.3 ± 0.6	35	Buchmann and Ehleringer (1997)
C3 → 2 yr C4			−15.0 ± 0.73	−22.8 ± 0.5	38	
C3 → 3 yr C4	−28.0	−15.0	−25.5	−17.0	85	Neill et al. (1996)
C3 → 13 yr C4			−22.0	−17.0	85	
C3 → 81 yr C4			−17.0	−12.9	≈100	
C3 → 1 yr C4		−12.1	−25.8	−18.4	54	Rochette and Flanagan (1997)

13.4 Conclusions

Thus far, only a few studies have measured Δ_e or provide the necessary data to calculate Δ_e values (Table 13.3). However, the Δ_e estimates presented here proved to be less variable than previously expected, and therefore offer the great potential to be widely used in biosphere/atmosphere modeling since year-to-year fluctuations may be small (Buchmann et al., 1997a,c). The Δ_e value is a temporal and spatial integration of foliage characteristics (past and current) within the entire ecosystem, and is influenced by litter and decomposition processes. A practical advantage of Δ_e estimates lies in the relatively simple data collection that can be easily performed in remote areas.

In Table 13.3, we compiled all of the Δ_e estimates available to the authors, spanning ecosystems from different biomes between 69° North and 42° South. We used the Lancaster (1990) dataset in combination with tropospheric data from Francey et al. (1995) to calculate Δ_e. However, uncertainties in the range 0.2–0.5‰ may arise because the Cape Grim data were used for all Lancaster sites. Δ_e estimates for boreal ecosystems ranged between 15.9‰ and 19.3‰, with an average value of 18.2‰. The most extensive boreal data set (Flanagan et al., 1996), collected during one growing

Table 13.3. Estimates of Δ_e for evergreen (e) and deciduous (d) ecosystems

Latitude	Biome	Species	Life Form	Δ_e [%][a]	Reference
69°N	boreal	tussock tundra	d	15.9	from Lancaster (1990)
64°N		Picea abies/Pinus sylvestris	e	16.3	from Högberg and Ekblad (1996)
61°N		tundra	d	18.5	from Lancaster (1990)
56°N		Picea mariana	e	19.0 ± 0.2	Flanagan et al. (1996)
		Pinus banksiana	e	18.9 ± 0.2	Flanagan et al. (1996)
		Populus tremuloides	d	18.8 ± 0.5	Flanagan et al. (1996)
54°N		Picea mariana	e	18.5 ± 0.1	Flanagan et al. (1996)
		Pinus banksiana	e	19.0 ± 0.1	Flanagan et al. (1996)
		Populus tremuloides	d	19.3 ± 0.6	Flanagan et al. (1996)
		Pinus ssp./Picea ssp.	e	18.1	from Lancaster (1990)
48°N	temperate	Pseudotsuga menziesii/ Tsuga ssp.	e	17.7[b]	from Keeling (1961)
47°N		Abies amabilis	e	19.2 ± 0.2	Buchmann et al. (1997c)
46°N		Pinus ssp.	e	16.7 ± 0.2	from Lancaster (1990)
45°N		Pinus resinosa	e	17.7 ± 0.4	from Berry et al. (1997)
44°N		Acer ssp./Alnus ssp.	d	20.3 ± 0.1	from Lancaster (1990)
41°N		Pinus contorta	e	18.3 ± 0.9	Buchmann et al. (1997b)
		Populus tremuloides	d	18.0 ± 0.7	Buchmann et al. (1997b)
		Acer ssp.	d	17.1 ± 0.4	Buchmann et al. (1997b)
		deciduous forest	d	16.1	from Lancaster (1990)
38°N		Pinus ssp.	e	18.2	from Lancaster (1990)
		Pinus ssp./Abies ssp.	e	16.6 ± 0.2[b]	from Keeling (1961)
36°N		Sequoia sempervirens	e	18.3 ± 0.5[b]	from Keeling (1961)
		Pinus ssp.	e	18.5[b]	from Keeling (1961)
33°N		scrub oak	d	19.4	from Lancaster (1990)
20°N	tropical	primary rainforest	d	20.2	from Lancaster (1990)
10°N		primary rainforest	d	17.3 ± 0.1	from Broadmeadow et al. (1992)
9°N		primary rainforest	d	21.1	from Sternberg et al. (1989)
		primary rainforest	d	20.7	from Lancaster (1990)
5°N		primary rainforest	d	20.4 ± 0.1	Buchmann et al. (1997a)
3°S		primary rainforest	d	20.3[b]	from Quay et al. (1989)
10°S		primary rainforest	d	19.5	from Lloyd et al. (1996)
42°S		primary rainforest	d	16.2[b]	from Francey et al. (1985)
Crop rotation, irrigated					
42°N	C_3	Medicago sativa	d	13.8	Buchmann and Ehleringer (1997)
	C_4	Zea mays	d	13.2	Buchmann and Ehleringer (1997)
Crops, irrigated					
32°N	C_3	Triticum aestivum	d	21.6	from Yakir and Wang (1996)
	C_3	Gossypium hirsutum	d	18.4	from Yakir and Wang (1996)
	C_4	Zea mays	d	12.3	from Yakir and Wang (1996)

[a] Original data or calculated from intercepts of the relationship of $1/[CO_2]$ versus $\delta^{13}C$ of canopy air given in the cited reference

[b] Samples analysed may have been contaminated by N_2O. Applying a N_2O correction factor of +0.2‰ would result in Δ_e 0.2‰ lower than the values given here.

season, provides Δ_e estimates that vary less, averaging 18.9‰ for both evergreen and deciduous boreal forests. The estimates for temperate forests were lower than those for boreal forests, averaging 18.0 ‰, and show a wide spread from 16.1‰ to 20.3‰. In contrast, the Δ_e estimates for tropical forests show generally a narrow range (19.5–21.1‰) and average 20.4‰, although two forests exhibited lower Δ_e estimates (Francey *et al.*, 1985 and Broadmeadow *et al.*, 1992). This may further indicate that most of these study sites were undisturbed by human impact such as historic or more recent land use changes (Bush and Colinvaux, 1994). Thus, the spread in the data for boreal and temperate forests is probably due to natural variations among sites. The most likely candidate may be climatic differences among sites, especially site moisture regime: wet coastal sites tended to have greater Δ_e than dry inland sites.

Results of studies where sites were sampled more than once indicated that the balance between photosynthesis and stomatal conductance (intrinsic WUE) also seems to influence carbon isotope discrimination of forest ecosystems. Δ_e of boreal *Picea mariana* and *Pinus banksiana* stands averaged 18.9‰ (54°–56° N; Flanagan *et al.*, 1996). Estimates for wet temperate *Abies amabilis* stands in the Pacific Northwest were on average 19.2‰ (47° N; Buchmann *et al.*, 1997c), while Δ_e estimates for dry temperate evergreen forests were lower: 16.7‰ for *Pinus* spp. stands in Montana (46° N; Lancaster, 1990), 18.3‰ for *P. contorta* stands in Utah (41° N; Buchmann *et al.*, 1997b), and 16.6‰ for *Pinus* spp./*Abies* spp. stands and 18.3‰ for *Sequoia sempervirens* in California (36°–38° N; Keeling, 1961). Deciduous forest ecosystems exhibited a similar trend with latitude. Boreal *Populus tremuloides* stands averaged 19.1‰ (54°–56° N; Flanagan *et al.*, 1996). Estimates for temperate deciduous forests decreased from 20.3‰ for *Acer* ssp./*Alnus* ssp. (44° N; Lancaster, 1990) to 17.1‰ for dry temperate *Acer* ssp. stands in Utah (41° N; Buchmann *et al.*, 1997b). Thus, on a global scale, decreasing Δ_e might reflect increased ecosystem WUE, under conditions when precipitation decreases and/or the evaporative demand increases. However, further studies of Δ_e are required for a wide variety of terrestrial ecosystems throughout the globe to validate this trend.

How do Δ_A estimates compare with Δ_e estimates? Whereas the model from Lloyd and Farquhar (1994) implies pronounced differences between life forms, the field data do not yet support this separation between evergreen and deciduous forests. The modeled Δ_A estimates for cool/cold conifer forests averaged 15.4‰ (Lloyd and Farquhar, 1994). In contrast, all available Δ_e estimates for evergreen forests were higher, ranging from 16.3‰ to 19.2‰. The absolute differences between the modeled Δ_A and the calculated Δ_e values were between 0.1‰ and 4.0‰ for natural ecosystems of both evergreen and deciduous life forms, and up to 9.9‰ for agricultural stands. Differences of less than 1‰ for natural ecosystems are well within the analytical errors of the factors used to calculate Δ_e (see Section 13.2.5). Differences of 1–2‰ might be due to specific site conditions, and are within the expected variability of the modeled values (J. Lloyd personal communication). However, once these differences exceed the 2‰ range, they might either indicate the influence of the soil compartment or that the model does not adequately describe the stand ecophysiology at a certain site.

The largest discrepancy between modelled Δ_A and calculated Δ_e estimates were found for the agricultural sites (2.0‰ to 9.9‰), which may be associated with site history. The crop rotation study illustrated very nicely that Δ_e is not just the weighted average or integral of all Δ_{leaf} values, but that $\delta^{13}C$ of soil organic matter has a great impact on ecosystem performance (Buchmann and Ehleringer, 1997). While Δ_{leaf} of corn plants ranged between 4‰ and 5‰, the estimate of Δ_e averaged 13.2‰ for two corn stands; Δ_{leaf} of alfalfa plants was about 20‰, whereas Δ_e was estimated as 13.8‰ for the alfalfa stand. Because Δ_e integrates not only over foliage carbon isotope discrimination, but also includes this 'memory effect' of past vegetation cover as reflected in soil organic matter, site history needs to be considered to estimate Δ_e values. This is not only true for systems that are currently heavily influenced by human activities such as agriculture or land use changes, but also for systems that are generally considered undisturbed mature forests as cautioned by Bush and Colinvaux (1994); they could demonstrate a 4000-year history of human disturbance in a so-called undisturbed remote tropical forest.

References

Arrouays, D., Balesdent, J., Mariotti, A. and Girardin, C. (1995) Modelling organic carbon turnover in cleared temperate forest soils converted to maize cropping by using ^{13}C natural abundance measurements. *Plant Soil*, **173**, 191–196.

Balesdent, J., Mariotti, A. and Guillet, B. (1987) Natural ^{13}C abundance as a tracer for studies of soil organic matter dynamics. *Soil Biol. Biochem*, **19**, 25–30.

Balesdent, J., Mariotti, A. and Boisgontier, D. (1990) Effect of tillage on soil organic carbon mineralization estimated from ^{13}C abundance in maize fields. *J. Soil Science* **41**, 587–596.

Bender, M., Ellis, T., Tans, P.P., Francey, R.J. and Lowe, D. (1996) Variability in the O_2/N_2 ratio of southern hemisphere air, 1991–1994: implications for the carbon cycle. *Global Biogeochem. Cycles* **10**, 9–21.

Berry, S.C., Varney, G.T. and Flanagan, L.B. (1997) Leaf $\delta^{13}C$ in *Pinus resinosa* trees and understory plants: variation associated with light and CO_2 gradients. *Oecologia* **109**, 499–506.

Bird, M.I., Chivas, A.R. and Head, J. (1996) A latitudinal gradient in carbon turnover times in forest soils. *Nature* **381**, 143–146.

Boden, T.A., Kaiser, R.J., Sepanski, R.J. and Stoss, F.W. (1994) Trends '93: A Compendium of Data on Global Change. ORNL/CDIAC-65. Carbon Dioxide Information Analysis Center, Oak Ridge National Laboratory, Oak Ridge, TN. 984 pp.

Broadmeadow, M.S.J., Griffiths, H., Maxwell, C. and Borland, A.M. (1992) The carbon isotope ratio of plant organic material reflects temporal and spatial variations in CO_2 within tropical forest formations in Trinidad. *Oecologia*, **89**, 435–441.

Brooks, J.R., Flanagan, L.B., Varney, G.T. and Ehleringer, J.R. (1997a) Vertical gradients in photosynthetic gas exchange characteristics and refixation of respired CO_2 within boreal forest canopies. *Tree Phys.* **17**, 1–12.

Brooks, J.R., Flanagan, L.B., Buchmann, N. and Ehleringer, J.R. (1997b) Carbon isotope composition of boreal plants: functional grouping of life forms. *Oecologia* **110**, 301–311.

Buchmann, N. and Ehleringer, J.R. (1997) CO_2 concentration profiles, carbon and oxygen isotopes in C_4 and C_3 crop canopies. *Agric. Forest Meteorol.* in press.

Buchmann, N. Kao, W.Y., Ehleringer, J.R. (1996) Carbon dioxide concentrations within forest canopies – Variation with time, stand structure, and vegetation type. *Global Change Biology* **2**, 421–432.

Buchmann, N., Guehl, J.M., Barigah, T.S. and Ehleringer, J.R. (1997a) Interseasonal comparison of CO_2 concentrations, isotopic composition, and carbon dynamics in an Amazonian rainforest (French Guiana). *Oecologia*, **110**, 120–131.

Buchmann N, Kao WY, Ehleringer JR. (1997b) Influence of stand structure on carbon-13 of vegetation, soils, and canopy air within deciduous and evergreen forests in Utah (USA). *Oecologia* **110**, 109–119.

Buchmann, N., Hinckley, T.M. and Ehleringer, J.R. (1997c) Carbon isotope dynamics in *Abies amabilis* stands in the Cascades. *Can. J. For. Res.* submitted.

Bush, M.B. and Colinvaux, P.A. (1994) Tropical forest disturbance: paleoecological records from Darien, Panama. *Ecology* **75**, 1761–1768.

Chabot, B.F. and Hicks, D.J. (1982) The ecology of leaf life spans. *Ann. Rev. Ecol. Syst.* **13**, 229–259.

Ciais, P., Tans, P.P., Trolier, M., White, J.W.C. and Francey, R.J. (1995) A large northern hemisphere terrestrial CO_2 sink indicated by the $^{13}C/^{12}C$ ratio of atmospheric CO_2. *Science*, **269**, 1098–1101.

Conway, T.J., Tans, P.P., Waterman, L.S., Thoning, K.W., Masarie, K.A. and Gammon, R.H. (1988) Atmospheric carbon dioxide measurements in the remote global troposphere, 1981–1984. *Tellus*, **40B**, 81–115.

Conway, T.J., Tans, P.P., Waterman, L.S., Thoning, K.W., Kitzis, D.R., Masarie, K.A. and Zhang, N. (1994) Evidence for interannual variability of the carbon cycle from the National Oceanic and Atmospheric Administration/Climate Monitoring and Diagnosis Laboratory Global Air Sampling Network. *J. Geophys. Res.* **99**, 22 831–22 855.

Denning, A.C., Fung, I.Y. and Randall, D. (1995) Latitudinal gradient of atmospheric CO_2 due to seasonal exchange with land biota. *Nature* **376**, 240–243.

Ehleringer, J.R. (1991) $^{13}C/^{12}C$ fractionation and its utility in terrestrial plant studies. In: *Carbon Isotope Techniques.* (eds. D.C. Coleman and B. Fry). Academic Press, San Diego, pp. 187–200.

Farquhar, G.D., Ehleringer. J.R. and Hubick, K.T. (1989) Carbon isotope discrimination and photosynthesis. *Ann. Rev. Plant. Phys. Plant Mol. Biol.* **40**, 503–537.

Flanagan, L.B., Brooks, J.R., Varney, G.T., Berry, S.C. and Ehleringer, J.R. (1996) Carbon isotope discrimination during photosynthesis and the isotope ratio of respired CO_2 in boreal forest ecosystems. *Global Biogeochem. Cycles* **10**, 629–640.

Francey, R.J., Gifford, R.M., Sharkey, T.D. and Weir, B. (1985) Physiological influences on carbon isotope discrimination in huon pine (*Lagarostrobus franklinii*). *Oecologia* **66**, 211–218.

Francey, R.J., Tans, P.P., Allison, C.E., Enting, I.G., White, J.W.C. and Trolier, M. (1995) Changes in oceanic and terrestrial carbon uptake since 1982. *Nature* **373**, 326–330.

Friedli, H. and Siegenthaler, U. (1988) Influence of N_2O on isotope analyses in CO_2 and mass spectrometric determination of N_2O in air samples. *Tellus* **40B**, 129–133.

Gash, J.H.C. (1986) A note on estimating the effect of a limited fetch on micrometeorological evaporation measurements. *Boundary Layer Meteorology* **35**, 409–413.

Grace, J., Lloyd, J., McIntyre, J., Miranda, A.C., Meir, P., Miranda, H.S., Moncrieff, J., Massheder, J., Wright, I.R. and Gash, J.H.C. (1995) Fluxes of carbon dioxide and water vapour over an undisturbed tropical forest in south-west Amazonia. *Global Change Biology* **1**, 1–12.

Gregorich, E.G., Ellert, B.H. and Monreal, C.M. (1995) Turnover of soi organic matter and storage of corn residue carbon estimated from natural ^{13}C abundance. *Can. J. Soil Sci.* **75**, 161–167.

Högberg, P. and Ekblad, A. (1996) Substrate-induced respiration measured *in situ* in a C_3-plant ecosystem using additions of C_4-sucrose. *Soil Biology Biochemistry* **28**, 1131–1138.

Houghton, R.A. (1995) Land-use change and the carbon cycle. *Global Change Biology* **1**, 275–287.

Houghton, J.T., Meira Filho, L.G., Callander, B.A., Harris, N., Kattenberg, A. and Maskell, K. (1996) *Climate Change 1995. The Science of Climate Change.* Cambridge University Press, Cambridge, 572 pp.

Keeling, C.D. (1958) The concentration and isotopic abundances of atmospheric carbon dioxide in rural areas. *Geochim. Cosmochim. Acta* **13**, 322–334.

Keeling, C.D. (1961a) The concentration and isotopic abundances of atmospheric carbon dioxide in rural and marine areas. *Geochim Cosmochim Acta*, **24**, 277–298.

Keeling CD. (1961b) A mechanism for cyclic enrichment of carbon-12 by terrestrial plants. *Geochim. Cosmochim. Acta* **24**, 299–313.

Keeling, C.D., Carter, A.F. and Mook, W.G. (1984) Seasonal, latitudinal, and secular variations in the abundance and isotopic ratios of atmospheric CO_2. 2. Results from oceanographic cruises in the tropical Pacific Ocean. *J. Geophys. Res.*, **89**, 4615–4628.

Keeling, R.F., Piper, S.C. and Heimann, M. (1996) Global and hemispheric CO_2 sinks deduced from changes in atmospheric O_2 concentration. *Nature* **381**, 218–221.

Keeling, C.D., Adams, J.A., Ekdahl, C.A. and Guenther, P.R. (1976) Atmospheric carbon dioxide variations at the South Pole. *Tellus* **28**, 552–564.

Körner, C. (1994) Scaling from species to vegetation: the usefulness of functional groups. In: *Biodiversity and Ecosystem Function* (eds E.D. Schulze and M.M. Caldwell). Springer, Berlin, pp. 117–140.

Komhyr, W.D., Gammon, R.H., Harris, T.B., Waterman, L.S., Conway, T.J., Taylor, W.R. and Thoning, K.W. (1985) Global atmospheric CO_2 distribution and variations from 1968–1982 NOAA/GMCC CO_2 flask sample data. *J. Geophys. Res.* **90**, 5567–5596.

Kruijt, B., Lloyd, J., Grace, J., McIntyre, J., Farqhuar, G.D., Miranda, A.C. and McCrakcen, P. (1996) Source and sinks of CO_2 in Rondonian tropical rainforest, inferred from concentrations and turbulence along a vertical gradient. In: *Amazonian Deforestation and Climate* (eds J.H.C. Gash, C.A. Nobre, J. Roberts, R.L. Victoria). Wiley, New York, pp. 331–351.

Lancaster, J. (1990) Carbon-13 fractionation in carbon dioxide emitting diurnally from soils and vegetation at ten sites on the North American continent. Dissertation. University of California, San Diego, 184 pp.

Lin, G. and Ehleringer, J.R. (1997) Carbon isotopic fractionation does not occur during dark respiration in C_3 and C_4 plants. *Plant Phys.* **114**, 391–394.

Lloyd, J. and Farquhar, G.D. (1994) ^{13}C discrimination during CO_2 assimilation by the terrestrial biosphere. *Oecologia* **99**, 201–215.

Lloyd, J., Kruijt, B., Hollinger, D.Y., Grace, J., Francey, R.J., Wong, S.C., Kelliher, F.M., Miranda, A.C., Gash, K.H.C., Vygodskaya, N.N., Wright, I.R., Miranda, H.S., Farquhar, G.D. and Schulze, E.D. (1996) Vegetation effects on the isotopic composition of atmospheric CO_2 at local and regional scales: theoretical aspects and a comparison between rain forest in Amazonia and a boreal forest in Siberia. *Aust. J. Plant Phys.* **23**, 371–399.

Marshall, J.D. and Zhang, J. (1994) Carbon isotope discrimination and water-use-efficiency in native plants of the north-central Rockies. *Ecology* **75**, 1887–1895.

Mook, W.G. and Jongsma, J. (1987) Measurement of the N_2O correction for $^{13}C/^{12}C$ ratios of atmospheric CO_2 by removal of N_2O. *Tellus* **39B**, 96–99.

Mook, W.G. and van der Hoek, S. (1983) The N_2O correction in the carbon and oxygen isotopic analysis of atmospheric CO_2. *Isotope Geoscience* **1**, 237–242.

Mook, W.G., Koopmans, M., Carter, A.F. and Keeling, C.D. (1983) Seasonal, latitudinal, and secular variations in the abundance and isotopic ratios of atmospheric CO_2. 1. Results from land stations. *J. Geophys. Res.* **88**, 10 915–10 933.

Neill, C., Fry, B., Melillo, J.M., Steudler, P.A., Moraes, J.F.L. and Cerri, C.C. (1996) Forest- and pasture-derived carbon contributions to carbon stocks and microbial respiration of tropical pasture soils. *Oecologia* **107**, 113–119.

Pales, J.C. and Keeling, C.D. (1965) The concentration of atmospheric carbon dioxide in Hawaii. *J. Geophys. Res.* **70**, 6053–6076.

Quay, P., King, S., Wilbur, D. and Wofsy, S. (1989) $^{13}C/^{12}C$ of atmospheric CO_2 in the Amazon Basin: Forest and river sources. *J. Geophys. Res.* **94**, 18 327–18 336.

Rochette, P. and Flanagan, L.B. (1997) Quantifying rhizosphere respiration in a corn crop under field conditions. *Soil. Sci. Soc. Am. J.* **61**, 466–474.

Schönwitz, R., Stichler, W. and Ziegler, H. (1986) $\delta^{13}C$ values of CO_2 from soil respiration on sites with crops of C_3 and C_4 type of photosynthesis. *Oecologia* **69**, 305–308.

Schuepp, P.H., Leclerc, M.Y., MacPherson, J.I. and Desjardin, R.L. (1990) Footprint prediction of scalar fluxes from analytical solutions of the diffusion equation. *Boundary Layer Meteorology* **50**; 355–373.

Schulze, E.D. (1982) Plant life forms and their carbon, water, and nutrient relations. *Plant Physiological Ecology* II. Springer, Berlin, pp 615–676.

Sobrado, M.A. and Ehleringer, J.R. (1997) Leaf carbon isotope ratios from a tropical dry forest in Venezuela. *Flora* **192**, 121–124.

Sokal, R.R. and Rohlf, F.J. (1981) *Biometry*. W.H. Freeman and Company, New York.

Sternberg, L.S.L. (1989) A model to estimate carbon dioxide recycling in forests using $^{13}C/^{12}C$ ratios and concentrations of ambient carbon dioxide. *Agric. Forest. Met.* **48**, 163–173.

Sternberg, L.S.L., Mulkey, S.S. and Wright, S.J. (1989) Ecological interpretation of leaf isotope ratios: influence of respired carbon dioxide. *Ecology* **70**, 1317–1324.

Tans, P.P., Bakwin, P.S. and Guenther, D.W. (1996) A feasible Global Carbon Cycle Observing System: a plan to decipher today's carbon cycle based on observations. *Global Change Biology* **2**, 309–318.

Tans, P.P., Fung, I.Y. and Takahasi, T. (1990) Observational constraints on the atmospheric CO_2 budget. *Science* **247**, 1431–1438.

Trumbore, S.E., Chadwick, O.A. and Amundson, R. (1996) Rapid exchange between soil carbon and atmospheric carbon dioxide driven by temperature change. *Science* **272**, 393–396.

Wedin, D.A., Tieszen, L.L., Dewey, B. and Pastor, J. (1995) Carbon isotope dynamics during grass decomposition and soil organic matter formation. *Ecology* **76**, 1383–1392.

Yakir, D. and Wang, X-F. (1996) Fluxes of CO_2 and water between terrestrial vegetation and the atmosphere estimated from isotope measurements. *Nature* **380**, 515–517.

14

Assessing sensitivity to change in desert ecosystems – a stable isotope approach

James R. Ehleringer, R. David Evans, David Williams

14.1 Introduction

Growth and gas exchange of plants in arid zones are driven primarily by periodic pulses of moisture and nutrient availability. Both seasonal and inter-annual changes in precipitation occur in response to long-term weather cycles causing direct impacts on productivity, competitive displacement, and mortality (Comstock and Ehleringer, 1992; Ehleringer, 1993c; Smith *et al.*, 1997). Deviations from mean climatic conditions might be expected to increase under predicted climate change scenarios, particularly given the disequilibrium between terrestrial and oceanic regions as radiative forces increase (Rind *et al.*, 1990). Surface disturbance caused by land-use change has resulted in significant changes in soil nitrogen levels of aridland ecosystems (Belnap, 1995; Evans and Belnap, 1998) and this may be further exacerbated by variations in monsoonal precipitation patterns.

In this chapter, we explore how analyses of stable isotope ratio patterns in arid land soils and plants help us better understand ecosystem dynamics in response to changes in moisture and nitrogen availability. Arid land ecosystems on the Colorado Plateau in the western United States provide an opportunity to examine plant response to the potential sensitivity of ecosystem components to variations in summer and winter moisture inputs along moisture clines, while at the same time exploring the impact of different land-use patterns on nitrogen availability.

To facilitate modelling approaches for estimating productivity it would be ideal if all vegetation components responded equally to seasonal variations in resource input. More realistically, it would be reasonable to expect that different vegetation components could be aggregated into functional groups, such as by life form, with members of the same life form responding similarly to seasonal pulses in moisture and nitrogen. While stable isotopes may not contribute directly to estimating productivity, stable isotope studies will contribute directly to understanding mechanistic aspects of the constraints on productivity in different species and/or functional groups, thereby

Stable Isotopes, edited by H. Griffiths.
© 1998 BIOS Scientific Publishers Ltd, Oxford.

complementing more traditional approaches (Ehleringer *et al.*, 1993). Carbon isotopes in organic material provide an estimate of the extent to which different gas exchange components affect productivity (Farquhar *et al.*, 1989). Hydrogen and oxygen isotopes in xylem waters contribute directly to quantifying use of monsoonal and winter water sources (Ehleringer and Dawson, 1992). Nitrogen isotope ratios provide a quantitative estimate of the balance between nitrogen inputs and losses from the soils and species-specific patterns of nitrogen use, particularly in response to disturbance (Evans and Ehleringer, 1993, 1994; Evans and Belnap, 1998). Together stable isotopes provide an opportunity to assess ecosystem dynamics and the sensitivity of these vegetation components to change and to complement more traditional gas exchange and biomass-assessment approaches.

14.2 Deserts as pulse-driven ecosystems

Desert ecosystems are generally characterised by low resource levels, with water representing the single most limiting resource constraining primary productivity through its influence on plant water status and soil nutrient availability (Noy Meir, 1973). Moisture in the upper soil layers is available as a series of brief pulses following intermittent precipitation events. High potential evapotranspiration during the growing season means that soil moisture does not persist long in the upper soil layers, irrespective of the extent of vegetative cover. The amount of plant-available nitrogen in soils is closely tied to these pulses of moisture because nitrogen mineralization is most rapid following precipitation (Burke, 1989; Evans and Ehleringer, 1994; Matson *et al.*, 1991; Zaady *et al.*, 1996). Productivity in deserts can then be thought of as a series of pulse events controlled primarily by soil moisture, and secondarily by nutrient availability following precipitation events (Figure 14.1). The high temporal variability between rainfall events on both an interannual and seasonal bases amplifies the impact of pulses of soil moisture availability on ecosystem dynamics (Noy Meir, 1973).

On the Colorado Plateau, precipitation comes either from winter storms generated in the Gulf of Alaska or from summer convection storms generated by the Arizona monsoon system. Understanding the current seasonal and regional patterns of precipitation inputs into an ecosystem has ramifications at several levels: on carbon and mineral cycling at the ecosystem level, on biodiversity at the community level, and on productivity and adaptation at the population and species levels. The interior deserts of Arizona, Nevada, and Utah represent the driest regions of western North America, resulting from a combination of rainshadow effects and either the southern limits of winter moisture input, the northern limits of summer moisture input, or both

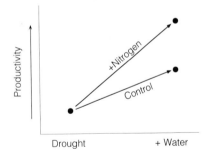

Figure 14.1. A hypothetical relationship of how water and soil nitrogen influence productivity in desert ecosystems, which are pulse driven.

(Comstock and Ehleringer 1992; Houghton 1979). Shifts in strengths of storm-generating conditions in the Pacific and in the Gulf influence both the magnitude and seasonality of soil moisture availability (Bryson and Lowry 1955; Mitchell 1976) and therefore constrain periods of primary productivity activity in these aridland ecosystems (Caldwell, 1985; Dobrowolski *et al.*, 1990). One major consequence predicted by global climate change scenarios is a change in monsoonal (summer) precipitation (Schlesinger and Mitchell, 1987; Mitchell *et al.*, 1990); it will increase in some areas and decrease in others. A second is increased soil temperatures and increased interior drought associated with ocean-land temperature disequilibrium (Rind *et al.*, 1990).

Southeastern portions of Utah (northwestern portions of the Colorado Plateau) form a broad northern border for the region influenced by the Arizona monsoonal system (Figure 14.2). Annual precipitation across Colorado Plateau ecosystems ranges from 100 to 400 mm (Houghton 1979), with these ecosystems experiencing low precipitation due to rain-shadow effects from winter-generated storms and the northern limits of monsoonal systems generated in the summer months. While on average approximately half the annual moisture is from summer moisture events, the year-to-year variability is high and depends on the intensity of the Arizona monsoon system that develops in a particular year (Adang and Gall, 1989; Houghton, 1979; Moore *et al.*, 1989). Pack rat midden data indicate that central and northern Utah had an extensive summer-precipitation climate several thousand years ago (Betancourt *et al.*, 1990; Cole, 1990), but the onset of regional summer drought is less clear. Variations in the intensities and predictability of summer rain should have significant impacts on primary productivity in these aridland ecosystems (Ehleringer and Mooney, 1983; Hadley and Szarek, 1981; Noy Meir, 1973; Smith and Nowak, 1990).

14.3 Water uptake patterns by aridland plants

Given that soil moisture levels limit productivity in deserts, how do plants respond to the pulses in moisture input? In particular, how do plants respond to variations in

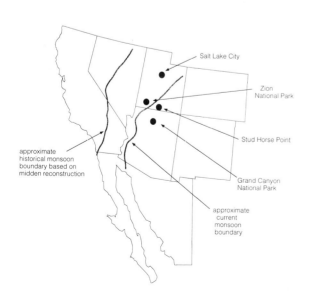

Figure 14.2. The current and historical distribution limits of the Arizona monsoon system in western North America. Also shown are major sites referred to in this chapter.

summer moisture events? Our initial hypothesis was that perennial plants would extract moisture from surface layers when it was available and then switch to moisture from deeper soil layers in between summer-precipitation pulses.

Stable isotopes in xylem water have been particularly useful in quantifying soil layers from which plants extract water, since roots do not fractionate during water uptake (reviewed in Ehleringer and Dawson, 1992; see Chapter 11). Moisture derived in winter, often as snow, penetrates to deeper layers because cold temperatures constrain biological activities. When the upper layers dry out in the spring and summer, moisture in the upper layers is recharged by precipitation events during the growing season. Differences in cloud temperatures during winter and summer precipitation events result in contrasting stable isotope ratio values in the precipitation (Figure 14.3). As a consequence, there is a layering of water in the soil profile with different stable isotope ratios. Analysis of xylem waters then allows us to quantify the extent to which species are using moisture derived from one zone to another and of the dynamics in switching between these two contrasting moisture zones.

Contrary to initial expectations, *Acer grandidentatum* and *Quercus gambelii* in the xeric woodlands north of the monsoon boundary did not use moisture in the upper layers at any point during the growing season (Figure 14.3). Instead, these dominant shrubs appeared to use winter-derived moisture that had penetrated to deeper soil depths. When monsoonal moisture fell in the latter part of the summer, neither species used appreciable amounts of this surface water source.

In the desert ecosystem of the Colorado Plateau, perennial plants used only a limited fraction of the summer moisture input (Figure 14.4). Moderate precipitation events in the early spring charged the profile and served as the water source from spring through early summer. The strong evaporative demand in this desert ecosystem resulted in some evaporative enrichment within the soil profile that was reflected in the isotopic composition of xylem water in early summer (July). Following strong

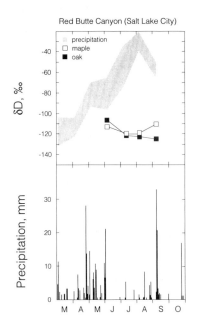

Figure 14.3. *Upper diagram, The hydrogen isotope ratio of precipitation in Salt Lake City (grey) and of xylem water from* Acer grandidentatum *(open square) and* Quercus gambelii *(closed square) for plants in Red Butte Canyon during the course of the year. Lower diagram. The seasonal dynamics of daily precipitation. Modified from Phillips and Ehleringer (1995).*

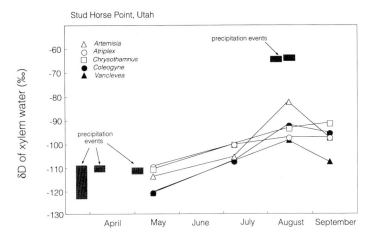

Figure 14.4. *The seasonal dynamics of hydrogen isotope ratio of water in xylem sap for five common desert shrubs in a Colorado Plateau desert ecosystem at Stud Horse Point, Utah. The species are* Artemisia filifolia, Atriplex canescens, Chrysothamnus nauseosus, Coleogyne ramosissima *and* Vanclevea stylosa. *Also shown are the hydrogen isotope ratios of precipitation events. Data are modified from Lin* et al. *(1996).*

precipitation events in early August (~ 60 mm), there were moderate changes in the xylem water isotopic composition of these shrubs, indicating a limited uptake of the moisture retained in the upper soil layers.

Ehleringer *et al.* (1991) had reported similar results for water uptake by different shrubs within this desert ecosystem. In that study, roots of annual and herbaceous species fully utilised the moisture in the upper soil layers (Figure 14.5). In contrast, there was a wide variation in the response of woody perennial shrubs to summer mois-ture inputs. On average, 54% of the xylem water in woody perennial plants was derived from summer precipitation, but within individual species this value ranged from 1% to 79% summer-derived moisture use.

In a follow-up study, Lin *et al.* (1996) used stable isotopes to follow uptake of mois-ture from artificial summer rain events of different magnitudes (Figure 14.6). Plants were monitored following the artificial precipitation events to see how long it took for moisture to be taken up. Typically, 2 to 3 days was required for maximum uptake. Since moisture in the uppermost soil layers only persists for 5–10 days following rain, a large fraction of this moisture is lost by evaporation and is not used by plants. Lin *et al.* (1996) observed that desert perennial shrubs differed in the extent to which they responded to monsoonal moisture inputs. On average, for a 25-mm precipitation event, less than 20% of the plant moisture was derived from the saturated upper soil layers (Figure 14.7). A greater response was seen in some species following a 50-mm precipitation event, but in no case did this upper-layer moisture account for more than 42% of the xylem water. Several of the species (*Atriplex canescens* and *Chrysothamnus nauseosus*) did not respond appreciably to either a 25-mm or a 50-mm precipitation event, with summer-derived moisture accounting for 7% or less of the water being taken up by the plants. Based on isotope ratio gradients within the soil profile, Thorburn and Ehleringer (1995) had concluded that these desert perennials were deriving most of their summer moisture from a depth of ~0.6 m.

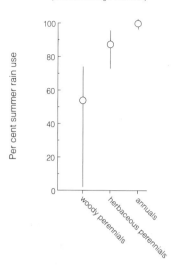

Summer rain use in August by different plants in
a Colorado Plateau desert shrub community
(data from Ehleringer et al. 1991)

Figure 14.5. The per cent
usage of summer rains by woody
perennials, herbaceous
perennials, and annuals in a
Colorado Plateau desert
ecosystem at Stud Horse Point,
Utah. Circles represent the mean
for each life form and the vertical
bars represent the entire range of
values for different species within
that life form. Based on data in
Ehleringer et al. (1991).

Some species at sites on the monsoon boundary show year-to-year variation in the
extent to which they take up summer moisture. For example, Gregg (1991) reported
that *Juniperus osteosperma*, a dominant tree of pinyon-juniper woodlands did not take
up appreciable amounts of summer rain on sites north of the Arizona monsoon
boundary. At Coral Pink (just east of Zion National Park and on the monsoon boundary), she reported that junipers did not take up summer moisture. Flanagan *et al.*
(1992) working at the same Coral Pink site reported that junipers did not take up
summer rain in one year, but did so in the following year. Evans and Ehleringer (1994)
also working at Coral Pink observed that some juniper individuals did take up summer
rains in yet a different year, but that others did not. At Tintic (north of the monsoon
boundary), Donovan and Ehleringer (1994) reported that junipers did not respond to
summer rain, specifically following a 21-mm summer precipitation event. *Artemisia
tridentata*, a widespread shrub whose distribution spans the monsoon boundary, has
been shown to take up summer moisture, irrespective of the location (Donovan and

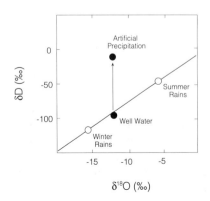

Figure 14.6. The meteoric
water line for Stud Horse Point,
Utah, showing the mean values
for winter precipitation, summer
precipitation, well water, and the
deuterium-spiked water used for
irrigation studies. Modified from
Lin et al. (1996).

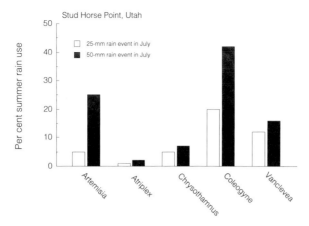

Figure 14.7. *The per cent usage of summer rains by five common woody perennials in a Colorado Plateau desert ecosystem at Stud Horse Point, Utah. Plants were provided with either a 25-mm rain event or a 50-mm rain event and then 3–5 days following this the xylem sap was sampled to determine the extent of the summer rain usage. The species are* Artemisia filifolia, Atriplex canescens, Chrysothamnus nauseosus, Coleogyne ramosissima, *and* Vanclevea stylosa. *Modified from Lin* et al. *(1996).*

Ehleringer, 1994; Evans and Ehleringer, 1994; Flanagan *et al.*, 1992). In contrast to *J. osteosperma* with a widespread distribution, *Pinus edulis* (the other common tree in the pinyon-juniper woodland) does not occur north beyond the monsoon boundary. Both Flanagan *et al.* (1992) and Evans and Ehleringer (1994) reported that *Pinus edulis* at Coral Pink used summer moisture inputs and was consistently more responsive than *J. osteosperma* to summer rain events. The factors contributing to a year-to-year variability in moisture uptake by plants along the monsoon boundary are unclear, but previous soil moisture stress, soil temperature, and the magnitude of the summer precipitation event are likely to be contributing factors.

14.4 Stomatal limitations as inferred from carbon isotope ratios

Long-term estimates of the ratio of intercellular to ambient CO_2 concentrations (C_i/C_a) in C_3 species can be derived from leaf carbon isotope ratios (Farquhar et al., 1989). Ehleringer (1993a, b) proposed that the C_i/C_a ratio reflected a metabolic set point, reflecting tradeoffs in the relative rates of CO_2 supply through stomata versus photosynthetic demand for that CO_2. Variations in the C_i/C_a ratio negatively correlate with plant longevity. On the basis of structure, species in these arid land ecosystems can be classified into four functional groups (tree, woody perennial shrub, herbaceous perennial, and annual) with carbon isotope discrimination negatively correlated with plant longevity (Ehleringer, 1993a,b; Ehleringer and Cerling, 1995; Ehleringer and Cooper, 1988; Schuster *et al.*, 1992). These patterns suggest that stomatal limitations in gas exchange become most prominent in long-lived functional groups. One possible tradeoff is that an increased stomatal constraint on gas exchange is associated with the reduced likelihood of xylem cavitation during drought periods. In response to increased resource levels from summer irrigation, all plants responded equally and as a result carbon isotope discrimination differences among species were maintained (Lin *et al.*, 1996). Changes in carbon isotope discrimination by C_3 and C_4 species closely parallel resource availability in both winter and summer-monsoon precipitation seasons, with carbon isotope discrimination increasing with stress in C_4 plants and decreasing with stress in C_3 plants (Evans and Ehleringer, 1994; Flanagan *et al.*, 1992; Lin *et al.*, 1996).

14.5 Nitrogen sources

Cryptobiotic crusts are biological soil crusts composed of cyanobacteria, lichens, mosses, green algae, and fungi. They are found in mesic environments, tropical and temperate deserts, and in polar regions (Eldridge and Greene, 1994), but they reach their best development in arid regions where they are ubiquitous on undisturbed arid and semi-arid soils. The absence of fire and grazing by large mammals has allowed the cryptobiotic crusts of the Colorado Plateau and Great Basin to become especially well developed (Mack and Thompson, 1982). The crusts can be as great as 10 cm deep and approach 100 % coverage in undisturbed ecosystems (Harper and Marble, 1988; Kleiner and Harper, 1972). The cryptobiotic crust is held together by cyanobacteria that exude a gelatinous sheath which binds soil particles and organisms (Campbell *et al.*, 1989; Belnap and Gardner, 1993). Surface disturbance disrupts the cohesiveness of the cryptobiotic crust causing loss of nitrogen fixation (Belnap, 1996; Belnap *et al.*, 1994). The consequences of surface disturbance are uncertain because the relative importance of nitrogen assimilated by the crust versus that deposited by atmospheric deposition is not known (West, 1990).

Evans and Ehleringer (1993) used nitrogen isotope ratios to show that these organisms were the primary nitrogen sources into Colorado Plateau ecosystems (Figure 14.8). The relationship between the isotopic composition and nitrogen content of soils followed a Raleigh distillation model (Fustec *et al.*, 1991; Mariotti *et al.*, 1981). Disturbance of the cryptobiotic crust resulted in net loss of nitrogen from the ecosystem causing an increase in soil nitrogen isotope composition (Evans and Ehleringer, 1993). Rates of nitrogen mineralisation depend on the availability of soil organic nitrogen (Binkley and Hart, 1989; Matson *et al.*, 1991) so mineralization potentials were also strongly correlated with soil nitrogen isotope ratios (Evans and Ehleringer, 1994). Fractionation does not occur during uptake of ammonium or nitrate (Evans *et al.*, 1996; Yoneyama and Kaneko, 1989) so plant nitrogen isotope composition becomes a reliable estimator of soil nutrient quality (Evans and Ehleringer, 1994). Plants growing on disturbed sites had higher nitrogen isotope ratios and lower leaf nitrogen contents than those on less disturbed sites (Evans and Ehleringer, 1994).

The sensitivity to change of the Colorado Plateau desert ecosystems, induced by either land-use practices or by shifting monsoonal conditions is great (Figure 14.8). Plant function, decomposition, and cryptobiotic function are all influenced by periodic moisture pulses, whether they come in the summer or in the winter. As with the reduced responsiveness of desert perennials to summer rain events, Belnap (1996) has shown that nitrogen-fixation rates by cryptobiotic crusts are less responsive to summer rain events than they are to winter-spring precipitation events. The long-term vegetation composition in these desert ecosystems will be a function of variation in summer moisture input, land-use pressures which select against some components of the herbaceous vegetation that can use the summer moisture inputs, and the susceptibility of these sites to invasions associated with either disturbance or changes in soil/climatic conditions.

	Undisturbed	Disturbed	Difference		Undisturbed	Disturbed	Difference
30 years of recovery following cryptobiotic soil crust disturbance in Canyonlands National Park				**Soil crusts with moderate grazing at Coral Pink Sand Dunes State Park** (data from Evans and Ehleringer 1994)			
soil δ¹⁵N (‰)	3.6 ± 0.4	5.1 ± 0.3	1.5 ‰		1.1	2.9	1.8 ‰
plant δ¹⁵N (‰)	1.1 ± 0.7	2.6 ± 0.3	1.5 ‰		1.1	2.7	1.6 ‰
soil N (mg/g)	0.41 ± 0.01	0.27 ± 0.03	66 %		0.44	0.26	69 %
plant N (mg/g)	- -	- -			2.9	2.4	21 %

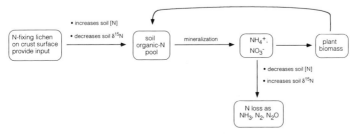

Figure 14.8. *Changes in the soil and plant nitrogen isotope ratio values for plants growing on soils that have intact or disturbed cryptobiotic crusts in Canyonlands National Park. Shown also is a model to account for the observed variations in nitrogen isotope ratio associated with crust disturbance. Modified from Evans and Ehleringer (1994) and Evans and Belnap (1998).*

14.6 Sensitivity of Colorado Plateau arid land ecosystems to invasions

The contributions of cryptobiotic crusts to nitrogen input in aridlands and its disruption by anthropogenic activities have important implications for the nitrogen cycle in arid regions (Evans and Ehleringer, 1993). The physical destruction of cryptobiotic crusts through land-use change (Figure 14.9) can eliminate the predominant source of nitrogen input. In the long term, removal of this nitrogen input source, coupled with continuous gaseous losses of nitrogen from the ecosystem, will ultimately decrease the amount of nitrogen available for plant growth. Crust recovery is slow (Belnap, 1991), and so in effect the nitrogen cycle is broken, with significantly reduced nitrogen inputs but continued and possibly accelerated rates of nitrogen loss from the system (Evans and Ehleringer, 1993). As indicated in Figures 14.8 and 14.10, this degradation leads to decreased fertility and ultimately to degradation of community structure and shifts in composition towards species that are either capable of nitrogen fixation or are tolerant of low nitrogen availability (Schlesinger *et al.*, 1990). Comparative studies of grazed and nearby pristine sites on the Colorado Plateau have already provided evidence for these changes in species composition (Kleiner and Harper, 1972).

Water-limited and nutrient-poor ecosystems are likely candidates for biological invasions (Mooney and Drake, 1986). The Colorado Plateau is no exception, in part due to both cryptobiotic crust damage and by land-use change (Kleiner and Harper, 1972; Loope *et al.*, 1988). Cryptobiotic crust damage makes the ecosystem more susceptible to invasion; one suggestion presented earlier is that disturbed soils may have less nitrogen available to plants than do pristine cryptobiotic soils.

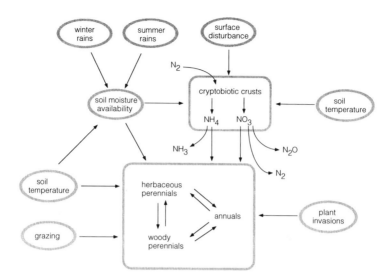

Figure 14.9. A model of the impacts of changes in precipitation patterns, soil temperature, grazing, and plant invasions, on the dynamics and interactions between plant components of a Colorado Desert ecosystem and the cryptobiotic crust.

This then raises the question of how much effect the increasing crust disturbance will have on the magnitude of the pulse precipitation event (Figure 14.11). Data from the previous section indicate that both soil and plant nitrogen contents have been decreasing on crust-disturbed ecosystems. Given that photosynthetic capacity is tightly correlated with leaf nitrogen content (Field and Mooney, 1986; Figure 14.11), one would expect that primary productivity by existing species should be decreased on disturbed sites.

At some point in the deterioration of the nitrogen levels within these desert ecosystems, there may be a shift toward favouring C_4 species over the predominantly C_3 native species. This will arise because C_4 plants are intrinsically more efficient in their nitrogen-use efficiency (NUE) than C_3 plants and can achieve greater photosynthetic rates per unit leaf nitrogen content. In Colorado Plateau deserts, invasive species are increasing in abundance; most of the weedy species are annuals. Winter weeds include the bromes (*Bromus* sp.) and mustards; summer annuals are primarily C_4 chenopods (*Bassia, Kochia, Halogeton, Salsola*). South African C_4 grasses are becoming more aggressive in the southern deserts (Anable *et al.*, 1992; Cox *et al.*, 1988), but thus far have had limited impact thus far on the Colorado Plateau ecosystems. By far the most noxious of the summer-time weeds is *Salsola*; this species appears to become established quickly on disturbed soil. In general, weedy species have a difficult time becoming established on undisturbed cryptobiotic soils.

These invading species are capable of using summer moisture inputs to a greater extent than can the native woody species (Ehleringer *et al.*, 1991; Lin *et al.*, 1996). If greater access to summer moisture results in greater productivity by invading species and greater acquisition of nutrients released by decomposition during periodic summer precipitation events, then invading species may have a competitive advantage over native perennial species in the long run.

Colorado Plateau desert

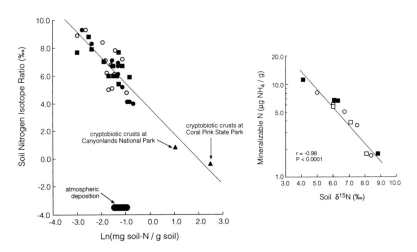

Figure 14.10. Left diagram. The relationship between soil nitrogen isotope ratio and the concentration of nitrogen in soils in a Colorado Plateau desert ecosystem. Right diagram. The relationship between mineralizable nitrogen and soil nitrogen isotope ration a Colorado Plateau desert ecosystem. Modified from Evans and Ehleringer (1993, 1994).

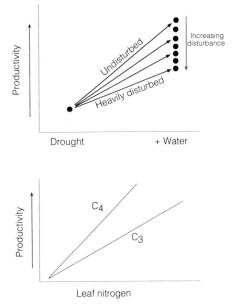

Figure 14.11. Top diagram. The hypothetical relationship of how water and soil nitrogen influence productivity in desert ecosystems, which are pulse driven, when crust disturbance is incorporated into the ecosystem interactions. Bottom diagram. The relationship between photosynthesis and leaf nitrogen content for C_3 and C_4 species.

14.7 Conclusions

Water and nitrogen limitations constrain plant productivity in the desert ecosystems of the Colorado Plateau. Moisture inputs are biseasonal, acting as pulse events with prolonged drought periods between major precipitation periods. There is no clear evidence for niche differentiation with respect to water source by perennial functional groups in the spring growing season. However, there are clear differences in the capacities of woody life forms to utilise summer moisture inputs. Differences in water source utilisation occur within the same life form, limiting a simplified niche-partitioning approach. Herbaceous annuals (including invasive species) utilise more of the monsoon precipitation than do woody perennials. Temperature may play a role in limiting the uptake of upper-layer moisture by some species, creating an imbalance in the capacities for different life forms to compete for monsoonal moisture input.

Variations in the soil nitrogen isotope ratios and nitrogen contents suggest a progressive loss of nitrogen from soils that fits a Rayleigh relationship. These disturbance-driven patterns are consistent with a decrease in nitrogen-fixing capacity relative to nitrogen loss; cryptobiotic crusts are implied as the primary nitrogen source in these ecosystems. Plant nitrogen isotope ratios then become a strong indicator of the quality of the soil resource following disturbance. Many of the invading species are C_4 plants with a lower intrinsic nitrogen requirement than the native vegetation.

These patterns suggest that the native vegetation of the Colorado Plateau may not fully utilise monsoonal moisture inputs and may be exposed to progressively eroding soil nitrogen availability associated with disturbance. Together these factors imply that such native ecosystems are unlikely to be responsive to changes in monsoonal moisture input (at least in their current composition) and are sensitive to change by invasive species. Fluctuations in the relative contributions of winter and monsoonal moisture and disturbance-driven decreases in soil nitrogen availability would both contribute to this effect.

References

Adang, T.C. and Gall, R.L. (1989) Structure and dynamics of the Arizona Monsoon boundary. *Mon. Weath. Rev.* **117**, 1423–1438.

Anable, M.E., McClaran, M.P. and Ruyle, G.B. (1992). Spread of introduced Lehmann lovegrass *Eragrostis lehmanniana* Nees. in Southern Arizona, USA. *Biological Conservation* **1992**, 181–188.

Belnap, J. (1995) Surface disturbances, their role in accelerating desertification. *Environ. Monitor. Assess.* **37**, 39–57.

Belnap, J. (1991) Effects of wet and dry pollutants on the physiology and elemental accumulation of cryptobiotic soil crusts and selected rock lichens. Dissertation, Brigham Young University.

Belnap, J. (1996) Soil surface disturbance in cold deserts: effects on nitrogenase activity in cyanobacterial-lichen soil crusts. *Biology and Fertility of Soils.* **23**, 362–367.

Belnap, J. and Gardner, J.S. (1993) Soil microstructure in soils of the Colorado Plateau: the role of the cyanobacterium *Microcoleus vaginatus*. *Great Basin Naturalist* **53**, 40–47.

Belnap, J., Harper, K.T., and Warren, S.D. (1994) Surface disturbance of cryptobiotic soil crusts: nitrogenase activity, chlorophyll content, and chlorophyll degradation. *Journal of Arid Soil Research and Rehabilitation* **8**, 1–8.

Betancourt, J.L., Van Devender, T.R., and Martin, P. (1990) *Packrat Middens, The Last 40,000 Years Of Biotic Change.* University of Arizona Press, Tucson.

Binkley, D. and Hart, S.C. (1989) The components of nitrogen availability assessments in forest soils. *Adv. Soil Sci.* **10**, 57–111.

Burke, I. C. (1989) Control of nitrogen mineralization in a Sagebrush Steppe landscape. *Ecology* **70**, 1115–1126.

Bryson, RA. and Lowry, R.P. (1955) Synoptic climatology of the Arizona summer precipitation singularity. *Bull. Am. Meteorol. Soc.* **36**, 329–339.

Caldwell, M.M. (1985) Cold deserts, *Physiological Ecology Of North American Plant Communities* (eds B.F. Chabot and H.A. Mooney). Chapman and Hall, New York. pp. 198–212.

Campbell, S.E., Seeler, J.S., and Glolubic, S. (1989) Desert crust formation and soil stabilization. *Arid Soil Res. and Reh.* **3**, 217–228

Cole, K. (1990) Late Quaternary zonation of vegetation in the eastern Grand Canyon. *Science* **217**, 1142–1145.

Comstock, J.P. and Ehleringer, J.R. (1992) Plant adaptation in the Great Basin and Colorado Plateau. *Great Bas. Nat.* **52**, 195–215.

Cox, J.R., Martin, M.H., Ibarra, F.A., Fourie, J.H., Retham, N.F.G. and Wilcox, D.G. (1988) The influence of climate and soils on the distribution of four African grasses. *J. Range Manage.* **41**, 127–138.

Dobrowolski J.P., Caldwell, M.M. and Richards, J.H. (1990) Basin hydrology and plant root systems, In: *Plant Biology Of The Basin And Range*. eds C.B. Osmond, L.F. Pitelka, and G.M. Hidy, Springer Verlag, New York. pp 243–292.

Donovan, L.A. and Ehleringer, J.R. (1994) Water stress and use of summer precipitation in a Great Basin shrub community. *Funct. Ecol.* **8**, 289–297.

Ehleringer, J.R. (1993a) Carbon and water relations in desert plants, an isotopic perspective, *Stable Isotopes and Plant Carbon–Water Relations* (eds J.R. Ehleringer, A.E. Hall, and G.D. Farquhar). Academic Press, San Diego. pp 155–172.

Ehleringer, J.R. (1993b) Gas exchange implications of isotopic variation in aridland plants, *Water Deficits, Plant Responses From Cell To Community* (eds J.A.C. Smith and H. Griffiths). BIOS Scientific Publ.,pp 265–284.

Ehleringer, J.R. (1993c) Variation in leaf carbon isotope discrimination in *Encelia farinosa*, implications for growth, competition, and drought survival. *Oecologia* **95**, 340–346.

Ehleringer, J.R. and Cerling, T.E. (1995) Atmospheric CO_2 and the ratio of intercellular to ambient CO_2 levels in plants. *Tree Physiol.* **15**, 105–111.

Ehleringer, J. R. and Cooper, T. A. (1988) Correlations between carbon isotope ratio and microhabitat in desert plants. *Oecologia* **76**, 562–566.

Ehleringer, J.R. and Dawson, T.E. (1992) Water uptake by plants, perspectives from stable isotope composition. *Plant Cell Environ.* **15**, 1073–1082.

Ehleringer, J.R., Hall, A.E. and Farquhar, G.D. (eds). 1993. *Stable Isotopes and Plant Carbon–Water Relations*. Academic Press, San Diego. 555 pp.

Ehleringer, J.R., Phillips, S.L., Schuster, W.F.S. and Sandquist, D.R. (1991) Differential utilization of summer rains by desert plants, implications for competition and climate change. *Oecologia* **88**, 430–434.

Ehleringer, J.R., and Mooney, H.A. (1983) Productivity of desert and mediterranean-climate plants. *Ency. Plant Physiol. N.S.* **12D**, 205–231.

Eldridge, D.J., and Greene, R.S.B. (1994) Microbiotic soil crusts: a review of their roles in soil and ecological processes in the rangelands of Australia. *Aust. J. Soil Res.* **32**, 389–415.

Evans, R. D. and Belnap, J. (1998) Long-term consequences of disturbance on nitrogen dynamics of an arid ecosystem. *Ecology*, in press.

Evans, R. D., Bloom, A.J., Sukrapanna, S.S. and Ehleringer, J.R. (1996) Nitrogen isotope composition of tomato (*Lycopersicon esculentum* Mill. cv. T-5) grown under ammonium or nitrate nutrition. *Plant Cell Environ.* **19**, 1317–1323.

Evans, R.D. and Ehleringer, J.R. (1993) Broken nitrogen cycles in aridlands, evidence from $\delta^{15}N$ of soils. *Oecologia* **94**, 314–317.

Evans, R.D. and Ehleringer, J.R. (1994) Water and nitrogen dynamics in an arid woodland. *Oecologia* **99**, 233–242.

Farquhar, G. D., Ehleringer, J. R. and Hubick, K. T. (1989) Carbon isotope discrimination and photosynthesis. *Ann. Rev. Plant Physiol. Mol. Biol.* **40**, 503–537.

Field, C.B. and Mooney, H.A. (1986) The photosynthesis–nitrogen relationship in wild plants. *On the Economy of Plant Form and Function* (ed T.J. Givnish). Cambridge University Press, Cambridge, pp. 1–16.

Flanagan, L.B., Ehleringer, J.R. and Marshall, J.D. (1992) Differential uptake of summer precipitation and groundwater among co-occurring trees and shrubs in the southwestern United States. *Plant Cell Environ.* 15, 831–836.

Fustec, E., Mariotti, A., Grillo, X. and Sajus, J. (1991) Nitrate removal by denitrification in alluvial ground water: role of a former channel. *J. Hydrol* 123, 337–354.

Gregg, J.W. (1991) The differential occurrence of the mistletoe *Phoradendron juniperinum* on its host, *Juniperus osteosperma*, in the western United States. Masters Thesis, University of Utah, Salt Lake City. 78 pp.

Hadley, N.F. and Szarek, S.R. (1981) Productivity of desert ecosystems. *BioScience* 31, 747–753.

Harper, K.T. and Marble, J.R. (1988) A role for nonvascular plants in management of arid and semiarid regions. *Vegetation Science Applications for Rangeland Analysis and Management* (ed P.T. Tueller) Kluwer Academic, Boston, pp. 135–169.

Houghton, J.G. (1979) A model for orographic precipitation in the north central Great Basin. *Mon. Weath. Rev.* 107, 1462–1475.

Kleiner, E.F. and Harper, K.T. (1972) Environment and community organization in grasslands of Canyonlands National Park. *Ecology* 53, 299–309.

Lin, G., Phillips, S.L. and Ehleringer, J.R. (1996) Monsoonal precipitation responses of shrubs in a cold desert community on the Colorado Plateau. *Oecologia* 106, 8–17.

Loope, L.L., Sanchez, P.G. Tarr, P.W., Loope, W.L. and Anderson, R.L. (1988) Biological invasions of arid land nature reserves. *Biol. Cons.* 44, 95–118.

Mack, R.N. and Thompson, J.N. (1982) Evolution in the steppe with few large, hooved mammals. *American Naturalist* 119, 757–773.

Mariotti, A., Germon, J.C., Hubert, P., Kaiser, P., Letolle, R., Tardieux, A. and Tardieus, P. (1981) Experimental determination of nitrogen kinetic isotope fractionation: some principles; illustration for the denitrification and nitrification processes. *Plant and Soil* 62, 413–430

Matson, P. A., Volkmann C., Coppinger K. and Reiners, W. A.. (1991) Annual Nitrous Oxide Flux and Soil Nitrogen Characteristics in Sagebrush Steppe Ecosystems. *Biogeochemistry* 14, 1–12.

Mitchell, V.L. (1976) The regionalization of climate in the western United States. *J. Appl. Meteorol.* 15, 920–977.

Mooney, H.A., and J.A. Drake (eds). 1986. *Ecology of Biological Invasions of North America and Hawaii.* Springer Verlag, New York.

Moore, T.J., Gall, R.L. and Adang, T.C. (1989) Disturbances along the Arizona Monsoon boundary. *Mon. Weath. Rev.* 117, 932–941.

Noy Meir, I. (1973) Desert ecosystems, environment and producers. *Ann. Rev. Ecol. Syst.* 4, 25–41.

Peterjohn, W.T. and Schlesinger, W.H. (1990) Nitrogen loss from deserts in the Southwestern United States. *Biogeochemistry* 10, 67–79.

Phillips, S.L. and Ehleringer, J.R. (1995) Limited uptake of summer precipitation by bigtooth maple (*Acer grandidentatum* Nutt) and Gambel's oak (*Quercus gambelii* Nutt). *Trees* 9, 214–219.

Rind, D., Goldenberg, R.,Hansen, J., Rosenzweig, C. and Ruedy, R. (1990) Potential evapotranspiration and the likelihood of future drought. *J. Geophys. Res.* 95, 9983–10 004.

Schlesinger, M.E. and Mitchell, J.B.F. (1987) Climate model simulations of the equilibrium climatic response to increased carbon dioxide. *Rev. Geophys.* 25, 760–798.

Schlesinger, W.H., Reynolds, J.F., Cunningham, G.L., Heunneke, L.F., Jarrell, W.M. Virginia, R.A. and Whitford, W.G. (1990) Biological feedbacks in global desertification. *Science* 247, 1043–1048.

Schuster, W.S.F., Sandquist, D.R., Phillips, S.L. and Ehleringer, J.R. (1992) Comparisons of carbon isotope discrimination in populations of aridland plant species differing in lifespan. *Oecologia* 91, 332–337.

Smith, S.D., Monson, R.K. and Anderson, J.E. (1997) *Physiological ecology of North America deserts.* Springer Verlag, Heidelberg.

Smith, S.D. and Nowak, R.S. (1990) Ecophysiology of plants in the intermountain lowlands, In: *Plant Biology Of The Basin And Range.* eds C.B. Osmond, L.F. Pitelka and G.M. Hidy Springer-Verlag, New York. pp 179–241.

Thorburn, P.J. and Ehleringer, J.R. (1995) Root water uptake of field-growing plants indicated by measurements of natural-abundance deuterium. *Plant Soil* **177**, 225–233.

West, N.E. (1990) Structure and function of microphytic soil crusts in wildland ecosystems. *Adv. Ecol. Res.* **20**, 179–223.

Yoneyama, T. and Kaneko, A. (1989) Variations in the natural abundance of ^{15}N in nitrogenous fractions of komatsuna plants supplied with nitrate. *Plant and Cell Physiology* **30**, 957–962.

Zaady, E., Groffman, P. M., and Shachak, M. 1996. Litter as a regulator of N and C dynamics in macrophytic patches in Negev desert soils. *Soil Biol. Biochem.* **28**, 39–46.

Carbon stable isotope fractionation in marine systems: open ocean studies and laboratory studies

A.M. Johnston and H. Kennedy

15.1 Introduction

The stable carbon isotopic composition of surface water particulate organic matter ($\delta^{13}C_{POM}$) has been shown to correlate with the concentration of dissolved carbon dioxide [$CO_{2(aq)}$]. The potential of this proxy has been exploited mainly by palaeoceanographers for the reconstruction of past variations in the ocean and atmospheric CO_2, derived from the analysis of the carbon isotopic composition of organic matter in marine sediments ($\delta^{13}C_{SOM}$). Although these correlations have been shown to be useful, a demonstration of an apparent empirical relationship between two parameters is not necessarily indicative of a directly coupled process and it is obvious that we need to develop a sound theoretical basis for the interpretation of any relationship. The main mechanism suggested to account for the observed covariations in $\delta^{13}C_{POM}$ and [$CO_{2(aq)}$] is not yet very well constrained. It is based on the assumption that in the majority of phytoplankton species, inorganic carbon acquisition and assimilation occurs solely by the diffusive entry of $CO_{2(aq)}$ and carboxylation by ribulose 1,5 bisphosphate carboxylase/oxygenase (RUBISCO). Under these circumstances the difference in isotopic composition between $CO_{2(aq)}$ ($\delta^{13}C_{CO2(aq)}$) in the surrounding sea water and the phytoplankton cell, ϵ_p, ($\delta^{13}C_{CO2(aq)} - \delta^{13}C_{POM}$) is an expression of the fractionation during cellular fixation of carbon as modified by the degree to which carbon transport may be rate limiting with respect to carbon fixation. When this expression is used in higher plant stable isotope analysis the difference is termed Δ, (($\delta^{13}C_{CO2} - \delta^{13}C_{ORG})/(1+\delta^{13}C_{ORG})$); the difference between ϵ_p and Δ at $\delta^{13}C_{CO2}$ and $\delta^{13}C_{ORG}$ values normally encountered tend to be negligible.

Subsequently examples have been reported where $\delta^{13}C_{POM}$ can vary independently of [$CO_{2(aq)}$] and other, mostly physiological factors have been proposed to cause a

Stable Isotopes, edited by H. Griffiths.
© 1998 BIOS Scientific Publishers Ltd, Oxford.

weakening of the initial correlation (Descolas-Gros and Fontugne 1985; Francois *et al.*, 1993; Laws *et al.*, 1995). Current research has focussed on gaining a fuller understanding of these physiological factors and the investigation of the examples where the correlation is not so strong may eventually lead to a better understanding of phytoplankton inorganic carbon acquisition. In a similar vein, Chisholm (1992), when reviewing the influence of size on phytoplankton physiology, wrote 'the simplicity of the general relationship serves as a stable background against which the exceptions can shine'.

In a number of ways the laboratory experimentalists have some advantages over the ship-based oceanographers. First, they can offer a single-species culture uncontaminated with other phytoplankton or bacteria, which if the experiments are carried out properly, contain little detrital material. Second, the culture conditions can be manipulated to determine the scale and effect that changing a single variable can have on the isotopic composition of a phytoplankton species. In the past, a lack of control in the experimental conditions and/or the incomplete analysis of the ambient chemical environment have led to measured changes in isotopic composition that cannot be fully interpreted. This is especially true of batch or closed cultures where progressive changes in the chemistry and isotopic composition of the medium are rarely monitored, and yet can often have a strong control on the isotopic composition of the culture. The alternative technique is to use continuous culture. While this allows a constant growth environment and simpler analysis of the data it can be argued that it is not as close to conditions found in the natural environment as are those in batch culture. From oceanographic and laboratory studies it is clear that CO_2 is a master variable in determining the isotopic composition of phytoplankton. There are three stages of fractionation involved in determining the net fractionation expressed by the cell (ϵ_p, Δ): the fractionation between CO_2 and HCO_3^- in the solution surrounding the cell; the fractionation that occurs as CO_2 is assimilated by the cell and the fractionation that occurs during the enzyme-catalysed reactions of carboxylation (Francois *et al.*, 1993; Goericke *et al.*, 1994, O'Leary, 1981). These steps can be greatly influenced by the environmental conditions to which the marine phytoplankter is exposed, and the extent of any genotypic and phenotypic control in determining ϵ_p (Δ) can be investigated much more rigorously through well-designed culture experiments than through synoptic studies in the open ocean.

15.2. Open ocean studies

15.2.1 *Observations on the variation of $\delta^{13}C_{POM}$ in the oceans*

The collection of oceanic surface particulate organic matter (POM) samples to delineate variations of $\delta^{13}C_{POM}$ led to the discovery that there was a distinct and coupled change of $\delta^{13}C_{POM}$ with latitude (Figure 15.1) and sea-surface temperature (Fontugne and Duplessy 1978, 1981; Sackett *et al.*, 1965). These correlations, however did not represent a direct cause and effect and the more relevant correlation was found to be between $\delta^{13}C_{POM}$ and $[CO_{2(aq)}]$ (Figure 15.2). The link between these parameters lay in the change of $[CO_{2(aq)}]$ in the surface ocean with temperature and hence latitude (Rau *et al.*, 1989, 1991). At a constant concentration of CO_2 in the atmosphere, the solubility of CO_2 increases with decreasing water temperature. This leads to the surface waters of the cold, high latitude oceans being about 2.5 times more enriched in CO_2

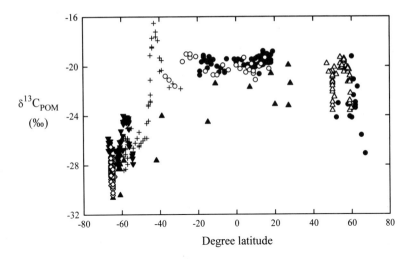

Figure 15.1. *Variation in $\delta^{13}C_{POM}$ plotted against latitude.* ● *Fontugne and Duplessy, 1981;* ▲ *Sackett et al., 1965;* ▼ *Kennedy and Robertson, 1995;* + *Francois et al., 1994;* ◇ *Rau et al., 1991;* ○ *Fontugne and Duplessy, 1978;* △ *Kennedy unpublished data.*

than those of the warm, low latitude oceans. Thus it was proposed that the isotopic composition of phytoplankton depends on the concentration of dissolved carbon dioxide in sea water. The more fundamental explanation for the interdependence of $\delta^{13}C_{POM}$ on $CO_{2(aq)}$ is that the isotopic composition of phytoplankton is an expression of the isotopic fractionation of carbon during cellular fixation as modified by the degree to which carbon transport to the cell may be rate limiting with respect to carbon fixation. Few concurrent measurements of $[CO_{2(aq)}]$ and $\delta^{13}C_{POM}$ have been made

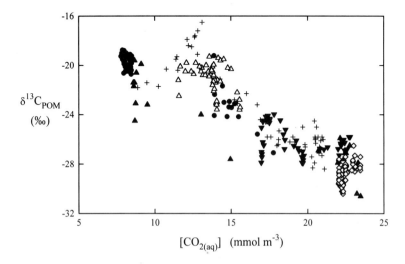

Figure 15.2. *Variation in $\delta^{13}C_{POM}$ plotted against $[CO_{2(aq)}]$ calculated for seawater in equilibrium with an atmosphere containing CO_2 at a concentration of 350 μmol mol⁻¹. Symbols the same as for Figure 15.1.*

and for the majority of the field data only location, temperature and $\delta^{13}C_{POM}$ have been recorded. Therefore, non-concurrent measurements of $[CO_{2(aq)}]$ or values calculated assuming equilibrium with the local atmosphere $[CO_{2(aq)}]_{equilib}$ have comprised the major part of data sets testing the co-variation of $\delta^{13}C_{POM}$ and $[CO_{2(aq)}]$. An early regression equation was derived from data collected in the south Atlantic and southern oceans (Rau et al., 1989), is given by $\delta^{13}C_{POM} = -0.8\,[CO_{2(aq)}]_{equilib} -12.6$.

A potential use for this correlation suggested by palaeoceanographers was in the reconstruction of past variations in $[CO_{2(aq)}]$ and the partial pressure of CO_2 in seawater $[CO_2]$ from the $\delta^{13}C$ record of the organic matter stored in ocean sediments (Freeman and Hayes, 1992; Hayes et al., 1989; Pedersen et al., 1991; Rau et al., 1989). To be able to reconstruct the $[CO_2]$ of ancient atmospheres from $\delta^{13}C_{POM}$, it must be assumed that the relationship between $\delta^{13}C_{POM}$ and $[CO_{2(aq)}]$ is constant and predictable and that the surface ocean is in equilibrium with the atmosphere. That there is scatter in this relationship is clear (Figure 15.2). Goericke and Fry (1994) suggested that the range of $\delta^{13}C_{POM}$ values measured at a single site might, in some cases, be as high as the whole latitudinal range of observed of $\delta^{13}C_{POM}$ values. A degree of truth in this statement can be seen in Figure 15.1. where a large range of $\delta^{13}C_{POM}$ values have been reported in some areas of the oceans for example 55 to 65°S and 47 to 60°N. When $\delta^{13}C_{POM}$ is plotted against $[CO_{2(aq)}]_{equilib}$ this variability is reduced but not removed and $\delta^{13}C_{POM}$ is often less negative than predicted by published regression equations (e.g. Rau et al., 1989). In these cases it could be because the actual $[CO_{2(aq)}]$ was low relative to local equilibrium so that the $\delta^{13}C_{POM}$ was less negative than predicted. In many regions, the oceanic surface waters can be low (or high) in $[CO_{2(aq)}]$ relative to local equilibrium (Tans et al., 1990). These changes can be pronounced in areas where there are strong seasonal changes in productivity. In Figure 15.2 where $\delta^{13}C_{POM}$ varies independently of $[CO_{2(aq)}]_{equilib}$, direct measurements have shown a deficit in surface water $[CO_{2(aq)}]$ relative to local equilibrium, especially during phytoplankton blooms. For example, in the North Atlantic, deficits relative to equilibrium of 1.4 to 3 μmol kg^{-1} (Rau et al., 1992), and in the Antarctic, 1 to 8 μmol kg^{-1} (Kennedy and Robertson, 1995) have been reported, and could lead to actual values of $\delta_{13}C_{POM}$ being 1‰ to 6‰ heavier than values predicted, assuming that the surface waters are in $_{equilib}$rium with the atmosphere.

The first concurrent measurements of $\delta^{13}C_{POM}$ and $CO_{2(aq)}$ were made at 47°N 20°W (Rau et al., 1992), and showed that even when $[CO_{2(aq)}]$ was below local equilibrium, due to biological utilisation, a correlation with $\delta^{13}C_{POM}$ existed although the regression equation, $\delta^{13}C_{POM} = -1.5\,[CO_{2(aq)}] -2.1$ was different from that previously reported using compilations of oceanographic data (Rau, 1989). Thus it was suggested that the primary association between $\delta^{13}C_{POM}$ and $[CO_{2(aq)}]$ still held and that the seasonal variations in $\delta^{13}C_{POM}$ at any one latitude reflected the modulation of surface water $[CO_{2(aq)}]$ by biological activity. In such regions $\delta^{13}C_{POM}$ would co-vary with surface ocean $[CO_2]$ but they would not represent suitable sites at which to reconstruct atmospheric pCO_2 from $[CO_{2(aq)}]$ or $\delta^{13}C_{POM}$. Nevertheless, isotopic studies in these regions can still provide some useful information and palaeoceanographic applications include the reconstruction of glacial/interglacial variations in the degree of local disequilibrium between surface water and the atmosphere (see Chapter 22 for equivalent terrestrially-based estimates). Müller et al. (1994) estimated surface water $[CO_{2(aq)}]$ using $\delta^{13}C_{SOM}$ and compared the surface water $[CO_2]$ with atmospheric $[CO_2]$ records, taken from the Vostok ice core (Barnola et al., 1987). They showed that for the last 160,000 years in the eastern Angola

Basin, the [CO_2] of the surface water had been consistently high relative to atmospheric levels. In addition, the data indicated that the degree of disequilibrium between surface water and atmosphere had diminished during glacial periods in this region and suggested that local productivity had been higher at this time than during the interglacial periods.

15.2.2 *Environmental and physiological factors that affect net isotope fractionation in phytoplankton*

Using a smaller compilation of concurrent $\delta^{13}C_{POM}$ and [$CO_{2(aq)}$] measurements (Francois *et al.*, 1993, Kennedy and Robertson 1995; Rau *et al.*, 1992), it can be seen that there still remains a degree of scatter in the data (Figure 15.3) and it has been proposed that additionally environmental and physiological factor(s) can cause phytoplankton $\delta^{13}C_{POM}$ to vary independently of [$CO_{2(aq)}$] (Francois *et al.*, 1993; Goericke *et al.*, 1994; Laws *et al.*, 1995).

Closed system effects. During phytoplankton blooms, CO_2 is assimilated faster than it can be resupplied by advection/turbulence from deeper waters or from the atmosphere, due to the slow rate of CO_2 gas exchange (Broecker and Peng, 1982). This means that during and after blooms, surface water [$CO_{2(aq)}$] may be low relative to local equilibrium and in this semi-closed environment the discrimination against ^{13}C during carbon assimilation by phytoplankton leads to ^{13}C enrichment of the remaining total dissolved inorganic carbon (DIC) pool. If isotopic discrimination remains constant, continued phytoplankton growth will result in progressive shifts of the isotopic composition of the DIC ($\delta^{13}C_{DIC}$) and $\delta^{13}C_{POM}$ to less negative values. The effect of a closed system on $\delta^{13}C_{DIC}$ and $\delta^{13}C_{POM}$ can be described using a Rayleigh distillation model (Mariotti *et al.*, 1981).

Closed system effects were first raised by Deuser *et al.*, (1968) who went on to make a generalised interpretation of the ^{13}C enriched phytoplankton values observed

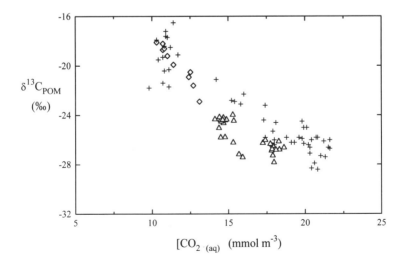

Figure 15.3. *Relationship between concurrent measurements of $\delta^{13}C_{POM}$ and [$CO_{2(aq)}$].*
+ Francois et al., *1994; ◇ Rau* et al., *1992; △ Kennedy and Robertson, 1995.*

in the Black Sea (Deuser, 1970). Bottle incubations, closed cultures and mesocosm experiments (Fry and Wainright, 1991; Thompson and Calvert, 1994) have successfully used the Rayleigh distillation model to predict net isotopic discrimination between DIC and phytoplankton. Analysis of DIC samples collected sequentially allows estimates to be made of the isotopic discrimination provided that the fraction of DIC remaining or utilised has been measured (Fry, 1996; Guy et al., 1989).

The isotopic changes that occur during a phytoplankton bloom cannot always be described by the simple Rayleigh distillation model. During the North Atlantic Bloom Experiment NABE, Rau et al. (1992) discounted closed system effects because the $\delta^{13}C_{DIC}$ changes that were predicted (0.035‰) using the Rayleigh model could not be reconciled with the observed variation (0.25‰) in the isotopic composition of particulate inorganic carbon, here used as a proxy for $\delta^{13}C_{DIC}$. In addition Nakatsaka et al. (1992) found that the assumption of a constant isotopic discrimination could only predict the $\delta^{13}C_{POM}$ during the early part of a phytoplankton bloom. The whole data set could only be modelled successfully when changes in isotopic discrimination with growth rate were taken into account (see Section 15.2.4 for a discussion of these effects).

Variations with species composition. Models of isotopic fractionation in phytoplankton (Francois et al., 1993; Goericke et al., 1994; Laws et al., 1995; Rau et al., 1992) show different levels of parameterisation with which to define the relationship between $\delta^{13}C_{POM}$ and $[CO_{2(aq)}]$. The underlying theory states that it is the interplay between the rate of transport of CO_2 across the plasmalemma relative to enzymatic fixation that plays the major role in specifying the degree of isotopic discrimination between DIC or $[CO_{2(aq)}]$ and phytoplankton POM. Modification of the external environment, changes in the physiology of the existing phytoplankton community or successional changes in phytoplankton composition may all affect the rate or nature of CO_2 supply and fixation and hence the relationship between $\delta^{13}C_{POM}$ and $[CO_{2(aq)}]$. This can occur to such an extent that in some field collections there is no observable correlation between $\delta^{13}C_{POM}$ and $[CO_{2(aq)}]$ (Fogel et al., 1992; Kopczynska et al., 1995). A better understanding of the physiological processes affecting $\delta^{13}C_{POM}$ could help determine the mechanism(s) and control(s) on carbon uptake by phytoplankton.

There are few publications that report $\delta^{13}C_{POM}$ and detailed descriptions of phytoplankton community composition (e.g. Kopczynska et al., 1995). Not surprisingly, no publications yet combine this type of information with detailed work on cell physiology. The definition of cause and effect is therefore very difficult. This is due to the level of complexity of factors that could affect $\delta^{13}C_{POM}$ and generally the lack of resources/expertise to measure all the possible variables concurrently. Most data that show any correlation between $\delta^{13}C_{POM}$ and another variable probably reflect the major control on $\delta^{13}C_{POM}$ at that particular site and time, and hence using this information some generalisations are possible.

Numerous authors have reported less negative $\delta^{13}C_{POM}$ values in diatoms relative to other phytoplankton groups (e.g. Fry, 1996). Generally, these values are representative of a lower isotopic discrimination (ϵ_p, Δ) during phytoplankton growth. Data from Prydz Bay Antarctica show that when diatoms represent more than 50% of the phytoplankton biomass $\delta^{13}C_{POM}$ becomes less negative with higher relative diatom abundance (Kopczynska et al., 1995). But at three inner shelf stations, where diatoms represented less than 50% of the biomass, the trend was reversed and Wainright and Fry (1994) have also found that diatom abundance does not always correlate with the least negative

$\delta^{13}C_{POM}$. In these cases it may be because the identifiable phytoplankton comprised only a small percentage of the total POM that the relation between diatom abundance and less negative $\delta^{13}C_{POM}$ values breaks down. Flagellates have been shown to exhibit more negative $\delta^{13}C_{POM}$ than co-existing diatoms (Gearing et al., 1984) and these differences are also seen during the seasonal succession of diatoms to flagellates with a concurrent increase in the degree of isotopic discrimination leading to more negative $\delta^{13}C_{POM}$ for flagellates relative to diatoms in Narragansett Bay (Gearing et al., 1984), the Delaware Estuary (Cifuentes et al., 1988; Fogel et al., 1992) and Woods Hole (Wainright and Fry, 1994). While cell size (large diatoms to small flagellates) can be one factor that affects the net isotopic fractionation expressed by phytoplankton, cell size alone may not be the sole cause. Growth rate as well as cell size is important. Analytical models show that small phytoplankton with slow growth rates should have the most negative $\delta^{13}C_{POM}$ values, while large rapidly growing phytoplankton would be predicted to have the most positive $\delta^{13}C_{POM}$ values (Francois et al., 1993; Goericke et al., 1994).

Variations with growth rate. The rapid growth of diatoms is well known, typically growing three times faster than dinoflagellates of equal size (Chisholm, 1992) and the possible dependence of $\delta^{13}C_{POM}$ on growth rate has received renewed interest (Laws et al., 1995). The differential growth rate argument is common to a number of other publications, whether the differences are inter- or intra-specific. Kopczynska et al. (1995) and Cifuentes et al. (1988) found good correlations of $\delta^{13}C_{POM}$ with POM ($r = 0.93$), and with estimates of areal productivity ($r=0.73$) respectively, suggesting that phytoplankton production/growth had a strong effect on the degree of isotopic discrimination. In the Antarctic, Fischer (1991) observed that two communities dominated by centric diatoms showed different degrees of discrimination. *Corethron criophilum* displayed a lower degree of discrimination and more negative values of $\delta^{13}C_{POM}$ (-32‰ to -35‰) than *Thalassiosira* spp (-23‰ to -27‰.). These differences were related to the higher specific growth rates for *Thalassiosira* spp than for *Corethron criophilum*. The data suggest that the determination of specific growth rate or production is important to the interpretation of $\delta^{13}C_{POM}$. All else being equal, if two sites have the same rate of primary production, but differing biomasses, there will be a corresponding difference in cell carbon demand. As the net isotopic discrimination reflects the balance between the supply of carbon to the cell and demand for carbon within the cell, this difference should be reflected in the isotopic composition of the phytoplankton.

15.2.3 *Mechanism of carbon acquisition*

An additional reason for a range of $\delta^{13}C_{POM}$ at a constant $[CO_{2(aq)}]$ is that it is due to the mechanism of carbon acquisition. Diatoms have been suggested to be able to actively assimilate bicarbonate and/or enhance fixation of carbon through the use of bicarbonate, while other phytoplankton, for example, flagellates, are assumed to be dependent solely on the diffusive supply of CO_2 to the cell (Falkowski, 1991; Fogel et al., 1992; Fry and Wainright, 1991; Thompson and Calvert 1994). The enzyme responsible for the carboxylation of phospho–*enol*pyruvate (PEP), using bicarbonate as a substrate, is phospho–*enol*pyruvate carboxylase (PEPC). Extremely high rates of ß-carboxylation have been observed transiently in nitrogen-starved chlorophyte cultures that have been enriched with ammonia (Guy et al., 1989), but conclusive physiological and isotopic data to support these findings have not yet been presented.

Goericke *et al.* (1994) has suggested that even if ß-carboxylation is occurs in natural phytoplankton populations it will only be responsible for a small fraction of the carbon fixed when averaged over the generation time of a phytoplankton. In addition, field evidence for high PEPC activities and low isotopic discrimination are not conclusive. Although Descolas-Gros and Fontugne (1985) reported a strong positive correlation between $\delta^{13}C_{POM}$ and chlorophyll-normalised PEPC activity, Kopczynska *et al.* (1995) found that the relatively positive $\delta^{13}C_{POM}$ sampled in Prydz Bay came from a period characterised by low specific nitrate and ammonium uptake rates, and hence, probably low PEPC activity.

In conclusion, from ship board studies of the isotopic composition of natural populations of phytoplankton in open-ocean systems it is clear that any simplistic mechanistic inferences from the relationship between $\delta^{13}C_{POM}$ and $[CO_{2(aq)}]$ can be criticised. Complementary laboratory studies that refine our understanding of the mechanism of carbon acquisition and the effects that environmental factors have on the net isotopic fractionation of carbon in phytoplankton will aid in our interpretation of the observed large variations in isotopic composition of surface and sedimentary POM.

15.3. Laboratory studies

15.3.1 *The assimilation and carboxylation of inorganic carbon*

Laboratory experiments have been conducted with cultures to study the mechanisms of carbon acquisition employed by marine and freshwater algae. The results of these experiments indicate a variety of different methods may be employed by different species or by the same species, at different times, depending on the physical, biological and chemical characteristics of the growth environment. The simplest mechanism by which inorganic carbon can be assimilated by photosynthetic organisms is diffusion. A concentration gradient is set up between the environment and chloroplast by the carboxylation of CO_2 by RUBISCO. However, the physiological photosynthetic characteristics of marine phytoplankton suggest that additional mechanisms of inorganic carbon assimilation can be operative in these algae. They have higher affinities for inorganic carbon (than would be expected for CO_2 diffusion and RUBISCO-catalysed carboxylation), low CO_2 compensation points and accumulate inorganic carbon internally to concentrations in excess of the bulk phase concentration (Burns and Beardall, 1987; Johnston and Raven, 1996). There are some weak points to this summary. A number of the experiments were performed at external inorganic carbon concentrations which were considerably less than the 2 mol m^{-3} normally found in seawater (see Johnston and Raven, 1996). A recent report (Colman and Rotatore, 1995) does give direct evidence to support the idea that some marine phytoplankton are able to actively transport CO_2 and HCO_3^-. It should be noted that there are few reports of marine phytoplankton that appear unable to utilise HCO_3^- ions and are dependent on CO_2 diffusion for their inorganic carbon supply. One such species is the non-calcifying strain of the coccolithophorid *Emiliania huxleyi* (Nimer *et al.*, 1992). Riesebell *et al.* (1993) suggested, following a theoretical approach to this problem, that marine phytoplankton are dependent on CO_2 diffusion and that their growth rates were limited by the concentrations of $CO_{2(aq)}$.

15.3.2. *Culture of marine phytoplankton for stable isotope analysis*

While the home-based experimenter has a number of advantages over the ship-based oceanographer, the cell physiologist is constrained by the limitations of the laboratory (see also Chapter 16). Few workers have enough space to be able to grow large volumes of cultures so that they can harvest sufficient quantities of cells at cell densities that reflect those in the natural environment, while experiments at high cell densities are not to be recommended because of the potentially unrealistic chemical changes leading to biological acclimation. This problem can be highlighted by the wide range of $\delta^{13}C_{POM}$ values reported for marine phytoplankton. While C_3 and C_4 plants can be distinguished on the basis of their $\delta^{13}C$ values, (C_3: –32‰ to –20‰, C_4: –17‰ to –9‰) marine phytoplankton have been reported as having values between -33‰ and -5‰ (O'Leary, 1981, see Chapter 8). This would suggest that either marine phytoplankton vary in capacity for extracting and fixing inorganic carbon, or that experimental procedure may greatly influence the isotopic composition inorganic carbon. Some critical aspects of the experimental procedures are outlined below.

Inorganic carbon sources for photosynthesis. Seawater in equilibrium with atmospheric CO_2 (369 μmol mol^{-1} CO_2) at pH 8.1, 10°C and 35‰ salinity has a [DIC] of 2.176 mol m^{-3} ($CO_{2(aq)}$: 0.016 mol m^{-3}; HCO_3^-; 2.0 mol m$_3^-$; CO_3^-:= 0.16 mol m^{-3}). In these conditions the isotopic fractionation between $CO_{2(aq)}$ and HCO_3^- is 10.05‰, so for atmospheric CO_2 with a $\delta^{13}C$ of -8.0‰, the isotopic composition of DIC ($\delta^{13}C_{DIC}$) will be 1.76‰ (vs. VPDB). The CO_2 diffusion coefficient in water is 10,000 times lower than in air and so potentially aquatic photosynthetic organisms may be limited by their ability to extract CO_2 from solution. Until recently it was common for $\delta^{13}C_{POM}$ values to be published without the corresponding source $\delta^{13}C_{DIC}$ or $\delta^{13}C_{CO2(aq)}$ values, and so it is difficult to compare the data of different research groups (see Section 15.2.1). Even if it is assumed that many of the earlier reports were based on studies in which the $\delta^{13}C_{DIC}$ was about 0‰ and $\delta^{13}C_{CO2(aq)}$ was about -8‰ (depending on the temperature) at the start of the culture period, the isotopic composition of the source inorganic carbon can change due to differential uptake of the lighter ^{12}C, thus making the remaining inorganic carbon heavier. This effect can be overcome if the culture media is sufficiently well aerated to ensure that the inorganic carbon removed photosynthetically is replaced from atmospheric CO_2. This is illustrated in Figure 15.4 where *Phaeodactylum tricornutum* was grown under different aeration regimes resulting in a range of $\delta^{13}C$ values (Johnston and Raven, 1992). This is possible for some types of marine algae which can tolerate a degree of agitation and aeration, namely diatoms, but is not possible with dinoflagellates (Flynn, 1996). This has led to an imbalance with $\delta^{13}C$ data related to diatoms greatly exceeding that of dinoflagellates. Although the $\delta^{13}C$ of atmospheric CO_2 is about -8‰, it is frequently much more negative in laboratories in which, with closed windows at the end of the working day, the concentration of CO_2 can increase to 600 μmol mol^{-1} (i.e. 60 Pa CO_2) CO_2 and the $\delta^{13}C_{CO2(aq)}$ can be as low as -11‰. Much research interest today is centred on the effect that changing CO_2 partial pressures will have on the levels of ^{13}C discrimination of marine phytoplankton. To do this work, partial pressures greater than the ambient 36.5 Pa CO_2 are used and it is often necessary to use cylinder gases with 1–5 kPa CO_2. The CO_2 in these cylinders is normally isotopically very light (-45‰ to -28‰). If these cylinders are to be used as sources of CO_2 it is vitally

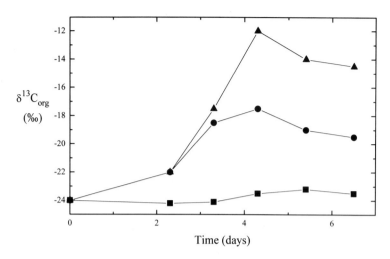

Figure 15.4. *The effect of aeration on the $\delta^{13}C$ values of* Phaeodactylum tricornutum *grown in f/2 media in batch culture at 20 °C and 95 μmol m^{-2} s^{-1} continuous light, aeration rates:* ■ *2 dm^{3} min^{-1},* ● *0.1 dm^{3} min^{-1},* ▲ *0 dm^{3} min^{-1}.*

important that the dissolved inorganic carbon in the culture media is in isotopic and chemical equilibrium with the 'light' atmospheric CO_2.

Batch culture versus continuous culture. There are two alternative ways in which phytoplankton can be cultured to study the environmental effects on the level of ^{13}C discrimination, namely, batch or continuous culture. Each technique has benefits and drawbacks. In batch culture, the growth medium is enriched with nutrients so that often the cells are in a nutrient-saturated environment. During growth, even though nutrients have been assimilated, there is, still, usually an excess in solution. This holds for most nutrients except the one central to this discussion, dissolved inorganic carbon. Unless the batch culture is sufficiently aerated, DIC can decrease with the result that the pH of the growth medium will rise. As the pH increases, the ratio of $CO_{2(aq)}$, HCO_3^- and CO_3^- will alter, as the equilibrium constants of carbonic acid are pH sensitive, with the effect that although the drawdown of CO_2 is relatively small compared to the initial DIC concentration, the reduction in CO_2 is relatively large compared to the initial $CO_{2(aq)}$. Combined with this 'pH drift' is the probable acclimation in photosynthetic physiology, since as the $CO_{2(aq)}$ and DIC concentrations decrease the algal cell can show an increasing affinity for inorganic carbon. An additional aspect, that needs to be considered, is the differential removal of the lighter ^{12}C by the photosynthesising cells. As DIC is removed from solution the $\delta^{13}C_{DIC}$ becomes less negative, and as the $\delta^{13}C_{POM}$ is an integration of the recent history of $\delta^{13}C_{DIC}$, transport and carboxylation processes, it is sometimes difficult to interpret the $\delta^{13}C_{POM}$ from such studies. Three approaches have been adopted when using batch culture. First, the closed system in which the change in $\delta^{13}C_{POM}$, $\delta^{13}C_{DIC}$ and the increase in biomass carbon have been calculated using a Rayleigh distillation, see Thompson and Calvert (1995). Second, Hinga *et al.* (1994) measured $\delta^{13}C_{DIC}$ of the initial growth medium and used large enough volumes so that the change in DIC concentrations and the level of pH drift was greatly reduced. Third, it is possible to use open systems of batch culture as long as the investigated species can tolerate the degree of aeration necessary to maintain air-equilibrium concentrations of DIC (Johnston and Raven, 1992).

Figure 15.5 shows the response of *Thalassiosira angulata* when cultured in a range of CO_2 concentrations in batch culture. It should be noted that at higher pCO_2 the growth rates were lower than those observed closer to atmospheric pCO_2 levels. While batch culture can have the theoretical advantage over steady-state continuous culture, as it more closely resembles the natural environment in terms of the effect of phytoplankton blooms, two points need to be made. First, because the volumes of most batch culture experiments are so small, normally less than 1 dm^{-3}, cell densities tend to be 3–4 orders of magnitude greater than found in the natural environment. Second, to date most batch cultures have very high levels of inorganic nutrients. Typical concentrations of nitrate and phosphate added to seawater for nutrient enrichment are 880 and 30 mmol m^{-3}, respectively, where as in the ocean concentrations above 35 and 3–5 mmol m^{-3} are rarely found. It is unknown what effect continued exposure of laboratory cultured algae to such high levels of nitrate and phosphate has had on their photosynthetic and general physiology. Given the problems of chemical and isotopic equilibria associated with batch culture it would seem the steady-state system of a continuous culture would be a desirable alternative and in many ways this is the case. Once set up a continual supply of cells in the same physiological state can be obtained. It is possible to measure a number of critical parameters to check the status of the inorganic carbon system – these include: pH, alkalinity, nutrient status, DIC concentration and $\delta^{13}C_{DIC}$. Again the type of phytoplankton species that can be grown successfully in continuous culture is restricted to those that can tolerate a degree of aeration and stirring. An additional problem associated with continuous culture is that some species, notably diatoms, attach themselves to the surfaces of the culture vessel and in the pores of the aerating stone resulting in patches of high biomass which will not rinse out.

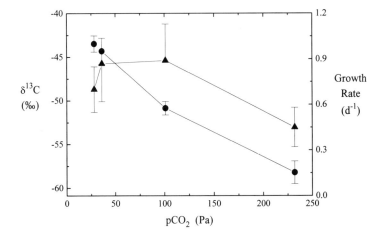

Figure 15.5. *Relationship between* CO_2 *and* $\delta^{13}C_{ORG}$ *(▲) and growth pH (●) of the diatom* Thalassiosira angulata *grown in f/2 media in batch culture at 20 °C and 95 μmol m^{-2} s^{-1} continuous light, aeration rate 1 dm^3 min^{-1}.*

15.3.3. *The effect of CO_2 concentration on carbon isotope fractionation*

The successful use of continuous culture in the study of carbon isotope fractionation caused by changes in growth rate by nitrate limitation over a range of $[CO_{2(aq)}]$ has recently been reported for the marine diatom *Phaeodactylum tricornutum* by Laws *et al.* (1995). While being rarely found in open oceans, it has been widely used as an experimental species, having many of the characteristics of a highly efficient HCO_3^- user, and can be thought of as an extreme case (see Johnston and Raven, 1996). The Laws *et al.* (1995) study showed that, by increasing the growth rate from 0.5 to 1.4 in a nitrate limited chemostat, there was an increase in the pH (7.83 to 8.11), a less negative $\delta^{13}C_{DIC}$ and a lower $[CO_{2(aq)}]$ (22 to 10 μmol kg^{-1}). What this study reveals is that, if the nutrient availability (NO_3^-) in a chemostat is increased, the inorganic carbon system cannot remain at equilibrium, resulting in a drawdown of CO_2. The effect is to change the two variables that can influence the level of ^{13}C discrimination, $[CO_{2(aq)}]$ and the rate of growth determined by the limiting nutrient, in this case nitrate. It should be noted that the observed decline of growth rate with increased $[CO_{2(aq)}]$ is a function of the decreased flow rate increasing the nitrate limitation, not a reflection of the $[CO_{2(aq)}]$. Using the same species but with well air-equilibrated batch culture rather than chemostats, Johnston (unpublished data) found that increasing $CO_{2(aq)}$ had little effect on the growth rate or the rate of photosynthesis of *P. tricornutum* but that the level of fractionation increased with increased $CO_{2(aq)}$ (Figures 15.6A and B) Both of these studies used isotopically light CO_2 gas cylinders as the source of atmospheric CO_2, Laws *et al.* (1995) 2.06 kPa CO_2 with $\delta^{13}C_{CO2}$ -45. 1‰, Johnston (unpublished) 5.0 kPa CO_2 with $\delta^{13}C_{CO2}$ -34.5‰. It is possible that in both of these studies the DIC may not have been in isotopic equilibrium with the headspace CO_2. Laws *et al.* (1995) report $\delta^{13}C_{DIC}$ as being between -15.9‰ and -8.8‰ and Johnston (unpublished data) recorded $\delta^{13}C_{CO2(aq)}$ of -20‰. Given the isotopic fractionation between $CO_{2(aq)}$ and HCO_3^- and the pH of the culture media, $\delta^{13}C_{DIC}$ would be expected to be 8‰ to 10‰ heavier than the atmospheric CO_2, as the bulk of inorganic carbon is in the form of HCO_3^-. It remains to be determined what effect such an isotopic disequilibrium would have on the calculation of the overall level of discrimination as it is probable that the $CO_{2(aq)}$ pool is being supplied from both the isotopically depleted atmospheric $CO_{2(g)}$ and enriched HCO_3^- pool.

Hinga *et al.* (1994) measured the level of discrimination in the diatom *Skeletonema costatum* and the non-calcifying form of the coccolithophorid *Emiliania huxleyi* using closed system batch culture. They ensured that the increase in biomass was small enough not to alter the $\delta^{13}C_{DIC}$. Their approach differs from that of Laws *et al.* (1995) in that the DIC concentration, and hence CO_2 concentration, was altered by adding $NaHCO_3$ of known isotopic composition to a maximum concentration of 10 mol m^{-3} at a constant pH. The ϵ_p (Δ) increased in response to increasing $CO_{2(aq)}$ but the range of the response was generally small (see Fry, 1996). What is noteworthy in this study is that the non-calcifying *E. huxleyi*, which is thought to be dependent on CO_2 diffusion for its inorganic acquisition, had lower levels of ^{13}C discrimination than *S. costatum*.

15.3.4 *The role of other environmental parameters on carbon isotope fractionation*

There is much interest in the relationship between the level of ^{13}C discrimination and the growth rate or photosynthetic rate of carbon assimilation (Goericke *et al.*, 1994;

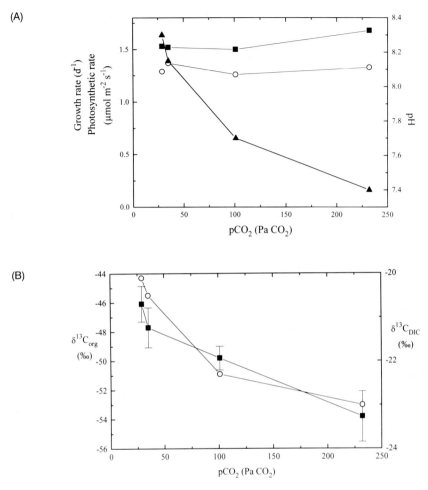

Figure 15.6. *Relationship between pCO$_2$ (A) photosynthetic rate (■), growth rate (○) and growth pH (▲); (B) δ^{13}C$_{org}$ (■) and δ^{13}C$_{DIC}$ (○) of the diatom* Phaeodactylum tricornutum *grown in f/2 media in batch culture at 20 °C and 95 μmol m^{-2} s^{-1} continuous light, aeration rate 1 dm^3 min^{-1}.*

Laws *et al.*, 1995; Rau *et al.*, 1996; Raven *et al.*, 1995). Such a relationship is evident from the analysis of the Farquhar model in the which the level of discrimination is controlled by the difference in the CO$_2$ concentration of the bulk phase (C$_b$) and the internal CO$_2$ concentration (C$_c$) (see Chapter 19).

Photon flux density. One of the main factors that can limit photosynthesis, and hence growth, is the photon flux density (PFD). Interestingly some studies have shown that the growth PFD appears to have little effect on the level of ^{13}C discrimination. Even though the growth rates followed the predicted response for increased growth PFD the level of discrimination only increased at low levels of PFD (5 μmol m^{-2} s^{-1}) in *Phaeodactylum tricornutum* (Johnston, 1996). Korb *et al.* (1996) found a similar response with *Ditylum brightwellii* and *Chaetoceros calcitrans*. These studies used batch and continuous culture respectively, and were air-equilibrated. However,

Thompson and Calvert (1995) reported that the level of discrimination by *Emiliania huxleyi* was primarily controlled by PFD and by the length of the light/dark cycle.

Temperature. The relationship between $\delta^{13}C_{POM}$ and $[CO_{2(aq)}]$ is greatly influenced by temperature as the solubility of CO_2 is primarily controlled by temperature. There are sparse laboratory data that separate the effect of temperature on solubility and on CO_2 fixation. Phytoplankton need to be cultured over a range of temperature at a constant $[CO_{2(aq)}]$ by altering the atmospheric $[CO_2]$. Johnston (1996) attempted this using *Phaeodactylum tricornutum* and showed that with increasing temperature the $\delta^{13}C_{POM}$ became less negative. It is interesting to note that the growth rate showed maxima at an intermediate temperature (16 °C). An alternative method is to saturate the inorganic carbon pool so that changes in temperature have little effect on the CO_2 and this can be done by culturing the algae in 5 kPa CO_2. Degens *et al.* (1968) reported that temperature had little effect on the $\delta^{13}C_{POM}$ of *Cyclotella nana* when compared to the $\delta^{13}C_{CO_2}$. The overall level of discrimination (18.6‰) is lower than expected for a CO_2-saturated system and this has been widely observed. Again caution is required when a high $[CO_2]$ is used because of this effect in their study, the pH was as low as 5.8 (Degens *et al.*, 1968).

Nutrients. Until recently the majority of photosynthetic physiological studies on marine phytoplankton have used nutrient-enriched culture media (see Chapter 16). While this has produced a large data set on nutrient replete cells we are still uncertain as to the true impact nutrient limitation has on the photosynthetic physiology of these species. Nutrient limitation will have a profound effect on growth rates and it is also likely to determine which mechanism of inorganic carbon assimilation the cell is to adopt. This in turn should have consequences for the level of carbon discrimination. Unfortunately there are very little data that considers the role of nutrient limitation directly. It is possible that the unusual result of Laws *et al.* (1995), whereby the growth rate decreased with increasing $CO_{2(aq)}$, could be related to the lower than normal concentration of NO_3^- (100 mmol m^{-3}) used in their study. In contrast the coccolithophorid *Emiliania huxleyi* needs to be kept in a low nutrient medium (less than 2 mmol m^{-3}) if it is to be kept in a high calcifying state. Yet when grown under the same conditions of light, pH and CO_2 the calcifying strain exhibits the same $\delta^{13}C_{POM}$ as the low or non-calcifying strain (Johnston, 1996).

pH. While not likely to be that important in the open ocean as a parameter that will influence the level of carbon discrimination, pH must be considered when discussing laboratory experiments. As mentioned previously, changes in $[CO_2]$ will alter the pH more than the DIC concentration. Most studies investigating the effect of $CO_{2(aq)}$ on carbon discrimination will need to consider the possible role of variations in pH. It is quite difficult to separate the cause and effect of changing $CO_{2(aq)}$ and pH. Two studies, adopting different approaches, have shown that pH can influence carbon discrimination. In the first, Johnston (1996) used batch culture where the pH was controlled by the use of buffers. When grown over a range of pH, in air-equilibrium concentrations of CO_2, *Phaeodactylum tricornutum* exhibited little variation in growth rates or in the level of carbon isotope discrimination. Under similar conditions, the calcifying form of *Emiliania huxleyi* was unable to grow below pH 7.8, above this pH the level of carbon isotope discrimination increased with increasing pH. Caution is required when using chemical buffers such as *Tris* as they can inhibit growth if the concentration used it too high. Hinga *et al.* (1994) adopted a different approach in their study of *Skeletonema costatum* by controlling the initial pH of the culture medium and altering the alkalinity and DIC concentration. They reported that pH appears to influence carbon discrimination above pH 7.885, the level of discrimination increasing with increased pH.

It is clear that the relationship between CO_2 concentration and growth rate/carbon demand is central to the level of ^{13}C discrimination. Other factors that may influence the level of ^{13}C discrimination may do so directly by altering the growth rate, for example nutrient resource limitation (light and nitrate). It is possible that other factors may act indirectly. Nutrient limitation may alter the mechanism of inorganic carbon acquisition. For example at low light, an algal cell may be less dependent on inorganic carbon accumulation, which would be reflected in higher levels of ^{13}C discrimination.

15.4 Recommendations for Future Research

This review has demonstrated that important insights into the environmental conditions under which carbon fixation occurs can potentially be provided through the analysis of $\delta^{13}C_{POM}$ and $\delta^{13}C_{SOM}$. However, in sediments, and to some extent in the water column, the interpretation of the isotope data can be complicated by the co-existence of many types of phytoplankton, zooplankton, bacteria and detritus. Recent advances in instrumentation developed the use of compound specific isotopic analyses and has shown that application of this approach could potentially yield information regarding class-specific phytoplankton processes. Compound specific analyses of chlorophylls (Bidigare *et al.*, 1991; Kennicutt *et al.*, 1992) or carotenoids, for example, may provide a more consistent measure of isotopic discrimination during carbon fixation. Analyses of sedimentary porphyrins (Hayes *et al.*, 1989, 1990) will provide a link to the historical records. More ubiquitous in the sedimentary record are lipids and determination of their isotopic compositions may provide a more practical link between the past and the present (Hayes *et al.*, 1990).

In laboratory studies there is a need for more careful studies on the relationship between $CO_{2(aq)}$ and the level of discrimination against both ^{13}C and ^{18}O for a wider range of species. Detailed measurements of the isotopic and chemical equilibria of the inorganic carbon system are also necessary if inter-laboratory comparisons of cultured algae are to be realised. At some stage the effect of resource limitation (PFD and nutrients) on the central relationship needs to be investigated. One of the major weaknesses of laboratory studies, is that most algae are maintained at remarkably constant conditions of PFD, temperature and nutrients. It is important that the interactions between these environmental conditions and the effect this has on carbon discrimination is addressed.

References

Barnola, J., Raynaud, D., Korotkevich, D. and Lorius, C. (1987) Vostok ice core provides 160,000 year record of atmospheric CO_2. *Nature* **329**, 408–414.

Bidigare, R.R., Kennicutt, M.C., Keeney-Kennicutt, W.L. and Macko, S.A. (1991) Isolation and purification of chlorophylls a and b for the determinations of stable carbon and nitrogen isotope compositions. *Anal. Chem.* **63**, 130–133.

Broecker, W. and Peng, T.-H. (1982) Tracers in the sea. Lamont–Doherty Geological Observatory, Palisades, NY, 690 pp.

Burns, B.D. and Beardall, J. (1987) Utilization of inorganic carbon by marine microalgae. *J. Exp. Mar. Biol. Ecol.* **175**, 75–86.

Chisholm, S.W. (1992) Phytoplankton size. *In Primary productivity and Biogeochemical cycles in the seas.* (eds. P.G. Falkowski, A.D. Woodheads). Plenum Press N.Y. pp 213–237.

Cifuentes, L.A., Sharp, J. and Fogel, M. (1988) Stable carbon and nitrogen biogeochemistry in the Delaware estuary. *Limnol. Oceanogr.* **33**, 1102–1115.

Colman, B. and Rotatore, C. (1995) Photosynthetic inorganic carbon uptake and accumulation in two marine diatoms. *Plant Cell Environ.* **18**, 919–924.

Descolas-Gros, C. and Fontugne, M.R.(1985) Carbon fixation in marine phytoplankton: carboxylase activities and stable isotopes ratios; physiological and palaeoclimatological aspects. *Mar. Biol.* **87**, 1–6

Degens, E.T., Guillard, R.R.L., Sackett, W.M. and Hellebust, J.A. (1968) Metabolic fractionation of carbon isotopes in marine plankton- I. Temperature and respiration experiments. *Deep Sea Res.* **68**, 1–9.

Deuser, W., Degens, E. and Guillard, R. (1968) Carbon isotope relationships between plankton and seawater. *Geochim. Cosmochim. Acta* **32**, 657–660.

Deuser, W.G. (1970) Isotopic evidence for diminishing supply of available carbon during diatom bloom in the Black Sea. *Nature* **225**, 1069–1071.

Falkowski, P.G. (1991). Species variability in the fractionation of ^{13}C and ^{12}C by marine phytoplankton. *J. Plankton Res.* **13**, 21–28.

Farquhar, G.D., O'Leary, M.H. and Berry, J.A. (1982) On the relationship between carbon isotope discrimination and the inter-cellular carbon-dioxide concentration in leaves. *Aust. J. Plant Physiol.* **9**, 121–137.

Fischer, G. (1991) Stable carbon isotope ratios of plankton carbon and sinking organic-matter from the Atlantic sector of the southern-ocean. *Mar. Chem.* **35**, 581–596.

Flynn, K. (1996) Toxic marine phytoplankton: some aspects of nutrient ecophysiology. *The Phycologist* **43**, 20.

Fogel, M.L., Cifuentes, A., Velinsky, D.J. and Sharp, J. (1992) Relationship of carbon availability in estuarine phytoplankton to isotopic composition. *Mar. Ecol. Prog. Ser.* **82**, 291–300.

Fontugne, M. and Duplessy, J.-C. (1978) Carbon isotope ratio of marine plankton related to surface water masses. *Earth Planet Sci. Lett.* **41**, 365–371.

Fontugne, M. and Duplessy, J.-C. (1981) Organic carbon isotopic fractionation by marine plankton in the temperature range -1 to 31°C. *Oceanol Acta* **4**, 85–90.

Francois, R., Altabet, M.A., Goericke, R., McCorkle, D.C., Brunet, C. and Poisson, A. (1993) Changes in the $\delta^{13}C$ of surface-water particulate organic matter across the subtropical convergence in the SW Indian ocean. *Global Biogeochemical Cycles* **7**, 627–644.

Freeman, K. and Hayes, J. (1992) Fractionation of carbon isotopes by phytoplankton and estimates of ancient CO_2 levels. *Global Biogeochemical Cycles* **6**, 185–198.

Fry, B. (1996) $^{13}C/^{12}C$ Fractionation in marine diatoms. *Mar. Ecol. Prog. Ser.* **134**, 283–294.

Fry, B. and Wainright, S.C. (1991) Diatom sources of ^{13}C-rich carbon in marine food webs. *Mar. Ecol. Prog. Ser.* **76**, 149–157.

Gearing, J. N., Gearing, P. J., Rudnick, D. T., Requejo, A. G. and Hutchins, M. J. (1984) Isotopic variability of organic carbon in a phytoplankton-based, temperate estuary. *Geochim, Cosmochim. Acta* **48**, 1089–1098.

Goericke, R. and Fry, B. (1994) Variations in marine plankton $\delta^{13}C$ with latitude, temperature and dissolved CO_2 in the world ocean. *Global Biogeochemical Cycles* **8**, 85–90.

Goericke, R., Montoya, J. and Fry, B. (1994) Physiology of isotope fractionation in algae and cyanobacteria. In *Stable isotopes in ecology.* (eds. K. Lajtha and B. Michener), Blackwell Scientific, Boston, pp. 187–221.

Guy, R., Vanlerberghe, G. and Turpin, D. (1989) Significance of phosphoenolpyruvate carboxylase during ammonium assimilation. *Plant Physiol.* **89**, 1150–1157.

Hayes, J. M., Popp, B., Takigiku, R. and Johnson, M. (1989) An isotopic study of biogeochemical relationships between carbonates and organic carbon in the Greenhorn Formation. *Geochim. Cosmochim. Acta* **53**, 2961–2972.

Hayes, J.M., Freeman, K.H., Popp, B.H. and Hoham, C.H. (1990) Compound specific isotope analyses, a novel tool for reconstruction of ancient biogeochemical processes. *Org. Geochem.* **16**, 1115–1128.

Hinga, K.R., Arthur, M.A., Pilson, M.E.Q. and Whitaker, D. (1994) Carbon-isotope fractionation by marine-phytoplankton in culture – the effects of CO_2 concentration, pH, temperature, and species. *Global Biogeochemical Cycles* **8**, 91–102.

Johnston, A.M. and Raven, J.A. (1992) Effect of aeration rates on growth rates and natural abundance $^{13}C/^{12}C$ ratio of *Phaeodactylum tricornutum. Mar. Ecol. Prog. Ser.* **87**, 295–300.

Johnston, A.M. and Raven, J.A. (1996) Inorganic carbon accumulation by the marine diatom *Phaeodactylum tricornutum. Eur. J. Phycology* **31**, 285–290.

Johnston, A. M. (1996) The effect of environmental variables on ^{13}C discrimination by 2 marine phytoplankton. *Mar. Ecol. Prog. Ser.* **132**, 257–263.

Kennedy, H. and Robertson, J.E. (1995) Variations in the isotopic composition of particulate organic carbon in surface water along an 88°W transect from 67°S to 54°S. *Deep-Sea Research Part II- Topical studies in Oceanography* **42**, 1109–1122.

Kennicutt, M.C., Bidigare, R.R., Macko, S.A. and Keeney-Kennicutt, W.L. (1992) The stable isotopic composition of photosynthetic pigments and related biochemicals. *Chemical Geology* **101**, 235–245.

Kopczynska, E.E., Goeyens, L., Semench, M. and Dehairs, F. (1995) Phytoplankton composition and cell carbon distribution in Prydz Bay, Antarctica: relation to organic particulate matter and its $\delta^{13}C$ values. *J. Plankton Res.* **17**, 685–707.

Korb, R.E., Raven, J.A., Johnston, A.M. and Leftley, J.W. (1996) Effects of cell-size and specific growth rate on stable carbon- isotope discrimination by 2 species of marine diatom. *Marine Ecol. Prog. Ser.* **143**, 283–288.

Laws, E. A., Popp, B. N., Bidigare, R. R., Kennicutt, M. C. and Macko, S. A. (1995) Dependence of phytoplankton carbon isotopic composition on growth rate and $[CO_2]$aq: theoretical considerations and experimental results. *Geochim. Cosmochim. Acta* **59**, 1131–1138.

Maberly, S. C., Raven, J. A. and Johnston, A. M. (1992). Discrimination between ^{12}C and ^{13}C by marine plants. *Oecologia* **91**, 481–492.

Mariotti, A., Germon, J., Hubert, P., Kaiser, P., Letolle, R., Tardieux, A. and Tardieux, P. (1981). Experimental determination of nitrogen kinetic isotopic fractionation: some principles: illustrations for denitrification and nitrification processes. *Soil Science* **62**, 413–443.

Müller, P.J., Schneider, R. and Ruhland, G. (1994) Late quaternary pCO_2 variations in the Angola current: evidence from organic carbon $\delta^{13}C$ and alkenone temperatures. In *Carbon cycling in the Glacial Ocean: Constraints on the Ocean's Role in Global Change.* (eds R. Zahn, T. F. Pedersen, M. A. Kaminski, and L. Labeyriem, pp 343–366.

Nakatsuka, T., Handa, N., Wada, E. and Wong, C. (1992) The dynamic changes of stable isotopic ratios of carbon and nitrogen in suspended and sedimented particulate organic matter during a phytoplankton bloom. *J. Mar. Res.* **50**, 267–298.

Nimer, N. A., Dixon, G. K. and Merrett, M. J. (1992) Utilization of inorganic carbon by the coccolithophorid *Emiliania huxleyi* (Lohmann) Kamptner. *New Phytol.* **120**, 152–158.

O'Leary, M. (1981) Carbon isotope fractionation in plants. *Phytochem.* **20**, 553–567.

Pedersen, T., Nielsen, B. and Pickering, M. (1991) Timing of late Quaternary productivity pulses in the Panama Basin and implications for atmospheric CO_2. *Palaeoceanography* **6**, 657–677.

Rau, G., Riebesell, U. and Wolf-Gladrow, D. (1996) A model of photosynthetic ^{13}C fractionation by marine phytoplankton based on diffusive molecular CO_2 uptake. *Mar. Ecol. Prog. Ser.* **133**, 275–285.

Rau, G.H., Takahashi, T., Des Marais, D.J. and Sullivan, C.W. (1991) Particulate organic matter ^{13}C variations in the Drake Passage. *J. Geophys. Res.* **96**, 15131–15135.

Rau, G.H., Takahashi, T., Des Maris, D.J., Repeta, D.J. and Martin, J.H. (1992) The relationship between $\delta^{13}C$ of organic matter and $[CO_2(aq)]$ in ocean surface water: Data from a JGOFS site in the Northeast Atlantic Ocean and a model. *Geochim. Cosmochim. Acta* **56**, 1413–1419.

Rau, G.H., Takahashi, T. and Marais, D.J.D. (1989) Latitudinal variations in plankton $\delta^{13}C$: implications for CO_2 and productivity in past oceans. *Nature* **341**, 516–518.

Raven J.A., Walker, D.I., Johnston, A.M., Handley, L.L. and Kübler, J.E. (1995) Implication of ^{13}C natural abundance measurements for photosynthetic performance by marine macro-phytes in their natural environment. *Mar. Ecol. Prog. Ser.* **123**, 193–205.

Riebesell, U., Wolf-Gladrow, D.A. and Smetacek, V. (1993) Carbon dioxide limitation of marine phytoplankton growth rates. *Nature* **361**, 249–251.

Robertson, J. E, Watson, A. J., Langdon. C, Ling. R.D. and Wood, J. W. (1993) Diurnal variation in surface pCO_2 and O_2 at 60°N, 20°W in the North Atlantic. *Deep Sea Research, Part II* **40**, 409–422.

Sackett, W. M., Eckelmann, W. R., Bender, M.L. and Be, A.W.H. (1965) Temperature depen-dence of carbon isotope composition in marine phytoplankton and sediments. *Science* **148**, 235–237.

Tans, P., Fung, Y. and Takahashi, T. (1990) Observational constraints on the global atmospheric CO_2 budget. Science **247**, 1431–1438.

Thompson, P. and Calvert, S. (1994) Carbon isotope fractionation by a marine diatom: the influence of irradiance, daylength, pH and nitrogen source. *Limnol. Oceanogr.* **39**, 1835–1844.

Thompson, P. and Calvert, S. (1995) Carbon isotope fractionation by *Emiliania huxleyi*. *Limnol. Oceanogr.* **40**, 673–679.

Wainright, S. and Fry, B. (1994) Seasonal variation of the stable isotopic compositions of coastal marine plankton from Woods Hole, Massachusetts and Georges Bank. *Estuaries* **17**, 552–560.

^{15}N and the assimilation of nitrogen by marine phytoplankton: the past, present and future?

N.J.P. Owens and L.J. Watts

16.1 Introduction

The World Ocean, covering approximately 70% of the surface of the planet, is the largest individual ecosystem on earth. The surface illuminated waters of this system, extending to a maximum of approximately 100 m, but more typically to less than 50 m, contributes about half of the global photoautotrophic-derived organic production (Whittaker and Likens, 1975). However, whilst areal production rates of the most productive marine waters compare or exceed those of the most productive terrestrial systems, the productivity of the marine system is typically lower than terrestrial systems and is highly variable, both temporally and spatially. This variability ranges over several orders of magnitude, across metre- to ocean-basin-spatial scales and diel- to decadal-temporal scales, (significantly longer if paleo-oceanographic scales are considered). The availability of light and nutrients are the critical factors underlying this variability. Excluding light availability (which clearly is critical when considering photoautotrophic processes) a major question arises as to which nutrient is the most important in controlling (limiting) marine photoautotrophic production? This debate has occupied marine scientists for as long as it has been recognised that nutrients are important. Notwithstanding the important current debates about the rôle of micronutrients, for example iron (Banse, 1991; de Baar *et al.*, 1990; Martin and Fitzwater, 1988; Martin *et al.*, 1990), attention has usually been focussed on the macronutrients phosphorus (P) and nitrogen (N). It is outside the scope of this review to discuss the arguments and evidence concerning the relative importance of P and N limitation. However, the balance of the evidence suggests that nitrogen is invariably the most important macro-nutrient controlling marine autotrophic production. ^{15}N studies have made an enormous contribution to this understanding.

Stable Isotopes, edited by H. Griffiths.
© 1998 BIOS Scientific Publishers Ltd, Oxford.

In common with terrestrial systems, nitrogen occurs in marine systems in both organic and inorganic, particulate and dissolved forms. Dissolved organic nitrogen usually dominates, typically representing over 95% of the total nitrogen pool (excluding di-nitrogen; N_2) in surface ocean waters, but particulate organic nitrogen, in the form of living, or recently living, biomass may also represent a dominant fraction on occasions. However, the various pools of N vary enormously in space and time (see Sharp (1983), for a comprehensive review of the distribution of N in the marine environment). Organic nitrogen has been relatively little studied in marine systems. In part this is due to the relative difficulty in identifying the huge variety of organic compounds in seawater. However, in addition, organic nitrogen compounds are generally considered to be unimportant in sustaining autotrophic (phytoplankton) production. Nevertheless, the rôle of organic nitrogen, particularly low molecular weight (LMW) compounds is beginning to receive more attention with the advent of new analytical techniques and the increasing evidence that some LMW compounds may be utilised by phytoplankton (see, for example, Antia et al., 1991; Bronk and Glibert, 1993). Nitrate and ammonium are without doubt the most significant inorganic nitrogen compounds supporting phytoplankton growth. Of these, nitrate is generally the most abundant, being found in surface seawater at concentrations between ~ 10 nmol l^{-1} to ~ 25–35 μmol l^{-1} . Ammonium concentrations, in contrast, generally do not exceed $\sim 2\mu$mol l^{-1} and more typically is found at concentrations between 10 nmol l^{-1} and 1μmol l^{-1} . For both compounds (and in particular ammonium) the accuracy of the concentration measurements at the lower bound of the range is usually poor because of analytical difficulties. Indeed, it is only in recent years, with the advent of new analytical techniques, that it has been possible to determine these low concentrations at all (for example, Garside, 1982; Jones, 1991). A third, combined-form of inorganic nitrogen utilised by phytoplankton, nitrite, is not thought to be important. However, nitrite has been implicated in the nitrogen economy of phytoplankton being 'excreted' under certain environmental conditions (Collos, 1982a; Flynn and Flynn, 1997) and some field studies have measured its uptake using ^{15}N techniques (McCarthy et al., 1984).

Di-nitrogen although inorganic, warrants particular attention since utilisation of N_2 is confined to nitrogen-fixing organisms, whereas the ability to utilise other inorganic forms appears to be ubiquitous in the phytoplankton. Although N_2 is the most abundant form of nitrogen in the oceans, typically being found in concentrations $\sim 800 – 1000$ μmol l^{-1}, surprisingly few (apparently) pelagic marine organisms utilise N_2 as a significant source of nitrogen. Several studies have quantified N_2-fixation but have concluded that it is not an important input of nitrogen globally. However, locally, N_2-fixation may be quantitatively important in supporting autotrophic production (Carpenter, 1983).

One of the most striking features of the marine environment is the spatial and temporal variations in the concentrations of the inorganic (and organic) forms of nitrogen. For nitrate, for example, concentrations may vary between > 20 μmol l^{-1} during late winter to < 100 nmol l^{-1} during summer in mid to high latitude ocean regions. Similarly, concentrations of inorganic nitrogen may range from ~ 100 nmol l^{-1} year round in the central, oligotrophic gyres of the Pacific Ocean, for example, to > 25 μmol l^{-1} in the Southern Ocean. The underlying mechanisms controlling these concentrations are oceanographic, both physical and biological.

Although there has been a consistent interest in phytoplankton production and its control over the past few decades, there has been renewed interest recently because of the possible importance of marine phytoplankton production in the global carbon cycle, and in particular its potential as a sink for anthropogenically derived carbon dioxide (with obvious implications for the 'greenhouse effect': see Chapters 15 and 24). Because of the surface area of the marine system, and the often high (and potentially greater?) rates of autotrophic production, it is not surprising that marine systems would come under scrutiny. Given the importance of nitrogen in supporting marine autotrophic production it is also not surprising that a large number of studies have focused on attempting to understand the details of the rôle of nitrogen in this cycle. The use of ^{15}N has been crucial in illuminating the role of nitrogen in this system.

In this mini-review the historical context of the use of ^{15}N in marine studies is briefly reviewed in relation to phytoplankton ecology and the origins of the concepts and techniques developed from these early studies. This is followed by a summary of some recent (past 2 to 3 decades) studies that have utilised ^{15}N. The review concludes with a view towards the future. Throughout, the emphasis of the review has been to concentrate on the ecological aspects of the subject with relatively little reference made to the substantial body of parallel work carried out on physiological aspects. Overall, the purpose of this review is to indicate how valuable ^{15}N has been, and continues to be, in helping to illuminate the role of nitrogen in phytoplankton nitrogen assimilation, ecology and biological oceanography.

16.2 The past – a short historical review

16.2.1 *Biological concepts*

The origins of phytoplankton ecology and its study using ^{15}N can be justifiably traced back to the pioneering work of Liebig in the mid-nineteenth century. Liebig, working in agricultural chemistry, formulated a number of 'laws', the most significant of which (within this context) is known as the 'Law of the Minimum' (see de Baar (1994) for an authoritative overview of Liebig's – and other – influential early studies). The central tenet of Liebig's Law of the Minimum is that a single nutrient ultimately limits plant growth. This principle has been hugely important in influencing agricultural, plant and marine sciences, and remains so today. According to de Baar, the principle of the Law of the Minimum was first introduced to the marine sciences by Brandt (1899) who suggested that nitrogen was likely to be the limiting nutrient for marine algae. Brandt followed this key work with a series of authoritative publications concerning the marine nitrogen cycle (Brandt, 1902; Brandt, 1920) – (see also de Baar (1994) for a more complete account). Most of Brandt's early ideas are still valid today. However, the importance of Brandt's work is that it provided the impetus for subsequent testing of his hypotheses by a number of hugely influential workers, most notably Brand, Rakestraw, Cooper and Harvey (e.g. Brand and Rakestraw, 1941a,b; 1941b; Brand *et. al.*, 1937, 1939, 1942; Cooper, 1933, 1935; Harvey, 1926, 1933, 1940; Harvey *et al.*, 1935). These workers, over a period of approximately 20 years, introduced ideas of marine nutrient assimilation, cycling and nutrient regeneration and clearly demonstrated their links with hydrography. This work provided the 'springboard' for the introduction of ^{15}N techniques.

16.2.2 *The introduction of ^{15}N*

Although the majority of the stable isotopes had been discovered and investigated in the early part of the 20th century (see Beynon (1960) and a brief history in Owens (1987)), it was not until 1955 (Hoering, 1955) that the first reported use of ^{15}N in the marine sciences was made, and then only as an 'in passing' note about natural variations of ^{15}N in 'seaweed' and 'clams'. The most important early marine natural abundance work appeared six years later with the work of Benson and Parker (1961) and Richards and Benson (1961) who utilised ^{15}N variations in N$_2$ (and N$_2$/Ar ratios) to demonstrate the importance of denitrification in marine systems. The first systematic study of natural variations of ^{15}N in marine samples was that of Miyake and Wada (1967). Thus, it is clear that although '^{15}N techniques' in general were introduced to the marine sciences by the late 1950s/early 1960s, it is difficult to establish exactly when ^{15}N was first used as a tracer or to identify the true precedents. However, the introduction of tracer techniques for marine phytoplankton growth studies by Steemann-Nielsen (1952), although using the radioactive isotope ^{14}C, probably provided the conceptual impetus for the use of ^{15}N as a deliberate tracer. Nevertheless, despite this fundamental step, and the clear need for direct rate measurements identified by the classical hydrographic studies of Cooper, Harvey and others, the first reported use of ^{15}N in phytoplankton work was not until 1961 when Dugdale *et al.* (1961) published their work on N$_2$-fixation in the Sargasso Sea; although this had been preceded by a similar study in freshwater (Dugdale *et al.*, 1959). Over the first half of the 1960s there was a 'stream' of papers from an active group all utilising the 'new' technique. The work not only concerned the measurement of N$_2$-fixation in lakes (Dugdale and Dugdale, 1962; Goering and Neess, 1964; Neess *et al.*, 1962) and marine waters (Dugdale *et al.*, 1964; Goering *et al.*, 1966), but also included studies of nitrogen assimilation (Goering *et al.*, 1964) and denitrification (Goering and Dugdale, 1966a,b. Interestingly, these early studies 'borrowed' the techniques and methods of soil/agricultural scientists, most notably those of Hauck *et al.* (1958), Rittenberg (1946) and Rittenberg *et al.* (1939), in parallel with the concepts similarly borrowed from agriculture a century earlier. Importantly, this work coincided with developments and concepts of tracer use (for example, Sheppard, 1962) and seawater analytical chemistry (Strickland and Parsons, 1960). Specific ^{15}N methods were also developed, for example the introduction of rapid Dumas combustion methods for sample preparation (Barsdate and Dugdale, 1965).

This early development phase culminated in probably the most influential (and certainly the most cited) of marine phytoplankton ^{15}N studies: Dugdale and Goering's conceptual paper on phytoplankton–nitrogen relationships in the upper ocean (Dugdale and Goering, 1967) – see below for a detailed discussion. Whilst this was undoubtedly a seminal paper (and included ^{15}N data from field experiments), a companion, theoretical paper published simultaneously by Dugdale (1967) was at least as important, since it combined the conceptual model of Dugdale and Goering (1967) with the notion of nutrient (nitrogen) limitation, *sensu* Liebig, and the possible utility of Michaelis–Menten kinetics in nitrogen uptake studies. Interestingly, this study is rarely cited.

Together, these two studies marked the end of the 'early phase' of ^{15}N–phytoplankton studies and introduced the concepts and hypotheses that have dominated subsequent studies. The work that immediately followed largely addressed the details of these models. For example, the validity of the Michaelis–Menten kinetic approach

(MacIsaac and Dudgale, 1969, 1972; multi-nutrient interactions (Conway, 1977; McCarthy, 1972; McCarthy *et al.*, 1977); nutrient-light interactions (MacIsaac and Dudgale, 1969, 1972); and the concept of nutrient preference (McCarthy *et al.*, 1977) – see also below. ^{15}N studies were also very quickly introduced into the investigation of all aspects of the marine nitrogen cycle, (see Harrison (1983) for a short review).

16.2.3 *Dugdale and Goering (1967) and 'new production'*

The conceptual model introduced by Dugdale and Goering (1967) is shown diagramatically in Figure 16.1. The euphotic zone is assumed to be a 'closed box' bounded at its lower level by the thermocline. The phytoplankton within this box may obtain nitrogen from two sources (indicated by I in Figure 16.1): 1. Inputs external to the box via lateral transport; upward vertical transport from beneath the thermocline; vertical downward transport through the upper surface that is from the atmosphere; 2. Inputs from the recycling of nitrogen (by zooplankton and bacteria), within the box. Nitrogen may be lost from the system (indicated by O in Figure 16.1) via the sinking of phytoplankton, zooplankton and zooplankton faecal pellets through the thermocline, and other exports for example, fishing etc. Dugdale and Goering proposed that the phytoplankton production 'supported' by the external inputs is 'new production', reflecting the 'new' inputs of nitrogen, and that 'supported' by recycled nitrogen is 'regenerated production'. They further proposed that to maintain steady-state the magnitude of the export of nitrogen (i.e. the sum of the 'O's) cannot exceed the sum of the inputs (the sum of the 'I's). Dugdale and Goering simplified the system further by assuming that the predominant form of 'new nitrogen' is nitrate and that of regenerated nitrogen is ammonium. It is important to note, however, that this assumption is only really applicable to open ocean situations where there are no significant inputs of reduced nitrogen. Thus, although the model of new and regenerated production is applicable for coastal waters, the assumption of the forms of nitrogen to support these is seriously violated. The 'validity' of the model was demonstrated by a comprehensive set of shipboard ^{15}N assimilation measurements, which introduced techniques that are largely still utilised in contemporary studies.

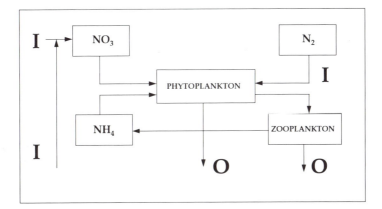

Figure 16.1. Diagrammatic representation of the nitrogen cycle in the upper ocean showing the concept of 'New' and 'Regenerated' nitrogen. (Based on Dugdale and Goering, 1967).

Whilst it is known that several of the assumptions of the model are not always valid, for example, 'new' and 'regenerated' nitrogen' are not always in the forms of nitrate and ammonium respectively, also steady-state is rarely achieved or sustained, this model forms the basis of contemporary biological oceanography. The fact that very few studies of phytoplankton ecology – nutrient relations have not cited Dugdale and Goering, is testament to the importance of this work. However, the 'chapter' on the foundation studies of ¹⁵N in biological oceanography would not be complete without reference to the important work of Eppley and Peterson (1979). Although, chronologically, this work appeared some considerable time following the 'groundbreaking' earlier studies, there is no doubt this was a 'milestone' paper. In essence Eppley and Peterson (1979) 'revisited' the original Dugdale and Goering model. Using (largely) data from ¹⁵N studies, Eppley and Peterson extended the model to suggest that 'new production', quantitatively approximated to the sinking flux of particulate organic matter, or the maximum amount of organic material available for transfer into higher trophic levels (i.e. export production), for example into fisheries. Eppley and Peterson also introduced the parameter 'f' – the proportion of the new production to total production. That is:

$$f = \frac{\text{New Production}}{\text{Total Production}} \qquad (16.1)$$

Theoretically f can vary between 0 and 1. Eppley and Peterson suggested, using data from ¹⁵N studies, that f largely varies between ~ 0.05 and 0.5, however, several more recent studies have shown f values up to at least 0.9 (see below). The importance of this work is that it introduced the notion of the fundamental linkage of nutrient supply and recycling, phytoplankton production and availability of 'production' for the support of all other ocean biological processes. Whilst this concept is sufficiently important in its own right, this paradigm has assumed significantly more importance in recent years in the context of the global carbon cycle, given the potential for marine export production to be quantitatively equal to the maximum production (i.e. carbon) that can be sequestered to the deep ocean. This is clearly of fundamental relevance to contemporary questions related to establishing the impact of anthropogenic activities on the global carbon cycle and subsequent questions concerning the 'greenhouse effect'. It is also important to note that although there are several methods available to estimate new production, ¹⁵N methods are by far the most widely used.

The publication of Eppley and Peterson's paper provides a natural break between the early and groundbreaking studies that utilised ¹⁵N techniques and all of those that have followed subsequently. It is interesting to note in reviewing this history, that the major advances were made through a combination of new concepts, for example those introduced by Liebig, Brand, Cooper, Harvey, Dugdale and Goering and new techniques, for example, those of Rittenberg, Steeman–Nielsen and Neess. The final section of this review suggests where the next developments in this subject may occur. However, it is important to consider the current state of understanding and this is covered in the next sections.

16.3 Natural abundance v. tracer studies

An interesting and valuable aspect of ¹⁵N is that it may be used to elucidate phytoplankton nitrogen dynamics through its use as both a deliberate tracer and by investigating variations in its natural abundance. Both uses have been applied to phytoplankton studies, although the number of studies that have used ^{15}N as a deliberate tracer far outnumber natural abundance studies.

However, addition of tracer, sample containment and other steps in the experimental procedure seriously perturb the natural conditions and may lead to potentially large inaccuracies in any rate measurements made (see Section 16.4). Conversely, samples for natural abundance studies may be collected with the minimum of handling procedures and thus this technique offers the possibility of examining *in vivo* or near *in vivo* conditions. However, whilst several natural abundance studies have proved valuable in illuminating various nitrogen cycling processes involving phytoplankton (for example, Miyake and Wada, 1967; Mariotti *et al.*, 1984; Altabet and McCarthy, 1985; Altabet *et al.*, 1986), the natural abundance technique essentially provides information of a 'state-variable' nature rather than the rates of processes. Although this can be valuable, for example, by providing qualitative information on the importance of oceanic di-nitrogen fixation (e.g. Wada *et al*, 1975; Wada and Haltori, 1976; Minagawa and Wada, 1986) no quantitative information can be obtained. Also, the basis of the interpretation of natural abundance field data is theoretical kinetic isotope fractionation (see Owens, 1987). There has been virtually no rigorous experimentation on the consequences of, for example, growth rate, nutritional status and temperature on the variations of ^{15}N in phytoplankton (compared with C: see Chapter 15). A further disadvantage of the technique is the difficulty in separating phytoplankton from the background mass of non-living suspended particulate material. Although the use of well-defined meshes and filters can provide some differentiation, this remains a practical difficulty and a serious limitation. This further diminishes the usefulness of the approach. Nevertheless, variations in natural abundance could offer considerable insight into the nitrogen dynamics/relations of marine phytoplankton and perhaps should be identified as a priority for research.

16.4 The present – methods and techniques

It is not the intention to provide a detailed account of all the ^{15}N methods used in estimating nitrogen assimilation by marine phytoplankton, but rather to provide a general account for completeness. Details of all the methods outlined below can be found by reference to the individual studies identified later in this section, and in general terms the use of ^{15}N in a biological oceanographic context can be conveniently divided into three sections that follow.

16.4.1 *Sampling and addition of the isotope*

Sampling from large oceanographic research ships is almost entirely carried out using conductivity, temperature, depth (CTD) sensor packages fitted with an array of sampling bottles (rosette). The sensors provide real-time information, via a conducting cable, about conditions in the ocean at the time of sampling. Modern instrumentation not only provides the 'basic' CTD suite but usually is deployed with a variety of other sensors (e.g. fluorometers, irradiance sensors, oxygen electrodes) to provide a comprehensive description of the physical and chemical environment. Although these packages are generally capable of deployment to full ocean depths (> 5000 m), deployments for phytoplankton studies are usually confined to the upper layers, typically the photic zone. The depth of the photic zone is, of course, variable depending upon local conditions, however, it is usually operationally defined as the depth of penetration of 1% (or 0.1%) of photosynthetically active radiation incident at the surface. The sample bottles (typical volumes 2–30 litres) attached to the rosette are capable of closure at any depth by controls from the ship, thus it is possible to collect the required sample volumes from any depth desired, concomitant with information about the physicochemical environment. Simpler sampling techniques (e.g. manually operated sample bottles), are of course possible, indeed are usual for near-shore or land-based studies but the method outlined above has become the 'norm' for most oceanographic studies.

The number and volume of the samples is dependent upon the intended measurements. In the majority of cases information about variability throughout the euphotic zone is required, thus samples are collected at intervals to provide the necessary coverage to the required depths (in many cases sufficient samples are collected to provide reliable information for subsequent depth integrations). In most cases between 2 and 10 discrete depths are sampled. The sample depths may be distributed equally or targeted to specific features (e.g. sub-surface chlorophyll maxima). In some studies it is more appropriate to collect a large volume of sample at a single discrete depth. In other cases the volume of sample(s) required is so large that it is necessary to 'pool' the contents of several sample bottles before further experimentation. This is often valuable if several biological or chemical variables require investigation simultaneously and where it is desirable to work with 'one sample'.

Pooling several samples reduces the natural variability inherent in CTD–rosette sampling but suffers from the drawback of introducing an additional handling step. After collecting the samples it is important to minimise the effects of 'on deck' surface temperature and light. In the case of experiments with photoautotrophs it is usual to collect samples during the hours of darkness (typically pre-dawn) to prevent light shock to the organisms. Whatever the number and volume of samples it is usual to begin the experiments as soon as possible after collection. It is also important to use non-contaminating techniques (e.g. contamination from metals and nutrients) during the entire sampling and subsequent handling procedures.

In the majority of cases it is necessary to divide the samples into a number of small volume vessels in which the experiments are subsequently carried out. Samples may be sub-divided without further manipulation or in some cases samples may be pre-filtered (for example, using a defined pore-sized mesh) to remove a particular component of the plankton, for example, larger zooplankton. The number, type and volume of bottles used will vary with the type of experiment, but typical volumes are 0.5 – 2.5 litres; consideration must be given to the optical and chemical properties of the bottles. Once the array of 'experimental' bottles has been prepared it is usual to add the ¹⁵N tracer(s). Typically, known quantities of an individual substrate labelled with ¹⁵N is added to each bottle. Although the amount and type of substrate added depends upon the investigation it is often the case that substrates are required to be added with the intention of not disturbing the natural conditions. Clearly, when dealing with a nitrogen substrate that is present in very low concentrations, and may be growth limiting, this is a difficult, if not impossible, task. A major disadvantage of the use of ¹⁵N isotopes (when compared with radio-isotopes) is their relative insensitivity, thus it is often necessary to add considerable quantities of substrate to enable subsequent detection. All ^{15}N studies suffer from the drawback of potential perturbation of the natural nutrient environment because of the addition of tracer. The magnitude of the perturbation depends upon specific conditions, however, the aim of many studies is to add nitrogen (in the form of ^{15}N tracer) at less than 10% of the ambient concentration of the substrate under investigation in the belief that this will not perturb the system. The '10% rule' is rather a pragmatic target, based loosely on Michaelis–Menten considerations, however, this target invariably has no basis since the Michaelis–Menten kinetic parameters are rarely known *a priori*. Notwithstanding this difficulty, it is usually important to have knowledge of the ambient concentrations of the nitrogen species under investigation prior to the addition of the ^{15}N substrate in order to target the addition.

16.4.2 *Incubation of samples*

The incubation of samples is entirely dependent upon the experiment. In many cases the prepared bottles are attached to metal/plastic frames suspended on ropes beneath buoys and incubated *in situ*. Samples are usually returned to the depth from which they were originally collected, and the incubation-rig allowed to 'free-float' away from the ship. Incubation times vary from a few hours to, usually, no longer than 24 hours. A typical experiment for example, might be a pre-dawn sample collection and preparation of 'experimental' bottles; deployment of *in situ* incubation rig at dawn; incubation for 24 hours with recovery of the incubation bottles the following dawn. An alternative scenario might be to recover the incubation rig at dusk and incubate the samples in darkness, at appropriate temperatures, onboard the ship for the dark period of the incubation. *In situ* type incubations may be carried out over any time period required, however, because of the logistical difficulties in ship operations involved in the deployment and recovery of *in situ* rigs, short-term incubations are normally restricted to simulated *in situ* incubations (see below). Despite the practical difficulties of *in situ* type incubations they remain a popular method of carrying out ^{15}N-phytoplankton experiments. Their main advantage is that the incubation conditions (particularly the light environment) are as near as possible those that the natural population would experience. However, a major disadvantage is that the experiments cannot be replicated. It is difficult (but not impossible) to carry out 'time course experiments' (see below) and there are innumerable practical difficulties involved in deploying and recovering *in situ* incubation rigs at sea.

Simulated *in situ* incubations are an alternative to *in situ* incubations and overcome several difficulties. For these incubations a variety of devices can be used to provide an environment that mimics natural conditions. Most 'incubators' utilise natural light as the light source, a range of light levels being achieved through the use of neutral density or wavelength dependent filters. Since these incubators are invariably placed on the open deck of ships it is necessary to regulate the temperature. This is usually achieved through the use of seawater pumped through containers in which the incubation bottles are suspended. In a fewer number of cases artificial light sources have been used to mimic the intensity and wavelength of light pertaining to particular water depths. Short-time incubations are much easier to undertake with these simulated *in situ* incubators and they offer the possibility for replicated experiments.

16.4.3 *Analysis*

Following the desired time interval incubations are terminated. In the majority of cases this is achieved by gentle filtration onto appropriate pore-size filters. Glass-fibre filters are popularly used since they provide a suitable support for subsequent analysis and potential contamination is easily removed. However, glass-fibre filters suffer from the drawback that they have poorly defined pore sizes. More accurately defined pore size filters (e.g. polycarbonate membranes, or nylon meshes) may be utilised to size-fractionate the phytoplankton sample in a series of sequential filtrations. The time taken for filtration is a potentially serious drawback of this technique. This could represent a significant fraction of very short incubation experiments. Filtrations also may result in lysis of cells and subsequent loss of label. Filters must also be rinsed of any possible ^{15}N-contaminating solution. Filters are then usually frozen for storage prior to subsequent landbased analysis, although with modern instrumentation it is possible to carry out ^{15}N analyses on board ship.

No details of ^{15}N measurement will be given since these are readily available. However, it is relevant to note the calculation of the uptake rate of nitrogen since there has been considerable debate about how this is achieved. Collos (1987) showed from first principles that many of the methods used for calculating nitrogen uptake rates lead to mathematical errors. Collos suggested the following equations that were valid for almost all conditions (but note the special caveats below):

$$\rho_N = N_f \, (C_p - C_{to})/(C_d - C_{to}) \, (\Delta t) \tag{16.2}$$

Where: ρ_N is the absolute uptake rate of dissolved ^{15}N label into particulate matter (phytoplankton); N_f is the concentration of particulate nitrogen at the end of the incubation period; C_p is the concentration of the label (in atom% ^{15}N) in the particulate phase after incubation; C_d is the concentration of the label (in atom% ^{15}N) in the dissolved phase at the start of the incubation (time zero); C_{to} is the concentration of the label (in atom% ^{15}N) in the particulate phase at time zero; Δt is the time interval of the incubation.

The corresponding equation for the specific uptake rate V is:

$$V = (C_p - C_{to})/(C_d - C_{to}) \, (\Delta t) \tag{16.3}$$

Collos (1987) noted that whilst the equation for the absolute uptake rate was insensitive to assimilation of multiple sources of nitrogen (usually the case in nature), the corresponding calculation for the specific uptake rate could lead to substantial underestimates where more that one nitrogen source was assimilated. This is a function of the fact that the ^{15}N technique relies on measuring the ratio of ^{14}N/^{15}N rather than absolute quantities of label (as it does, for example, in ^{14}C incubations) (see Collos and Slawyk, 1985). There is no practical solution to the problem.

As Collos (1987) showed, these equations assume ideal conditions where C_d does not alter over the time course of the incubation. However, all incubations involving natural waters suffer from the potentially significant drawback that a variety of nitrogen cycling processes occur simultaneously within the incubation bottles and these result in an alteration (dilution) of the ^{15}N label in the dissolved fraction (C_d). The most serious source of this error in most cases is in estimating ammonium assimilation rates, since ammonium production rates (ammonification) are frequently rapid, often of the same order as the assimilation rates (see for example, Glibert, 1982; Harrison, 1978; Koike *et al.*, 1986).

This can lead to serious errors in the value of C_d. Estimates of the assimilation of organic nitrogen will be similarly influenced. Similar difficulties arise with the estimation of assimilation of nitrate, but, because nitrate production rates (nitrification) are typically slow in ocean waters, relative to nitrate assimilation, this error will be small. A practical solution to these difficulties is to measure C_d at least at the end of the incubation and to apply an isotope dilution correction. A single end-point estimate of C_d and correction necessarily implies a linear rate of dilution; a more accurate correction should employ a time series of measurements of C_d or the application of a temporal isotope dilution model. These are rarely employed. The question of isotope dilution in ^{15}N incubations is particularly serious and has been the subject of much discussion (see for example, (Garside, 1984; Garside and Glibert, 1984; Glibert and Garside, 1989; Glibert et al., 1982c, 1985; Harrison and Harris, 1985; Laws, 1985).

Other difficulties arise with these incubations, for example, imprecision in the estimates of ambient nutrient concentrations; the exhaustion of labelled substrate; effects of containment (bottle effects); stress on the organisms from sampling; storage of samples. These and others are outside the scope of this review (see Harrison (1983) for details of many of the difficulties), but careful consideration must be given to the design and implementation of assimilation experiments.

16.5 The present – a discussion of results from recent studies

There are many examples of contemporary ^{15}N studies that could be given as examples. However, for the purposes of this review two areas of current interest have been selected: global variation in f ratio and preference of phytoplankters for a particular nitrogen substrate.

16.5.1 *Global variation in f ratio*

As noted above, Eppley and Peterson (1979) introduced the parameter of the f ratio as a measure of the proportion of the total production available for export from the upper ocean. This exportable production supports not only all subsequent elements of the food chain (for example fisheries) but also drives the potential sedimentation and ultimate sequestration of organic carbon to deep-sea sediments. The latter is particularly significant for the estimation of the global carbon cycle. It is not surprising therefore that a considerable effort has been expended in attempting to estimate the levels of new production (f ratio) globally. Figure 16.2 summarises the results of 37 studies (representing 74 estimates world-wide) that have measured the f ratio using ^{15}N techniques. The compilation is not exhaustive but it is representative. A simple arithmetic average of the data suggests a global f ratio of 0.37. Clearly this is far too simplistic an approach since it does not take into account the relative areas of the different ocean regions represented or any seasonal effects.

In order to provide a more representative global estimate of the f ratio, the data from Figure 16.2 have been sub-divided into six broad ocean regions (Figure 16.3). The most striking feature of the results is the extreme variability exhibited within each of the different regions, with no clear pattern emerging. The one exception perhaps being the 'equatorial' region which typically exhibits low f ratios. (For this study 'equatorial' has been defined as the oceanic gyres found to each side of the Equator in all the major oceans – strictly, these regions extend into the sub-tropics and thus are

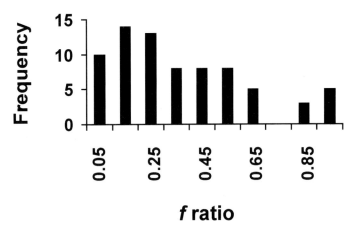

Figure 16.2. Frequency of f ratios from 74 estimates from 37 studies. The f ratios used represent integrated values over the depth of the photic zone. Where not explicitly cited in the individual studies integrated f ratios were calculated if sufficient data were available. Data sources: (Bury et al., 1995; Cochlan, 1986; Cota et al., 1992; Dugdale and Goering, 1967; Dugdale and Wilkerson, 1991; Dugdale et al., 1992; Eppley and Peterson, 1979; Eppley et al., 1977; Glibert et al., 1982a; Goering et al., 1970; Harrison et al., 1987; Jacques, 1991; Knauer et al., 1990; Kristiansen et al., 1994; Kristiansen et al., 1992; McCarthey, et al., 1996; Owens et al., 1986; Owens et al., 1993; Owens et al., 1991; Owens et al., 1990; Pena, Harrison et al., 1992; Pena et al., 1994; Probyn, 1985; Probyn and Painting, 1985; Probyn et al., 1996; Rees et al., 1995; Ronner et al., 1983; Sambrotto et al., 1993; Sathyendranath et al., 1991; Smith, 1990; Smith, 1991; Vezina, 1994; Waldron et al., 1995; Watts and Owens, 1997; Wheeler, 1993; Wilkerson and Dugdale, 1992)

not exclusively equatorial). These large ocean gyres are ultra-oligotrophic and typically exhibit nitrogen concentrations in the nanomolar range and low phytoplankton biomass (e.g. Banse, 1991; Dugdale and Goering, 1967; Eppley and Peterson, 1979; Knauer *et al.*, 1990; McCarthy *et al.*, 1996; Owens *et al.*, 1993; Pena *et al.*, 1992, 1994; Wilkerson and Dugdale, 1992). These systems are permanently thermally stratified thus are effectively isolated from nutrient sources in the deep ocean apart from the relatively poor supply from upward diffusion. They are also distant from terrestrial/riverine inputs thus it is not surprising that new production rates are typically low. The wide variations in f ratio found in all the other regions is caused by a variety of factors, most importantly seasonality. For example, temperate and high latitude regions typically exhibit high f ratios during the spring and early summer, but if thermal stratification persists the supply of new nitrogen gradually becomes exhausted and these systems then tend towards low f ratio systems (Glibert *et al.*, 1982b; Kristiansen *et al.*, 1994; Owens *et al.*, 1986; Rees *et. al.*, 1995; Sambrotto *et al.*, 1993). Upwelling systems also show a wide range in f ratio because there are usually pronounced upwelling periods (Owens, *et al.*, 1993; Watts and Owens, 1997) providing a large input of new nitrogen (resulting in high f ratios) after which the system tends to 'run down' towards a largely regenerated production system (lower f ratios) (Dugdale *et al.*, 1990).

A re-calculated global f ratio based on these regions is shown in Table 16.1 and provides an estimate of 0.33. A third estimate (details not shown) based on regional primary production data and the regional f ratios from Figure 16.3 and Table 16.1

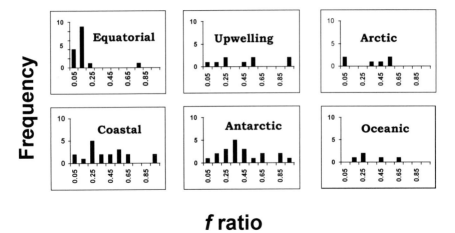

f ratio

Figure 16.3. Frequency of f ratios (as in Figure 16.2) sub-divided into ocean 'regions'. 'Equatorial' samples include sub-tropical gyres. 'Upwelling' samples refer to major coastal upwelling sites – for example, Arabian Sea. 'Oceanic' samples refer to all samples that do not fit into any of the other categories.

provides a global new production of ~ 21 Gt Carbon year $^{-1}$ which when compared to a total global phytoplankton production rate of ~ 53 Gt Carbon year^{-1} (Martin *et al.*, 1987; Platt, 1994) yields a global *f* ratio of ~ 0.40. The convergence of these estimates is perhaps surprising given the rather simple approaches used. However, these estimates are somewhat higher than several others derived from a variety of alternative approaches. For example 0.18 – 0.20 (Eppley and Peterson, 1979), 0.15 – 0.22 (Sarmiento *et al.*, 1993), 0.23 – 0.28 (Najjar *et al.*, 1992). The simple scaling approaches used here, and the failure to take account of seasonal effects may result in our apparent overestimates of the global *f* ratio.

However, it is interesting to speculate how much the estimate may be in error through the use of the ^{15}N technique itself, and two aspects in particular. First the influence of isotope dilution during the incubations (as noted above), and second, the failure to take account of forms of regenerated nitrogen other than ammonium. Certainly, the isotope dilution errors can be considerable and could easily account for overestimates of the *f* ratio in excess of 50% by underestimating the rate of ammonium assimilation (see references cited above), yet many studies do not account for it. Similarly, it is well known that phytoplankton assimilate a wide range of forms of 'regenerated nitrogen' other than ammonium for example, urea, amino acids and other organic compounds (Antia *et al.*, 1991) but very few field studies take account of this. In large part this is due to the difficulty in estimating the ambient concentrations of the compounds which is a pre-requisite for calculating the uptake rate. Urea assimilation has been measured in a few studies (for example Bury *et al.*, 1995; Eppley *et al.*, 1977; Kristiansen *et al.*, 1994; Kristiansen *et al.*, 1992; McCarthy, 1972; Ronner *et al.*, 1983; Watts and Owens, 1997) and the assimilation rates are often similar to rates of ammonium assimilation. It is possible, however, that some of this observed uptake is due to organisms other than phytoplankton. Nevertheless, the absence of assimilation data for organic nitrogen compounds will lead to an overestimation of the *f* ratio.

Table 16.1. *Estimate of global* f *ratio (new production) from* 15N *incubations conducted in a variety of ocean regions*

Ocean Region[a]	Mean f ratio[b]	Area (× 10^6 km^2) (%)	f ratio scaled to area[c]	% contribution of area to global f ratio
Equatorial	0.18	114 (32%)	0.058	17.6
Upwellings	0.46	10 (2.8%)	0.013	4.0
Arctic	0.33	15 (4.2%)	0.014	4.3
Coastal	0.42	50 (13.9%)	0.058	17.6
Southern Ocean	0.44	80 (22.3%)	0.098	29.8
Ocean	0.35	90 (25.1%)	0.088	26.7
Totals	0.37	359 (100%)	0.33	100%

[a]Areas defined as in Figure 16.3. [b]From Figure 16.3. [c]Weighted by % area.

Temporal variability in phytoplankton production is an important feature of many marine environments. This variability is frequently induced by changes in the availability of nutrients, thus marked temporal changes in f ratio are to be expected. Yet, there are very few comprehensive 15N studies that have examined temporal variability on the appropriate time-scales. Failure to take account of this variability undoubtedly diminishes the usefulness of the 15N studies to date in attempting to estimate global f ratios. Several 15N studies have measured temporal changes in nitrogen assimilation (for example, Cochlan, 1986; Glibert, 1982; Glibert, Goldman *et al.*, 1982b; Knauer, Redalje *et al.*, 1990; Kristiansen *et al.*, 1994; McCarthy *et al.*, 1977; Owens *et al.*, 1986; Paasche and Kristiansen, 1982; Rees *et al.*, 1995; Sambrotto *et al.*, 1993; Sarmiento *et al.*, 1993; Smith, 1990) but many of these are on very coarse time scales, e.g. only two seasons. This is unlikely to be sufficient given the typical time scale of variability. It is probable that the estimates of global f ratio attempted above, based on the extant 15N data, will be seriously in error.

An example of a detailed seasonal 15N study is shown in Figure 16.4. In this study nitrogen assimilation was measured monthly, together with total productivity (Rees *et al.*, 1995). The magnitude of the variability in f ratio is readily apparent with the spring bloom (May) clearly exhibiting very high f ratios, with the lower levels of production during summer (June–September) exhibiting much lower f ratios. A simple arithmetic average of the monthly data suggests an annual mean f ratio of ∼ 0.36 and it is easily appreciated how this would be different if only data from selected months were available. However, if new production is computed for each month (i.e. f ratio x primary production rate) and integrated over the year (the more appropriate calculation), an annual f ratio of ∼ 0.5 results. Clearly, if a similar 'correction' is applied to the data used above to calculate the global f ratio, our estimates would be even more disparate from those obtained by other techniques. However, there is no basis for assuming that this is a valid correction for the other studies. Nevertheless, this example serves to demonstrate the magnitude of the errors in f ratio estimation if temporal variability is not accounted for. However, even the resolution of nitrogen assimilation over monthly time intervals is unlikely to provide a full appreciation of the temporal variability since this occurs over a

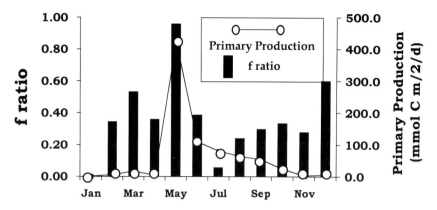

Figure 16.4. Seasonal variation of f ratio and primary production measured in a fjordic sea-loch. Adapted from Rees et al. (1995).

range of scales, even minutes to hours (for example, Glibert and Goldman, 1981; Goldman and Glibert, 1982). It is doubtful whether ^{15}N techniques alone will ever be sufficient to provide the data necessary for global modelling purposes. Rather, the technique is probably more appropriate in resolving details of the nitrogen ecophysiology and ecology of the plankton which can then be utilised in informing the global modelling efforts. However, possible techniques for providing more appropriate temporal and spatial extrapolations are considered in Section 16.6.

16.5.2. *Preference for different forms of nitrogen*

One of the findings from the earliest field ^{15}N studies was the apparent preference for the reduced forms of nitrogen, for example ammonium and urea, over nitrate. The most cited cause for this preference is the relatively higher energetic costs of utilisation of nitrate than ammonium due to the necessity for the reduction of nitrate (Syrett, 1981). These early studies assumed that the uptake and subsequent assimilation of nitrate were closely linked and that ammonium inhibited nitrate uptake. However, it is now known that there is frequently an uncoupling of nitrate uptake and assimilation (Collos, 1982a,b, 1984; Dortch *et al.*, 1979) and that the physiological interaction of nitrate and ammonium assimilation is highly complex (see Dortch, 1990; Flynn *et al*, 1997; Raven *et al.*, 1992), so that the basis of the apparent preference for ammonium may involve other factors (e.g. metal availability – Raven (1992)) in addition to the relative energetic costs of assimilation. A further complication is that preference for ammonium may be linked to an inhibition of uptake of nitrate, thus the interaction of nitrate and ammonium assimilation is actually a combination of preference and inhibition (Dortch, 1990). Notwithstanding the complex nature of the interactions there have been many field and laboratory studies that have utilised ^{15}N to examine the problem.

A formal method of examining relative preference nitrogen substrates was introduced by McCarthy *et al* (1977). This is the relative preference index (RPI) and utilises data from field-based ^{15}N incubations. Theoretically, the RPI provides a measure of the relative uptake rates (preference) of two or more forms of nitrogen relative to their respective ambient concentrations (= availability). It is calculated thus:

$$\text{RPI}_{\text{ammonium}} = \frac{\dfrac{\rho_{\text{ammonium}}}{\Sigma \rho_{\text{N}}}}{\dfrac{[\text{ammonium}]}{[\Sigma \text{N}]}} \qquad (16.4)$$

where: ρ_{ammonium} is the ammonium uptake rate; $\Sigma \rho_{\text{N}}$ is the sum of the uptake rates of all the nitrogen sources; [ammonium] is the ambient concentration of ammonium; [ΣN] is the sum of the concentrations of all the nitrogen sources. The equivalent RPI of nitrate (or any other nitrogen substrate) can be calculated by substituting for ammonium. Values of RPI$_{\text{ammonium}}$ >1 indicate a preference for ammonium, values <1 indicate a preference for an alternative substrate.

Since its introduction, many studies have utilised the RPI as an ecophysiological indicator (for example, Cota *et al.*, 1992; Glibert *et al.*, 1982a, b; Lancelot *et al.*, 1986; Owens *et al.*, 1986; Owens *et al.*, 1991; Probyn, 1985; Probyn and Painting, 1985; Rees *et al.*, 1995; Smith, 1990). In almost all cases these studies have shown a preference for ammonium, and this has been interpreted in various ways as being of importance in an ecological context, for the environment under study. Although McCarthy *et al.* (1977) argued that the RPI was a valuable indicator of the nitrogen status of the environment, its use and validity have been questioned (Dorch, 1990; Paasche, 1988; Stolte and Riegman, 1996), since its value may be strongly biased by the ambient nutrient concentrations rather than having any physiological basis.

Dorch (1990) suggested that a more meaningful indication of substrate preference is to compare the Michaelis–Menten parameters of $K_{1/2}$ and V_{max}, where lower $K_{1/2}$ and/or higher V_{max} indicate preference for that substrate. Table 16.2 summarises the findings from a large number of ^{15}N field studies where this has been done. It can be seen that although the mean of the quotients indicate a strong preference for ammonium, the range of observations suggest that ammonium is not always preferred relative to nitrate. These data somewhat contradict the evidence from RPI studies that invariably indicate a preference for ammonium over nitrate. Interestingly, the ratios of both $K_s\text{NO}_3$, $K_s\text{NH}_4$ and $V_{\text{max}}\text{NO}_3$: $V_{\text{max}}\text{NH}_4$ suggest that the preference for ammonium is less pronounced in high nutrient (largely coastal) environments. Invariably,

Table 16.2. Relative preference of ammonium and nitrate substrates for phytoplankton in natural populations determined by the use of ^{15}N techniques. Values of $K_{1/2}\text{NO}_3/K_{1/2}\text{NH}_4$ >1 and $V_{\text{max}}\text{NO}_3/V_{\text{max}}\text{NH}_4$ < 1 indicate preference for ammonium

Region	$K_{1/2}\text{NO}_3/K_{1/2}\text{NH}_4$	$V_{\text{max}}\text{NO}_3/V_{\text{max}}\text{NH}_4$
Oligotrophic		
Range	0.17 – 7	0.1 – 0.59
Mean (*n*)[a]	2.14 (8)	0.29 (9)
High nutrient		
Range	0.46 – 32.8	0.3 – 2.18
Mean (*n*)[a]	7.4 (11)	0.79 (16)

Data Source Dorch (1990). [a]*n* = number of individual studies.

high nutrient, coastal environments are high nitrate environments and these data support an increasing body of information that suggests that the phytoplankton population in these environments is capable of taking advantage of the high nitrate environment by maximising nitrate uptake rates.

In the context of the observation above, it is also interesting to note that the size composition of the phytoplankton community may be influenced by the relative supplies of nitrogen substrates and that small cells appear to have a greater preference for ammonium than larger cells (Owens et al., 1990; Probyn, 1985; Stolte et al., 1994; Stolte and Riegman, 1995). Figure 16.5 shows a clear dominance of uptake of ammonium by phytoplankters < 20 μm diameter in a ^{15}N study from the Antarctic. Because of the nature of the study it is not possible to determine whether there was any physiological basis to the observations, however, if applicable generally, this would have a profound effect on the composition of the entire planktonic community and the subsequent food-chain dynamics and nutrient cycling of that environment. The possible interaction between nutrient status and phytoplankton community is demonstrated further by the data in Figure 16.6 which shows that the ambient ammonium concentration may influence the composition of the phytoplankton community. Given the implications of the size composition of the plankton community on food-chain dynamics, sedimentation, carbon export, it is important to determine the interactions of nutrient concentrations, nitrogen ecophysiology and phytoplankton composition. [15]N studies provide the ideal tool for carrying out this analysis.

16.6 The future – possible developments for ^{15}N studies in biological oceanography

There is no doubt that the ^{15}N technique has been hugely influential in developing

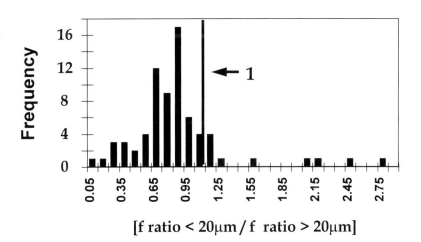

Figure 16.5. The frequency of the quotients of f ratios measured in the < 20 μm and >20 μm plankton size fractions around South Georgia, Southern Ocean. The vertical line indicates where the f ratios of the two size fractions are equal. The distribution of the quotients is significantly (Students t test: p < 0.05) less than one indicating that the smaller size fraction exhibited a lower f ratio than the larger fraction. Reproduced from Owens et al. (1991).

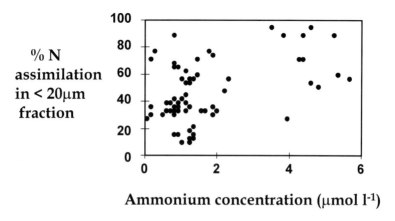

%N assimilation in < 20µm fraction

Ammonium concentration (µmol l⁻¹)

Figure 16.6. Proportion of phytoplanktonic inorganic nitrogen assimilation attributable to the < 20 µm diameter size fraction as a function of ambient ammonium concentration around South Georgia, Southern Ocean. Reproduced from Owens et al.(1991).

knowledge and understanding of the marine nitrogen cycle. The present status of the technique has been achieved through the introduction of new conceptual models and the development of new analytical techniques, many of which have benefitted from parallel developments from other disciplines. What is the future of the technique and where will the new developments occur? There is no doubt that there are innumerable research needs but whether these will be realised is open to question.

In the area of techniques, the introduction of continuous flow isotope ratio mass spectrometry (CF – IRMS) (Preston, 1992; Preston and Barrie, 1991; Preston and Owens, 1983) has been important in improving the speed and precision of analysis of ¹⁵N. The improved sample throughput has enabled a level of experimentation (e.g. replication) to be achieved that was prohibitive before its introduction. Also, the relatively small size of the instrumentation allows analysis to be carried out on board ships (Owens, 1988) which overcomes potential problems of sample storage, but more importantly allows for immediate 'feedback' of results that can be used to refine the study. Whilst these techniques now enable sub-microgram quantities of N to be analysed with sufficient precision, a major problem for marine studies is the often very small mass of sample N available for analysis. This causes problems with experimental design. Thus a marked improvement (at least an order of magnitude) in sensitivity would be highly desirable. Allied to this need is an improvement in the speed and precision of analysis of dissolved nitrogen substrates at the often extremely low concentrations available in seawaters (typically nmole l⁻¹). Improvements in 'wet chemical' analysis have been achieved in recent years that now offer reliable estimation of nitrate, ammonium and nitrite at appropriate concentrations but they are relatively slow and difficult to automate. A highly desirable development would be 'solid state' sensors to provide 'real time' estimation of concentrations. Whilst these might be possible for inorganic substrates, the analysis of organic substrates is a major difficulty holding back understanding. Current developments in analysis of dissolved amino-acids for subsequent ¹⁵N analysis (Preston *et al.*, 1996) are a significant step forward, as will be the more widespread use of gas chromatography coupled to CF-IRMS (GC-

CF-IRMS); the latter enabling the ^{15}N label in individual cellular constituents to be analysed.

In terms of concepts, it is unlikely that the biological oceanographic paradigms of new production and export production (Dugdale and Goering, 1967; Eppley and Peterson, 1979) will be replaced in the foreseeable future, however, new understanding is required about how the system operates under non-steady state conditions (the more common state). Better understanding of the interaction of competing nitrogen substrates and how this influences, and is influenced by, the phytoplankton community, is also required. This can be achieved, at least in part, by continued experimentation along lines similar to those currently, but by paying particular attention to measuring the full range of possible nitrogen substrates and, simultaneously, their recycling. This will need to be carried out with smaller additions of ^{15}N to reduce the perturbations of adding substrates. All of these are becoming more achievable with the method developments above. The field approach will be considerably enhanced with more detailed understanding of the physiology of organisms. A very promising development is the introduction of models that combine detailed physiology with the needs of field based studies (for example, Flynn et al., 1997). This type of modelling approach allows specific hypotheses to be tested.

Whilst future developments outlined above will be beneficial, there remains the major difficulty of extrapolating the limited information available from direct experimentation to relevant temporal and spatial scales. Major developments in the use of remote sensing (airborne and satellite sensors), coupled with ecophysiological and biophysical models have led to considerable advances in assessing primary production over suitable scales (Platt, 1994; Platt et al., 1988, 1991; Platt and Sathyendranath, 1988, 1995; Sathyendranath and Platt, 1988; Sathyendranath et al., 1989, 1996) similar developments are needed to extend this approach to include nitrogen cycling.

In an attempt to extend the remote sensing technique to estimating regional and global f ratios Sathyendranath et al. (1991) introduced the concept of 'compound remote sensing'. This is a multi-step approach to overcome the difficulty that phytoplankton have no known electromagnetic signature that can be measured remotely, that indicates the form of nitrogen assimilated. It is possible that such a signature may be detected in the future, for example, through subtle changes in pigment composition, but until such time an indirect approach is the only alternative.

Figure 16.7 shows diagramatically one protocol that might be used to determine f ratio over large space scales. The key to the technique is the availability of satellite data of ocean surface temperature (SST). The satellite data may be used to generate vertical temperature profiles that in turn may be used to derive a temperature nitrate concentration relationship. Both relationships would be based on empirical data from a range of ocean areas. Similarly, empirically (^{15}N) derived nitrate $-f$ ratio relationships would allow f ratios to be calculated for each pixel in the original satellite image. The individual steps involved in this protocol have been shown to be valid (Harrison, 1987; Sathyendranath et al., 1991; Watts and Owens, unpublished data). This suggested protocol is similar to (Sathyendranath et al., 1991) but differs in that the ambient nitrate vertical structure is employed. An alternative approach currently under investigation by Watts et al. is to utilise empirical (^{15}N) f ratio – primary production relationships. In this approach basin-scale primary production is derived using an established light-dependent model (Platt et al., 1991) and these data, in turn, are used to derive the corresponding f ratios.

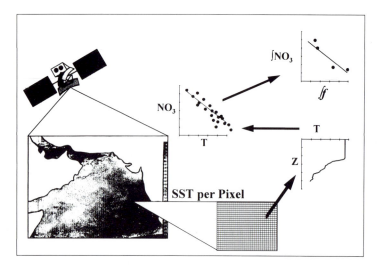

Figure 16.7. Diagrammatic representation of a protocol for the use of satellite remote sensing for the estimation of f ratios over large spatial scales. See text for explanation.

The major difficulty with all these approaches is defining the spatial and temporal limits within which individual models and empirical relationships may be used. This difficulty is currently overcome by utilising the concept of 'biogeochemical provinces'. In brief, biogeochemical provinces are spatial areas which exhibit similar ecological and biogeochemical properties within a certain time-frame/s. Thus in this context, to be useful, one would expect to find 'province-specific' ^{15}N relationships with say, primary production. Since these properties are essentially dynamic in space and time, it is likely that the province boundaries will also be dynamic. Whilst it might be possible then to define biogeochemical provinces entirely by geographic areas, this is not ideal since the province boundaries will be static. Ideally the definition of a province should incorporate a property that can be measured remotely so that the boundaries can be defined at any one time. Sathyendranath *et al.* (1995) provides a good example of a similar approach for the North Atlantic. Much work remains to be done in identifying biogeochemical provinces and the empirical (or mechanistic) relationships suitable for the remote sensing protocols outlined above. However, this approach is a significant conceptual development for the use of ^{15}N in marine studies that will continue to provide new understanding of the marine environment for the foreseeable future.

16.7 Summary and conclusions

^{15}N studies have been immeasureably influential in marine science and phytoplankton ecology for nearly three decades. Building on ideas, concepts and techniques extending well into the 19th century, there have been 'bursts' of activity following the introduction of new ideas, punctuating periods of consolidation. The use of ^{15}N has provided significant understanding of the ecophysiology and temporal and spatial variability of phytoplankton nitrogen assimilation that would otherwise have been impossible. Furthermore, it was through the use of ^{15}N techniques that the current paradigms of biological oceanography were based. In parallel with the use of the techniques there has been steady progress in measurement technology, and recently introduced developments appear set to continue this trend. In addition, new concepts of understanding how the oceans may be sub-divided are currently being developed, in part, alongside the ^{15}N techniques, which are likely to provide a major step forward for biological oceanography.

References

Altabet, M.A. and McCarthy, J.J. (1985) Temporal and spatial variations in the natural abundance of ^{15}N in PON from a warm-core ring. *Deep-Sea Res.* **32**, 755–772.

Altabet, M.A., Robinson, A.R. and Walstead, L.J. (1986) A model for the vertical flux of nitrogen in the upper ocean: stimulating the alteration of isotopic ratios. *J. Mar. Res.* **44**, 203–225.

Antia, N.J., Harrison, P.J. and Oliveira, L. (1991) The role of dissolved organic nitrogen in phytoplankton nutrition, cell biology and ecology. *Phycologia* **30**, 1–89.

Banse, K. (1991). Iron availability, nitrate uptake and exportable new production in the subarctic Pacific. *J. Geophys. Res.* **96**, 741–748.

Barsdate, R.J. and Dugdale, R.C. (1965) Rapid conversion of organic nitrogen to N_2 for mass spectrometry: An automated Dumas procedure. *Anal. Biochem.* **13**, 1–5.

Benson, B.B. and Parker, P.D.M. (1961) Nitrogen/argon and nitrogen isotope ratios in aerobic sea water. *Deep-Sea Res.* **7**, 237–253.

Beynon, J.H. (1960). *Mass Spectrometry and its Applications to Organic Chemistry*. Elsevier, Amsterdam.

Brand, T.v. and Rakestraw, N.W. (1941a) Decomposition and regeneration of nitrogenous organic matter in seawater. II. Influence of temperature and source and condition of water. *Biol. Bull., Woods Hole.* **79**, 231–236.

Brand, T.v. and Rakestraw, N.W. (1941b) Decomposition and regeneration of nitrogenous organic matter in seawater. IV. Interrelationship of various stages, influence of concentration and nature of particulate matter. *Biol. Bull., Woods Hole.* **81**, 63–69.

Brand, T.v., Rakestraw, N.W. and Renn, C.E. (1937) The experimental decompositon and regeneration of nitrogenous organic matter in sea water. *Biol. Bull., Woods Hole.* **72**, 165–175.

Brand, T.v., Rakestraw, N.W. and Renn, C.E. (1939) Further experiments on the decompositon and regeneration of nitrogenous organic matter in sea water. *Biol. Bull., Woods Hole.* **77**, 285–296.

Brand, T.v., Rakestraw, N.W. and Zabor, J.W. (1942) Decomposition and regeneration of nitrogenous organic matter in sea water. V. Factors influencing the length of the cycle; observations upon the gaseous and dissolved organic nitrogen. *Biol. Bull., Woods Hole.* **83**,273–282.

Brandt, K. (1899) Ueber den stoffwechsel im meere (Rektoratsrede). *Wissenschaftliche Meeresuntersuchungen, Abteilung Kiel, Neue Folge.* **4**, 215–230.

Brandt, K. (1902). Ueber den stoffwechsel im meere. 2. Abhandlung. *Ueber den stoffwechsel im meere (Rektoratsrede).* **6**, 23–79.

Brandt, K. (1920) Ueber den stoffwechsel im meere. 3. Abhandlung. *Wissenschaftliche Meeresuntersuchungen, Abteilung Kiel, Neue Folge.* **18**, 185–229.

Bronk, D.A. and Glibert, P.M. (1993) Application of a ^{15}N tracer method to the study of dissolved organic nitrogen uptake during spring and summer in Chesapeake Bay. *Marine Biol.* **115**, 501–508.

Bury, S.J., Owens, N.J.P. and Preston, T. (1995). ^{13}C and ^{15}N uptdake by phytoplankton in the marginal ice zone of the Bellingshausen Sea. *Deep-Sea Res. II* **42**, 1225–1252.

Carpenter, E.J. (1983) Nitrogen fixation by Marine *Oscillatoria* (Trichodesmium) in the world's oceans. In Nit*rogen in the Marine Environment.* (eds E.J. Carpenter and D.G. Capone) Academic Press, New York, pp. 65–103.

Cochlan, W.P. (1986) Seasonal study of uptake and regeneration of nitrogen on the Scotian Shelf. *Continental Shelf Res.* **5**, 555–557.

Collos, Y. (1982a) Transient situations in nitrate assimilation by marine diatoms. 2. Changes in nitrate and nitrite following a nitrate perturbation. *Limnol. and Oceanogr.* **27**, 528–535.

Collos, Y. (1982b) Transient situations in nitrate assimilation by marine diatoms. III. Short-term uncoupling of nitrate uptake and reduction. *J. Exp. Marine Biol. and Ecology* **62**, 285–295.

Collos, Y. (1984) Transient situations in nitrate assimilation by marine diatoms. V. Interspecific variability in biomass and uptake during nitrogen starvation and resupply. *Marine Ecol. Progr. Series* **17**, 25 – 31.

Collos, Y. (1987) Calculations of ¹⁵N uptake rates by phytoplankton assimilating one or several nitrogen sources. *Int. J. Appl. Radiat. and Isotopes* **38**, 275 – 282.

Collos, Y. and Slawyk, G. (1985) On the compatibility of carbon uptake rates calculated from stable and radioactive isotope data: implications for the design of experimental protocols in aquatic primary productivity. *J. Plankton Res.* **7**, 595 – 603.

Conway, H.L. (1977) Interactions of inorganic nitrogen in the uptake and assimilation by marine phytoplankton. *Marine Biol.* **39**, 221–232.

Cooper, L.H.N. (1933) Chemical constituents of biological importance in the English Channel. Part 1. Phosphate, silicate, nitrate, nitrite, nitrite, ammonia. *J. Marine Biol. Assoc. U.K.* **18**, 677–728.

Cooper, L.H.N. (1935) The rate of liberation of phosphate in sea water by the breakdown of plankton organisms. *J. Marine Biological Assoc. U.K.* **20**, 197–200.

Cota, G.F., Smith, W.O.J., Nelson, D.M., Muench, R.D. and Gordon, L.I. (1992) Nutrient and biogenic particulate distributions, primary productivity and nitrogen uptake in the Weddell-Scotia Sea marginal ice zone during winter. *J. Marine Rese.* **50**, 155–181.

de Baar, H.J.W. (1994) von Liebig's law of the minimum and plankton ecology (1899–1991). *Prog. Oceanogr.* **33**, 347–386.

de Baar, H.J.W., Buma, A.G.J., Nolting, R.F., Cadee, G.C., Jacques G. and Treguer P.J. (1990) On iron limitation of the Southern Ocean: experimental observations on the Weddell and Scotia Seas. *Marine Ecol. Progr. Ser.* **65**, 105–122.

Dortch, Q. (1990) The interaction between ammonium and nitrate uptake in phytoplankton. *Marine Ecol. Progr. Ser.* **61**, 183 – 201.

Dortch, Q., Ahmed S.I. and Packard, T.T. (1979) Nitrate reductase and glutamate dehydrogenase activities in *Skeletonema costatum* as measures of nitrogen assimilation rates. *J. Plankton Res.* **1**, 169 – 186.

Dugdale, R.C. (1967) Nutrient limitation in the sea: dynamics, identification, and significance. *Limnonol. and Oceanogr.* **12**, 685–695.

Dugdale, R.C., Dugdale V., Neess, J. and Goering, J. (1959) Nitrogen fixation in lakes. *Science* **130**, 859–860.

Dugdale, R.C. and Goering, J.J. (1967) Uptake of new and regenerated forms of nitrogen in primary productivity. *Limnol. Oceanogr.* **12**, 196–206.

Dugdale, R.C., Goering, J.J. and Ryther, J.H. (1964) High nitrogen fixation rates in the Sargasso Sea and the Arabian Sea. *Limnol. Oceanogr.* **9**, 507–510.

Dugdale, R.C., Menzel, D.W. and Ryther, J.H. (1961) Nitrogen fixation in the Sargasso Sea. *Deep-Sea Res.* **7**, 298–300.

Dugdale, R.C. and Wilkerson, F.P. (1991) Low specific nitrate uptake rate: A common feature of high-nutrient, low-chlorophyll marine ecosystems. *Limnol. Oceanogr.* **36**, 1678–1688.

Dugdale, R.C., Wilkerson, F.P., Barber, R.T. and Chavez, F.P. (1992) Estimating new production in the Equatorial Pacific Ocean at 150°W. *J. Geophys. Res.* **97**, 681–686.

Dugdale, R.C., Wilkerson, F.P. and Morel, A. (1990) Realization of new production in coastal upwelling areas: A means to compare relative performance. *Limnol. Oceanogr.* **35**, 822–829.

Dugdale, V.A. and Dugdale, R.C. (1962) Role of nitrogen fixation in Sanctuary Lake, Pennsylvania. *Limnol. Oceanogr.* **7**, 170–177.

Eppley, R.W. and Peterson, B.J. (1979) Particulate organic matter flux and planktonic new production in the deep ocean. *Nature* **282**, 677–680.

Eppley, R.W., Sharp, J.H., Renger, E.H., Perry, M.J. and Harrison, W.G. (1977) Nitrogen assimilation by phytoplankton and other micro-oganisms in the surface waters of the central North Pacific Ocean. *Marine Biol.* **39**, 111–120.

Flynn, K.J. and Flynn K. (1997) The release of nitrite by marine dinoflagellates – development of a mathematical simulation. *Marine Biol.* **in press.**

Flynn, K.J., Fasham, M.J.R. and Hipkin, C.R. (1997) Modelling the interactions between ammonium and nitrate uptake in marine phytoplankton. *Philo. Trans. Roy. Soc., London B,* **in press.**

Garside, C. (1982) A chemiluminescent technique for the determination of nanomolar concentrations of nitrate, nitrite, or nitrite alone in seawater. *Marine Chem.* 11; 159–167.

Garside, C. (1984) Apparent ^{15}N uptake kinetics resulting from remineralization. *Limnol. Oceanogr.* **29**, 204 – 210.

Garside, C. and Glibert P.M. (1984) Computer modeling of ^{15}N uptake and remineralization experiments. *Limnol. Oceanogr.* **29**, 190 – 204.

Glibert, P.A., Biggs, D.C. and McCarthy, J.J. (1982a) Utilization of ammonium and nitrate during austral summer in the Scotia Sea. *Deep-Sea Res.* **29**, 837–850.

Glibert, P.M. (1982) Regional studies of daily, seasonal and size fraction variability in ammonium remineralization. *Marine Biol.* **70**, 209–222.

Glibert, P.M. and Garside, C. (1989) Discussion on 'Spring recycling rates of ammonium in turbid continental shelf waters off the southeastern United States'. *Continental Shelf Res.* **9**, 197–200.

Glibert, P.M. and Goldman, J.C. (1981) Rapid ammonium uptake by marine phytoplankton. *Marine Biol. Lett.* **2**, 25–31.

Glibert, P.M., J.C. Goldman and E.J. Carpenter (1982b) Seasonal variations in the utilization of ammonium and nitrate by phytoplankton in Vineyard Sound, Massachusetts, USA. *Marine Biology* **70**, 237–249.

Glibert, P.M., Lipschultz, F., McCarthy, J.J. and Alatabet, M.A. (1985) Has the mystery of the vanishing ^{15}N in isotope dilution experiments been resolved? *Limnol. Oceanogr.* **30**, 444–447.

Glibert, P.M., Lipschultz, F., McCarthy, J.J. and Altabet, M.A. (1982c) Isotope dilution models of uptake and remineralization of ammonium by marine plankton. *Limnol. Oceanogr.* **27**, 639–650.

Goering, J.J. and Dugdale, R.C. (1966a) Denitrification rates in an island bay in the Equatorial Pacific Ocean. *Science* **154**, 505–506.

Goering, J.J., Dugdale, R.C. and Menzel, D.W. (1964) Cyclic diurnal variations in the uptake of ammonia and nitrate by photosynthetic organisms in the Sargasso Sea. *Limnol. Oceanogr.* **9**, 448–451.

Goering, J.J., Dugdale, R.C. and Menzel, D.W. (1966) Estimates of *in situ* rates of nitrogen uptake by *Trichodesmium* sp. in the tropical Atlantic Ocean. *Limnol. Oceanogr.* **11**, 614–620.

Goering, J.J. and Dugdale V.A. (1966b) Estimates of the rates of denitrification in a subarctic lake. *Limnol. Oceanogr.* **11**, 113–117.

Goering, J.J. and Neess. J.C. (1964) Nitrogen fixation in two Wisconsin lakes. *Limnol. Oceanogr.* **9**, 530–539.

Goering, J.J., Wallen, D.D. and Nauman, R.M. (1970) Nitrogen uptake by phytoplankton in the discontinuity layer of the eastern subtropical Pacific Ocean. *Limnol. Oceanogr.* **15**, 789–796.

Goldman, J.C. and Glibert, P.M. (1982) Comparitive rapid ammonium uptake by four species of marine phytoplankton. *Limnol. Oceanogr.* **27**, 814 – 817.

Harrison, W.G. (1978) Experimental measurements of nitrogen remineralization in coastal waters. *Limnol. Oceanogr.* **23**, 684 – 694.

Harrison, W.G. (1983) Nitrogen in the marine environment. IV.2. Use of isotopes. In: *Nitrogen in the Marine Environment.* (eds E.J. Carpenter and D.G. Capone) Academic Press, New York, pp. 763–807.

Harrison, W.G. and Harris, L.R. (1985) Isotope dilution and its effects on measurements of nitrogen and phosphorus uptake by oceanic microplankton. *Marine Ecol. Prog. Ser.* **27**, 254–261.

Harrison, W.G., Platt, T. and Lewis, M.R. (1987) *f*-Ratio and its relationship to ambient nitrate concentration in coastal waters. *J. Plankton Res.* **9**, 235–248.

Harvey, H.W. (1926) Nitrate in the sea. *J. Marine Biol. Assoc. U.K.* **14**, 71–88.

Harvey, H.W. (1933) On the rate of diatom growth. *J. Marine Biol. Assoc. U.K.* **19**, 253–276.

Harvey, H.W. (1940) Nitrogen and phosphorus required for the growth of phytoplankton. *J. Marine Biol. Assoc. U.K.* **24**, 115–123.

Harvey, H.W., Cooper, L., LeBour, M.V. and Russell, F.S. (1935) Plankton production and its control. *J. Marine Biol. Assoc. U.K.* **20**, 407–441.

Hauck, R.D., Melsted, S.W. and Yankwich, P.E. (1958) Use of N – isotope distribution in nitrogen gas in the study of denitrification. *Soil Science* **86**, 287–291.

Hoering, T. (1955) Variations of nitrogen-15 abundance in naturally occurring substances. *Science* **122**, 1233–1234.

Jacques, G. (1991) Is the concept of new production-regenerated production valid for the Southern Ocean. *Marine Chemisty* **35**, 273–286.

Jones, R.D. (1991) An improved fluorescence method for the determination of nanomolar concentrations of ammonium in natural waters. *Limnol. Oceanogr.* **36**, 814–819.

Knauer, G.A., Redalje. D.G., Harrison, W.G. and Karl, D.M. (1990) New production at the VERTEX time-series site. *Deep-Sea Res.* **37**, 1121–1134.

Koike, I., Holm-Hansen, O. and Biggs, D.C. (1986) Inorganic nitrogen metabolism by Antarctic phytoplankton with special reference of ammonium cycling. *Marine Ecol. Progr. Ser.* **30**, 105 -116.

Kristiansen, S., Farbrot, T. and Wheeler, P.A. (1994) Nitrogen cycling in the Barents Sea – Seasonal dynamics of new and regenerated production in the marginal ice zone. *Limnol. Oceanogr.* **39**, 1630–1642.

Kristiansen, S., Syvertsen, E.E. and Farbrot, T. (1992) Nitrogen uptake in the Weddell Sea during late winter and spring. *Polar Biology* **12**, 245–251.

Lancelot, C., S. Mathot and Owens, N.J.P. (1986) Modelling protein synthesis, a step to an accurate estimate of net primary production: *Phaeocystis pouchetii* colonies in Belgian coastal waters. *Marine Ecol. Progr. Ser.* **32**, 193- 202.

Laws, E. (1985) Analytic models of NH_4^+ uptake and regeneration experiments. *Limnol. Oceanogr.* **30**, 1340 – 1350.

MacIsaac, J.J. and Dudgale, R.C. (1972) Interactions of light and inorganic nitrogen in controlling nitrogen uptake in the sea. *Deep-Sea Res.* **19**, 209–232.

MacIsaac, J.J. and Dugdale, R.C. (1969) The kinetics of nitrate and ammonia uptake by natural populations of marine phytoplankton. *Deep-Sea Res.* **16**, 45–57.

Martin, J.H. and Fitzwater, S.E. (1988) Iron deficiency limits phytoplankton growth in the north-east Pacific Subarctic. *Nature* **331**, 341–343.

Mariotti, A., Lancelot, C. and Billen, C.G. (1984) Natural isotopic composition of nitrogen as a tracer of origin for suspended organic matter in the Scheldt estuary. *Geochim. Cosmochim Acta* **48**, 549–555.

Martin, J.H., Fitzwater, S.E. and Gordon, R.M. (1990) Iron deficiency limits phytoplankton growth in Antarctic water. *Global Biogeochem. Cycles* **4**, 5–12.

Martin, J.H., Knauer, G.A., Karl, D.M. and Broenkow, W.W. (1987) VERTEX: carbon cycling in the northeast Pacific. *Deep-Sea Res.* **34**, 267 – 285.

McCarthy, J.J. (1972) The uptake of urea by natural populations of marine phytoplankton. *Limnol. Oceanogr.* **17**, 738–748.

McCarthy, J.J., Garside, C., Nevins, J.L. and Barber, R.T. (1996) New production along 140ºW in the equatorial Pacific during and following the 1992 El Nino event. *Deep-Sea Res. II* **43** no. 4–6, 1065–1093.

McCarthy, J.J., Kaplan, W. and Nevins, J.L. (1984) Chesapeake Bay nutrient and plankton dynamics. 2. Sources and sinks of nitrite. *Limnol. Oceanogr.* **29**, 84 – 98.

McCarthy, J.J., Taylor, W.R. and Taft, J.L. (1977) Ntrogenous nutrition of the plankton in the Chesapeake Bay. 1. Nutrient availability and phytoplankton preferences. *Limnol. Oceanogr.* **22**, 996–1011.

Minigawa, M. and Wada, E. (1986) Nitrogen isotope ratios of red organisms in the East China Sea: a characterization of biological nitrogen fixation. *Mar. Chem.* **19**, 245–259.

Miyake, Y. and Wada, E. (1967) The abundance ratio of $^{15}N/^{14}N$ in marine environments. *Rec. Oceanographic Works Jpn.* **9**, 32–53.

Najjar, R.G., Sarmiento, J.L. and Toggweiler, J.R. (1992) Downward transport and fate of organic matter in the oceans: simulations with a general circulation model. *Global Biogeochemical Cycles* **6**, 45 – 76.

Neess, J.C., Dugdale, R.C., Dugdale, V.A. and Goering, J.J. (1962) Nitrogen metabolism in lakes. I. Measurement of nitrogen-fixation with N^{15}. *Limnol. Oceanogr.* **7**, 163–169.

Owens, N.J.P. (1987) Natural variations in ^{15}N in the marine environment. *Advances in Marine Biol.* **24**, 389–451.

Owens, N.J.P. (1988) Rapid and total automation of shipboard ^{15}N analysis: Examples from the North Sea. *J. Exp. Marine Biol. Ecol.* **122**, 163–171.

Owens, N.J.P., Mantoura, R.F.C., Burkil, P.H., Howland, R.J.M., Pomroy, A.J. and Woodward, E.M.S. (1986) Nutrient cycling studies in Carmarthen Bay: Phytoplankton production, nitrogen assimilation and regeneration. *Marine Biol.* **93**, 329–342.

Owens, N.J.P., Mantoura, R.F.C., Burkill, P.H., Woodward, E.M.S., Bellan, I.E. and Aiken, J. (1993) Size fractionated primary production and nitrogen assimilation in the northwestern Indian Ocean. *Deep-Sea Res. II* **40**, 697–709.

Owens, N.J.P., Priddle, J. and Whitehouse, M.J. (1991) Variations in phytoplanktonic nitrogen assimilation around South Georgia and in the Bransfield Strait (Southern Ocean) *Marine Chem.* **35**, 287–304.

Owens, N.J.P., Woodward, E.M.S., Aiken, J.K., Bellan, I.E. and Rees, A.P. (1990) Primary production and nitrogen assimilation in the North Sea during July 1987. *Netherlands J. Sea Res.* **25**, 143–154.

Paasche, E. (1988) Pelagic primary production in nearshore waters. In: *Nitrogen cycling in coastal marine environments.* (eds T.H. Blackburn and J. Sorensen) Wiley, New York, pp. 33–57.

Paasche, E. and Kristiansen, S. (1982) Nitrogen nutrition of the phytoplankton in the Oslofjord. *Est. Coastal Shelf Sci.* **14**, 237 – 249.

Pena, M.A., Harrison, W.G. and Lewis, M.R. (1992) New production in central equatorial Pacific. *Marine Ecol. Progr. Ser.* **80**, 265–274.

Pena, M.A., Lewis, M.R. and Cullen, J.J. (1994) New production in the warm waters of the tropical Pacific Ocean. *J. Geophys. Res.* **99**, 14255–14268.

Platt, T. (1994) Global Marine Primary Productivity: Report to the European Space Agency, Bedford Institute of Oceanography.

Platt, T., Caverhill, C. and Sathyendranath, S. (1991) Basin-scale estimates of oceanic primary production by remote sensing: The North Atlantic. *J. Geophys. Res.* **96**, 15147–15159.

Platt, T. and Sathyendranath, S. (1988) Oceanic primary production: Estimation by remote sensing at local and regional scales. *Science* **241**, 1613 – 1620.

Platt, T. and Sathyendranath, S. (1995) Latitude as a factor in the calculation of primary production. In: *Ecology of Fjords and Coastal Waters.* (eds H.R. Skjoldal, C. Hopkins, K.E. Erikstad and H.P. Leinaas) Elsevier, Amsterdam, pp. 3–13.

Platt, T., Sathyendranath, S., Caverhill, C.M. and Lewis, M.R. (1988) Ocean primary production and available light: Further algorithms for remote sensing. *Deep-Sea Res.* **35**, 855–879.

Preston, T. (1992) The measurement of stable isotope natural abundance variations. *Plant, Cell Environ,* **15**, 1091 – 1097.

Preston, T., Bury, S., McMeekin, B. and Slater, C. (1996) Isotope dilution analysis of combined nitrogen in natural waters: *II:* Amino Acids. *Rapid Comm. Mass Spectrom.* **10**, 965 – 968.

Preston, T. and Barrie, A. (1991) Recent progress in continuous flow isotope ratio mass spectrometry. *International Laboratory*, 31–34.

Preston, T. and Owens, N.J.P. (1983) Interfacing an automatic elemental analyser with an isotope ratio mass spectrometer: the potential for fully automated total nitrogen and nitrogen-15 analysis. *The Analyst* **108**, 971 – 977.

Probyn, T.A. (1985) Nitrogen uptake by size-fractionated phytoplankton populations in the southern Benguela upwelling system. *Marine Ecol. Progr. Ser.* **22**, 249–258.

Probyn, T.A. and Painting, S.J. (1985) Nitrogen uptake by size-fractionated phytoplankton populations in Antarctic surface waters. *Limnol. Oceanogr.* **30**, 1327–1332.

Probyn, T.A., Waldron, H.W., Searson, S. and Owens, N.J.P. (1996) Diel variability in nitrogenous nutrient uptake at photic and sub-photic depths. *J. Plankton Res.* **18**, 2063–209.

Raven, J.A., Wollenweber, B. and Handley, L.L. (1992) A comparison of ammonium and nitrate as nitrogen sources for photolithotrophs. *New Phytologist* **121**, 19–32.

Rees, A.P., Owens, N.J.P., Heath, M.R., Plummer, D.H. and Bellerby, R.S. (1995) Seasonal nitrogen assimilation and carbon fixation in a fjordic sea loch. *J. Plankton Res.* **17**, 1307–1324.

Richards, F.A. and Benson, B.B. (1961) Nitrogen/argon and nitrogen isotope ratios in two anaerobic environments, the Cariaco trench in the Caribbean Sea and Dramsfjord, Norway. *Deep-Sea Res.* **7**, 254–264.

Rittenberg, D. (1946) The preparation of gas samples for mass spectrographic isotope analysis. In: *Preparation and measurement of isotopic tracers.* (eds D.W. Wilson, A.O.C. Nier and S.P. Reimann) Edwards, Ann Arbor, Michigan.

Rittenberg, D., Keston, A.S., Rosebury, F. and Schoenheimer, R. (1939) Studies in protein metabolism. II. The determination of nitrogen isotopes in organic compounds. *J. Biol. Chem.* **127**, 291–299.

Ronner, U., Sorensson, F. and Holm-Hansen, O. (1983) Nitrogen assimilation by phytoplankton in the Scotia Sea. *Polar Biology* **2**, 137–147.

Sambrotto, R.N., Martin, J.H., Broenkow, W.W., Carlson, C. and Fitzwater, S.E. (1993) Nitrate utilization in surface waters of the Iceland Basin during spring and summer of 1989. *Deep-Sea Res. II* **40**, 441–457.

Sarmiento, J.L., Slater, R.D., Fasham, M.J.R., Ducklow, H.W., Toggweiler, J.R. and Evans, G.T. (1993) A seasonal three-dimensional ecosystem model of nitrogen cycling in the North Atlantic euphotic zone. *Global Biogeochemical Cycles.* **7**, 417 – 450.

Sathyendranath, S. and Platt, T. (1988) The spectral irradiance field at the surface and in the interior of the ocean: A model for applications in oceanography and remote sensing. *J. Geophys. Res,* **93**, 9270–9280.

Sathyendranath, S., Platt, T., Caverhill, C.M., Warnock, R.E. and Lewis, M.R. (1989) Remote sensing of oceanic primary production: computations using a spectral model. *Deep-Sea Res.* **36**, 431–453.

Sathyendranath, S., Platt, T., Horne, E.P.W., Harrison, W.G., Ulloa, O., Outerbridge, R. and Hoepffner, N. (1991) Estimation of new production in the ocean by compound remote sensing. *Nature* **353**, 129–133.

Sathyendranath, S., Platt, T., Stuart, V., Irwin, B.D., Veldhuis, M.J.W., Kraay, G.W. and Harrison, W.G. (1996) Some bio-optical characteristics of phytoplankton in the NW Indian Ocean. *Marine Ecol. Progr. Ser.* **132**, 299 – 311.

Sathyendranath, S., Longhurst, A., Caverhill, C.M. and Platt, T. (1995) Regionally and seasonally differentiated primary production in the North Atlantic. *Deep Sea Res. I* **42**, 1773–1802.

Sharp, J.H. (1983) The distribution of inorganic nitrogen and dissolved and particulate organic nitrogen in the Sea. In: *Nitrogen in the Marine Environment.* (eds E.J. Carpenter and D.G. Capone) Academic Press, New York, pp. 1–35.

Sheppard, C.W. (1962) *Basic principles of the tracer method.* Wiley, New York.

Smith, W.O.J. (1990) Phytoplankton growth and new production in the Weddell Sea marginal ice zone in the ausral spring and autumn. *Limnol. Oceanogr.* **35**, 809–821.

Smith, W.O.J. (1991) Nutrient distributions and new production in polar regions: parallels and contrasts beween Arctic and Antarctic. *Marine Chem.* **35**, 245–257.

Steemann-Nielsen, E. (1952) The use of radioactive carbon (^{14}C) for measuring organic production in the sea. *J. Consiel Int. Expl. de la Mer* **18**, 117–140.

Stolte, W., McCollin, T., Noordeloos, A.A.M. and Riegman, R. (1994) Effect of nitrogen source on the size distribution within marine phytoplankton populations. *J. Exp. Marine Biol. Ecol.* **184**, 83 – 97.

Stolte, W. and Riegman, R. (1995) The effect of phytoplankton cell size on transient state nitrate and ammonium uptake kinetics. *Microbiology* **141**, 1221–1229.

Stolte, W. and Riegman, R. (1996) The relative preference index (RPI) for phytoplankton nitrogen use is only weakly related to physiological preference. *J. Plankton Res.* **18**, 1041–1045.

Strickland, J.D.H. and Parsons, T.R. (1960) *A manual of sea water analysis*. Bulletin of the Fisheries Research Board of Canada.

Syrett, P.J. (1981) Nitrogen metabolism of microalgae. In: *Physiological bases of phytoplankton ecology. Bulletin No. 210.* (ed. T. Platt) Canadian Government, Quebec, pp. 182–210.

Vezina, A.F. (1994) Mesoscale variability in nitrogen uptake rates and the f-ratio during a coastal phytoplankton bloom. *Limnol. Oceanog.* **39**, 854–868.

Wada, E., Kadonaga, T. and Matsuo, S. (1975) ^{15}N abundance in nitrogen of naturally occurring substances and assessment of denitrification from isotopic standpoint.

Wada, E. and Hattori, A. (1976) Natural abundance of ^{15}N in particulate organic matter in the North Pacific Ocean. *Geochem. Cosmoclimica Acta* **12**, 97–102.

Waldron, H.N., Attwood, C.G., Probyn, T.A. and Lucas, M.I. (1995) Nitrogen dynamics in the Bellingshausen Sea during the Austral spring of 1992. *Deep-Sea Res. II* **42**, 1253–1276.

Watts, L.J. and Owens, N.J.P. (1997) Nitrogen assimilation and the f-ratio in the northwest Indian Ocean during an intermonsoon period. *Deep-Sea Res. II,* **in press.**

Wheeler, P.A. (1993) New production in the sub-Arctic Pacific Ocean: Net changes in nitrate concentration, rates of nitrate assimilation and accumulation of particulate nitrogen. *Progr. Oceanogr.* **32**, 137–161.

Whittaker, R.H. and Likens, G.E. (1975) The biosphere and man. In: *Primary Productivity of the Biosphere.* (eds H. Lieth and R.H. Whittaker) Springer Verlag, Berlin, pp. 305–328.

Wilkerson, F.P. and Dugdale, R.C. (1992) Measurements of nitrogen productivity in the Equatorial Pacific. *J. Geophys. Res.* **97**, 669–679.

Archaeological reconstruction using stable isotopes

A.M. Pollard

17.1 Introduction

Archaeological reconstruction – the inference of past human behaviour from a study of material remains – has long employed the chemical investigation of artefacts as a key element of methodology, and more recently this has included both light and heavy isotopic measurements. This emerging research area has been called 'isotope archaeology' (e.g., Stos-Gale, 1995), to emphasise the parallel which exists between it and the well-established subject of isotope geology. Table 17.1 lists some of the areas where isotopic measurements have been used in archaeology, excluding those principally employed as dating techniques. This chapter focusses on the use of isotopes in human palaeoecology. The first section considers the importance of isotope palaeoclimatology in archaeology, as a means of establishing the 'backdrop' against which human society has developed, although most would say that environment and society have been far more interactive than is implied by this in the (geologically) recent past. The next section reviews the dietary interpretation of isotopic measurements made on human bone collagen, bone mineral and dental enamel – a large area of research, from which it is only possible to select certain key features. The review ends with some more recent studies aimed at reconstructing human migration patterns using stable isotopes plus the measurement of $^{87}Sr/^{86}Sr$ in bone mineral and dental enamel.

17.2 Isotope archaeology

17.2.1 *Historical background*

By the mid-19th century, it had been suggested that correlations in the elemental concentrations of inorganic archaeological artefacts (primarily metals and pottery) could

Stable Isotopes, edited by H. Griffiths.
© 1998 BIOS Scientific Publishers Ltd, Oxford.

Table 17.1. Isotope systems used archaeologically, other than those used primarily for dating (compiled from Ambrose (1993) and Stos-Gale (1995), with additions)

Element	Isotope ratio	Natural isotopic abundance (%)	Substances studied	Application
Hydrogen	$^2H/^1H$	$^1H = 99.985$ $^2H\,(=D) = 0.015$	Water, organic matter (cellulose, collagen, lipids, chitin, peat)	Climate, plant water metabolism
Carbon	$^{13}C/^{12}C$	$^{12}C = 98.89$ $^{13}C = 1.11$	Organic matter, carbonates, biomineralised tissue, soil, CO_2	Diet, plant water-use efficiency, climate and habitat, provenance (ivory, marble)
Nitrogen	$^{15}N/^{14}N$	$^{14}N = 99.633$ $^{15}N = 0.366$	Organic matter, soil, dissolved NO_3^- and NH_4^+, groundwater	Diet, nitrogen fixation pathways, animal water use, climate, groundwater pollution
Oxygen	$^{18}O/^{16}O$	$^{16}O = 99.759$ $^{17}O = 0.037$ $^{18}O = 0.204$	Water, biomineralised carbonates and phosphates, sedimentary phosphates and carbonates, silicates, organic matter	Climate, plant and animal water metabolism, ocean temperature, provenance (marble), chronostratigraphy
Sulphur	$^{34}S/^{32}S$	$^{32}S = 95.00$ $^{33}S = 0.76$ $^{34}S = 4.22$ $^{36}S = 0.014$	Organic matter, hydrocarbons, sulfates, sediments	Diet, pollution
Strontium	$^{87}Sr/^{86}Sr$	$^{84}Sr = 0.56$ $^{86}Sr = 9.86$ $^{87}Sr = 7.02$ $^{88}Sr = 82.56$	Bone and dental mineral, sediments, seawater	Diet, trophic levels, provenance, dating
Lead	$^{208}Pb/^{206}Pb$, $^{207}Pb/^{206}Pb$, $^{206}Pb/^{204}Pb$	$^{204}Pb = 1.4$ $^{206}Pb = 24.1$ $^{207}Pb = 22.1$ $^{208}Pb = 52.4$	Geological material, metals, silicates, biominerals, environmental samples	Provenance, pollution sourcing

Notes: Isotopes in **bold** are radiogenic
The isotope ratios given for lead are those usually reported in the archaeological literature, and differ from those reported geologically

be used to identify the source of these archaeological materials (the postulate of provenance), and even to provide relative dates for manufacture and use (Pollard and Heron, 1996). It was natural, therefore, that when mass spectrometry became available, this in turn should be applied to archaeological material. Thus, in the 1960s and 70s, the pioneering work of Brill established the use of lead isotope determinations (measuring ^{204}Pb, ^{206}Pb, ^{207}Pb and ^{208}Pb as three ratios) to address questions of provenance, initially of lead-containing glass, but later and much more usefully for objects of silver, lead and copper (Brill and Wampler, 1967). More recent studies have extended the list of materials susceptible to isotopic provenancing to include marble, obsidian, and gypsum using other isotope systems such as $\delta^{13}C$, $\delta^{18}O$ and $^{87}Sr/^{86}Sr$ (Stos-Gale, 1995). Although controversy has occasionally broken out about how some of these data (particularly lead isotopes) may be interpreted archaeologically (e.g., Budd *et al.*, 1993, 1995), such applications of isotopic methods are still very much valued and likely to increase in future.

17.2.2 *Isotope palaeoclimatology and archaeology*

At about the same time as the work of Brill, archaeologists were being introduced to the potential applications of the lighter stable isotopes – initially the use of oxygen isotopes for temperature reconstruction from marine cores (Emiliani, 1969) and marine molluscs (Shackleton, 1969). This archaeological interest in isotopic climatology has continued to the present day, and is one of the growing areas of interaction between archaeologists and other palaeoscientists (see Chapters 16, 20, 21 and 22). Of particular interest archaeologically has been the development of dendroclimatology and isotope dendroclimatology, for the obvious reason that climatic signals derived from trees are more likely to reflect the conditions directly experienced by past human societies, rather than the perhaps more remote (in human terms) record in the marine sediments. The fact that the dendroclimatalogical record is essentially Holocene, and potentially offers an annual scale of climatic reconstruction, also makes it far more relevant to the majority of archaeologists in their task of interpreting and explaining the post-glacial development of the human species. Recent advances in isotope dendroclimatology have been reviewed, for example by Trimborn *et al.* (1995) on Black Forest silver fir (*Abies alba*) and Switsur *et al.* (1995) on English oak (*Quercus robur*), and are also presented elsewhere in this volume (Chapter 18).

  This potential annual resolution is of particular importance when considering the possible impact on human society of large scale climatic events such as those which might have followed major volcanic activity, although dendro-isotopic studies through the relevant 'narrow ring' events associated with volcanic activity have yet to be fully published. Baillie (1995) has postulated a very strong link between presumed climatic downturns following major volcanic events and significant events in human history, such as the end of the Minoan civilisation on Crete in the middle of the second millennium BC, tentatively (and controversially – see Baillie and Munro, 1988) linked with a volcanic event identified as Santorini in 1628 BC. Preliminary measurements of $\delta^{13}C$ and δD have been made through the narrow ring sequences dated to 1628 BC in two Irish oak trees (Baillie, Pollard and Ramesh, unpublished data). They are difficult to interpret because of the relatively coarse sampling strategy – ten year blocks from 1650 – 1590 BC for one tree (Q5392), and ten year blocks covering 1650 – 1630 BC and 1620 – 1590 BC, but with two year blocks between 1630 and 1620 BC for another (Q1276) (see Figure 17.1). Clearly, a higher temporal resolution is needed, but one tree at least (Q5392) shows a significant increase in δD through this decade which might indicate a change in climatic conditions. As always, further work is required, and analytical techniques are improving all the time.

  To date, most terrestrial isotope palaeoclimatology studies have been carried out by measuring δD, $\delta^{13}C$ or $\delta^{18}O$ in peat cores, lake sediments or tree-rings (see Chapters 18 and 22). A relatively recent development has been the attempt to interpret stable oxygen isotope data from fossil mammalian bone and tooth enamel in terms of climatic evidence. Measurements of $\delta^{18}O$ in modern human bone phosphate mineral have yielded convincing positive correlations with $\delta^{18}O$ in local meteoric water (Longinelli, 1995), but environmental temperature reconstructions from similar measurements in medieval

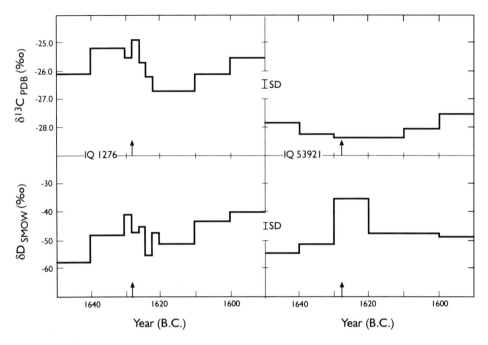

Figure 17.1. *Measurements of δ¹³C and δD from the cellulose of two Irish oak trees growing between 1650 and 1590 BC, showing narrow ring features identified with the volcanic eruption of Santorini (unpublished data from Baillie, Pollard and Ramesh).*

human tooth enamel from Greenland (Fricke *et al.*, 1995) have proved more controversial. Fricke and co-workers observed a reduction in δ¹⁸O of 3‰ interpreted as a temperature downturn corresponding to 'the Little Ice Age' in enamel samples dating to *c.* 1400 AD from the Julianehaab Bay region of Greenland. Bryant and Froelich (1996) noted a lack of correspondence between these data and the equivalent measurements from the adjacent Dye 3 ice core, concluding that the variation observed in the time series of human dental enamel isotopic measurements may have a physiological or social explanation rather than the climatic model put forward by Fricke *et al.* (1995).

It has recently been reported that δ¹³C measurements on Holocene bone collagen, charcoal and wood samples from across Europe show a trend from north to south and east to west which can be linked to latitudinal differences and oceanic influences on climate (van Klinken *et al.*, 1994). In bone collagen, for example, the average value of δ¹³C for archaeological samples from Spain is -18.9 ‰ (vs. V-PDB) compared with -20.8‰ for Sweden. Similar trends are reported for charcoal. Although somewhat preliminary, these observations are based on large data sets accumulated during the routine measurement of δ¹³C for the correction of radiocarbon dates, and clearly warrant further investigation. Interestingly, the same publication shows plots of δ¹³C from bone collagen, charcoal and wood against modern day mean July temperatures, sunshine, humidity and precipitation which show significant correlations (e.g., $r = 0.84$ between bone collagen δ¹³C and total July sunshine), which appear by eye to be as good as (if not better than) those reported elsewhere from tree-rings! This demonstrates the wealth of untapped proxy palaeoclimatic data which resides in various archaeological data banks, and the need to develop analytical techniques. Most recently, the development and application of compound-specific isotope analysis (GCC-IRMS: Chapters 1, 2 and 22) have provided more detailed insights into human palaeo-dietary preferences, whether from $\delta^{13}C$ of lipid residues in potsherds (Evershed et al., 1994) or bone collagen and tooth enamel (Stott and Evershed, 1996; Stott et al., 1997).

17.3 Dietary reconstruction using stable isotopes

Although palaeoclimatic reconstruction is of great importance in archaeology, the most immediate impact of 'isotope archaeology' has been in the area of human dietary reconstruction. The well-known saying 'you are what you eat' has been taken literally in archaeology for the last 20 years, and dietary reconstruction has been attempted using trace element levels in bone mineral and stable isotope studies on collagen and mineral in both bone and teeth (reviewed by, for example, Pate, 1994). It soon became apparent that inorganic trace element studies were potentially bedevilled by post-mortem diagenetic effects, the magnitude and significance of which continue to be debated (e.g., Sandford, 1993; Radosevich, 1993). Isotopic studies have been far less controversial and, for Holocene material at least, appear to avoid most of the diagenetic problems encountered with trace elements. Most authors conclude that if collagen survives, then the isotopic signal measured is unchanged from that which would have been measured in vivo. Collagen is conventionally deemed to have 'survived' if it has an atomic C/N ratio of between 2.9 and 3.6, consistent with measurements on modern bone collagen (Ambrose, 1990; DeNiro, 1985; Nelson et al., 1986), although some authors (e.g., Katzenberg et al., 1995) have carried out amino acid analysis on extracted protein to demonstrate the efficacy of their procedures. Reviews of dietary reconstruction using isotopic techniques include those of Ambrose (1993), DeNiro (1987), Schwarcz and Schoeninger (1991) and van der Merwe (1992) and see also Chapter 21.

17.3.1 North American human ecology

By the late 1970s it had been established that changes in the carbon isotopic signal at the base of the food web were indeed reflected in the signal obtained from ancient human bone collagen (Vogel and van der Merwe, 1977). In a classic study of the spread of maize (Zea mays) agriculture from Central America into eastern and central North America in pre-Columbian times, van der Merwe and Vogel (1977) demonstrated its arrival in the lower Illinois Valley by the end of the Late Woodland period (c. 1000 AD: see Figure 17.2). This conclusion was possible because maize is a C_4 plant with a $\delta^{13}C$ value of typically -15‰, compared to -25‰ in C_3 plants (see Chapter 8). Established archaeological opinion prior to this result was that maize had been introduced into this area by 400 AD or perhaps earlier (van der Merwe, 1992). This piece of work is now regarded as one of the triumphs of modern scientific archaeology. Because of the essentially ephemeral nature of the direct evidence for agricultural practices, it actually helped to solve a question which, on archaeological grounds alone, was relatively intractable, and established isotopic human palaeodietary studies as a legitimate research area.

The methodology, extended to include measurements of $\delta^{15}N$, was rapidly applied to

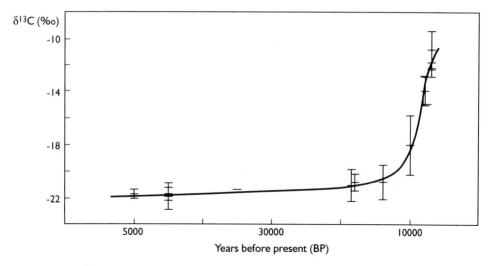

Figure 17.2. *Changes in the δ¹³C values in human bone collagen showing the adoption of maize agriculture in the North American Woodlands after 800 AD (van der Merwe, 1992; © The British Academy 1992. Reproduced by permission from* Proceedings of the British Academy, Vol. 77, New Developments in Archaeological Science).

bone collagen from elsewhere in the world to study of human (palaeo-)ecosystems, with varying degrees of success. The most comprehensive work to date has been carried out in southern and eastern Africa, where climatic conditions (resulting in a predictable mix of C_3 and C_4 plants at the base of the food web) and, in some cases, a hunter–gatherer lifestyle, have combined to give what might be regarded as an ideal system to study using this approach (Ambrose and DeNiro, 1986; Sealy and van der Merwe, 1985). This success can be attributed in part of the suitability of the methodology to the ecosystem, but also to the practice of attempting to reconstruct as many levels of the food web as possible, by analysing bone and tissue from a wide range of plants and animals with known or assumed predator–prey relationships. In less complex societies, it appears possible to simply regard humans as the top predators in the ecosystem, and apply the techniques of isotope ecology used elsewhere in zoology (Wada *et al.*, 1991).

Several isotopic studies of bone collagen have been carried out in the USA since the original work of van der Merwe and his colleagues. In combination with pollen analysis and coprolite analysis, Matson and Chisholm (1991) demonstrated the spread of maize subsistence during the Basketmaker II phase (*c.* 1st century BC) to the Cedar Mesa area of Utah, in contradiction to the previously expected date of 800–1000 AD. Buikstra and Milner (1991) investigated the degree of dependence of humans on maize in the central Mississippi and Illinois river valleys between 1000 and 1400 AD, showing a complex pattern of regional and temporal variation. Schurr (1992) made isotopic measurements on bone in order to elucidate the relationship between social position (as inferred from burial practice) and diet in the Middle Mississippian (1200 – 1450 AD) of eastern North America. Perhaps the most comprehensive recent study has been that of Larsen *et al.* (1992), who investigated human remains covering four prehistoric (range 1000 BC – 1450 AD) and two historic periods from the Georgia Bight, south-eastern coastal US. Less negative δ¹³C values were interpreted as increased utilisation of maize during the prehistoric period. This was confirmed by an accompanying decrease in $\delta^{15}N$ values, which allowed the authors to discount an alternative possible explanation for this phenomenon – the increased exploitation of marine resource exploitation. Further evidence for increasing maize consumption was afforded by the observation of increased dental caries – maize, being high in sugar, is more cariogenic in general than C_3 diets. Interestingly, however, one site (Irene) dated to the very end of the prehistoric period (1300 – 1450 AD) showed a marked decrease in $\delta^{13}C$ values, interpreted as a reversion to C_3 subsistence in the late agricultural phase. This is tied in with the archaeological evidence for a marked decline in social organisation in Mississippian cultures in the early fifteenth century – a phenomenon linked with climatic deterioration and a reduction in rainfall in the region.

Katzenberg et al. (1995) considered the isotopic evidence ($\delta^{13}C$, $\delta^{15}N$) for maize agriculture in collagen from human bone recovered in southern Ontario, Canada, dating to the period 400 – 1500 AD. They concluded that there was evidence for a gradual increase in the importance of maize in the diet from 650 – 1250 AD. Perhaps more importantly, they compiled published $\delta^{13}C$ data from human bone collagen from north-eastern North America (the Mississippi, Illinois and Ohio valleys, and the Great Lakes region) between 500 and 1300 AD (about 130 measurements). This shows $\delta^{13}C$ becoming less negative from around -20‰ in 600 AD up to around -10‰ by 1200 AD, but with regional variations which are counter-intuitive to the simple model for the spread of maize agriculture from the south. Southern states such as Missouri, Arkansas and Tennessee show little increase in maize utilisation until near the end of the period in question, whereas Illinois shows early and continued exploitation of maize. This evidence, combined with that provided by Larsen et al. (1992) for a reversion to more traditional agricultural resources during times of social or environmental stress, serves to underscore the complexity of reconstructing human palaeoecology, but also the value of stable isotope measurements in such studies.

17.3.2 South and Central America

Given the interest in the spread of maize into North America, it is perhaps surprising how relatively little dietary reconstruction has been done using bone collagen from Central America, and the lack of parallel studies of the spread of maize southwards into South America. White and Schwarcz (1989) measured trace elements in bone mineral and stable isotope values in human bone collagen from the Lowland Maya site of Lamanai, Belize, covering a date range of pre-Classic (1250 BC – 250 AD) to Historic (1520 – 1670 AD). From the $\delta^{13}C$ values of the collagen they estimated a dietary maize input of 50% at the beginning of the pre-Classic period, falling to 37% at the end of the this period, rising to 70% in post-Classic times. The observed decline in $\delta^{13}C$ (and an associated decline in the incidence of dental caries) at the end of the pre-Classic is taken to suggest a reduction in maize production, perhaps as a result of social or climatic change. $\delta^{15}N$ is constant in human collagen (at around +10‰) throughout the periods studied, with the exception of one individual – a man, buried together with a female in a high status Early Classic (250–400 AD) tomb burial. His $\delta^{15}N$ value of +13.2‰ was taken to indicate a higher consumption of imported coastal dietary resources by the elite.

In South America, van der Merwe et al. (1981) established that maize was a significant dietary component along the Orinoco River in Venezuela between 800 BC and 400 AD, in contrast to the previously held assumptions about tropical Amazonian ecosystems. These authors concluded that by around 400 AD the diet consisted of 80% or more of maize, and that this shift in subsistence strategy allowed a 15-fold increase in population density. This conclusion was based on an isotopic shift in $\delta^{13}C$ from -26‰ in pre-maize humans to -10‰ in those exploiting maize. The unusually negative average value of -26‰ (compared with -21.5‰ for prehistoric North American humans with a C_3 diet) was attributed to a depletion in ^{13}C as a result of carbon recycling in the tropical forest ecosystem (van der Merwe, 1992; but see Chapter 8). On the other side of the Andes, Ubelaker *et al.* (1995) used isotope studies to assess the dietary relationship between high status individuals and those sacrificed in the same tombs. The remains were from six shaft tombs excavated at La Florida, Quito, Ecuador, and dated to the Chaupicruz phase(*c.* 100 – 450 AD). Although ethnographic evidence from the 16th C. AD suggested that there might be a higher consumption of animal protein by the elite, this was not observed in the isotopic data. The only difference suggested analytically was a higher consumption of maize by the elite class (average $\delta^{13}C$ = -10.3‰ in the elite group, -11.6‰ in those of lower rank), attributed to a higher consumption of *chicha* (maize beer) by the elite.

17.3.4 *Prehistoric Europe*

Away from subsistence regimes with a terrestrial C_4 plant component, progress has been somewhat less spectacular but will perhaps advance rapidly with the development of GCC-IRMS techniques (see Chapters 1 and 2). Isotopic work carried out on human collagen from prehistoric Europe has focused on detecting differences in the balance between terrestrial C_3 subsistence and exploitation of marine resources, either temporally or geographically. The first work of this kind was Tauber (1981), who measured $\delta^{13}C$ in human remains dating from the Mesolithic (*c.* 5200–4000 BC) through to the end of the Iron Age (*c.* 1000 AD) in Denmark, and some historic Innuit samples from Greenland. This work showed a clear transition from the hunter–gatherer lifestyle of the Mesolithic ($\delta^{13}C$ approximately -11 to -15‰) to the agricultural lifestyle of the Neolithic and later populations ($\delta^{13}C$ approximately -13 to -27‰). In the absence of any C_4 terrestrial dietary component, the unusually high $\delta^{13}C$ in the Mesolithic is attributed to a subsistence pattern heavily dependent on marine resources. This is supported by similar figures for three pre-contact Innuit from Greenland, with an assumed diet dominated by marine food. In a similar study involving measurements of $\delta^{13}C$ and $\delta^{15}N$, combined with dental evidence, Lubell *et al.* (1994) demonstrated a gradual shift from marine exploitation to terrestrial food resources in Portugal at the Mesolithic/Neolithic transition – here dated to around 5000 BC. In this case $\delta^{13}C$ decreases from a range of -15 ‰ to -19‰ in the Mesolithic to between -19 and -20.5‰ in the Neolithic. The interpretation of this as being due to reduced marine input is supported by a concomitant change in $\delta^{15}N$ from greater than +10‰ in the Mesolithic to around +9‰ in the Neolithic.

Some work carried out on British material during the 1980s was aimed at demonstrating differences between two contrasting communities: Late Neolithic/Early Bronze Age Wessex (*c.* 3000–2000 BC) in southern England, representing an assumed inland (i.e. terrestrial C_3) population (Antoine *et al.*, 1988a; Pollard *et al.*, 1991), and Neolithic Orkney (a group of islands off the north coast of Scotland), with an expected marine-influenced diet (*c.* 3000 BC; Antoine *et al.*, 1988 a,b). This yielded significant isotopic differences between these groups, as summarised in Table 17.2. As

Table 17.2. *Summary of isotopic differences in human bone collagen between sites in Neolithic Orkney and Neolithic/Early Bronze Age Wessex (Antoine et al., 1988a, b; Pollard et al. 1991).*

	$\delta^{13}C(\%o)$	$\delta^{15}N(\%o)$
South of England		
Shrewton ($n = 19$)	-21.9 ± 0.8	4.6 ± 2.8
Wor Barrow ($n = 6$)	-22.0 ± 0.7	7.0 ± 0.4
Hambledon Hill ($n=25$)	-21.3 ± 2.6	6.7 ± 4.1
Fussell's Lodge ($n = 95$)	-23.2 ± 1.3	12.2 ± 3.6
Irthlingborough ($n = 70$)	-22.9 ± 1.1	11.6 ± 3.3
Orkney		
Isbister Cairn ($n = 30$)	-20.7 ± 0.6	10.6 ± 2.1
Holm of Papa Westray		
($n = 22$)	-20.4 ± 1.2	9.7 ± 3.5
Quanterness		
($n = 45\ \delta^{13}C, n = 55\ \delta^{15}N$)	-20.4 ± 0.6	8.3 ± 2.9
Point of Cott, Westray ($n = 10$)	-20.1 ± 0.3	9.0 ± 3.9

would be predicted, the data do not show the large differences in $\delta^{13}C$ which accompanied the Mesolithic/Neolithic transition, but there is still a significant difference between average $\delta^{13}C$ values for Orkney (-20.5‰) and southern England (overall figure -22.6‰). The $\delta^{15}N$ values for Orkney (+9.2‰) are higher than most of the English sites studied (average for Shrewton, Hambledon Hill and Wor Barrow +5.9‰), but, completely unexpectedly, two English sites (Fussell's Lodge and Irthlingborough) have a higher $\delta^{15}N$ value (average +11.9‰) than that from Orkney (see Figure 17.3). Although the authors have expressed some concern over the quality of these nitrogen isotope measurements (Pollard *et al.* 1991), it is noteworthy that the anomalous nitrogen figures are accompanied by more negative mean $\delta^{13}C$ values (average for these two sites -23.0‰, compared to -21.6‰ for the other English sites). The data were obtained from a wide range of skeletal elements, and there may also be some systematic variation in the isotope signal between different bones which has contributed to the larger than expected ranges – this point is being further investigated. Irthlingborough is geographically separate from the other sites, being situated near Peterborough in the east of England, whereas the other English sites (including the apparently isotopically similar Fussell's Lodge) are located in Wiltshire and Dorset. It is also an exceptional group of Late Neolithic/Early Bronze Age (Beaker) burials, because one of the burial chambers was found to be covered by the skulls of more than 100 wild and domesticated oxen (Halpin, 1987). However, as yet no simple explanation for these isotopic differences has emerged and confirmation of these unusual signatures is awaited. The effect of this large number of analyses on the overall picture for prehistoric Europe can be seen by plotting these British values onto the well-known diagram of DeNiro (1987; see Figure 17.4). This work at least has succeeded in demonstrating that isotopic differences in bone collagen do exist between groups of humans expected to have substantially different diets, despite the fact that the C_3/C_4 isotopic shift so central to New World isotopic human ecology is not present in prehistoric Europe.

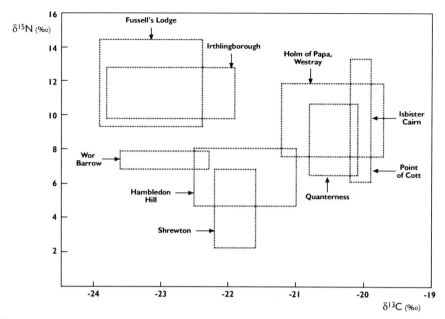

Figure 17.3. $\delta^{13}C$ and $\delta^{15}N$ in human collagen from various prehistoric British sites plotted as interquartile ranges around the medians (from Pollard, 1993).

17.3.5 *Carbon isotopes in mineralised tissue*

The vast majority of the work described above was carried out on collagen extracted from human bone – strictly, on the acid-insoluble organic matter contained in bone: there has been some debate about whether this is entirely collagen, and, in archaeological material, it is highly likely that it is not (Ambrose, 1990). In order to extend the method into 'deep time' (i.e., beyond the time over which collagen survives in bone), attention has been paid to the measurement of $\delta^{13}C$ in the carbonate incorporated into mineralised tissue, and especially dental enamel, since this is usually the most resilient mammalian tissue. This has allowed estimation of diet from $\delta^{13}C$ in a wide range of fossil animals (and to reconstruct the emergence of C_4 plants in Pakistan and Kenya between 10 and 15 Ma: Morgan *et al.*, 1994; see Chapter 21), but, more important archaeologically, in a series of fossil hominids stretching back to *Australopithecus robustus* at Swartkrans (Lee-Thorp *et al.*, 1994), dating to the range 1.8 – 1 Ma. Comparison of $\delta^{13}C$ in the dental enamel from eight hominid samples (average value - 8.5‰ ± 1.0‰) with measurements from the teeth of known C_4 grazers (average value *c.* 0‰) and C_3 browsers (average -11‰ to -12‰) from Swartkrans suggested that the Australopithecines had a significant component (25%–30%) of C_4 plant material in their diet, either obtained by eating C_4 grasses or, more likely, from consuming grazing animals. This suggest an omnivorous diet, in contradiction to the previously-held belief (based on dental studies) that Australopithecines were vegetarian. This shows that isotopic studies on fossil dental enamel have a significant role to play in understanding human evolution.

   In recent years it has become common, even in studies of human bone where collagen survives, to compare $\delta^{13}C$ values in collagen with those obtained from the bone

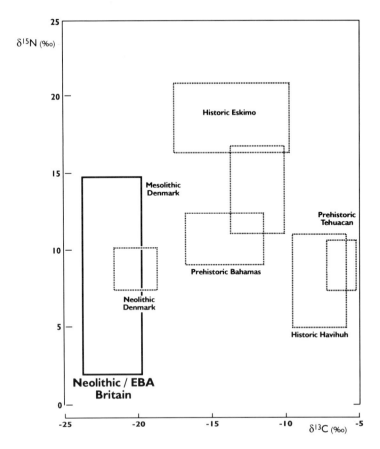

Figure 17.4. Approximate interquartile range of $\delta^{13}C$ and $\delta^{15}N$ in human collagen from various prehistoric British sites compared to the summary published by DeNiro (1987).

mineral (variously described as apatite, hydroxypatite, bioapatite, dahlite, carbonate apatite, etc.). Sullivan and Kreuger (1981) first demonstrated that there was a systematic difference between $\delta^{13}C$ values in the organic and mineral fraction of bone. They calculated this to be an average of +7.98‰ (apatite minus 'gelatin'), and observed that the isotope enrichment between diet and collagen was approximately +5‰ (tissue minus diet) for pure C_3 consumers and +6‰ for C_4 grazers. Since in each case the apatite is further enriched by +8‰, they were able to conclude that $\delta^{13}C$ from apatite was also highly correlated with dietary values. This relatively simple picture has been somewhat complicated by questions of 'isotope routing', whereby consideration is given to how the various tissues are biosynthesised (Schwarcz, 1991). Collagen is made up of amino acids, some of which are 'essential', in that they are not biosynthesised and are incorporated directly from the diet, and may therefore have a 'non-dietary' isotope signal. The non-essential amino acids are biosynthesised, and should be in isotopic equilibrium with blood glucose. The overall isotopic signal in collagen is therefore a weighted average of that from the essential and the non-essential amino acids, and might in some circumstances give a biased view of dietary inputs (see Chapter 2). However, bone and tooth mineral carbonate should be in isotopic equilibrium with dissolved blood CO_2. This has led to suggestions that the mineral $\delta^{13}C$ values in bone should reflect that in the dietary carbohydrate, whilst $\delta^{13}C$ in collagen represents that of the protein source (Kreuger and Sullivan, 1984). Experimental $\delta^{13}C$ studies using laboratory rats on isotopically controlled diets have suggested that dietary protein is indeed routed to collagen, but that the isotope enrichment between diet and collagen varies considerably if the protein and non-protein components of the diet have significantly different values (Ambrose and Norr, 1993). These authors conclude that the collagen isotope signal may substantially underestimate the non-protein component of the diet. This is not the case with bone carbonate values, which are said to give reliable estimates of the isotopic composition of the whole diet. From this it can be seen that the recent trend towards improved measurement techniques, such as routine measurement of both mineral and collagen $\delta^{13}C$ values, and the application of GCC-IRMS to separate individual amino acids and lipids from human collagen, are likely to have an increasing impact on these debates, and undoubtedly represent the key to future developments.

As was mentioned above (Section 17.2.2) the analysis of specific lipid fractions using GCC-IRMS has already been used for more precise palaeodietary reconstruction. The analysis of plant and animal residues in unglazed cookery pots has led to the identification of epicuticular waxes from *Brassica* and ketones from animal fats (Evershed *et al.*, 1994; 1995a). In addition, the $\delta^{13}C$ signal of cholesterol (and associated diagenetic products) has provided complimentary analysis to that of collagen in whole fossils (Stott *et al.*, 1997), validating the distinction between marine and terrestrial food sources made using this method (Evershed *et al.*, 1995b; Stott and Evershed, 1996).

17.4 Isotopes and human mobility

It would be inappropriate to leave a review such as this without discussing briefly the important new isotopic methodologies now being brought to bear on the important question of human mobility in antiquity. As with dietary studies, this methodology has largely grown out of isotopic studies of other mammals – in this case, the need to trace the movement of elephants in Africa, and to be able to provenance traded ivory (Koch *et al.*, 1995; van der Merwe *et al.*, 1990; Vogel *et al.*, 1990). A key factor in this approach is the knowledge that different elements of the mammalian skeleton represent different periods during the life history of an individual. Bone tissue, which, because of metabolic turnover, is constantly remodelled, will reflect the latter part of an individual's life – the exact period is not known, and will vary from bone to bone depending on the rate of turnover, which in turn depends on age, but in humans is likely to range up to 10 years (Sealy *et al.*, 1995). Human dental tissue is formed early in life, and is essentially unaltered during subsequent life, especially the enamel. Stable isotope measurements on enamel and dentine are therefore likely to reflect environmental factors relevant to the early stages of human life (perhaps the first 20 years), whereas the same measurements on bone mineral and collagen will be determined by conditions in the last few years of life. Any significant change in diet or environment may be expected to show up as an isotopic difference. A significant addition to the methodology has been the measurement of $^{87}Sr/^{86}Sr$ ratio in mineralised tissue, as first suggested by Ericson (1985). This ratio is characteristic of the local underlying geology, and passes unmodified through the food chain. Migrations which result in the last years of life being spent in a region of different geology to that in which the early years were spent should show up in a different strontium isotope ratio.

Sealy *et al.* (1995) reported the measurement of $\delta^{13}C$, and $\delta^{15}N$ in dentine, enamel, rib and femur (or humerus) from five individuals excavated in South Africa. Two were pre-historic Khoisan hunter–gatherers, whose isotopic signatures showed similar values in all tissues, suggesting no significant movement during their lifetime. Another one was an adult male from Cape Town Fort, probably a European who died at the Cape during the 17th C. AD. His isotopes also showed no variation, and it was assumed that he was newly-arrived in the Cape when he died. A fourth individual, another European from the Dutch East India Company station at Oudespost (occupied between 1673 and 1732 AD), did show significant differences between dental and bone measurements – less negative $\delta^{13}C$ values and higher $\delta^{15}N$ figures – a pattern attributed to greater consump-tion of sea food in adult life. His strontium isotope ratio was similar for all tissues, and indicated a coastal habitation throughout life. The fifth individual was an African women assumed to be a slave, and dating to the 18th C. AD. She showed a marked shift in all ratios measured. In early life through to early adulthood she showed $\delta^{13}C$ and $\delta^{15}N$ measurements consistent with a C_4 biome – interpreted as a tropical or sub-trop-ical inland area. At the end of her life, she had increased $\delta^{15}N$ values consistent with a more marine diet. Her strontium ratios dropped from 0.732 in canine dentine to 0.717 in the rib, suggesting a move to a coastal environment, consistent with where she was found and the increase in seafood input. It would appear that the whole of an individ-ual's life history is isotopically encoded in the skeleton.

Perhaps more significant archaeologically is the work of Price *et al.* (1994) which focussed purely on strontium isotope ratios in teeth and bone from eight individuals from Bell Beaker burials in Bavaria. The 'Beaker folk' are traditionally seen as a group of people identified by their burial practices (including the characteristic ceramic 'beakers') who spread through central and western Europe at the end of the Neolithic (*c.* 2500 BC), and who may have brought with them the knowledge of metalworking. What is not clear archaeologically is whether the 'beaker folk' themselves moved, or whether 'beaker ideas' moved and were simply adopted by indigenous populations. Hence the interest in an isotopic method of detecting the movement of individuals or groups of individuals. Price *et al.* plotted total Sr concentration in the bone and teeth against the $^{87}Sr/^{86}Sr$ ratio. Of the eight individuals measured, two were claimed to show clear evidence of having moved significantly between adulthood and childhood. This pilot study was felt to at least show that the method had promise in resolving a long-standing archaeological question. Inherent in all these methodologies which measure either absolute Sr levels or strontium isotope ratios in mineralised tissue is the assumption that diagenesis has not altered the signal since death. This is a matter of some debate (e.g., Nelson *et al.*, 1986), but the consensus of current opinion is that the repeated acid washing procedures used removes any diagenetic mineral because it has a higher solubility than biomineralised apatite (Sillen and Sealy, 1995). This seems to be a crucial point common to several palaeobiological studies, and one which deserves further examination.

17.5 Summary

It is clear from the above that isotopic methods over the last 20 years have been increasingly used to answer questions about prehistoric human environments, diets, and, in the last few years, migrations. Many of the methods used are common with those employed elsewhere in palaeobiology, but interpretation is always more difficult in the archaeological context because of the additional factor present in human ecology – the social or 'ritual' dimension. It might be argued that such endeavours are essentially the application of established techniques of isotope biogeochemistry to questions of archaeological interest, and, as such, no more than an interesting diversion. This is not the case. Apart from the unique scientific challenges provided by attempting to solve questions of past human behaviour from a study of material remains, there is a considerable contribution to be made by archaeology to the other palaeosciences. Throughout the Holocene, and often back into the Late Glacial, archaeological sequences provide a wealth of material with exceptional chronological control. This material is not just artefactual – stone tools, ceramics, metals, etc. – it is rich in faunal assemblages and sedimentary sequences containing a wide variety of environmental information. It has all been carefully excavated and reported, but it is not always published in the accessible literature. Vast amounts of information reside in site archives, but, perhaps more importantly, these archives also contain physical remains of great scientific potential. To take an obvious example, insect assemblages exist from throughout the Holocene, from well-defined geographical and chronological locations, which may provide additional climatic information by stable isotopic examination of their chitin. The potential therefore exists to add significantly to the post-glacial climatic picture, by exploiting in new ways already existing material. Within archaeology, 'isotope archaeology' is now well established as a valuable addition to the scientific armoury of problem-solving techniques. The archaeological resource is yet to be exploited to its full potential for historical environmental and ecological reconstructions.

References

Ambrose, S.H. (1990) Preparation and characterization of bone and tooth collagen for isotopic analysis. *J. Archaeological Sci.* **17**, 431–451.

Ambrose, S.H. (1993) Isotopic analysis of paleodiets: methodological and interpretative considerations. In *Investigations of Ancient Human Tissue* (ed M.K. Sandford). Gordon and Breach Science Publishers, Langhorne, Pennsylvania, USA, pp. 59–130.

Ambrose, S.H. and DeNiro, M.J. (1986) Reconstruction of African human diet using bone-collagen carbon and nitrogen isotope ratios. *Nature* **319**, 321–324.

Ambrose, S.H. and Norr, L. (1993) Experimental evidence for the relationship of the carbon isotope ratios of whole diet and dietary protein to those of bone collagen and carbonate. In *Molecular Archaeology of Prehistoric Human Bone* (eds J. Lambert and G. Grupe), Springer, Berlin, pp. 1–43.

Antoine, S.E., Dresser, P.Q., Pollard A.M. and Whittle A.W.R. (1988a) Bone chemistry and dietary reconstruction in Prehistoric Britain: examples from Wessex. In *Science and Archaeology Glasgow 1987* (eds E.A. Slater and J.O. Tate). British Archaeological Reports British Series Vol. 196, Oxford, pp. 369 – 380.

Antoine, S.E., Pollard, A.M., Dresser, P.Q. and Whittle A.W.R. (1988b) Bone chemistry and dietary reconstruction in Prehistoric Britain: examples from Orkney, Scotland. In *Proceedings of the 26th International Archaeometry Symposium, Toronto* (eds R.M. Farquhar, R.G.V. Hancock and L.A. Pavlish), Archaeometry Laboratory, Department of Physics, University of Toronto, Toronto, pp. 101 – 106.

Baillie, M.G.L. (1995) *A Slice Through Time*. B.T. Batsford Ltd., London.

Baillie, M.G.L. and Munro, M.A.R. (1988) Irish tree-rings, Santorini and volcanic dust veils. *Nature* 332, 344–346.

Brill, R.H. and Wampler, J.M. (1967) Isotope studies of ancient lead. *Am. J. Archaeology* 71, 63–77.

Bryant, J.D., Froelich, P.N. (1996) Oxygen-isotope composition of human tooth enamel from medieval Greenland – linking climate and society – comment. *Geology* 24, 477–478.

Budd, P., Gale, D., Pollard, A.M., Thomas, R.G. and Williams, P.A. (1993) Evaluating lead isotope data: further observations. *Archaeometry* 35, 241–263.

Budd, P., Haggerty, R., Pollard, A.M., Scaife, B. and Thomas, R.G. (1995) Rethinking the quest for provenance. *Antiquity* 70, 168–174.

Buikstra, J.E. and Milner, G.R. (1991) Isotopic and archaeological interpretations of diet in the central Mississippi valley. *J. Archaeological Sci.* 18, 319–329.

DeNiro, M.J. (1985) Postmortem preservation and alteration of *in vivo* bone collagen isotope ratios in relation to palaeodietary reconstruction. *Nature* 317, 806–809.

DeNiro, M.J. (1987) Stable isotopy and archaeology. *Am. Sci.* 75, 182–191.

Emiliani, C. (1969) The significance of deep-sea cores. In *Science in Archaeology* (eds D. Brothwell and E. Higgs). 2nd edn. Thames and Hudson, London, pp. 109–117.

Ericson, J.E. (1985) Strontium isotope characterization in the study of prehistoric human-ecology. *J. Human Evolution* 14, 503–514.

Evershed, R.P., Arnst, K.I., Collister, J., Eglinton, G. and Charters, S. (1994) Application of isotope ratio monitoring gas chromatography/mass spectrometry to the analysis of organic residues of archaeological interest. *Analyst* 119, 909–914.

Evershed, R.P., Stott, A.W., Raven, A., Dudd, S.N., Charters, S. and Leyden A. (1995a) Formation of long-chain ketones in ancient-pottery vessels by pyrolysis of acyl lipids. *Tetrahedron Lett.* 36, 8875–8878.

Evershed, R.P., Turner-Walker, G., Hedges, R.E.M., Tuross, N. and Leyden A. (1995b) Preliminary results for analysis of lipids in ancient bone. *J. of Archaeological Science* 22, 277–290.

Fricke, H.C., O'Neil, J.R. and Lynnerup, N. (1995) Oxygen-isotope composition of human tooth enamel from medieval Greenland – linking climate and society. *Geology* 23, 869–872.

Halpin, C. (1987) Irthlingborough, *Current Archaeology* No. 106, pp.331–332.

Katzenberg, M.A., Schwarcz, H.P., Knyf, M. and Melbye F.J. (1995) Stable isotope evidence for maize horticulture and paleodiet in southern Ontario, Canada. *Am. Antiquity* 60, 335–350.

Koch, P.L., Heisinger, J., Moss, C., Carlson, R.W., Fogel, M.L. and Behrensmeyer A.K. (1995) Isotopic tracking of change in diet and habitat use in African elephant. *Science* 267, 1340–1343.

Kreuger, H.W. and Sullivan C.H. (1984) Models for carbon isotope fractionation between diet and bone. In *Stable Isotopes in Nutrition* (eds J. Turnland and P. Johnson), American Chemical Society Symposium Series 258, Washington, pp. 205–220.

Larsen, C.S., Schoeninger, M.J., van der Merwe, N.J., Moore, K.M. and Lee-Thorp J.A. (1992) Carbon and nitrogen stable isotopic signatures of human dietary change in the Georgia Bight. *Am. J. Physical Anthropology* 89, 197–214.

Lee-Thorp, J.A., van der Merwe, N.J. and Brain C.K. (1994) Diet of *Australopithecus-robustus* at Swartkrans from stable carbon isotopic analysis. *J. Human Evolution* 27, 361–372.

Longinelli, A. (1995) Stable isotope ratios in phosphate from mammal bone and tooth as climatic indicators. In *Problems of Stable Isotopes in Tree-Rings, Lake Sediments and Peat-bogs as Climatic Evidence for the Holocene* (ed B. Frenzel). European Science Foundation, Strasbourg, pp. 57–70.

Lubell, D., Jackes, M., Schwarcz, H., Knyf, M. and Meiklejohn, C. (1994) The Mesolithic Neolithic transition in Portugal – isotopic and dental evidence of diet. *J. Archaeological Sci.* 21, 201–216.

Matson, R.G. and Chisholm, B. (1991) Basketmaker II subsistence – carbon isotopes and other dietary indicators from Cedar-Mesa, Utah. *Am. Antiquity* 56, 444–459.

Morgan, M.E., Kingston, J.D. and Marino, B.D. (1994) Carbon isotope evidence for the emergence of C_4 plants in the Neogene from Pakistan and Kenya. *Nature* 367, 162–165.

Nelson, B.K., DeNiro, M.J., Schoeninger, M.J., De Paolo, D.J. and Hare P.E. (1986) Effects of diagenesis on strontium, carbon, nitrogen and oxygen concentration and isotopic composition of bone. *Geochim. Cosmochim. Acta* 50, 1941–1949.

Pate, F.D. (1994) Bone chemistry and paleodiet. *J. Archaeological Method Theory*, 1, 161–209.

Pollard, A.M. (1993) Tales told by dry bones. *Chemistry and Industry* 10, 359–362.

Pollard, A.M., Antoine, S.E., Dresser, P.Q. and Whittle A.W.R. (1991) Methodological study of the analysis of bone. In *Archaeological Sciences 1989* (eds P. Budd, B. Chapman, C. Jackson, R. Janaway and B. Ottaway), Oxbow Monograph 9, Oxford, pp. 363–372.

Pollard, A.M. and Heron, C. (1996) *Archaeological Chemistry*, Royal Society of Chemistry, Cambridge, UK.

Price, T.D., Grupe, G. and Schrötter P. (1994) Reconstruction of migration patterns in the Bell Beaker period by stable strontium isotope analysis. *Appl. Geochem.* 9, 413–417.

Radosevich, S.C. (1993) The six deadly sins of trace element analysis: a case of wishful thinking in science. In *Investigations of Ancient Human Tissue* (ed M.K. Sandford). Gordon and Breach Science Publishers, Langhorne, Pennsylvania, USA, pp. 269–332.

Sandford, M.K. (1993) Understanding the biogenic-diagenetic continuum: interpreting elemental concentrations of archaeological bone. In *Investigations of Ancient Human Tissue* (ed. M.K. Sandford). Gordon and Breach Science Publishers, Langhorne, Pennsylvania, USA, pp. 3–57.

Schurr, M.R. (1992) Isotopic and mortuary variability in a middle Mississippian population. *Am. Antiquity* 57, 300–320.

Schwarcz, H.P. (1991) Some theoretical aspects of isotope paleodiet studies. *J. Archaeological Sci.* 18, 261–275.

Schwarcz, H.P., and Schoeninger M.J. (1991) Stable isotopes in human nutritional ecology. *Yearbook of Physical Anthropology* 34, 283–321.

Sealy, J., Armstrong, R. and Schrire C. (1995) Beyond lifetime averages: tracing life histories through isotopic analysis of different calcified tissues from archaeological human skeletons. *Antiquity* 69, 290–300.

Sealy, J.C. and van der Merwe N.J. (1985) Isotope assessment of Holocene human diets in the southwestern Cape, South-Africa. *Nature* 315, 138–140.

Shackleton, N.J. (1969) Marine mollusca in archaeology. In *Science in Archaeology* (eds D. Brothwell and E. Higgs). 2nd edn. Thames and Hudson, London, pp. 407–414.

Sillen, A. and Sealy J.C. (1995) Diagenesis of strontium in fossil bone – a reconsideration of Nelson *et al.* (1986). *J. Archaeological Sci.* 22, 313–320.

Stos-Gale, Z.A. (1995) Isotope archaeology – a review. In *Science and Site* (eds J. Beavis and K. Barker). Bournemouth University School of Conservation Sciences, Poole, UK, pp. 12–28.

Stott, A.W. and Evershed R.P. (1996) $\delta^{13}C$ analysis of cholesterol preserved in archarological bones and teeth. *Anal. Chem.* 68, 4402–4408.

Stott, A.W., Evershed R.P. and Tuross N. (1997) Compound-specific approach to $\delta^{13}C$ analysis of cholesterol in fossil bones. *Org. Geochem.* 26, 99–103.

Sullivan, C.H. and Kreuger H.W. (1981) Carbon isotope analysis of separate phases in modern and fossil bone. *Nature* 292, 333–335.

Switsur, R., Waterhouse, J.S., Field, E.M., Carter, T. and Loader N. (1995) Stable isotope studies in tree rings from oak – techniques and some preliminary results. In *Problems of Stable Isotopes in Tree-Rings, Lake Sediments and Peat-bogs as Climatic Evidence for the Holocene* (ed B. Frenzel). European Science Foundation, Strasbourg, pp. 129–140.

Tauber, H. (1981) ^{13}C evidence for dietary habits of prehistoric man in Denmark. *Nature* 292, 332–333.

Trimborn, P., Becker, B., Kromer B., and Lipp, J. (1995) Stable isotopes in tree-rings: a palaeoclimatic tool for studying climatic change. In *Problems of Stable Isotopes in Tree-Rings, Lake*

Sediments and Peat-bogs as Climatic Evidence for the Holocene (ed B. Frenzel). European Science Foundation, Strasbourg, pp. 163–170.

Ubelaker, D.H., Katzenberg, M.A. and Doyon, L.G. (1995) Status and diet in precontact highland Ecuador. *Am. J. Physical Anthropology* **97**, 403–411.

van der Merwe, M.J. (1992) Light stable isotopes and the reconstruction of prehistoric diets. In: *New Developments in Archaeological Science* (ed A.M. Pollard). Proceedings of the British Academy 77, Oxford University Press, Oxford, UK, pp. 247–264.

van der Merwe, N.J., Lee-Thorp, J.A., Thackeray, J.F., Hall-Martin, A., Kruger, F.J., Coetzee, H., Bell, R.H.V and Lindeque, M. (1990) Source-area determination of elephant ivory by isotopic analysis, *Nature* **346**, 744–746.

van der Merwe, N.J., Roosevelt, A.C. and Vogel, J.C. (1981) Isotopic evidence for prehistoric subsistence change at Parmana, Venezuela. *Nature* **292**, 536–538.

van der Merwe, N.J. and Vogel, J.C. (1977) ^{13}C content of human collagen as a measure of prehistoric diet in woodland North America. *Nature* **276**, 815–816.

van Klinken, G.J., van der Plicht, H. and Hedges, R.E.M. (1994) Bone ^{13}C/^{12}C ratios reflect (palaeo-) climatic variation. *Geophys. Res. Lett.*, **21**, 445–448.

Vogel, J.C. and Eglington B. and Auret J.M. (1990) Isotope fingerprints in elephant bone and ivory. *Nature* **346**, 747–749.

Vogel, J.C. and van der Merwe, N.J. (1977) Isotopic evidence for early maize cultivation in New York State, *Am. Antiquity* **42**: 238–242.

Wada, E., Mizutani, H. and Minagawa M. (1991) The use of stable isotopes for food web analysis, *Crit. Rev. Food Sci. Nutrition* **30**, 361–371.

White, C.D. and Schwarcz, H.P. (1989) Ancient Maya diet – as inferred from isotopic and elemental analysis of human-bone. *J. Archaeological Sci.* **16**, 451–474.

Stable isotopes in tree ring cellulose

Roy Switsur and John Waterhouse

18.1 Introduction

Much effort is being expended in the reconstruction of surrogate climatic and environmental data over several time scales. The history of climate is of exceptional interest and importance and has been investigated through traditional methods such as floral and faunal analysis, pollen analysis, tree ring widths and tree ring X-ray analysis. However, it is probable that stable isotope research provides the best method currently available for the study and reconstruction of past climates and environments (see Chapters 20, 21 and 22). Efforts are being made to investigate and interpret many natural systems, including ice cores and lake and ocean sediments, through these agencies. It is accepted that high resolution information is vital importance for modelling climatic change and the prediction of possible global warming and its concomitant consequences and as such has been adopted as a core project of the International Geosphere–Biosphere Programme. Of the various methods available, tree ring isotopic analysis offers the possibility of the most detailed climatic information from continental temperate zones, as opposed to ice caps or the deep sea, with the highest time resolution. Trees form the most extensive ecosystem of the temperate zones of the earth and preserve an annually datable record extending for many thousands of years. The most generally useful record is that contained in growth rings where the potential is available for the recovery of information with annual and seasonal resolution extending through much of the Holocene using several tree species.

It was Urey (1947) who first suggested the possibility that there might be a climatic signal in cellulose noting that 'plant compounds synthesized at different temperatures may contain varying amounts of C-13'. Later Epstein *et al.* (1976) postulated that since 'temperature variations over the earth's surface are correlated with D/H and $^{18}O/^{16}O$ variations in meteoric water, natural systems which record variations in D/H and $^{18}O/^{16}O$ should preserve a record of past temperature variations'. Hence, since wood is comprised primarily of hydrogen, carbon and oxygen, the isotopic analysis of the growth rings should provide a record of past climate related to the time of synthesis of

Stable Isotopes, edited by H. Griffiths.
© 1998 BIOS Scientific Publishers Ltd, Oxford.

the compounds. The growth rings thus constitute a natural data acquisition and storage system that may be accessed by several methods. Their isotopes can not only help to elucidate important environmental parameters such as temperature and rainfall but also details of the variations of past atmospheric carbon dioxide concentrations. In addition stable isotope analyses can assist in our investigations of the mechanism of photosynthesis and other processes in trees. Reviews are by Leavitt (1990), Long (1982), Ramesh et al. (1986b), White (1988) and Yakir (1992) and Chapter 3.

Isotope values in the tree ring cellulose are expressed in the conventional 'δ' notation as deviations in permil (‰) from the international standards; PDB (for carbon), Vienna-SMOW (for hydrogen and oxygen); such that $\delta_{sample} = (R_{sample}/R_{standard} -1) \times 1000$, where R is the $^{13}C/^{12}C$, D/H or $^{18}O/^{16}O$ ratio in sample and standard. Typical analytical uncertainties necessary in the isotopic determinations for palaeoenvironmental research are ±0.05 (‰) for $\delta^{13}C$, ± 0.2 (‰) for $\delta^{18}O$, ±3 (‰) for δD.

18.2 Atmospheric carbon dioxide and carbon stable isotope ratios

18.2.1 *Global cycle*

The sole source of carbon in the growth rings of trees may be assumed to be derived from the carbon dioxide of the atmosphere and the ring constituents actively record the variations of atmospheric $\delta^{13}C$. Leavitt and Long (1988), Peng et al. (1983), Stuiver (1978), Stuiver et al. (1984), have used $\delta^{13}C$ chronologies derived from tree rings as surrogate chronologies of $\delta^{13}C$ of atmospheric carbon dioxide. These contain the effects of both fossil fuel and biosphere inputs. The fossil-fuel effect may be estimated by the use of radiocarbon dilution measurements and hence an idea of the biospheric input may be obtained. The isotopic ratios are also governed by variations in geophysical parameters that affect the photosynthetic process. There is an important requirement for increased knowledge of atmospheric carbon dioxide concentrations in the past and of the global carbon cycle; the isotopic composition of the growth ring cellulose as described above can help significantly with these aims. The carbon cycle is characterized by various sources and sinks, or reservoirs (see also Chapter 24). Bowen (1991) estimates these as comprising the atmosphere (600 Gigatonnes carbon), biosphere (600 to 970 Gt.), fossil fuels (*ca.* 5000 Gt.), soils (100–300 Gt.), oceans (36000 Gt.) and multifarious sediments (10×10^6 Gt.). The world's forests contain up to 500 Gt. carbon. Approximately 103 Gt. are absorbed annually by the terrestrial biota during photosynthesis of which 100 Gt. are returned during respiration and decomposition (Leggett, 1990). In the ocean / atmosphere cycle about 92 Gt. carbon are absorbed by phytoplankton as bicarbonate ions but 90 Gt. are returned to the atmosphere by respiration (see Chapter 15). These carbon fluxes appear to be in approximate equilibrium and Freyer (1979) suggests that any net absorption might be balanced by carbon dioxide generated by hot spring, fumarole and volcanic action. The combustion of fossil fuels and forest trees during the Industrial Revolution has disturbed the natural cycle and Berner and Lasaga (1989) have calculated that the carbon input to the atmosphere during the past 200 years is equivalent to the total mass of carbon in living material on the planet.

Prior to the middle of the present century the predominant source of anthropogenic input of carbon dioxide into the atmosphere was through deforestation and agricultural clearance but this has been overtaken by fossil fuel combustion and cement manufacture; the annual carbon dioxide release from these sources grew from 250 M tonnes in 1860 to 1250 Mt. in 1930; after the depression of the 1930s the output increased and from 1945 to the present the rate has increased by a roughly constant 112 Mt. per year to the present value of about 6000 Mt. Currently, slightly less than 9 Mt. of biomass are burned annually. Of this carbon dioxide production 30%–40% is absorbed by the oceans. The global carbon cycle is not yet well understood and arguments continue as to whether the atmosphere constitutes a source or a sink and annual budget estimates vary by about one million tonnes (see Chapter 24). Investigations using tree growth-rings could aid in making the decision. Sink models assume that photosynthesis is stimulated by the increasing carbon dioxide content, from around 270 μmol mol^{-1} in the late 19th century to over 350 μmol mol^{-1} at present, the rate increasing by 1–1.5 μmol mol^{-1} per annum. The oscillations in carbon dioxide concentration have been measured at stations in both the northern and southern hemispheres and seasonal oscillations of about 0.45% are observed, thus reflecting the annual photosynthetic and respiration cycles of the biosphere (Keeling *et al.*,1979; Mook *et al.*,1983). Plants in the northern hemisphere take up 4 Gt of carbon dioxide through photosynthesis between April and September and this corresponds to lowest concentrations of atmospheric carbon dioxide. The mean stable isotopic composition of the atmospheric carbon dioxide been altered by the releases from cement manufacture and fossil fuel combustion. The carbon dioxide from fossil fuel is considerably depleted in the ^{13}C isotope than the general atmospheric value of around -8 ‰ so that the mean value is slowly decreasing. Measured values (Francey *et al.*, 1996) of global atmospheric behaviour over the past decade indicate a gradual decrease in δ^{13}C between 1982 and 1994 from values of about -7.5‰ to -7.6‰ to between -7.7‰ and -7.8‰ with a pronounced flattening from 1988 to 1990. Andres (1996) has computed the global δ^{13}C value of carbon dioxide from fossil fuel combustion and cement manufacture from 1896 to 1992; the value changes smoothly from about -24.2‰ to -25.2‰ between 1860 and 1940 after which there is a more rapid decrease to -28.3‰ by 1970 and thereafter a change of only 0.2‰ during the last 20 years. The δ^{13}C chronologies from tree cellulose show a fall of 1‰ to 2‰ from pre-industrial times to the present. This may be compared with δ^{13}C chronologies obtained from ice cores (Friedli *et al.*, 1986) which indicate a decline of 1.4‰ during a similar period.

18.2.2 *Methods for determining δ^{13}C*

In their studies of isotopes in tree rings early workers such as Craig (1954), Farmer and Baxter (1974) and Libby and Pandolfi (1974) tended to work with samples of contiguous blocks of between five and twenty rings and based their isotopic analyses on whole wood. These strategies had the disadvantages of not exploiting the unique annual resolution of the ring chronologies and of using chemically and isotopically inhomogeneous material. Current investigations are now tending towards the study of single, absolutely-dated rings with the analyses of pure α-cellulose extracted from them.

   In palaeoenvironmental research normal procedure is to undertake a site survey of the study area recording the edaphic conditions, competing tree species and other vegetation necessary to specify the site where possible local meteorological records are obtained. Then follows a dendrochronological study of the trees under investigation and the assignment of calendar dates to the growth rings. Next the rings are removed from the sample cores or tree sections under study. Prior to the chemical procedures it is necessary to reduce the wood to small size. Whilst this may be done with a ball or knife mill, there is a risk of charring and affecting the early and late wood differentially and also powdered wood tends to clog filters leading to losses unacceptable for careful isotopic work. It has been found expedient to cut the rings into thin (20 μm) serial sections using a sledge microtome or scalpel (Switsur et al., 1995) since these allow thorough penetration of the reagents into the wood in the preparation of α-cellulose. The translucent sections also make it very easy to differentiate and to separate the early spring wood from the main summer growth, which latterly is frequently the preferred material for analysis (Epstein and Yapp, 1976; Lipp et al., 1992, 1993, 1994; Switsur et al., 1995).

The extraction of α-cellulose usually follows the method described by Green (1963) as modified by Mullane (1988) and by Tans and Mook (1980). This is time-consuming and efforts to increase the rate of sample throughput have been described by Leavitt and Danzer (1993). This technique does not produce pure α-cellulose, but rather a holocellulose which contains non-glucose sugars as impurities. The technique is based on delignification of the wood in small fragile glass-fibre packets suspended in a beaker of oxidant and is capable of processing about 10 samples simultaneously; however a new technique of Loader et al. (1996) is fast and produces pure α-cellulose. Here 2 mg of thin wood slivers are contained in a small glass thimble (as used in miniature Soxhlet extraction) having a sintered glass plate a few millimeters from the end. Up to 40 of these are immersed in hot (70 °C) sodium chlorite – acetic acid oxidant in an ultrasonic bath for 4 h using the degas facility, with four regular additions of further oxidant. This produces a homogeneous mixture and the wood structure is no longer visible. The samples are vacuum filtered, washed and returned to the ultrasonic bath containing 10% sodium hydroxide at 90 °C. This is repeated using 17% sodium hydroxide at room temperature to ensure the complete removal of impurities. After washing with hydrochloric acid and copious water the pure α-cellulose is vacuum dried. This technique has the advantage of making all the operations in the same vessel ensuring quantitative processing. The α-cellulose is used for isotopic analysis of hydrogen, carbon and oxygen.

For measurement of $\delta^{13}C$, oxidation of the α-cellulose follows the methods of Sofer (1980) or Boutton (1991) in which the sample is mixed with an excess of copper oxide, with or without copper and silver, in an evacuated pre-dried borosilicate glass tube. This is heated at 450 °C for 18 h. The carbon dioxide produced is thoroughly dried and used for isotopic analysis in a mass spectrometer. Being a batch process, the throughput is limited by available equipment, usually to about 20 to 30 samples per day for one operator. However, on-line methods have been developed for rapid processing of carbon isotope samples in which the α-cellulose is weighed into a miniature tin boat and oxidised in a pulse of oxygen in an elemental analyser attached to the input of the mass spectrometer. Use of a carousel attached to the equipment enables many samples may be loaded simultaneously and the isotopic ratios of roughly 200 samples may be measured during a 24 h period. A new method for determining the $\delta^{18}O$ of organic material is described in Chapter 3.

18.2.3 *Climate relationships with $\delta^{13}C$.*

Preliminary studies (eg. Lipp et al.,1991, 1993; Ramesh et al., 1985, 1986; Saurer and Siegenthaler, 1989) have demonstrated high correlation coefficients of $\delta^{13}C$ with temperature and relative humidity over periods of about 30 years for several tree species. Time series of a century and longer have been investigated (Loader, 1995; Loader and Switsur, 1996; Sonninen and Jungner, 1995; Switsur et al., 1994; Switsur et al.,1995; Switsur et al.,1996) and similarly high correlations have been found for *Quercus robur* and *Pinus sylvestris*. The general trend of the data are similar in that the $\delta^{13}C$ values remain at around -24‰ from earlier times up to about 1880 and then decrease for about 30 years, followed by a recovery until roughly 1940, since which time there has been a continuous decrease to the present. These general trends have been observed in other tree ring studies (Freyer and Belacy, 1983; Leavitt and Long, 1989; Stuiver et al., 1984).The pattern, however, is not exhibited by all trees. The data over 100-year periods or longer for Finland, England and Scotland have been used to calibrate and to verify climatic variables using records from local meteorological stations. The reconstructions, with some exceptions, agree closely with the original climate records (see Figure 18.1).

A few high resolution studies of $\delta^{13}C$ variation across individual growth rings have been made. Wilson and Grinsted (1977) analysed two very wide rings of *Pinus radiata* and suggested that the variations were due to changes in leaf temperature. In their study of 1982, Leavitt and Long analysed a total of 7 slices across two rings of *Juniperus depperanea* which showed a rapid increase and decrease of $\delta^{13}C$ values during the growing season but the cause was uncertain. Their later work (1991) similarly studied rings from 19 *Pinus ponderosa* trees and here a relationship was found between $\delta^{13}C$ and temperature/soil moisture. Ogle and McCormac (1994) studied holocellulose from multiple slices from individual wide rings selected from the ring-width

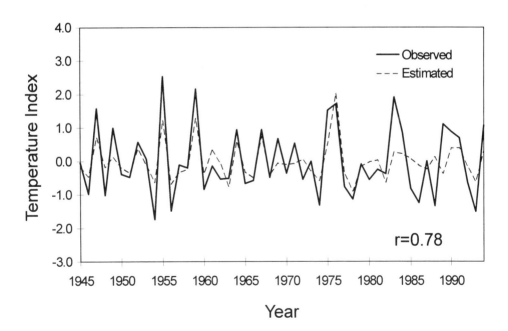

Figure 18.1. *Climate relationships in $^{13}C/^{12}C$ ratios of tree ring cellulose from oak in east England (Sandringham, UK). Reconstruction of the mean temperature for July and August from measurements of $\delta^{13}C$ in cellulose calibrated against meteorological observations during the previous 50 years, and applied to values measured between 1945 and 1994. In the figure these are compared with the temperature index for the equivalent period, with a correlation constant of 0.78.*

chronologies of *Quercus petraea*. Two thirds of the samples exhibited a 'spring deple-tion' of $\delta^{13}C$ values with varying depths and lengths primarily in the early growth but occasionally crossing the early/late wood boundary. However, the investigation with the highest resolution (Loader *et al.*, 1995) used α-cellulose extracted from 200 serial sections each, 30 μm thick, cut by microtome from across two consecutive growth rings of *Quercus petraea*. Five of these slivers were combined to make up 40 samples across the 2 rings. Each sample thus represented only a few days of growth. The detailed $\delta^{13}C$ values showed a high correlation with both the meteorological monthly relative humidity and weekly ambient temperature records. It has been possible to cal-culate relative humidity and temperature coefficients from the data and interpret them in terms of the carbon isotope model described below.

The supply of CO_2 for reduction during photosynthesis is regulated by stomata, and coupled to environmental conditions and water loss. The internal CO_2 concentra-tion (C_i) is related to both water use and the extent of carbon isotope discrimination, providing the indirect association between Δ and water-use efficiency (WUE) which has been central to tree-ring studies. (Farquhar 1980; Farquhar *et al.*, 1982; Francey and Farquhar 1982; see Chapter 9). Details of discrimination by differing photosyn-thetic pathways, and the calculation of the CO_2 concentration within the chloroplast, C_c, in relation to discrimination against ^{13}C and ^{18}O are described in earlier chapters (see Chapters 3, 8, 9 and 10). To date, tree-ring studies have used the simplified model of discrimination and C_i (Farquhar *et al.*, 1982; see Chapter 8):

$$\delta^{13}C_p = \delta^{13}C_a - a + (b-a) \times \frac{C_i}{C_a} \qquad (18.1)$$

with gas exchange related through Fick's law:

$$C_i = C_a - A/g \qquad (18.2)$$

and hence to transpired water and photosynthetic water use efficiency, *WUE* (see Chapter 9, Equation 9.2):

$$WUE = \frac{A}{E} = \frac{1}{1.6} \frac{C_a(1 - C_i/C_a)}{(e_i - e_a)} \qquad (18.3)$$

where a is discrimination against ^{13}C due to diffusion ($\approx 4.4‰$); b is isotopic dis-crimination during carboxylation ($\approx 27‰$); E is evaporation rate; C_i is internal car-bon dioxide concentration; C_a is atmospheric carbon dioxide concentration; $\delta^{13}C_p$ is delta carbon value in plant tissue; $\delta^{13}C$ is delta carbon value of atmospheric carbon dioxide; A is carbon assimilation rate; g is stomatal conductance for carbon dioxide; C_a, C_i is vapour pressures in air and leaf, respectively.

The stomatal size, density and assimilation rate are known to be affected by climatic parameters, for example at higher temperatures the values of g decreases, so that the resul-tant carbon isotope ratio in the plant tissue would become less negative. The carbon iso-topic composition of plants thus commonly preserves strong signals related to moisture or temperature (Figure 18.1; Dupouey *et al.*, 1993; Dupouey, 1995; Farquhar *et al.*, 1989; Livingstone and Spittlehouse, 1993). Climatic changes control A and g variations and hence C_i. The models are heavily dependent upon atmospheric source isotope data, which as described in the carbon cycle above, have (unfortunately), limited geographical and temporal coverage but see Chapter 24. Nevertheless the models are important steps in the interpretation of the $\delta^{13}C$ signals in tree ring time series, for example, as noted, changes of A/g (reflecting C_i/C_a) have been used (Leavitt and Long, 1988; Stuiver *et al.*, 1984) to correct tree ring $\delta^{13}C$ chronologies. The corrections over-estimate the effect expected from fossil fuels alone, the residuals being the contribution from the biospheric material to the atmosphere (see Figure 18.2). Besides being affected by climatic parameters, the stomatal size is affected by pollutants such as sulphur dioxide and ozone, so that $\delta^{13}C$ in tree rings may be used as a pollution monitor (Freyer 1979; Martin and Sutherland 1990; Martin *et al.*, 1988).

18.3 Hydrogen and oxygen stable isotope ratios

Local precipitation is the dominant source of both the oxygen and hydrogen atoms in the cellulose of tree rings, with a detailed review given in Chapter 11. The δD and $\delta^{18}O$ values of meteoric water are known to depend upon ambient temperature (see Chapter 23); hence if the δD and $\delta^{18}O$ values cellulose of tree rings reflects those of the local precipitation, tree rings should contain a record of temperature. The route from water to tree ring cellulose, however, is not direct, and modification to the isotope ratios of the source water used by the tree can occur at several stages. In particular, transpiration enriches leaf water in the heavier isotopes of both hydrogen and oxygen (see Chapters 10 and 11). The degree of enrichment depends upon humidity; hence local values of humidity can also be

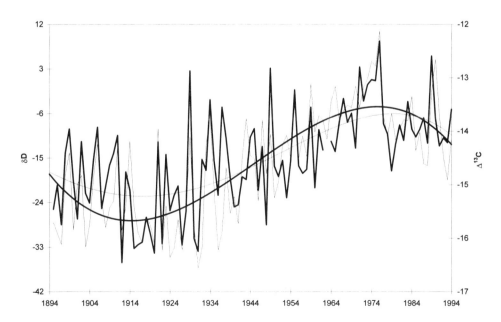

Figure 18.2. Carbon isotope discrimination ($\Delta^{13}C$; fine line) and hydrogen isotope composition (δD; thick line) are shown for cellulose of Pinus sylvestris L. from Bedfordshire, UK. The annual values are the mean of two trees with long-term trends shown by a sixth-order polynomial fit to each series. Carbon isotope discrimination is calculated using the changing $\delta^{13}C$ of the ambient carbon dioxide during this period. Such a correction then brings the carbon isotope signal back into synchronisation with δD.

expected to influence to isotope ratios in tree rings. Further modifications to the isotope ratios (so-called biological fractionation) occur during photosynthesis and ensuing chemical reactions in the leaves and trunk (see Chapters 3 and 4).

18.3.1 *Methods for determining δD and δ^{18}O*

Methods have been developed for measuring δD and δ^{18}O values with high accuracy. For δD measurements, the most commonly used method requires nitration of the α-cellulose, so as to remove the hydroxyl hydrogens, which are readily exchangeable with environmental water (Heuser, 1944). The carbon-bound hydrogens are non-exchangeable (Epstein *et al.*, 1976; Yapp and Epstein, 1977). Pyrolysis of the nitrated cellulose with copper oxide yields water, which upon reduction with metallic reducing agents such as zinc (Coleman *et al.*, 1982; Hayes and Johnson, 1988) or uranium (Bigeleisen *et al.*, 1952) gives hydrogen gas used for measurement of the isotopic ratio of the carbon-bound hydrogens. A second method makes use of the isotopic exchange that occurs between the hydroxyl hydrogens and water in order to calculate the hydrogen isotope ratio of the carbon-bound hydrogens of cellulose. The technique was first described by Wilson and Grinsted (1975). Exchange is carried out with water of known isotopic composition. If the exchange reaches equilibrium, the δD value of the carbon-bound hydrogens of cellulose can be calculated knowing the δD value of the water, the δD value of the exchanged cellulose and the equilibrium fractionation factor for exchange. The method has been recently improved by Feng *et al.* (1993). They found that treatment of the cellulose with sodium hydroxide solution prior to reaction with water allowed the exchange to reach equilibrium more rapidly. Despite the precision achievable with these methods, the rate of measurement of δD values of cellulose samples is very slow compared with that now available for the on-line method for δ^{13}C determinations. Current attempts to develop an on-line system for hydrogen isotope analysis (Gehre *et al.*, 1996) as well as the developments described in Chapter 1, promise improvements for the future.

For oxygen isotope measurements, one of two different methods is generally used. Both rely on the incorporation of oxygen atoms of cellulose into carbon dioxide for isotope ratio measurement in a mass spectrometer. The first method, originally developed by Rittenberg and Pontecorvo (1956), involves the pyrolysis of cellulose with mercury(II) chloride. The resulting process of oxidative degradation converts the oxygen of cellulose into a mixture of carbon monoxide and carbon dioxide. As the carbon monoxide produced was found to be isotopically heavier by about 0.5‰ than the carbon dioxide (Thompson and Gray, 1977), the carbon monoxide is converted to carbon dioxide and carbon by an electric discharge (Agget *et al.*, 1965). All the oxygen in the cellulose is therefore quantitatively converted to carbon dioxide. The pyrolysis with mercury(II) chloride also produces hydrogen chloride. This can be removed by the use of weak bases such as quinoline (Dunbar and Wilson, 1983; Rittenberg and Pontecorvo, 1956), isoquinoline (DeNiro and Cooper, 1989; Yakir and DeNiro, 1990) and 5,6-benzoquinoline (Rittenberg and Pontecorvo, 1956). These compounds are inconvenient to use owing to their toxicities and the difficulty of removing the carbon dioxide from them. Two materials more suitable for the removal of hydrogen chloride have been recently reported. One is a weakly basic macroreticular ion exchange resin containing tertiary amine groups (Field *et al.*, 1994); the other is zinc (Sauer and Sternberg, 1994).

The second method for determination of δ^{18}O of cellulose involves pyrolysis of cellulose in a nickel vessel. The procedure was first used by Thompson and co-workers

(Epstein *et al.*, 1977; Gray and Thompson, 1976; Thompson and Gray, 1977). Cellulose powder is heated in a sealed nickel vessel to around 1000 °C, at which temperature the nickel walls are porous to hydrogen. Removal of hydrogen ensures that oxygen is incorporated entirely into carbon monoxide and carbon dioxide. Disproportionation of CO to CO_2 is accomplished via an electric discharge. The method was modified by Brenninkmeijer and Mook (1981) to allow more rapid processing. Addition of nickel powder to the reaction vessel catalysized the disproportionation reaction of CO to CO_2 at a temperature of 350 °C, thus eliminating the need for the use of an electric discharge. A disadvantage of both these procedures is the welding necessary to seal the nickel vessels. A more convenient and rapid procedure has been recently developed by Edwards *et al.* (1994) who use re-sealable nickel vessels. They report that the CO and CO_2 formed on pyrolysis have the same isotopic composition. This could result from the freezing of the isotopic equilibrium between the gases at the pyrolysis temperature (950 °C) as the vessel cools. Disproportionation of CO to CO_2, using either nickel powder of an electric discharge, is therefore unnecessary.

As in the case for hydrogen isotope ratios, there is a need for an on-line method for oxygen isotopes. Werner *et al.* (1996) describe a method by which cellulose is pyrolysed in the presence of excess carbon to give carbon monoxide, which is passed directly into the mass spectrometer. A revised method which can analyse organic material "contaminated" with N, has recently been described for the preparation and analysis of ^{18}O in CO (Farquhar *et al.*, 1997; see Chapter 3).

18.3.2 *Climate relationships*

Several studies have found that there is indeed a direct relationship between isotope ratios in tree ring cellulose and in local precipitation or environmental water (Epstein and Yapp, 1976; Yapp and Epstein, 1977, 1982a; Lipp *et al.*, 1994; White *et al.* 1994). Yapp and Epstein (1982a) derived equation 18.4 for a variety of plant species growing in differing local climates.

$$\delta D_{CN} = 0.87\delta D_{EW} + 11 \qquad\qquad (18.4)$$

where δD_{CN} is the hydrogen isotope ratio of nitrated cellulose and δD_{EW} is the hydrogen isotope ratio the local environmental water.

Significant correlations have also been found between δD and $\delta^{13}C$ of tree ring cellulose and other environmental parameters (see Figure 18.2). Relationships have been reported for δD and mean annual temperature (Gray and Song, 1984; Yapp and Epstein, 1982b), growing season temperature (Lipp *et al.*, 1992, 1993, 1994; Ramesh *et al.*, 1986), seasonal relative humidity (Lipp *et al.*, 1992, 1993, 1994; Ramesh *et al.*, 1986a) and precipitation amounts (Dubois, 1984; Epstein and Yapp, 1976; Lawrence and White, 1984; Ramesh *et al.*, 1986a; Yapp and Epstein, 1985). For $\delta^{18}O$, relationships have been similarly reported for mean annual temperature (Burk and Stuiver, 1981; Gray and Thompson, 1976), growing season temperature (Switsur *et al.* 1996), and seasonal relative humidity (Figure 18.3; Ramesh *et al.*, 1986a; Switsur *et al.* 1996; see Chapter 3). The scale of these studies has varied in range from the analysis of homogenized blocks of annual rings from trees across continental distances (Burk and Stuiver, 1981; Gray and Song, 1984; Yapp and Epstein, 1982b), to the late wood of annual rings within individual trees (Lipp *et al.*, 1992, 1993, 1994; Switsur *et al.* 1995,

1996), to isotope measurements across individual annual rings (White *et al.*, 1994). The observed relationships between isotope ratios and environmental parameters across such a range of scales provides great confidence in the use of tree rings for climatic reconstruction (see Figure 18.3). Epstein and Krishnamurthy (1990) used the reported relationship between temperature and cellulose δD (Yapp and Epstein, 1982b) in their analysis of 23 trees from around the world. They interpreted the increasing δD values in 21 of the trees as evidence for an increase in global temperature over the past century. The values from a bristlecone pine suggested that temperatures have been rising for at least the past 1000 years.

Two studies from trees from quite different geographical areas – silver fir from the Kashmir valley in India (Ramesh *et al.*, 1985) and oak from Northern Ireland (Robertson *et al.*, 1995) – have investigated the degree of coherence between isotope ratios along different radii of the same tree. Both studies showed that although values of δD and δ¹⁸O from the same ring vary around the circumference, the pattern of year to year variation is consistent between radii. It is therefore important that samples from along the same radius be used for climatic studies.

Much progress has been made in understanding the mechanisms of isotopic fractionation that can occur during the complex processes between water entering the tree and the incorporation of hydrogen and oxygen in the cellulose of tree rings. No significant

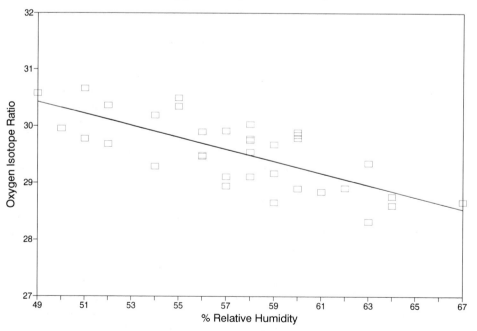

Figure 18.3. *The correlation between oxygen isotopes and climate, shown as the relationship between ¹⁸O/¹⁶O in tree ring cellulose (for a single oak tree from Wennington Wood, UK) and relative humidity in July for the years 1887–1922. The regression equation is δ¹⁸O =-0.12(RH) + 36.4‰, with a correlation coefficient of 0.6 (see also Chapter 3).*

fractionation occurs during the uptake of water in the roots and its passage up to the leaves (White et al., 1985; see Chapter 11). A model for isotopic fractionation in the leaf has been developed (Dongmann et al., 1974; Farris and Strain, 1978; Flanagan et al., 1991) from one originally derived for evaporation of water from the ocean (Craig and Gordon, 1965). If leaf water and water vapour in the leaf are assumed to be in isotopic equilibrium, then at the steady state the isotope composition of leaf water (R_l) can be expressed as:

$$R_l = \alpha_k \alpha_e (1-h) R_s + R_a \alpha_e h \qquad (18.5)$$

where R_s is the isotope composition of the source water, R_a is the isotope ratio of atmospheric water vapour, h is the relative humidity, α_k is the kinetic fractionation factor between liquid and vapour, and α_e is the equilibrium fractionation factor between liquid and vapour.

R_{cell}, the isotope ratio of carbon-bound hydrogen or oxygen in the cellulose of tree rings, can be expressed as follows:

$$R_{cell} = \alpha_k \alpha_e \alpha_b (1-h) R_s + \alpha_e \alpha_b h R_a \qquad (18.6)$$

Where α_b is the net biochemical fractionation factor involved in the processes from leaf water to cellulose defined by the term R_{cell}/R_l.

Equation 18.6 has been modified assuming that the source water used by the plant is in isotopic equilibrium with water vapour in the atmosphere and that R_s has the same value as R_{MW}, the isotope ratio in local precipitation (Yapp and Epstein, 1982a; Edwards et al., 1985; Edwards and Fritz, 1986). This produces equation 18.7:

$$R_{cell} = \alpha_k \alpha_e \alpha_b (1-h) R_{MW} + \alpha_b h R_{MW} \qquad (18.7)$$

For oxygen isotopes, a consistent value of around +27‰ is found for the biochemical isotope effect (α_b = 1.027) (Yakir and DeNiro, 1990). Epstein et al. (1977) accounted for this figure by assuming that CO_2 equilibrates with leaf water prior to fixation, leading to an enrichment of around 41‰. Addition of a water molecule to carbon dioxide during photosynthesis would give an overall value of 27‰ higher than leaf water. An alternative explanation (Sternberg and DeNiro, 1983) is that the oxygen atoms in cellulose have at some stage been in carbonyl groups of sugars. Enzyme-mediated reversible hydration of the carbonyl groups will lead to exchange of oxygen with metabolic water. In support of this, the isotope effect for a model exchange reaction using acetone was found to be +27‰ (Sternberg and DeNiro, 1983). A detailed discussion of factors controlling the ^{18}O signal in organic material and exchange of carbonyl-O is given in Chapter 3.

The situation with hydrogen is more complex. The overall biological fractionation factor seems to be determined by two processes. Reduction reactions involving NADPH during the Calvin cycle produce a high negative fractionation effect; whereas exchange with leaf water of carbon-bound hydrogens adjacent to carbonyl groups in sugars (via keto-enol tautomerism) produces a high positive fractionation (Yakir and DeNiro, 1990). In their studies on Lemna gibba, Yakir and DeNiro (1990) measured values of -171‰ and +158‰ respectively for the two fractionation effects under autotrophic conditions. The differing balance between these two effects leads to differing observed values for the overall biological fractionation factor (see Chapters 1 and 4).

A further question to be addressed is the extent to which further exchange of hydrogen and oxygen isotopes occurs during synthesis of cellulose in the cambium from sucrose translocated from the leaf. Equations 18.6 and 18.7 ignore this possibility. Several studies have shown, however, that this could be a significant factor (see also Chapter 3). Sternberg and Cooper (1986) grew carrot tissue in the dark using sucrose as the only source of nutrient (heterotrophic conditions).They found that nearly half of the oxygen atoms of sucrose were exchanged with water during the conversion of sucrose to cellulose. A similar result was reported by Yakir and DeNiro (1990) for both oxygen and hydrogen exchange. They showed that under heterotrophic conditions 40% of both the oxygen and carbon-bound hydrogen of sucrose fed to *Lemna gibba* exchanged with water prior to cellulose formation. In a study using potatoes, DeNiro *et al.* (1989) showed that the $\delta^{18}O$ of shoot cellulose was influenced not only by the water in the leaves, but also water in the shoots and roots. In particular, $\delta^{18}O$ values of shoot cellulose are related to the values of $\delta^{18}O$ of tuber water and not to those of tuber starch, which is the source of sucrose used in the formation of shoot cellulose. They suggested that this was caused by the breakdown of hexoses to trioses prior to the formation of shoot cellulose. Hill *et al.* (1994) demonstrated that this process is also relevant to trees. They showed that the oxygen atoms of starch, stored from the previous growing season, undergo exchange with the water in the trunk during the formation of early wood in oak trees. Evidence for the role of trioses in the exchange process was provided by the distribution of C-14 in labeled fructose extracted from cores of wood that had been incubated with [1-^{14}C] and [6-^{14}C] glucose. There is no corresponding direct evidence for the exchange of carbon-bound hydrogens with water in the trunk. If, however, the above mentioned value of +158‰ for the fractionation that occurs when *Lemna gibba* is grown in darkness with sucrose as the only carbon source is applicable to formation of trunk cellulose from translocated sucrose, then any exchange with trunk water would result in a δD value of trunk cellulose significantly higher than that of the source water. This is clearly not the case. Reported values of δD of trunk cellulose are not greatly different from (usually slightly less than) those for the source water, as is exemplified by Equation 18.4.

A further complication for the application of Equations 18.5 to 18.7 is the apparent isotopic inhomogeneity of water in the leaf. A gradient of isotopic enrichment exists in the leaf between the incoming stem water and water at the sites of evaporation (Flanagan *et al.*, 1991). A more complete discussion on this can be found in Yakir (1992) and Chapters 3 and 10 in this volume.

In the 'δ' notation of Craig, Equation 18.7 can be written in the forms 18.8 and 18.9 for hydrogen and oxygen isotopes respectively (Edwards and Fritz, 1986):

$$\delta D_{cell} = A \, \delta D_{MW} + 1000 \, (A-1) \tag{18.8}$$

$$\delta^{18}O_{cell} = B \, \delta^{18}O_{MW} + 1000 \, (B-1) \tag{18.9}$$

where $A = {}^{2}\alpha_b {}^{2}\alpha_e {}^{2}\alpha_k - {}^{2}\alpha_b ({}^{2}\alpha_e {}^{2}\alpha_k - 1) \times h$ and $B = {}^{18}\alpha_b {}^{18}\alpha_e {}^{18}\alpha_k - {}^{18}\alpha_b ({}^{18}\alpha_e {}^{18}\alpha_k - 1) \times h$. α_b, α_e, α_k and h are defined as above and superscript numbers 2 and 18 refer to hydrogen and oxygen respectively.

The good correlations reported between tree ring δD and $\delta^{18}O$ with temperature (via meteoric water) and humidity (via evapotranspiration) are consistent with this model. Equations 18.8 and 18.9 allow the calculation of the isotope ratios of meteoric water and humidity if it is assumed that δD_{MW} and $\delta^{18}O_{MW}$ at the site are related through Equation 18.10 (Craig, 1961).

$$\delta D_{MW} = 8\delta^{18}O_{MW} + 10 \qquad\qquad (18.10)$$

Edwards and co-workers have used this model to reconstruct climatic history of sites in southern Ontario in time scales ranging from centuries (Buhay and Edwards, 1993, 1995) to millennia (Edwards and Fritz, 1986). Changing climatic patterns are reconstructed using $\delta^{18}O_{MW}$ as a surrogate for temperature and calculated values of humidity.

The site conditions, in particular the hydrological setting, seem to play an important role in determining the relationship between environmental parameters and δD and $\delta^{18}O$ of tree ring cellulose. For example, Lipp *et al.* (1992, 1993, 1994) observed significant correlations of δD values of late wood of spruce trees from middle Franconia, Germany with July/August temperatures. This was as expected from a dry site such as this, where temperature-related summer precipitation is used for the formation of late wood. In a related study, the absence of any clear correlation between climatic parameters and δD values of fir trees from the Black Forest (Lipp *et al.*, 1991) was explained by the wetter conditions at this site. The trees here do not sample precipitation directly but sample well-mixed groundwater with no clear isotopic signature (Lipp *et al.*, 1994). The observations of Lipp and co-workers are in agreement with White *et al.* (1985), who showed that trees on dry sites use only summer rain with no evidence that groundwater contributes to the source water; however, trees on wetter sites use both summer rain and groundwater, and this leads to δD values that are more difficult to interpret. A study of white spruce from Alberta, Canada (Gray and Song, 1984) showed a significant correlation between cellulose δD and winter temperature, probably resulting from the contribution of winter snow to the source water used by the tree. A detailed discussion of the effect of differing water sources on stem water isotopic composition is given in Chapter 11.

A study of white pine growing in southeast New York state (White *et al.*, 1994) showed that δD values in the tree rings responded most strongly to changes in δD values of the source water, with changing humidity having little observed effect. The insensitivity to changing humidity was interpreted in terms of the local conditions, whereby water vapour from soil and upper-air cancel out leaf enrichment effects with temperature. The empirical relationship observed between δD of tree ring cellulose and source water for this site is nearly identical to Equation 18.6 obtained by Yapp and Epstein (1982a). The Equation derived by White *et al.* (1994), relating δD values of tree ring cellulose and source water under conditions of non-equilibrium, has a slope that was in agreement with the observed slope of the graph of δD(cellulose)/δD(source water). In contrast, the slope predicted from Equation 18.7 predicts a slope closer to 1.0. It therefore seems that models derived from Equation 18.7 are not valid at this site or at others where the condition of isotopic equilibrium does not apply.

A model based on the interplay between changing isotope signals in precipitation (due to changes in temperature) and in leaf water (due to changing humidity) shows promise for the recreation of past climatic patterns by comparing variation in the δD and $\delta^{18}O$ signals in individual trees (see Figures 18.2 and 18.3). Under climatic regimes characterized by cold/moist-warm/dry patterns, climatic signals in precipitation and leaf water reinforce each other, leading to co-variance in the δD and $\delta^{18}O$ values of tree rings. Climatic regimes characterized by cold/dry-warm/moist patterns can lead to anti-variance in the δD and $\delta^{18}O$ signals, owing to the different sensitivities of hydrogen and oxygen to fractionation in precipitation and in leaf water. An intermediate situation may lead to climatic signals in δD values of tree rings, but complacency in the $\delta^{18}O$ values. Such considerations have aided Edwards and co-workers in their attempts to reconstruct changing patterns of climate for sites in southern Ontario (Buhay and Edwards, 1993, 1995; Edwards and Buhay, 1996). Finally the observation that the ^{13}C and ^{18}O signal in organic material may distinguish co-limitation by photosynthetic and environmental factors is timely and important (see Chapter 3).

The increasing interest in the stable isotope ratios within tree rings reflects the potential importance of the unique record of information that they hold. Further investigations are necessary, however, before their potential can be fully realised. Work is in progress in Great Britain and Germany to study the effect of site moisture conditions on the hydrogen, oxygen and carbon isotope values in the annual growth rings oak trees from Finland, England and Ireland during the past century. A further investigation, also with annual resolution, is being made to discover how the three isotopes in different tree species, oak, pine and beech, react under the same site conditions and to what extent each correlates with local meteorological records for the past century. These are the first intensive studies that have been made over such a lengthy period with annual growth rings and using more than one isotope; they will provide a basis for further detailed research into past climates. This research effort is worthwhile; for stable isotope ratios in tree rings potentially hold not only records of environmental information with annual or even finer resolution, but also provide the means for a more detailed understanding of biochemical processes within trees.

Acknowledgements

We acknowledge the kindness of our colleagues, Neil Loader, Tony Carter, Iain Robertson, Alison Barker, Debbie Hemming and Mike Hall for their helpful discussions during the writing of this paper. We thank the NERC and the EC for research grants that have enabled us to carry out work in this interesting field.

References

Agget, J., Bunton, C.A., Lewis, T.A., Llewllyn, D.R., O'Connor, C. and Odell, A.L. (1965) The isotopic analysis of oxygen in organic compounds and in co-ordination compounds containing organic ligands. *Int. J. Appl. Radiat. Isotopes* **16**, 165–170.

Appleby, R.F., Davies, W.J. (1983) A possible evaporation site in the guard cell wall and the influence of leaf structure on the humidity response by stomata of woody plants, *Oecologia* **56**, 30–40.

Andres, R. (1996) Global and Latitudinal Estimates of $\delta^{13}C$ from Fossil-Fuel Consumption and Cement manufacture. *CDIAC Commun.* **22**, 9–10.

Berner, R.A. and Lasaga, A.J. (1989) Modelling the geochemical carbon cycle. *Scientific Am.* **260**, 54–61.

Bigeleisen, J., Perlman, M.L., Prosser, H.C. (1952) Conversion of hydrogenic materials to hydrogen for isotopic analysis. *Anal. Chem.* **24**, 1356–1357.

Boutton, T.W. (1991) Stable carbon isotope ratios of natural materials. 1. Sample preparation and mass spectrometric analysis. *In: Carbon Isotope Techniques.* (D.C. Coleman and B. Fry eds.) Academic Press, New York, pp. 155–171.

Bowen, R. (1991) *Isotopes and Climates*. Elsevier Applied Science, London, 483 pp.

Brenninkmeijer, C.A.M, Mook, W.G. (1981) A batch process for direct conversion of organic oxygen and water to CO_2 for $^{18}O/^{16}O$ analysis. *Int. J. Appl. Radiat. Isotopes* **32**, 137–141.

Buhay, W.M. and Edwards, T.W.D. (1993) Reconstruction of Little Ice Age climate in southwestern Ontario, Canada, from oxygen and hydrogen isotope ratios in tree rings. In: *Proceedings of an International Symposium on the Application of Isotope Techniques in the Study of Past and Current Environmental Change in the Hydrosphere and Atmosphere*, IAEA-SM-329/37, pp. 407–417.

Buhay, W.M. and Edwards, T.W.D. (1995) Climate in southwestern Ontario, Canada, between AD 1610 and 1885 inferred from oxygen and hydrogen isotopic measurements of wood cellulose from trees in different hydrological settings. *Quaternary Res.* **44**, 438–446.

Burk. L. and Stuiver, M. (1981) Oxygen isotope ratios in trees reflect mean annual temperature and humidity. *Science* **211**, 1417–1419.

Coleman, M.L., Shepherd, T.J., Durham, J.J., Rouse, J.E., Moore, G.R. (1982) Reduction of water with zinc for hydrogen isotope analysis. *Anal. Chem.* **54**, 993–995.

Craig, H. (1954) Carbon-13 in sequoia rings and the atmosphere. *Science* **119**, 141–143

Craig H. (1961) Isotopic variations in meteoric water. *Science,* **133**, 1702–1703.

Craig H. and Gordon L. (1965) Deuterium and oxygen-18 variation in the ocean and marine atmosphere. In: *Proceedings of a Conference on Stable Isotopes in Oceanography Studies and Palaeotemperatures* (ed. E. Tongiorgi). Pisa, Italy, pp. 9–130.

DeNiro, M.J. and Cooper L.W. (1989) Post-photosynthetic modification of oxygen isotope ratios of carbohydrates in the potato: Implications for paleoclimate reconstruction based upon isotopic analysis of wood cellulose. *Geochim. Cosmochim. Acta* **53**, 2573–2580.

Dongmann, G., Nurnberg H.W., Förstel H. and Wagener K. (1974) On the enrichment of $H_2^{18}O$ in the leaves of transpiring plants. *Radiat. Environ. Biophys.* **11**, 41–52.

Dubois, A.D. (1984) On the climatic interpretation of the hydrogen isotope ratios in recent and fossil wood. *Bulletin Soc. Belge. Geol.* **93**, 267–270.

Dunbar, J. and Wilson A.T. (1983) Re-evaluation of the $HgCl_2$ pyrolysis technique for oxygen isotope analysis. *Int. J. Appl. Radiat. Isotopes.* **48**, 932–934.

Dupouey, J.L. Leavitt, S.W., Choisnel, E., Jourdain, S. (1993) Modelling carbon isotope fractionation in tree rings based on effective evapotranspiration and soil water status. *Plant, Cell Environ.* **16**, 939–947.

Dupouey, J.L. (1995) Using $\delta^{13}C$ in tree rings as a bio-indicator of environmental variations and ecophysiological changes in tree functioning. In: *Paläoklimaforschung: Problems of stable isotopes in tree-rings, lake sediments and peat-bogs as climatic evidence for the Holocene* (B. Frenzel ed.) Vol. 15, pp. 97–104.

Edwards, T.W.D. (1993) Interpreting past climate from stable isotopes in continental organic matter. *In: Climate Change in Continental Isotope Records*. (P.K. Swart, K.C. Lohmann, J. McKenzie, S. Savin eds.). Am. Geophys. Union, Monograph, Vol. 78, pp. 333–341.

Edwards, T.W.D. and Fritz, P. (1986) Assessing meteoric water composition and relative humidity from ^{18}O and ^{2}H in wood cellulose: paleoclimatic implications for southern Ontario, Canada. *Appl. Geochem.* **1**, 715–723.

Edwards, T.W.D., Buhay, W.M., Elgood, R.J. and Jiang, H.B. (1994) An improved nickel-tube pyrolysis method for oxygen isotope analysis of organic matter and water. *Chem. Geol. (Isotope Geosci. Sect.)* **114**, 179–183.

Edwards, T.W.D. and Buhay, W.M. (1997) Paleoclimatic significance of oxygen and hydrogen isotope correlations in tree-rings. *Palaeogeogr. Palaeoclimatol. Palaeoecol.* (in press).

Epstein, S., Yapp, C.J. and Hall, J.H. (1976) The determination of the D/H ratio of non-exchangeable hydrogens in cellulose extracted from aquatic and land plants. *Earth Plan. Sci. Lett.* **30**, 241–251.

Epstein, S., Thompson, P. and Yapp C.J. (1977) Oxygen and hydrogen isotopic ratios in plant cellulose. *Science* **198**, 1209–1205.

Epstein, S. and Krishnamurthy R.V. (1990) Environmental information in the isotopic record of trees. *Phil. Trans. Roy. Soc. London* A 330, 427–439.

Epstein, S. and Yapp, C.J. (1976) Climatic implications of the D/H ratio of hydrogen in C-H groups in tree cellulose. *Earth Plan. Sci. Lett.* 30, 252–261.

Farmer, J.G. and Baxter, M.S. (1974) Atmospheric CO_2 levels as indicated by the stable isotope record in wood. *Nature* 247, 273–274

Farris, F. and Strain, B.R. (1978) The effect of water-stress on leaf $H_2^{18}O$ enrichment. *Radiat. Environ. Biophys.* 15, 167–202.

Farquhar, G.D. (1980) Carbon Isotope Discrimination by Plants: effects of carbon dioxide concentration and temperature via the ratio of intercellular and atmospheric CO_2 concentrations. *In: Carbon Dioxide and Climate: Australian Research.* (G.J. Pearman ed.) Australian Academy of Science, Canberra, pp. 105–110.

Farquhar, G.D., O'Leary, M.H. and Berry, J.A. (1982) On the relationship between carbon isotope discrimination and intercellular carbon dioxide concentration in leaves. *Austr. J. Plant Physiol.* 9, 121–137.

Farquhar, G.D., Ehleringer, J.R. and Hubick, K.T. (1989) Carbon isotope discrimination and photosynthesis. *Ann. Rev. Plant Phys. Plant Mol. Biol.* 40, 503–537.

Farquhar, G.D., Henry, B.K. and Styles, J.M. (1997) A rapid on-line technique for determination of oxygen isotope composition of nitrogen-containing organic matter and water. *Rapid Commun. Masspectrom.* 11, 1554–1560.

Feng, X., Krishnamurthy, R.V. and Epstein, S. (1993) Determination of D/H ratios of non-exchangeable hydrogen in cellulose: A method based on the water -cellulose exchange reaction. *Geochim. Cosmochim. Acta* 57, 4249–4256.

Field, E.M., Switsur, V.R. and Waterhouse, J.S. (1994) Improved technique for the determination of oxygen stable isotope ratios in wood cellulose. *Appl. Radiat. Isotopes.* 45, 177–181.

Flanagan, L.B., Bain, J.F. and Ehleringer, J.R. (1991) Stable oxygen and hydrogen isotope composition of leaf water in C_3 and C_4 plant species under field conditions. *Oecologia* 88, 394–400.

Flanagan L.B., Copmstock, J.P. and Ehleringer, J.R. (1991) Comparison of modelled and observed environmental influences on stable oxygen and hydrogen isotope composition of leaf water in *Phaseolus vulgaris* L. *Plant Physiol.* 96, 588–596.

Francey, R.J. and Farquhar, G.D. (1982) An explanation of $^{13}C/^{12}C$ in tree rings. *Nature* 297, 28–31.

Francey, R.J., Allison, C.E, Welch, E.D., Enting, I.G. and Goodman, H.S. (1996) *In situ* Carbon 13 and Oxygen 18 Ratios of Atmospheric CO_2 from Cape Grim. *CDIAC Commun.* 22, 10–11.

Freyer, H.D. (1979) Variations in the Atmospheric CO_2 content. In: *The Global Carbon Cycle*, (B. Bolin, E.T. Degens, S. Kempe and P. Ketner eds.), John Wiley and Sons, New York. pp. 79–99.

Freyer, H.D. and Belacy, N. (1983) $^{13}C/^{12}C$ records in northern Hemispheric trees during the past 500 years – anthropogenic impact and climatic superpositions. *J. Geophys. Res.* 88, 6844–6852.

Friedli, H., Lotscher, H., Oeschger, H., Siegenthaler, U. and Stauffer B. (1986) Ice core record of the $^{13}C/^{12}C$ ratio of atmospheric CO_2 in the past two centuries. *Nature* 324, 327–328.

Gehre, M., Hofling, R. and Kowski, P. (1996) Methodological Studies for D/H Analysis – a new technique for the direct coupling of sample preparation to an IRMS. *Isotopes in Environmental Health Studies* 32, 335–340.

Gray, J. and Song, S.J. (1984) Climatic implications of he natural variations of D/H ratios in tree ring cellulose. *Earth Planet Sci. Lett.* 70, 129–138.

Gray J. and Thompson P. (1976) Climatic information from $^{18}O/^{16}O$ ratios of cellulose in tree rings. *Nature* 262, 481–482.

Green, J.W. (1963) Wood cellulose. *In: Methods in Carbohydrate Chemistry III.* (R.L. Whistler ed.), Academic Press, New York, pp. 9–21.

Hayes, J.M. and Johnson, M.W. (1988) *Reagent and procedures for the preparation of H₂for hydrogen-isotopic analysis of water*. Ph.D. Thesis, Indiana University.

Heuser, E. (1944) *Cellulose Chemistry*. John Wiley, Chichester

Hill, S.A., Waterhouse, J.S., Field, E.M., Switsur, V.R. and ap Rees, T. (1994) Rapid recycling of triose phosphates in oak stem tissue. *Plant, Cell Environ*. 18, 931–936.

Keeling, C.D., Mook, W.G. and Tans, P.P. (1979) Recent trends in the ¹³C/¹²C ratio of atmospheric carbon dioxide. *Nature* 277, 121–123.

Lawrence, J.R. and White, J.W.C. (1984) Growing season precipitation from the D/H ratios of Eastern White pine. *Nature* 311, 558–560.

Leavitt, S.W. (1992) Isotopes and trace elements in tree rings. In: *Tree Rings and Environment*. (T.S. Bartolin, B.E. Berglund, D. Eckstein, F.H. Schweingruber eds.) Lund University, pp. 182–190.

Leavitt, S.W. and Long, A. (1988) Stable carbon isotope chronologies from trees in the south western United States. *Global Biogeochem. Cycles* 2, 189–198.

Leavitt, S.W. and Long A. (1989) The atmospheric δ¹³C record as derived from 56 pinyon trees at 14 sites in the southwestern United States. *Radiocarbon* 31, 469–474.

Leggett, J. (1990) *Global Warming*, The Greenpeace Report, Oxford University Press, Oxford, 554 pp.

Libby, L.M. and Pandolfi, L.J. Temperature dependence of isotopic ratios in tree rings. *Proc. Natl. Acad. Sci*. 71, 2482–2486

Lipp, J., Trimborn, P., Fritz, P., Moser, H., Becker, B. and Frenzel, B. (1991) Stable isotopes in tree ring cellulose and climatic change. *Tellus* 43B, 322–330.

Lipp, J., Trimborn, P., Becker, B. (1992) Rhythmic δD fluctuations in the tree-ring latewood cellulose of spruce trees (*Picea albes* L.). *Dendrochronologia* 10, 9–22.

Lipp, J., Trimborn, P., Graff, W. and Becker, B. (1993) Climatic significance of D/H ratios in the cellulose of late wood in tree rings from spruce (*Picea albes* L.). In: *Isotope Techniques in the study of past and current environmental changes in the hydrosphere and atmosphere*, International Atomic Energy Agency, Vienna, 1993, pp. 395–405.

Lipp, J., Trimborn, P., Graff, W. and Becker, B. (1994) A δ²H and δ¹³C chronology from tree rings of spruce (*Picea abies* L.) during the 20th century and climatic implications. In: *Proceedings of the Workshop Modelling of Tree-ring Development – Cell Structure and Environment* (H. Spiecker and P. Kahle eds.), Freiburg, Germany, pp. 24–38.

Livingstone, N.J., Spittlehouse, D.L. (1993) Carbon isotope fractionation in tree rings in relation to the growing season water balance. In: *Stable Isotopes and plant carbon-water relations*. (J.R., Ehleringer, A. E., Hall and G.D., Farquhar eds.) Academic Press, New York, pp. 141–153.

Loader, N.J. (1995) *The stable isotope dendroclimatology of* Pinus sylvestris *from Northern Britain. Ph.D Thesis,* Cambridge University, 356 pp.

Loader, N.J., Switsur, V.R. and Field, E.M. (1995) High resolution stable isotope analysis of tree rings: implications of 'microdendroclimatology' for palaeoenvironmental research. *The Holocene* 5, 457–460.

Loader, N.J. and Switsur, V.R. (1996) Reconstructing past environmental change using stable isotopes in tree rings. *Bot. J. Scotland*. 48, 65–78.

Long, A. (1982) Stable Isotopes in Tree Rings. In: *Climate from Tree Rings*. (M.K., Hughes, P.M. Kelly, J.R. Pilcher and V.C. LaMarche eds.) Cambridge University Press, pp. 12–17.

Martin, B., Bytnerowicz, A. and Thorstenson, Y.R. (1988) Effects of air pollutants on the composition of stable carbon isotopes, δ¹³C, of leaves and wood, and on leaf injury. *Plant Physiol*. 88, 213–218.

Martin, B. and Sutherland, E.K. (1990) Air pollution in the past recorded in width and stable isotope composition of annual growth rings of Douglas-fir. *Plant. Cell Environ*. 13, 839–844.

Mook, W.G., Koopmans, M., Carter, A.F. and Keeling C.D. (1983) Seasonal, latitudinal and secular variations in the abundance and isotopic ratios of atmospheric carbon dioxide. *J. Geophys. Res*. 88, 10915–10933.

Mullane, M.V., Waterhouse, J.S. and Switsur, V.R. (1988) On the development of a novel method for the determination of stable oxygen isotope ratios in cellulose. *Appl. Radiat. & Isotopes* **39**, 1028–1035.

Ogle, N. and McCormac, F.G. (1994) High resolution δ¹³C measurements of oak show a previously unobserved spring depletion. *Geophys. Res. Lett.* **21**, 2373–2375.

Peng, T.H., Broecker, W.S., Freyer, H.D. and Trumbore, S. (1983) A deconvolution of the tree ring based δ¹³C record. *J. Geophys. Res.* **88**, 3609–3620.

Ramesh, R., Bhattacharya, S.K. and Gopalan, K. (1985) Dendroclimatological implications of isotope coherence in trees from Kashmir Valley, India. *Nature* **317**, 802–804.

Ramesh, R., Bhattacharya, S.K. and Gopalan, K. (1986a) Climatic correlations in the stable isotope records of silver fir (*Abies pindrow*) trees from Kashmir, India. *Earth Planet. Sci. Lett.* **79**, 66–74.

Ramesh, R., Bhattacharya, S.K. and Gopalan, K. (1986b) Stable isotope systematics in tree cellulose as palaeoenvironmental indicators – a review. *J. Geo. Soc. India* **27**, 154–167.

Rittenberg, D. and Pontecorvo, L. (1956) A method for the determination of the ¹⁸O concentration of oxygen of organic compounds. *Int. J. Appl. Radiat. Isotopes.* **1**, 208–214.

Robertson, I., Field, E.M., Heaton, T.H.E., Pilcher, J.R., Pollard, M., Switsur, V.R. and Waterhouse, J.S. (1995) Isotope coherence in oak cellulose. In: *Paläoklimaforschung, Problems of stable isotopes in tree-rings, lake sediments and peat-bogs as climatic evidence for the Holocene* (ed. B Frenzel), Vol. 15, pp. 141–155.

Sauer, E.P. and Sternberg, L.d.S.L.O. (1994) Improved method for the determination of the oxygen isotopic composition of cellulose. *Anal. Chem.* **66**, 2409–2411.

Saurer, M. and Siegenthaler, U. (1989) ¹³C/¹²C isotope ratios in trees are sensitive to relative humidity. *Dendrochronologia* **7**, 9–13.

Schleser, G.H. (1991) Carbon isotope fractionation during CO_2 fixation by plants. In: *Modern Ecology: Basic and applied aspects*, (G. Esser and D. Overdieck eds.). Elsevier, pp. 603–622.

Sofer, Z. (1980) Preparation of carbon dioxide for stable isotope analysis. *Anal. Chem.* **52**, 1389–1391.

Sonninen, E. and Jungner, J. Stable carbon isotopes in tree rings of a Scots pine (Pinus sylvestris L.) from northern Finland. In: *Paläoklimaforschung, Problems of stable isotopes in tree-rings, lake sediments and peat-bogs as climatic evidence for the Holocene* (B. Frenzel ed.). Vol 15, pp. 121–128.

Sternberg, L.d.S.L. and DeNiro, M.J. (1983) Biogeochemical implications of the isotopic equilibrium fractionation factor between oxygen atoms of acetone and water. *Geochim. Cosmochim. Acta* **47**, 2271–2274.

Sternberg, L.d.S.L., DeNiro, M.J. and Savidge R.A. (1986) Oxygen isotope exchange between metabolites and water during biochemical reactions leading to cellulose synthesis. *Plant Physiol.* **82**, 423–427.

Stuiver, M. (1978) Atmospheric carbon dioxide and carbon reservoir changes. *Science* **99**, 253–258.

Stuiver, M., Burk, R.L. and Quay, P.D. (1984) ¹³C/¹²C ratios and the transfer of biospheric carbon to the atmosphere. *J. Geophys. Res.* **89**, 11731–11748.

Switsur, R., Waterhouse, J., Field, E., Carter, T., Hall, M., Pollard, M., Robertson, I., Pilcher, J. and Heaton, T. (1994) Stable isotope studies in tree rings of oak from the English Fenland and Northern Ireland. In: *Palaeoclimate of the Last Glacial/Interglacial Cycle,* (B.M. Funnell, R.L.F. Kay eds.) NERC Earth Sciences Directorate, Special publication 94/2, 67–73.

Switsur, V.R., Waterhouse, J.S., Field, E.M., Carter, T. and Loader, N. (1995) Stable isotope studies in tree rings from oak – techniques and some preliminary results In: *Paläoklimaforschung, Problems of stable isotopes in tree-rings, lake sediments and peat-bogs as climatic evidence for the Holocene* (B. Frenzel ed.). Vol. 15, pp. 129–140.

Switsur, V.R., Waterhouse, J.S., Field, E.M. and Carter, A.H.C. (1996) Climatic signals from stable isotopes in oak trees from East Anglia, Great Britain. In: *Tree Rings, Environment and Humanity* (J.S. Dean, D.M. Meko and T.W. Swetnam eds.), *Radiocarbon* pp. 637–645.

Tans, P.P. and Mook, W.G. (1980) Past Atmospheric CO_2 levels and the $^{13}C/^{12}C$ ratios in tree rings. *Tellus* **32**, 268–283.

Thompson, P. and Gray J. (1977) Determination of the $^{18}O/^{16}O$ ratios in compounds containing C,H, and O. *Int. J. Appl. Radiat. Isotopes* **28**, 411–415.

Trimborn, P., Becker, B., Kromer, B. and Lipp, J. (1995) Stable isotopes in tree rings: a palaeoclimatic tool for studying climatic change. In: *Paläoklimaforschung: Problems of stable isotopes in tree-rings, lake sediments and peat-bogs as climatic evidence for the Holocene* (B. Frenzel ed.). Vol. 15, pp. 163–170.

Urey, H.C. (1947) The Thermodynamic Properties of Isotopic Substances. *J. Chem. Soc.* 562–581.

Werner, R.A., Kornexl, A., Rossmann, A. and Schmidt, H-L. (1996) On-line determination of $\delta^{18}O$-values of organic substances. *Anal. Chim. Acta* **319**, 159–164.

White, J.W.C. (1988) Stable hydrogen isotopes in plants: a review of current theory and some potential applications. In: *Stable Isotopes in Ecological Research* (P.W. Rundel, J.R. Ehrlinger, and K.A. Nagy eds.), Springer, Berlin, pp. 144–162.

White, J.W.C., Cook, E.R., Lawrence, J.R. and Broecker, W.S. (1985) The D/H ratios of sap in trees: Implications for water sources and tree ring D/H ratios. *Geochim. Cosmochim. Acta* **49**, 237–246.

White, J.W.C., Lawrence, J.R. and Broecker, W.S. (1994) Modelling and interpreting D/H ratios in tree rings: a test case of white pine in the northeastern United States. *Geochim. Cosmochim. Acta* **58**, 851–862.

Wilson, A.T. and Grinsted, M.J. (1975) Palaeotemperatures from tree rings and the D/H ratio of cellulose as a biochemical thermometer. *Nature* **257**, 387–388.

Yakir, D. and De Niro, M.J. (1990) Oxygen and hydrogen isotope fractionation during cellulose metabolism in *Lemna gibba* L. *Plant Physiol.* **93**, 325–332.

Yakir, D. (1992) Variations in the natural abundance of oxygen-18 and deuterium in plant carbohydrates. *Plant, Cell and Environ.* **15**, 1005–1020.

Yapp, C. and J. Epstein, S.J. (1977) Climatic implications of meteoric water over North America (9,500–22,000 BP) as inferred from ancient wood cellulose C-H hydrogen. *Earth Planet. Sci. Lett.* **34**, 333–350.

Yapp, C. and Epstein S.J. (1982a) A re-examination of the cellulose carbon-bound hydrogen δD measurements and some factors affecting plant-water D/H relationships. *Geochim. Cosmochim. Acta* **46**, 955–965.

Yapp, C. and Epstein S.J. (1982b) Climatic significance of hydrogen isotope ratios in tree cellulose. *Nature* **297**, 636–639.

Yapp, C. and Epstein S.J. (1985) Seasonal contributions to the climatic variations recorded in tree ring deuterium/hydrogen data. *J. Geophys. Res.* **90**, 3747–3752.

Phylogeny, palaeoatmospheres and the evolution of phototrophy

J.A. Raven

19.1 Introduction

Data from palaeontology, from other earth sciences, and from molecular genetics and cladistics, has enabled a tentative timetable to be established for the evolution of photosynthetic organisms and for changes in the atmospheric (and aquatic and terrestrial) environments (Holland, 1984; Knoll, 1994; Raven, 1995). Measurements of the natural abundance of the stable isotopes of carbon have contributed to our understanding of this area. This chapter outlines the data which have been obtained and the conclusions which have been drawn from them, explores the extent to which stable carbon isotope studies have contributed to our understanding, and finally considers some possibilities for further contributions from work on stable carbon isotopes to this field.

19.2 The phylogeny of O_2-producing phototrophs and their carboxylases

19.2.1. *Endosymbiotic origin of plastids*

The prokaryotic cyanobacteria (including the chloroxybacteria or prochlorophytes) are the only O_2-evolving prokaryotes. O_2 evolution by eukaryotes can always be traced to an endosymbiotic event involving a unicellular, phagotrophic, non-O_2-evolving eukaryote ingesting a cyanobacterial cell. This cell did not suffer the usual fate of digestion and absorption but was retained as a photosynthetic organelle (plastid) which has lost the capacity for independent existence: some genes needed for independent growth were lost while many of those needed for chloroplast function were transferred to the eukaryotic (host) genome (see Allen and Raven, 1996; Bhattacharya and Medlin, 1995). Such a 'simple' ingestion of a cyanobacterium by a phagotrophic eukaryote accounts for the plastids (and nuclear, host genome) of the red algae and,

Stable Isotopes, edited by H. Griffiths.
© 1998 BIOS Scientific Publishers Ltd, Oxford.

through the loss of phycobilins and gain of chlorophyll *b* the green algae (Chlorophyta) and higher plants (Embryophyta).

The plastids of other algal taxa arose by a second round of endosymbioses which involved a eukaryotic O_2-evolving unicell, with a subsequent sequence of events similar to that detailed above for cyanobacteria to yield plastids with three or four membranes around them rather than the two found in red algae, green algae and higher plants. Such an endosymbiotic event involving a green alga and a non-photosynthetic euglenoid flagellate led to the photosynthetic euglenoids, and another event involving a green alga and an amoeboflagellate yielded chlorarachniophytes (Bhattacharya and Medlin, 1995; Palmer and Delwiche, 1996). Other endosymbiotic events involving four different higher taxa of phagotrophic non-photosynthetic flagellates gave rise to the Cryptophyta, Heterokontophyta, Haptophyta and Dinophyta; the exact number of endosymbiotic events involved here is still unclear (Cavalier-Smith *et al.*, 1996; despite the well-based arguments of Cavalier-Smith and Chao, 1996, that the Heterokontophyta should be referred to as the Ochrophyta, we retain the (possibly!) more familiar name). In all of these cases some pigment changes occurred; the cryptophytes altered the intracellular location of phycobilins and gained chlorophyllide c_2, while the other three divisions lost phycobilins and gained chlorophyllide(s) *c*. Phylogenetic analyses based on nucleotide sequences show that some of the eukaryote 'hosts' of these secondary endosymbioses branched earlier from the eukaryotic lineage than the 'hosts' of the red and green algae and higher plants so that these eukaryotes (e.g. the photosynthetic euglenoids) must have spent longer as non-photosynthetic eukaryotes than have the green and red algae.

19.2.2. *Anaplerotic carboxylases*

The implications of these phylogenetic considerations for stable C isotope studies relate directly mainly to the carboxylases other than ribulose-1,5-bisphosphate carboxylase-oxygenase (RUBISCO). Activity of these carboxylases accounts for less than 5% of the harvestable organic C in most O_2-evolvers (those with C_3 biochemistry), since they catalyse anaplerotic reactions which supply intermediates in biosynthesis which cannot be derived from the products of RUBISCO activity alone (Raven, 1996, 1997a). Most of this anaplerotic CO_2 fixation involves so-called $(C_3 + C_1)$ carboxylases which catalyse the addition of CO_2 or HCO_3- to pyruvate (pyruvate carboxylase, PC) or phosphoenolpyruvate (phosphoenolpyruvate carboxylase, PEPC or phosphoenolpyruvate carboxykinase, PEPCK). The occurrence of any one of these carboxylases as the major $(C_3 + C_1)$ carboxylase in a taxon varies among major taxa (see Table 19.1; Raven, 1997a), and at least some of these differences may be related to the main $(C_3 + C_1)$ carboxylase used by the eukaryotic host, or the prokaryotic (cyanobacterial) or photosynthetic eukaryotic endosymbiont. Further molecular genetic analyses are needed to test surmises as the origin of the major $(C_3 + C_1)$ carboxylase in these various taxa. As will be considered later, the implications of these variations in the predominant $(C_3 + C_1)$ carboxylase among major taxa studies of the natural abundance of stable isotopes relate to the different discriminations between ^{13}C-inorganic C and ^{12}C-inorganic C expressed in terms of a source inorganic value relative to CO_2; this discrimination is high for PEPCK, low for PEPC; and poorly characterised (but probably low) for PC (Table 19.2). The implications of the occurrence of a given $(C_3 + C_1)$ carboxylation involving exogenous inorganic C (followed by a $(C_4 - C_1)$ decarboxylation in the vicinity of RUBISCO) intervenes between exogenous inorganic C and the activity of RUBISCO. Such plants

Table 19.1. Predominant $C_3 + C_1$ carboxylases in O_2-evolving Photolithotrophs (from Raven, 1997a)

Taxon	$C_3 + C_1$	Immediate Inorganic C source
Prokaryotes		
Cyanobacteria	Phosphoenolpyruvate carboxylase	HCO_3^-
Eukaryotes		
Rhodophyta	Phosphoenolpyruvate carboxylase	HCO_3^-
Cryptophyta	Phosphoenolpyruvate carboxylase	HCO_3^-
Haptophyta	Phosphoenolpyruvate carboxykinase **or**	CO_2
	Phosphoenolpyruvate carboxylase	HCO_3^-
Heterokontophyta		
Bacillariophyceae	Phosphoenolpyruvate carboxykinase	CO_2
Phaeophyceae	Phosphoenolpyruvate carboxykinase	CO_2
Raphidophyceae	Phosphoenolpyruvate carboxykinase	CO_2
Dinophyta	Pyruvate carboxylase **or**	HCO_3^-
	Phosphoenolpyruvate carboxylase	HCO_3^-
	Phosphoenolpyruvate carboxykinase	CO_2
Chlorophyta	Phosphoenolpyruvate carboxylase **or**	HCO_3^-
	Phosphoenolpyruvate carboxykinase	CO_2
Bryophyta	Phosphoenolpyruvate carboxylase	HCO_3^-
Tracheophyta	Phosphoenolpyruvate carboxylase	HCO_3^-

are mainly the vascular plants exhibiting C_4 or CAM metabolism (at least 10% of vascular plant species), and the relatively few macroalgae with C_4- or CAM-like behaviour (Griffiths, 1992; Peisker and Henderson, 1992; Raven, 1991, 1997a).

19.2.3. *Ribulose bisphosphate carboxylase-oxygenase (RUBISCO)*

The structure of RUBISCO, the core carboxylase through which 95% or more of the total organic C in harvestable plant material has been routed, shows phylogenetic variations in nucleotide sequence within the (monophyletic) gene family. These variations can most readily be explained by two separate acquisitions of RUBISCO genes by eukaryotes other than via the endosymbiotic cyanobacterium which gave rise to all plastids by one or two rounds of endosymbiosis. The RUBISCO of Chlorophyta, Euglenophyta and Embryophyta (and, presumably, the Chlorarachniophyta) has been shown by molecular genetic studies to have been acquired directly from a cyanobacterium as part of the endosymbiosis which gave rise to all plastids (Bhattacharya and Medlin, 1995; Paul and Pichard, 1996; Palmer and Delwiche, 1996). The RUBISCO of Rhodophyta and thus (by secondary endosymbiosis) of Cryptophyta, Heterokontophyta and Haptophyta was derived from a RUBISCO of a β-proteobacterium (Bhattacharya and Medlin, 1995; Paul and Pichard, 1996; Palmer and Delwiche, 1996). It is not known if this β-proteobacterium was the one which gave rise to the mitochondria of these (and other) eukaryotes. The kinetic characteristics of the cyanobacterial RUBISCO and its derivatives (the enzyme of the Chlorophyta, Embryophyta and Euglenophyta) vary with the organism in which they are found. Thus, the affinity for CO_2, and the selectivity for CO_2 over O_2 (denoted by S_{rel},

Table 19.2. *Values of* α *(rate constant* $^{12}CO_2$/*rate constant with* $^{13}CO_2$*) for various reactions of* CO_2, *including carboxylation processes (from compilations in Raven, 1996, 1997a; Raven et al., 1994)*

Process	α
Gaseous $CO_2 \rightarrow$ dissolved CO_2	1.0010
Diffusion of dissolved CO_2 in solution	1.0007–1.0009
Flux of CO_2 or HCO_3- through membranes	1.00?
Uncatalysed conversion of dissolved CO_2 to HCO_3^-	1.013
Uncatalysed conversion of HCO_3- to dissolved CO_2	1.022
Carbonic anhydrase – catalysed conversion of dissolved CO_2 to HCO_3- (*in vitro*)	1.0001
Carbonic anhydrase – catalysed conversion of HCO_3- to dissolved CO_2 (*in vitro*)	1.0101
Fixation of dissolved CO_2 by RUBISCO from	
freshwater β-proteobacteria (*in vitro*)	1.018
freshwater cyanobacteria (*in vitro*)	1.022–1.025
freshwater and marine cyanobacteria (*in vivo*)	≤1.025
freshwater Rhodophyta (*in vivo*)	≤1.032
marine Rhodophyta (*in vivo*)	≤1.025
freshwater Bacillariophyceae (*in vivo*)	1.024
marine Phaeophyceae (*in vivo*)	≤1.029
freshwater Chlorophyceae (*in vivo*)	≤1.028
freshwater Bryophyta (*in vivo*)	≤1.033
terrestrial Tracheophyta (*in vitro*)	1.029
Fixation of HCO_3- (expressed in terms of dissolved CO_2) by phosphoenolpyruvate carboxylase (*in vitro*)	1.0047
Fixation of HCO_3- (expressed in terms of dissolved CO_2) by carbamylphosphate synthetase (*in vivo*)	1.001?
Fixation of CO_2 by phosphoenolpyruvate carboxykinase (*in vitro*)	1.024–1.040

In vitro values may be over estimates if ^{13}C-enriched C is lost from the organisms as CO_2 (e.g. in lipogenesis) or as organic C, and are under estimates when ^{13}C-enriched C is gained by anaplerotic CO_2/HCO_3- fixation (phosphoenolpyruvate carboxylase; carbamylphosphate synthetase).

the CO_2/O_2 selectivity factor, defined as $(V_{m(CO2)}.K_{m(O2)})/V_{m(O2)}.K_{m(CO2)})$, where V_m is the maximum velocity of the enzyme, and K_m the half-saturation value, for the substrate denoted in brackets), are lowest in the cyanobacteria, and highest in land plants with dif-fusive CO_2 entry to RUBISCO, with intermediate values for eukaryotes with CO_2 concentrating mechanisms (see Raven, 1996; Tabita, 1995; Uemura *et al.*, 1996). Importantly for the present volume is the finding that ^{13}C/^{12}C discrimination for CO_2 when there is no restriction by CO_2 diffusion is greater for the eukaryotic enzymes than for the cyanobac-terial RUBISCOs (Table 19.2). The β-proteobacterial RUBISCOs in eukaryotes have not thus far been shown to have CO_2 affinities as high as the highest values in chlorophyte and embryophyte RUBISCOs, but they do seem to have higher CO_2/O_2 selectivity factors than any of the cyanobacterially-derived RUBISCOs examined thus far. The $^{13}CO_2$/$^{12}CO_2$

discrimination factor (unconstrained by diffusion) α, defined as the rate constant for $^{12}CO_2$ fixation divided by the rate constant for $^{13}CO_2$ fixation, for the eukaryotic RUBISCOs derived from β-proteobacteria are similar to those for the eukaryotic RUBISCOs derived from cyanobacteria (Raven, 1996, 1997a; Table 19.2).

All of the RUBISCOs considered so far are of Form 1, that is have 8 large (catalytic) subunits and 8 small subunits in each hexadecameric enzyme molecule. The other version of RUBISCO is Form II, with 2 large (catalytic) subunits per enzyme molecule. Form II RUBISCO was thought to be confined to prokaryotes until a Form II enzyme related to that in α-proteobacteria was found as the sole functional RUBISCO, in those members of the Dinophyta with the most common pigmentation (chlorophyll a – chlorophyllide c_2 – peridinin): (Rowan et $al.$, 1996 and references cited therein). The kinetics of the dinophyte Form II RUBISCO have not yet been characterised due to instability of the extracted enzyme (Whitney and Yellowlees, 1995); this also precludes estimation of in $vitro$ values of $^{13}CO_2/^{12}CO_2$ discrimination for this enzyme. Data on the prokaryotic Form II enzyme show lower $^{13}CO_2/^{12}CO_2$ discrimination, CO_2 affinity and CO_2/O_2 selectivity than for any of the Form I enzymes (Raven, 1996; Robinson and Cavanaugh, 1995).

Thus, both kinetics and $^{13}CO_2/^{12}CO_2$ discrimination vary within the best-investigated subfamily of RUBISCOs (i.e. the cyanobacteria – Chlorophyta – Euglenophyta – Embryophyta lineage) both kinetics and $^{13}CO_2/^{12}CO_2$ discrimination vary. At least for the kinetics, the variation can be related to some extent to the habitat and to the presence or absence of a CO_2 concentrating mechanism (Raven, 1991, 1996, 1997a,b). Some kinetic differences seem to occur within the β-proteobacterial subfamily of eukaryotic RUBISCOs, although the available data are not adequate to decide if there is also variation in $^{13}CO_2/^{12}CO_2$ discrimination within this subfamily.

In the context of present-day habitats, we conclude from this discussion of the phylogeny of O_2-producing phototrophs and their carboxylases, that the genetic distance among the nucleotide sequences for both nuclear (e.g. 18S rRNA) and plastid (e.g. 16S rRNA) genes is greater for aquatic, and especially marine, O_2-evolvers than for terrestrial O_2-evolvers. Thus, extant terrestrial O_2-evolvers are cyanobacteria (free-living and lichenised), Chlorophyta (free-living and lichenised) and overwhelmingly, Embryophyta. In addition to these higher taxa, extant aquatic habitats have representatives of the Euglenophyta, Rhodophyta, Cryptophyta, Heterokontophyta, Haptophyta, Dinophyta and Chlorarachniophyta. The greatest diversity of higher taxa of O_2-evolvers occurs in marine habitats, while alga with diversity decreasing in the order freshwater, then soil, then more fully terrestrial (soil surface, rocks, surface of plant shoots) habitats (Raven, 1995). The greater diversity (in terms of genetic distance) in aquatic habitats is also seen for the major (C_3 + C_1) carboxylases in a given Division and the subfamilies of RUBISCO represented therein. Therefore, the range of maximum in $vivo$ $^{13}C/^{12}C$ discrimination values attainable during photosynthesis may be greater for aquatic than for terrestrial habitats. This will be considered again after the time-scale of the evolution of the various taxa and of the occurrence of terrestrial O_2-evolvers has been discussed in the context of the changes in atmospheric composition over the last 4.6 Ga.

19.3 Timing of origin of taxa of O_2-evolving phototrophs in relation to changes in atmospheric composition

Figure 19.1 shows the times at which various higher taxa of O_2-evolvers are thought to

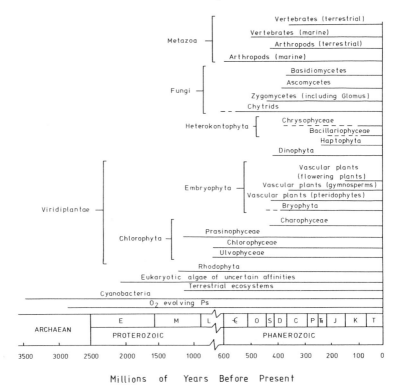

Figure 19.1. *Major events in the evolution of aquatic and terrestrial O_2-evolvers and of certain chemoorganotrophs. Note change of scale between Precambrian and Phanerozoic. C = Cambrian, O = Ordovician, S = Silurian, D = Devonian, C = Carboniferous, P = Permian, TR = Triassic, J = Jurassic, K = Cretaceous, T = Tertiary. Modified from Raven (1997b).*

have originated, and the timing of the first occurrence of terrestrial biota, based on fossil and palaeogeochemical evidence rather than on at least some interpretations of the 'molecular clock' (Doolittle *et al.*, 1995; Hagesewara *et al.*, 1996; Raven, 1995, 1997a,b). Figure 19.2 shows variations in atmospheric CO_2 and O_2 over the last 3.5 Ga. Comparison of the data in Figures 19.1 and 19.2 yield the following conclusions.

19.3.1. *Evidence for the development of phototrophy in relation to CO_2 and O_2*

Cyanobacteria, presumably with O_2 evolution, have existed for 3.45 Ga (Schopf, 1993). This considerably predates the occurrence of O_2 in the ocean or atmosphere at more than the $\leq 10^{-8}$ of the present level which could be produced abiologically by the photodissociation of H_2O (Figure 19.2; Holland, 1984; Schidlowski, 1988). Any free O_2 before ~2.5 Ga was consumed by oxidation of reductants (Fe^{2+}, S^{2-}) in the ocean and on the land surface which has been exposed to the atmosphere for at least 3.5 Ga (Buick *et al.*, 1995; Hoffman, 1995).

Fossil evidence for eukaryotes, in the form of an alga (*Grypanea spiralis*) of unknown affinity, are known from deposits 2.1 Ga old (Han and Runnegar, 1992 and, more convincingly, 1.7 Ga old deposits (Shixing and Huineng, 1995) as a cellularly preserved macroalga, again of unknown affinity.

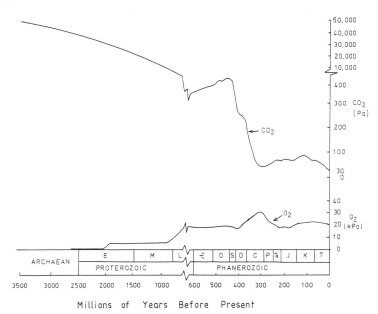

Figure 19.2. Atmospheric O_2 and CO_2 over the last 4 billion years. Note change of time scale between Precambrian and Phanerozoic, and also change of scale for CO_2. Modified from Raven (1997b); based on Berner (1990, 1993, 1994), Berner and Canfield (1989), Canfield and Teske (1996) and Holland (1994).

Geochemical evidence for terrestrial photosynthetic ecosystems, based on higher weathering rates than are expected from atmospheric CO_2 concentrations has been suggested for deposits 1.2 Ga old ago (Horodyski and Knauth, 1994).

Fossil evidence from marine deposits 0.7–1.4 Ga old suggests the presence of members of the (eukaryotic) algal divisions Chlorophyta, Rhodophyta and Heterokontophyta (Butterfield *et al.*, 1990; Knoll, 1994; cf. Kooistra and Medlin, 1996).

There is fossil evidence of the presence of three classes of the Chlorophyta (Prasinophyceae, Chlorophyceae *sensu lato* and Ulvophyceae *sensu lato*) in the Neoproterozoic. However, fossil evidence of the class Charophyceae, which are closest to the ancestors of the Embryophyta, is not known prior to about 428 Ma ago, that is at about the same time as the earliest Embryophyta (Graham, 1993; Raven, 1995; Taylor, 1995a,b). The earliest embryophytes in the Ordovician were probably like certain extant liverworts (Edwards *et al.*, 1995; Taylor, 1995a,b), while both the sporophyte and gametophyte phase of vascular plants are known from Upper Silurian/Lower Devonian strata (Raven, 1995). At least the sporophyte phase of these plants would have been capable of homoiohydry, that is retain hydration of the tissues despite low soil, water availability and high evaporative demand of the atmosphere (Raven, 1993, 1995). Yapp and Poths (1994) use $Fe(CO_3)OH$ in goethite in Ordovician palaeosols as a proxy for soil CO_2 partial pressure, which they relate to organic C input from primary productivity with subsequent chemoorganotrophic activity. Their conclusion is that primary productivity in the Ordovician approached extant values for moderately productive extant terrestrial ecosystems (Yapp and Poths, 1994). Raven (1993, 1995) points out that, even with high CO_2 partial pressures in the Ordovician (Figure 19.2), the productivities suggested by

Yapp and Poths (1994) require a larger area for CO_2 absorption per m² of ground than is provided by a simple microbial mat and that, at the liverwort level of organisation, this could be achieved by several layers of unistratose leaves or a ventilated foliose thallus, for example Marchantiales.

Subsequent evolution of terrestrial vegetation led to the tree life form (late Devonian), the gymnospermous reproductive mechanism (early Carboniferous) and the angiospermous reproductive mechanism (early Cretaceous), among the vascular plants (Figure 19.1; Willis, 1996). Evidence (including stable C isotope evidence) on the evolution of C_4 photosynthesis does not give conclusive indications until the mid Miocene (see Chapter 21), while that for CAM does not appear till the Pleistocene (Raven and Spicer, 1996). This late fossil evidence is despite low CO_2, and low CO_2/O_2, in the Upper Carboniferous, some 250 Ma before first clear indications of C_4 and CAM with low CO_2 and CO_2/O_2 in the Neogene (Miocene – Pliocene – Pleistocene) as likely selective forces favouring these pathways (see Chapter 21).

19.3.2. *The biogeochemical control of atmospheric CO_2 and O_2*

The Phanerozoic values for O_2 and CO_2 in Figure 19.2 derive from chemical mass balances of oxidized and reduced C, Fe, and S (O_2; Berner and Canfield, 1989) and from mass balances of inorganic and organic C (CO_2; Berner, 1990, 1993, 1994). Proterozoic and Archaean values for these gases are not so exact. O_2 values are deduced from such evidence as the survival of O_2-sensitive minerals such as uraninite and the occurrence of 'red beds' and 'banded iron formation', as well as the occurrence of organisms whose size and (putatively) aerobic metabolism demands a certain O_2 partial pressure for growth (Holland, 1984) and phylogenetic and sulphur-isotope studies (Canfield and Teske, 1996). For CO_2 we rely on arguments from the greenhouse effects, such that a larger greenhouse effect was needed earlier in the earth's history to explain the continuous occurrence of unfrozen oceans, the lower total radiant output of the 'weak young sun' (Newman, 1979), as well as or evidence from the mature of carbonate minerals and the weathering rate of terrestrial rocks.

The role of biota in bringing about Archaean, Proterozoic and Phanerozoic O_2 and CO_2 changes is known at least in semi-quantitative terms. Since photodissociation of water vapour could only account for less than 10^{-8} of the present O_2 partial pressure, the remainder is absolutely dependent on photolithotrophic O_2 production (Holland, 1984; Schidlowski, 1988). The quantity of O_2 left effectively represents a 'fossil' of previous photosynthesis in a given geological epoch, depending on the quantity of organic C (originally produced in parallel with O_2) which has been sedimented and not subsequently reoxidised, as well as that O_2 used in oxidising Fe(II) and S(II). Thus, photosynthetic O_2 production has been modulated by chemo-organotrophic, geological and geochemical processes (Berner and Canfield, 1989). As for CO_2, the net conversion of CO_2 to (sedimented) organic C caused a drawdown of CO_2 but the situation is complicated by other reactions, for example the precipitation of $CaCO_3$ in the ocean from the (presently) supersaturated solution, which is (presently) largely under biological control.

It is important to note that precipitation of $CaCO_3$ *per se* is a CO_2-producing reaction (Frankignoule, *et al.*, 1993). Weathering of silicates on land can, however, remove CO_2 from the atmosphere:

$$CaSiO_3 + CO_2 \rightarrow CaCO_3 + SiO_2$$

These considerations show that photolithotrophs had major impacts on atmospheric composition in terms of O_2 before the evolution of terrestrial ecosystems as well as subsequently (Figures 19.1 and 19.2).

19.3.3. *Effect of atmospheric O_2 on phototroph-based nutrient cycles*

The transition to an oxidising, O_2-containing atmosphere from a neutral or slightly reducing atmosphere had implications for availability of other resources. The oxidation of Fe(II) to Fe(III), which (with S(II) oxidation) delayed the accumulation of free O_2 until well after the onset of O_2-evolving photosynthesis, had major impacts on the availability of not only Fe but also of P. Thus, the oxidation of Fe(II) to Fe(III) leads to the precipitation of Fe(III) as oxide/hydroxide, thus minimizing free Fe(II) in solution and leading to the evolution of siderophore and plant surface reduction mechanisms which facilitate Fe acquisition for the numerous Fe-requiring catalysts which evolved early in the history of life (see Couture *et al.*, 1994; Hardison, 1993; Raven, 1997b). The occurrence of Fe(III) solid phases can lead to the binding of orthophosphate-P in oxidised aquatic and terrestrial sediments. The build-up of atmospheric O_2 leads to the occurrence of a stratospheric O_3 layer which acts as a (partial) screen for UV-B (reviewed by Rozema *et al.*, 1997). Such a screen would be especially important for any emerging terrestrial biota, since all natural waters attenuate UV-B to a greater extent than photosynthetically active radiation so the aquatic phototrophs are protected from UV-B relative to their terrestrial counterparts (Raven, 1995, 1997b; Raven and Sprent, 1989). The O_3-based UV-screen would also reduce the Fe(II) availability resulting from Fe(III) photoreduction in water bodies using UV-B. O_2 could also have acted as a feedback control on stratospheric O_3 by the use of O_2 in the peroxidative (using a V-containing peroxidase) production of halocarbons by marine biota (mainly algae) which, upon transfer to the stratosphere and breakdown to Cl, Br, OCl and OBr radicals, can catalyse O_3 destruction.

Other influences of O_2 on resource availability include the provision of combined N as NO_x via lightning (whose fossilized traces can be seen in fused sandstones) which makes up in part for decreased abiogenic production of combined N in the earlier, O_2-free atmosphere and the O_2 inhibition of biological N_2 fixation. A further influence of O_2 on resource supply concerns $(CH_3)_2S$ production by marine plants; this volatile compound is oxidized by atmospheric $^{\bullet}OH$ (produced photochemically from O_2) to produce, in part, SO_x which can supply S to S-deficient terrestrial habitats (Raven, 1995). A final influence of free O_2 on photolithotrophs is via the intracellular production of damaging O species, including free radicals and singlet O, in part offsetting the decreased UV-B flux resulting from the O_3 screen. While the phylogenetically earliest enzymes which scavenge such damaging O species as H_2O_2 and $O_2^{\bullet}$-. contained Fe (catalase, most peroxidases, some superoxide dismutases) or Mn (some superoxide dismutases), those which evolved later (e.g. a form of superoxide dismutase in Charophyceae and Embryophyta) has Cu and Zn, metals whose availability is increased as O_2 increases (Raven, 1995, 1997b).

Somewhat at variance with the argument developed above are phylogenetic arguments based on molecular genetic analyses of such respiratory catalysts as the cytochrome *b*–cytochrome c_1-Rieske non-heme iron complex, cytochrome oxidase, and F- and V-type H^+ ATPases which show that all of these components predate O_2-evolving photosynthesis (reviewed by Schäfer *et al.*, 1996). This means that aerobic respiration, which is one of the more 'obvious' biological consequences of O_2-build-up, predates O_2-evolving photosynthesis as source of the O_2 used as the substrate for the cytochrome oxidase and as the means of making available the Cu which is an essential component of all cytochrome oxidases.

19.3.4. *The effect of O_2 on RUBISCO activity in the aquatic environment*

These considerations show that the presence of free O_2 impacts on the availability of both resources such as N, P, S and Fe and on the potential for damage by O_2 free radicals and by UV-B. What have not been considered thus far are direct impacts on CO_2 assimilation, although clearly all of the influxes considered above will have an indirect effect on CO_2 fixation and hence, potentially, on $^{13}C/^{12}C$ discrimination. A direct effect on CO_2 fixation occurs whenever the O_2/CO_2 concentration ratio at the site of RUBISCO is low enough for significant oxygenase activity of RUBISCO to occur. With the high CO_2 levels envisaged for the Proterozoic as O_2 build-up began (Figure 19.2), the CO_2 concentration and CO_2/O_2 ratio at the site of RUBISCO activity would be relatively high even with diffusive entry of CO_2 and dissipation of O_2. This high $CO_2 : O_2$ would mean very substantial suppression of RUBISCO oxygenase, even with the low CO_2 affinities and low CO_2/O_2 selectivities which characterise the RUBISCOs of extant phototrophic proteobacteria and cyanobacteria. The argument developed above applies even when taking into account the attenuation of incident radiation by photosynthetic pigments in bulky objects like stromatolites or more planar microbial mats with their own significant diffusion boundary layer as well as internal diffusion distances which tend to restrict CO_2 entry and O_2 loss. This has clear implications (see later) for $^{13}C/^{12}C$ discrimination.

As CO_2 levels fell and O_2 stayed constant or rose in the late Proterozoic and early Phanerozoic the force of the argument made in the previous paragraph is decreased, at least for benthic macroscopic phototrophs. For planktonic phototrophs, and especially picoplanktonic organisms with diffusion boundary layers only some 1–2 μm thick, the late Proterozoic and early Phanerozoic CO_2 and O_2 concentrations in surface seawater would still be adequate to saturate the carboxylase functioning of RUBISCO and minimize the oxygenase activity. This would also be the case for the more complex terrestrial macrophytes, and especially the vascular plants from the late Silurian into the Devonian. The aqueous-phase diffusion distance for CO_2 to the site of RUBISCO is also only 1–2 μm, courtesy of very strict control of the position of the gas–water interface on the outer plant surface and in the intercellular spaces (Evans & von Caemmerer, 1996; Laisk & Loreto, 1996; Raven, 1991; Sültemeyer & Rinast, 1996). The benthic microscopic phototrophs in microbial mats underwater or on land might have become CO_2-limited due to the decreased CO_2 concentration and the significant diffusion distance (Raven, 1993, 1995). The same may apply to bulky aquatic macrophytes in the late Proterozoic and early Phanerozoic.

The onset of CO_2 limitation and the consequent selection pressure for the direct or indirect use of HCO_3- means that the enzyme carbonic anhydrase would have become essential uncatalysed HCO_3–CO_2 interconversion was not to limit photosynthesis. The three independently evolved families of carbonic anhydrase all use Zn (or Cd or Co) as an essential catalytic component; the availability of Zn increased as O_2 increased (Raven, 1997a).

The low CO_2 level in the Carboniferous, and the (probably) high O_2 level at that time, would have meant that even terrestrial C_3 vascular plants and picoplankton relying on diffusive CO_2 entry with short (a few μm) aqueous-phase diffusion distances would not saturate RUBISCO with CO_2 and would permit oxygenase activity of RUBISCO. For most of the Phanerozoic the CO_2 level, and CO_2/O_2 ratio, was higher than the present value, with CO_2 and CO_2: O_2 values in the glacial episodes in the Pleistocene which were lower than present values. In the Mesozoic the high CO_2 and CO_2/O_2 ratio would have saturated RUBISCO with CO_2 and made the very high temperatures compatible with C_3 photosynthesis (Raven and Spicer, 1996).

19.3.5. *Selective driving forces for the evolution of CO_2 concentrating mechanisms*

The discussion above shows that low CO_2, and CO_2/O_2, in photosynthesising cells, could have arisen in microbial mats and in bulky aquatic or terrestrial macrophytes with no internal ventilation in the late Proterozoic and early Phanerozoic due to the long CO_2 and O_2 diffusion pathlengths between the bulk phase and RUBISCO and the sites of O_2 evolution, even with high environmental CO_2 and CO_2/O_2. Organisms with much shorter CO_2 and O_2 diffusion paths would also have low intercellular CO_2 and CO_2/O_2 in the low CO_2/high O_2 environments of the Carboniferous. In both cases the steady-state CO_2 concentrations, and CO_2/O_2 ratios, at the site of RUBISCO activity would have been so low as to not saturate the enzyme with CO_2, and would permit oxygenase activity, even with the lowest known $K_{m(CO2)}$ and highest known S_{rel} for extant RUBISCOs (Raven, 1996). These extremes of CO_2 deprivation and low CO_2/O_2 would presumably have provided a selective pressure for CO_2 concentrating mechanisms which would have increased not only the maximum rate of light, N, and (on land) H_2O-saturated photosynthesis, but could also increase the CO_2 fixed per unit light absorbed or H_2O transpired, and the rate of CO_2 fixation per unit tissue N (see Raven, 1991).

There are three main categories of CO_2 concentrating mechanisms. One is the active transport of CO_2 or HCO_3- across a membrane or membranes; this is widespread today in primarily and secondarily aquatic phototrophs, many terrestrial algae and cyanobacteria, both free-living and lichenised, and the terrestrial embryophytic Anthocerotae: (Raven, 1991; Smith and Griffiths, 1996a,b). A second is the C_4 pathway, restricted in its classic form to terrestrial flowering plants, and involving localization of biochemical reactions in different cell types. The third is CAM, involving dark CO_2 fixation by a mechanism which has very little energy cost, with refixation of this CO_2 in the light phase, and mainly found in terrestrial vascular plants (Winter and Smith, 1996). The timing of the evolution of these various mechanisms is a matter of considerable contention; it is generally held that all three of the mechanisms are polyphyletic (Apel, 1994; Raven, 1991; Sinna and Kellog, 1996; Winter and Smith, 1996). At least for the C_4 pathway, and for the CAM flowering plants, they cannot be older than the angiosperms themselves, which at least on fossil evidence are less than 200 Ma old (Figure 19.1). For primarily aquatic plants it is likely that at least some of the major eukaryotic clades are well over 600 Ma old (Figure 19.1), while cyanobacteria are probably 3.5 Ga old; all of these organisms thus evolved well before the CO_2 level, and CO_2/O_2 ratio in the environment would have given large selective pressures favouring CO_2 concentrating mechanisms, and the likelihood of a polyphyletic origin of the CO_2 concentrating mechanisms based on transmembrane active transport of CO_2/HCO_3- (Raven, 1991, 1997a). However, it is likely that the carboxysomes (aggregations of RUBISCO in a proteinaceous shell in autotrophic prokaryotes using the photosynthetic carbon reduction cycle which are involved in CO_2 concentrating mechanisms) are monophyletic (Raven, 1997a). Furthermore, pyrenoids (aggregations of RUBISCO in the plastid stroma of some, mainly aquatic, eukaryotic phototrophs, associated with CO_2 concentrating mechanisms but not needed for all eukaryotic CO_2 concentrating mechanisms) are apparently ancestral features of photosynthetic eukaryotes and that their absence is due to (multiple) losses (Raven, 1997a). While Schopf (1968) has suggested that pyrenoid-like structures occur in putative eukaryotes from 1 Ga this identification, and thus the implication of a CO_2 concentrating mechanism, is open to question.

19.4 The contribution of $^{13}C/^{12}C$ measurements to our understanding of the evolution of phototrophy

The timing of the origin of these various pathways is one of the contributions which natural abundance studies of C isotopes can contribute to the study of the evolution of phototrophs. Granted a knowledge of the pathway of CO_2 fixation, $^{13}C/^{12}C$ ratios can be used to estimate the extent of diffusive limitation of photosynthesis in C_3 plants. $^{13}C/^{12}C$ ratios can also be related to the extent of the 'biological pump' in aquatic systems (drawdown of surface water inorganic C as organic C is sedimented).

19.4.1. *Derivation of carbon isotope discrimination terminology*

The interpretation of $^{13}C/^{12}C$ ratios in fossil organic C (or biotically influenced localized inorganic C deposits) in terms of photosynthetic pathways requires that the maximum extent of discrimination by the carboxylases involved in CO_2/HCO_3^- fixation, as well as the $^{13}C/^{12}C$ of the bulk phase CO_2/HCO_3^-. The reason for needing to know both of these parameters is shown by considering the equation which describes the $^{13}C/^{12}C$ discrimination by diffusion of CO_2 and of the subsequent carboxylation reaction(s), thus (Farquhar *et al.*, 1989):

$$\Delta = \alpha_d \frac{(C_b - C_c)}{(C_b)} + \alpha_c \frac{(C_c)}{(C_b)} - 1 \tag{19.1}$$

where C_c = CO_2 concentration in plastids
$\quad C_d$ = CO_2 concentration in the bulk medium (C_a in air)
$\quad \alpha_c$ = discrimination factor between $^{12}CO_2$ and $^{12}CO_2$ in carboxylation
$\quad \alpha_d$ = discrimination factor between $^{13}CO_2$ and $^{12}CO_2$ in diffusion
$\quad \Delta$ = difference between $^{13}C/^{12}C$ in organic C in the plant and $^{13}C/^{12}C$ in source CO_2, where Δ is defined in equation (19.2):

$$\Delta = \frac{\delta^{13}C_{source} - \delta^{13}C_{plant}}{1 + \delta^{13}C_{plant}} \tag{19.2}$$

where subscripts "source" and "plant" refer to the source CO_2 and to the plant organic C respectively, and $\delta^{13}C$ is defined by equation (19.3):

$$\delta^{13}C = \frac{(^{13}C/^{12}C)_{sample}}{(^{13}C/^{12}C)_{standard}} - 1 \tag{19.3}$$

where the subscripts "sample" and "standard" refer to the sample (source or plant in equation (19.2)) and the standard (carbonate from the Pee-Dee Belemnite, or PDB) respectively. The need for a knowledge of the source CO_2 value comes from equation (19.2) which defines Δ in terms of $\delta^{13}C$ values.

19.4.2. *Constraints on the interpretation of Δ values of aquatic and terrestrial organisms*

The use of Equation (19.1) in the study of extant (or herbarium) plants is that, granted that the plant is known to be terrestrial and vascular, a value of Δ in excess of 20‰ means C_3 (O'Leary, *et al.*, 1992), $C_3 - C_4$ intermediate (von Caemmerer, 1992) or the C_3 mode of facultative CAM plants (Griffiths, 1992). This prediction is mechanistically based on the weighted mean of the fractionation factor, α_d, for gas-phase diffusion through boundary layers and stomata, and of the fractionation factor, α_c, for RUBISCO with some (5%) parallel fixation by PEPC, CPS, etc. The processes regulating stomatal opening and photosynthetic biochemistry constrain the value of C_c/C_a and hence that of Δ (Equation (19.1)). The C_4 and CAM pathways have much lower Δ values (≤ 15‰), resulting from the use of PEPC as the initial carboxylase with subsequent decarboxylation and refixation by RUBISCO with some leakage of CO_2 (Griffiths, 1992; Peisker & Henderson, 1992). Granted that the Δ value in excess of 20‰ defines the plant as C_3 or $C_3 - C_4$ intermediates, variations in Δ within the range 20–26‰ can be accounted for by variations in C_c/C_a and in α_c (Equation (19.1); Farquhar *et al.*, 1989; Raven and Farquhar, 1990) as a function of genotype and environment, affecting the fraction of anaplerotic C fixation and the ratio of stomatal to biochemical conductance.

For aquatic plants, and poikilohydric land plants, diffusive CO_2 entry and C_3 biochemistry can yield much lower Δ values since diffusive limitation of photosynthesis can be much higher in water or with a water layer over the plant than in a C_3 vascular land plant where aqueous phase diffusion pathlengths are minimal. A lower Δ value thus cannot distinguish diffusive entry of CO_2 with C_3 biochemistry from operation of a CO_2 concentrating mechanism. However, a CO_2 concentrating mechanism based on HCO_3^- use can yield negative Δ values when Δ is referred (Equations (19.1)–(19.3) to exogenous CO_2, since HCO_3^- has a higher $^{13}C/^{12}C$ (and hence less negative $\delta^{13}C$) than does equilibrium CO_2 in solution or the gas phase. Thus, extreme values of Δ in aquatic plants can indicate exclusively diffusive CO_2 entry ($\Delta > 22$‰) or HCO_3^- use ($\Delta < 0$‰), although intermediate values do not rule out either pathway (or active influx of CO_2: see Raven, 1997a; Raven *et al.*, 1995). Interpretation of Δ values in C_3 aquatic plants using Equation (19.1) requires not only use of a lower α_d than for gas-phase diffusion, but also that the phylogenetic diversity of $^{13}CO_2/^{12}CO_2$ discrimination known for RUBISCO and in the anaplerotic carboxylases with different inorganic C $^{13}C/^{12}C$ discrimination (Raven, 1997a; Raven *et al.*, 1994, 1995).

Hence the interpretation of the $\delta^{13}C$ values of organic C in extant plants demands, as a minimum, a knowledge of the carboxylases involved, their inorganic C isotopic discrimination, the source CO_2 $\delta^{13}C$ value and whether the organism is terrestrial or aquatic. In addition to these requirements, the interpretation of fossil material requires consideration of diagenetic effects, that is the changes which occur in the conversion of live material to the fossil as found. Differential loss of chemical components with specific $^{13}C/^{12}C$ ratios is the main impact of diagenesis on $\delta^{13}C$ values. Cuticular components are frequently preferentially preserved, and such components (like all lipids) are more depleted in ^{13}C than bulk plant material, with differences between C_3 (large lipid-whole plant difference in $^{13}C/^{12}C$) and C_4 and CAM plants (smaller difference in $\delta^{13}C$: Collister *et al.* 1994). Such effects must be borne in mind in interpreting fossil $\delta^{13}C$ values.

19.4.3. *Applications of $^{13}C/^{12}C$ in tracing the evolution of phototrophy*

These sources of variability do not, however, mean that nothing can be made of $^{13}C/^{12}C$ measurements in interpreting the evolution of phototrophy. The examples that will be discussed are (1) the Archaean and Proterozoic $\delta^{13}C$ in sedimentary organic C; (2) the Δ value of terrestrial C_3 plants over the last 400 Ma or so; and (3) the timing of the evolution of the various CO_2 concentrating mechanisms.

Precambrian $^{13}C/^{12}C$ on Earth and Mars. Dealing first with the $^{13}C/^{12}C$ in the Precambrian, Schidlowski (1988) points out that there has been a disproportionation of ^{13}C and ^{12}C from the mantle-derived C with a $\delta^{13}C$ of –5‰, and cartonaceous components of the lithosphere and the biosphere for at least 3.5 Ga, giving a $\delta^{13}C$ of sedimentary organic C of –27 ± 7‰ and a $\delta^{13}C$ of sedimentary inorganic C (carbonate) of 0.4 ± 2.6‰. This has been interpreted by Schidlowski (1988) in terms of a relatively constant discrimination against ^{13}C in the photosynthetic production of organic C over the last 3.5 Ga, allowing for negligible discrimination during respiration, fire and chemical weathering which reconvert reduced C to inorganic, oxidized C. Furthermore, the data are consistent with a relatively time-invariant balance of reduction of inorganic C and oxidation of reduced C without, of course, any necessary implication that these two rates have been constant over the last 3.5 Ga. The more negative organic C values are within the range –27 ± 7‰ in the interval 3.5 Ga – 2 Ga ago when cyanobacteria were the only O_2-evolving phototrophs. Given the *in vitro* estimates of $^{13}CO_2/^{12}CO_2$ discrimination by RUBISCO from extant cyanobacteria even with negligible diffusive limitation of CO_2 fixation, these values would then be even more difficult to explain if there were a CO_2 concentrating mechanism, which is unlikely in view of the high CO_2 levels: (Figure 19.2, and Rye *et al.*, 1995). The possibility of preferential preservation of cell components with $^{13}C/^{12}C$ ratios lower than the mean cell value, or of atmospheric CO_2 depleted in $^{13}CO_2$ relative to today's value (or that expected from carbonate rocks of $\delta^{13}C = 0.4 ± 2.6‰$) cannot be ruled out. However, the lipid components of cyanobacteria are not as recalcitrant as are cuticles of higher plants, and global C isotopic mass balance considerations limit the extent of long-term depletion of ^{13}C from atmospheric CO_2.

The interpretations of Schidlowski (1988) as to the significance of the $^{13}C/^{12}C$ record of inorganic and organic C over the last 3.5 Ga has recently been extended back to 3.85 Ga (Hayes, 1996; Holland, 1997; Mojzsis *et al.*, 1996). This extension was made possible by the discovery of unmetamorphosed particles (surrounded by apatite) in the metamorphic material which characterizes sedimentary rock more than 3.5 Ga old (Hayes, 1996; Holland, 1997; Mojzis *et al.*, 1996). These $\delta^{13}C$ data are consistent with the operation of RUBISCO back to 3.85 Ga. The $\delta^{13}C$ values are not consistent with the operation of the reversed tricarboxylic acid cycle (found today, among phototrophs, in the non-O_2-evolving Chlorobiaceae) or of the β-hydroxypropionate cycle (found today, among phototrophs, in the non-O_2-evolving

Chloroflexaceae), both of which have very small $^{13}C/^{12}C$ discrimination (Sirevåg, 1995). Evidence of Chlorobiaceae-derived organic C has been found in later (Devonian) strata, based on the high $^{13}C/^{12}C$ of Chlorobiaceae-specific carotenoids (Summons and Powell, 1986), showing that isotope signals of the Chlorobiaceae can be detected in the fossil record. The interpretations of Schidlowski (1988) of ancient ^{13}C-depleted organic C as an indicator of RUBISCO activity should be viewed in the context of the $^{13}C/^{12}C$ of the organic products of simulated prebiotic syntheses (Chang et al., 1993). While these values can mimic the $^{13}C/^{12}C$ of RUBISCO-derived organic C, the magnitude of organic C accumulation is apparently more consistent with a biological origin.

Even more distant in space, and possibly in time, are the events leading to the $^{13}C/^{12}C$ ratios of organic materials and carbonates in meteorites which resulted from bolide impacts on mass. The $^{13}C/^{12}C$ data, like that derived from other sources, is not definitive in indicating the occurrence of life on Mars at the time (1.4 Ga? ago) at which the meteorite (of rock up to 4.6 Ga old) was dislodged from Mars (Anders et al., 1996; Grady et al., 1996; McKay et al., 1996). The organic material has a $\delta^{13}C$ of −22‰, while the source CO_2 (the Martian atmosphere) is probably at +36 ± 10‰ (Romanek et al., 1994). Grady, Wright and Pillinger (1996) point out that such a difference would be attributed on earth to methanogens, although Arnelle and O'Leary (1992) point out that α for PEPCK – catalysed carboxylation can be as high as 1.040, while the 'theoretical' value for carboxylases is as high as 1.060. Thus, the Martian values are not outwith the possible range for carboxylation, as opposed to methanogenic, reactions.

Embryophyte $^{13}C/^{12}C$. The second aspect of $^{13}C/^{12}C$ values in the palaecological context relates to the $\delta^{13}C$ values attributable to organic C from terrestrial plants (mainly vascular plants) over the last 400 Ma. The great predominance of data on organic C gives $^{13}C/^{12}C$ values which are consistent with C_3 biochemistry, granted the constancy of $^{13}C/^{12}C$ discrimination by carboxylases within the vascular plants over 400 Ma, relatively small diagenetic effects on $\delta^{13}C$ of organic matter, and relative constancy in the $\delta^{13}C$ of atmospheric CO_2 (France-Lanord and Derry, 1994; Troughton et al., 1974; Wedin et al., 1995). This conclusion is independent of whether bulk fossil C (coal, lignite) or individual organs of known plant species, even with known bases in $\delta^{13}C$ in some cases due to preferential preservation of cuticular components, are considered (Troughton et al., 1974; cf. Collister et al., 1994). The constancy of the $\delta^{13}C$ of atmospheric CO_2 has been challenged by Mora et al., (1996) for the middle to late Palaeozoic with evidence for atmospheric $\delta^{13}C$-CO_2 values as high as −1‰ which are not reflected in contemporary organic (coal) $\delta^{13}C$ values. A further complication (or opportunity!) in the interpretation of $\delta^{13}C$ values of C_3 plants over the last 400 Ma concerns the variations in CO_2 content of the atmosphere (Figure 19.2), stomatal density (McElwain and Chaloner, 1995; Raven, 1993; cf. Poole et al., 1996) and possibly in stomatal behaviour (Robinson, 1994). Can the relative constancy of C_3 plant $\delta^{13}C$ values tell us anything (via Equation (19.1)) of how stomatal optimization (Cowan, 1986) operated in the past?

Timing of the evolution of CO_2 concentrating mechanisms. The final question relating to palaeo $\delta^{13}C$ values and the evolution of phototrophy concerns the timing of the evolution of CO_2 concentrating mechanisms. Perhaps the simplest case is that of terrestrial C_4

plants which, in their typical form, have a unique combination of anatomical and $\delta^{13}C$ characteristics, i.e. Kranz anatomy and a less negative $\delta^{13}C$; some variant of Kranz anatomy and a more negative $\delta^{13}C$ means $C_3 - C_4$ intermediate metabolism, while an absence of Kranz anatomy means C_3 (low $\delta^{13}C$) or CAM (high $\delta^{13}C$) carbon acquisition. There is no anatomical feature common to CAM plants analogous to Kranz anatomy for C_4 plants (Raven and Spicer, 1996). In the absence of anatomical data the occurrence of a high $\delta^{13}C$ in organic C demand from terrestrial vascular plants cannot distinguish C_4 from CAM. The best examples of an (extant) plant whose anatomical and $\delta^{13}C$ characteristics unequivocally indicates C_4 photosynthesis is the grass *Tomlinsonia thomassonii* from the Miocene (~ 12 Ma ago) of California (Tidwell and Nambudiri, 1989; see Chapter 21). The anatomical and $\delta^{13}C$ data equally unequivocally show that the Pliocene (- Pleistocene?) *Tomlinsonia stichkania* was a C_3 plant (Tidwell and Nambudiri, 1990). The occurrence of an anatomically and isotopically recognizable C_4 grass species 12 Ma ago is a satisfying prelude to the evidence from $^{13}C/^{12}C$ of organic C in soils and, especially inorganic C derived from respired CO_2 in palaeosol $CaCO_3$ and in the enamel of horses teeth to indicate a large expansion in many warmer parts of the world of communities dominated by plants with high $^{13}C/^{12}C$ (C_4 from the palaeoecological context) in Upper Miocene some 7 Ma ago (McFadden and Cerling, 1996b; see Chapter 21). CAM plants, by contrast, have no such palaeontological documentation from $\delta^{13}C$ measurements (Raven and Spicer, 1996). Both C_4 and CAM are polyphyletic traits (see Sinna and Kellog, 1996; Winter and Smith, 1996).

Pre-Miocene $\delta^{13}C$ evidence of terrestrial C_4 or CAM plants is much less abundant and there is no evidence of Kranz anatomy indicative of C_4 photosynthesis. Bocherens *et al.* (1988) found a high $\delta^{13}C$ in collagen from the Cretaceous Hadrosaur *Anatosaurus* indicative of a diet of C_4 or CAM terrestrial plants, the context making aquatic plants with high $\delta^{13}C$ an unlikely food source. Wright and Vanstone (1991) found evidence in palaeosol $CaCO_3$ $\delta^{13}C$ values suggestive of respiration of C_4- or CAM-derived organic C in the Carboniferous (cf. Mora *et al.* (1996) for higher atmospheric source CO_2 $\delta^{13}C$ at that time). However, the plant organic C data in the Palaeozoic and Mesozoic, whether from bulk material (coal, lignite) or from identifiably plant species shows no evidence of significant C_4 or CAM metabolism, although an uppermost Cretaceous lignite sample has a $\delta^{13}C$ of -17.3‰ which might indicate C_4 or CAM (Bocherens *et al.*, 1993; Raven and Spicer, 1996; Troughton *et al.*, 1974). Consistency of $\delta^{13}C$ within plant species indicates limited diagenetic effects on $\delta^{13}C$ (Bocherens *et al.*, 1993). The $\delta^{13}C$ data do not give unequivocal support to the occurrence of C_4 or CAM plants on land before the Tertiary, despite the likely occurrence of low CO_2 and low CO_2/O_2 in the Carboniferous atmosphere (Figure 19.2).

The $^{13}C/^{12}C$ record is of less help than for aquatic than for terrestrial plants. Aquatic phototrophs rarely leave macroscopic organic remains recognizable to a species, although sedimentary molecular markers of higher taxa can be analysed from $\delta^{13}C$ (e.g. Hayes, 1993). The non-isotope biogeochemistry of such markers can be related to that of the bulk C of the parent organisms via calibrations with extant organisms (Hayes, 1993). However, the least equivocal evidence of a CO_2 concentrating mechanism based on $^{13}C/^{12}C$ ratios of organic C comes from $\delta^{13}C$ values which are less than 8‰ (at 20 °C) more negative than associated carbonates whose formation is uninfluenced by respiratory CO_2 since high $^{13}C/^{12}C$ ratios in organic matter indicate HCO_3^- use (Raven, 1997a). Even here the correlation between the removal of HCO_3^- from the medium faster than uncatalysed external conversion to CO_2 permits and a CO_2

concentrating mechanism may not be perfect (Raven, 1997a). It is clear that earlier attempts to pinpoint the time at which a shift toward a greater fraction of organisms with 'CO$_2$ pumps' in an aquatic ecosystem occurred on the basis of an increase in organic δ^{13}C (Raven and Sprent, 1989) may be flawed since a likely cause of the increased δ^{13}C within the range where either diffusive CO$_2$ entry or a CO$_2$ pump could explain the δ^{13}C in the organic sediment (Raven, 1997a) is that of a greater 'biological pump' activity. The 'biological pump' involves sedimentation of organic C resulting from 'new' primary productivity, so that a greater net loss of organic C involves a greater drawdown of surface water inorganic C and of ^{12}C relative to ^{13}C, leaving a relatively ^{13}C-enriched inorganic C pool and hence, from subsequent productivity, organic C with a higher ^{13}C/^{12}C ratio, and vice versa (e.g. Jenkyns, 1996; Knoll et al., 1996; Wignall and Twitchett, 1996; Visscher et al., 1996; see Chapter 15). A further aspect of this sort of analysis of ^{13}C/^{12}C ratios is that the possibility of palaeothermometry or palaeobarometry (of CO$_2$) on the basis of ^{13}C/^{12}C ratios in sedimentary organic C, depending ultimately on some version of Equation (19.1), is mechanistically if not empirically undermined by the occurrence of CO$_2$ concentrating mechanisms (Raven et al., 1993; Hayes, 1993; Fry, 1996) and by any diagenetic changes in δ^{13}C (Galimov, 1995).

19.5 Is the ^{13}CO$_2$/^{12}CO$_2$ discrimination factor of RUBISCO subject to direct natural selection?

Raven (1996) suggested that it was unlikely that the α value of RUBISCO (or other carboxylases) is subject to direct (evolutionary) selection, but is rather the result ('emergent property') of selection for other, for example, kinetic, characters such as K_m (CO$_2$) and S_{rel}. The (implicit) argument used by Raven (1996) was that, while the outcomes of variations in the values of α might be of selective significance, the potential magnitude of such variations is such that they would have very little significance.

Some of the potential outcomes of variations in α values for RUBISCO in a C$_3$ plant, or of the difference in ^{13}C/^{12}C of the products of terrestrial C$_3$ versus C$_4$ or CAM plants, are as follows:

(1) Differences in the mechanical properties of such structural polymers as cellulose or lignin depending on the ^{13}C/^{12}C ratio, with greater mechanical strength for a higher ^{13}C/^{12}C ratio. While greater mechanical strength per atom of C, and also per unit of energy used in CO$_2$ fixation, might be achieved there might not be more mechanical strength per unit mass of polymer so that there might be no net advantage in plant shoots of a higher ^{13}C/^{12}C.

(2) Differences in density of cell components, with a lower density for components of a given chemical composition with a lower ^{13}C/^{12}C. This relates not only to consideration (1) above, but also to the role of lipids in the flotation of marine diatoms and other organisms and of carbohydrates as ballast in many phytoplankton cells (Raven, 1997c; Richardson and Cullen, 1995; Richardson et al., 1996).

(3) Differences in diffusion coefficients through water (and lipid bilayers?) with higher diffusion coefficients for solutes with lower ^{13}C/^{12}C ratios; this could speed up diffusion-limited processes. Differences in the ^{13}C/^{12}C of membrane lipids might themselves alter diffusion coefficients of solutes within the membrane which could interact with the effects of variations in ^{13}C/^{12}C of the solute.

(4) Differences in the rate of enzyme-catalysed reactions where kinetic isotope effects can be expressed, with reactants with higher $^{13}C/^{12}C$ being transformed more slowly. Other things being equal, growth related processes resulting from post-RUBISCO reactions will be slower with higher $^{13}C/^{12}C$ ratios, an effect which would be exacerbated if the catalytic activity of ^{13}C-enriched proteins was less than that of corresponding ^{13}C-depleted proteins (see Schmidt *et al.*, 1995).

(5) Related to (4) above is the perception of sweetness, and thus, potentially, the attractiveness of sugars of different $^{13}C/^{12}C$ ratio, but at the same concentration in nectar, to pollinators. If the ^{13}C-depleted sugar is sweeter, then pollination might involve less energy for C input to sugar in nectar in a C_3 than a C_4 plant if sweetness determines the quantity of nectar ingested; if the pollinator relies on energy intake to regulate ingestion, then sweetness (and hence $^{13}C/^{12}C$) would have no impact (but see (6) below).

(6) Related to (5) above, the mass flow of concentrated solutions of organic compounds (e.g. in phloem, or into pollinators from nectar) can be regarded as energetically optimized in that the concentration of the organic solutes maximizes the material moved (mol C or J energy (m² sieve tube transverse area)$^{-1}$ s^{-1}) relative to the minimum energy input needed to overcome frictional and gravitational forces (Kingsolver and Daniel, 1979; Lang, 1978; Passioura, 1976). Clearly a change in the $^{13}C/^{12}C$ ratio of the solutes would alter the cost (minimum energy input): benefit (C or energy moved) relationship depending on whether the benefit was expressed in C or energy terms, since a lower $^{13}C/^{12}C$ means a higher content of metabolically recoverable energy per unit C moved.

These six interconnected examples shows that there are potentially selectable attributes of different $^{13}C/^{12}C$ ratios in the downstream products of CO_2 fixation, and thus of the $^{13}C/^{12}C$ discrimination by the carboxylation reactions generating the organic C. However, the observed differences in $\delta^{13}C$ between terrestrial C_3 and C_4 or CAM plants is only some 15‰, while the maximum difference in $\delta^{13}C$ of organic material between carboxylases expressing the theoretical maximum discrimination ($\alpha = 1.06$) and no discrimination ($\alpha = 1.0$) is some 60‰ (Arnelle and O'Leary, 1992) in both cases assuming a constant source $\delta^{13}C$. In absolute terms (Ziegler, 1995) the change from 0‰ to –60‰ involves a decrease in $^{13}C/^{12}C$ from 0.011237 to 0.010563, so differences in (for example) density (item (2) above) are unlikely to be large. A further complicating feature in the analysis are differences, as a function of C fixation pathway and habitat, on the δD and $\delta^{18}O$ of organic compounds (Ziegler, 1995). Any variations in the α value for carboxylases as a selective feature is likely to involve changes in the kinetics which may be subject to much greater selective control (see first sentence of Section 19.5).

19.6 Conclusions and prospects

$^{13}C/^{12}C$ natural abundance studies have contributed to our understanding of the evolution of phototrophy by studies of Precambrian photosynthetic processes, demonstrating dominance of C_3 plants on land through most of the last 400 Ma, and defining the time of the widespread contribution of C_4 plants to global productivity, if not the evolution of C_4 plants.

Future work would involve the following.

(1) Determination of $^{13}CO_2/^{12}CO_2$ discrimination by RUBISCO from a wider range of extant phototrophs;

(2) Characterisation the kind of (C_3 + C_1) carboxylases, and their $^{13}C/^{12}C$ discrimination, in a wider phylogenetic range of organisms;

(3) Increasing use of $^{13}C/^{12}C$ studies on individual molecular species which can show the $\delta^{13}C$ of particular higher taxa (fatty acids; sterols (steranes); chlorophyll derivatives);

(4) Further investigation of diagenesis and its impact on $\delta^{13}C$ values of fossil organic C;

(5) Focussing $\delta^{13}C$ studies related to timing of origins of C_4, CAM to organic C from epiphytic vascular plants (for CAM), and organisms with any trace of Kranz anatomy (for C_4);

(6) There is little possibility of direct natural selection for discrimination between $^{13}CO_2$ and $^{12}CO_2$ in the reaction of RUBISCO and other carboxylases.

Acknowledgements

I am most grateful to my colleagues who have corrected my errant thoughts on phylogeny, palaeobiogeochemistry and isotopy. It is a particular pleasure to acknowledge the late Mrs Thake, my boyhood neighbour in Wimbish, Essex, U.K., who held that she could distinguish cane sugar, which she preferred, to the 'new' beet sugar. As an arrogant schoolboy I dismissed this claim; while not now suggesting that she could detect $^{13}C/^{12}C$ directly, or via effects discussed in Section 19.5, different plant chemistry and structure, and processing, could have permitted such a distinction.

References

Allen, J.F. and Raven, J.A. (1996). Free-radical-induced mutation *vs* redox regulation: costs and benefits of genes in organelles. *J. Mol. Evoln.* **42**, 482–492.

Anders, E., Shearer, C.K., Papike, J.J., Bell, J.F., Clemett. S.J., Zare., R.N, McKay, D.S., Thomas-Keprta, K.L., Romanek, C.S., Gibson, E.K. Jr, Vali, H., Gibson, E.K. Jr, McKay, D.S., Thomson-Keprta, K. and Romanek, C.S. (1996). Evaluating the evidence for past life on Mars. *Science* **274**, 2119–2125.

Apel, P. (1994). Evolution of the C_4 photosynthetic pathway: a physiologists point of view. *Photosynthetica* **30**, 495–502.

Arnelle, D.R. and O'Leary, M.H. (1992). Binding of carbon dioxide to phosphoenolpyruvate carboxykinase from carbon kinetic isotope effect. *Biochemistry* **31**, 4363–4368.

Berner, R.A. (1990). Atmospheric carbon dioxide levels over Phanerozoic time. *Science* **249**, 1382–1386.

Berner, R.A. (1993). Palaeozoic atmospheric CO_2: importance of solar radiation and plant evolution. *Science* **261**, 68–70.

Berner, R.A. (1994). 3 Geocarb II: a revised model of atmospheric CO_2 over Phanerozoic time. *Am. J. Sci.* **294**, 56–91.

Berner, R.A. and Canfield, D.E. (1989). A new model for atmospheric oxygen over Phanerozoic time. *Am. J. Sci.* **289**, 333–361.

Bhattacharya, D. and Medlin, L. (1995). The phylogeny of plastids: a review based on comparisons of small-subunit ribosomal RNA coding regions. *J. Phycol.* **31**, 489–498.

Bocherens, H., Fizet, M., Cuif, J-P, Jaeger, J-J, Michard, J-G and Mariotti A (1988). Premieres mesures d'abundance isotopiques naturelles on ^{13}C et ^{12}N de la matière organique fissile de Dinosaure. Application à l'étude du regime alimentaire du genre *Anatosaurus* (Ornithischia, Hadrosauridae). *Compt. Rend. Acad. Sci. Paris D.* **306**, Series II, 1521–1525.

Bocherens, H., Friis, E.M., Mariotti, A. and Pedersen, K.R. (1993). Carbon isotope abundances in Mesozoic and Coenozoic fossil plants: Palaeoecological implications. *Lethaia* **26**, 347–358.

Buick, R., Thornett, J.R., McNaughton, N.J., Smith, J.B., Barbey, M.E. and Sarage, M. (1995). Record of emergent continental crust ~3.56 billion years ago in the Pilbara Craton of Australia. *Nature* **375**, 574–577.

Butterfield, N.J., Knoll, A.H. and Swett, K. (1990). A bangiophyte red alga from the Proterozoic of Arctic Canada. *Science* **250**, 104–107.

Canfield, D.E. and Teske, A. (1996). Late Proterozoic rise in atmospheric oxygen concentration confirmed from phylogenetic and sulphur-isotope studies. *Nature* **382**, 127–132.

Cavalier-Smith, T. and Chao, E.E. (1996). 18S rRNA sequence of *Heterosigma carterae* (Raphidophyceae), and the phylogeny of heterokont algae (Ochrophyta). *Phycologia* **35**, 500–510.

Cavalier-Smith, T., Couch, J.A., Thorsteinsen, K.E., Gilson, P., Deane, J.A., Hill, D.R.A. and McFadden, G.I. (1996). Cryptomonad nuclear and nucleomorph 18S rRNA phylogeny. *Eur. J. Phycol.* **31**, 315–338.

Chang, S., Des Marais, D., Mack, R., Miller, S.L. and Strathearn, G. (1993). Prebiotic organic synthesis and the origin of life. In *Earth's Earliest Biosphere* (ed J.W. Schopf), Princeton University Press, pp. 53–92.

Collister, J.W., Rieley, G., Stern, B., Eglington, G. and Fry, B. (1994). Compound specific ^{13}C analyses of leaf lipids from plants with differing carbon dioxide metabolisms. *Org. Geochem.* **21**, 619–627.

Couture, M., Chamberland, H., St-Pierre, B. and Guertin, M. (1994). Nuclear genes encoding chloroplast hemoglobins in the unicellular green alga *Chlamydomonas eugametos*. *Mol. Gen. Genet.* **243**, 185–197.

Cowan, I.R. (1986). Economics of carbon fixation in higher plants. *On the Economics of Plant Form and Function* (ed T.J. Givnish). Cambridge University Press, pp. 133–170.

Doolittle, R.F., Feng, D.F., Tsang, S., Cho, G. and Little, E. (1996). Determining divergence times of the major kingdoms of living organisms with a protein clock. *Science* **271**, 470–477.

Edwards, D., Duckett, J.G. and Richardson, J.B. (1995). Hepatic characters in the earliest land plants. *Nature* **374**, 635–636.

Evans, J.R. and von Caemmerer, S. (1996). Carbon dioxide diffusion inside leaves. *Pl. Physiol.* **110**, 339–346.

Farquhar, G.D., Ehleringer, J.R. and Hubick, K.T. (1989). Carbon isotope discrimination and photosynthesis. *Ann. Rev. Pl. Physiol. Pl. Mol. Biol.* **40**, 503–537.

France-Lanord, C. and Derry, L.A. (1994). δ^{13}C of organic carbon in the Bengal Fen: source evolution and C_3 and C_4 plant carbon to marine sediments. *Geochim. Cosmochim. Acta* **58**, 4809–4814.

Frankignoule, M., Canon, C. and Gattuso, J-P. (1993). Marine calcification as a source of carbon dioxide: positive feedback of increasing atmospheric CO_2. *Limnol. Oceanogr.* **39**, 458–462.

Fry, B. (1996). ^{13}C/^{12}C fractionation by marine diatoms. *Mar. Ecol. Progr. Ser.* **134**, 283–294.

Galimov, E.M. (1995). Fractionation of carbon isotopes on the way from living to fossil organic matter. *Stable Isotopes in the Biosphere* (eds E. Wada, T. Yoneyama, M. Minagawa, T. Ando and B.D. Fry). Kyoto University Press, Kyoto, pp. 133–170.

Grady, M., Wright, I. and Pillinger, C. (1996). Opening a Martian can of worms? *Nature* **382**, 575–577.

Graham, L.E. (1993). *Origin of Land Plants*. John Wiley and Sons Inc, New York.

Griffiths, H. (1992). Carbon isotope discrimination and the integration of carbon assimilation pathways in terrestrial CAM plants. *Plant Cell Environ.* **15**, 1051–1062.

Hagesawa, M., Fitch, W.M., Gogarten, J.P., Olendzenski, C., Hilario, E., Simon, C., Holsinger, K.E., Doolitle, R.F., Feng, D.F., Tsang, S., Cho, G. and Little, E. (1996). Dating the cenancester of organisms. *Science* **274**, 1750–1753.

Han, T-M. and Runnegar, B. (1992). Megascopic eukaryotic algae from the 2.1 billion year-old Negaunee Iron-Formation, Michigan. *Science* **257**, 232–235.

Hardison, R.C. (1993). A brief history of hemoglobin: plant, animal, protist and bacteria. *Proc. Natl. Acad. Sci. Wash.* **93**, 5675–5679.

Hayes, J.M. (1993). Factors controlling ^{13}C contents of sedimentary organic compounds: principles and evidence. *Mar. Geol.* **113**, 111–125.

Hayes, J.M. (1996). The earliest memories of life on earth. *Nature* **384**, 21–22.

Hoffman, P.F. (1995). Oldest terrestrial landscape. *Nature* **375**, 537–538.

Holland, H.D. (1984). *The Chemical Evolution of the Atmosphere and Oceans*. Princeton University Press, Princeton, NJ.

Holland, H.D. (1997). Evidence for life on earth more than 3850 million years ago. *Science* **275**, 38–39.

Horodyski, R.J. and Knauth, L.P. (1994). Life on land in the Precambrian. *Science* **263**, 494–498.

Jenkyns, H.C. (1996). Relative sea-level change and carbon isotopes: data from Upper Jurassic (Oxfordian) of Central and Southern Europe. *Terra Nova* **8**, 75–85.

Kingsolver, J.G., Daniel, T.L. (1979). On the mechanics and energetics of nectar feeding in butterflies. *J. Theoret. Biol.* **76**, 167–179.

Knoll, A.H. (1994). Proterozoic and Early Cambrian protists: evidence for increasing evolutionary tempo. *Proc. Natl. Acad. Sci. Wash.* **91**, 6743–6750.

Knoll, A.H., Bambach, R.K., Canfield, D.E. and Grotzinger, J.P. (1996). Comparative earth history and late Permian mass extinction. *Science* **273**, 452–457.

Kooistra, W.H.C.F. and Medlin, L.K. (1996). The origin of pigmented heterokonts and the diatoms. *J. Phycol.* **32**: (**suppl**) 25.

Laisk, A. and Loreto, F. (1996). Determining photosynthetic parameters for leaf CO_2 exchange and chlorophyll fluorescence. Ribulose-1,5-bisphosphate carboxylase/oxygenase specificity factor, dark respiration in the light, excitation distribution between photosystems, alternative electron transport rate, and mesophyll diffusion resistance. *Pl. Physiol.* **110**, 903–912.

Lang, A. (1978). A model of mass flow in the phloem. *Austr. J. Plant Physiol.* **5**, 535–546.

McElwain, J.C. and Chaloner, W.G. (1995). Stomatal density and index of fossil plants track atmospheric carbon dioxide in the Palaeozoic. *Ann. Bot.* **76**, 389–395.

McFadden, B.J. and Cerling, T.E. (1996). Fossil horses, carbon isotopes and global change. *TREE* **9**, 481–486.

McKay, D.S., Gibson, E.K. Jr, Thomas-Keptra, K.L., Vali, H., Romanek, C.S., Clemett, S.J., Chillier, X.D.F., Meechling, C.R. and Zare, R.N. (1996). Search for past life on Mars: possible relic biogenic activity in Martian Meteorite ALH 84001. *Science* **273**: 924–930.

Mojzsis, S.J., Arrhenius, G., McKeegan, K.D., Harrison, T.M., Nutman, A.P. and Friend, C.R.L. (1996). Evidence for life on earth before 3,800 million years ago. *Nature* **384**, 55–59.

Mora, C.L., Driese, S.G. and Colarusso, L.A. (1996). Middle to Late Palaeozoic atmospheric CO_2 levels from soil carbonate and organic matter. *Science* **171**, 1105–1107.

Newman, M.J. (1979). The evolution of the solar constant. *Origins of Life* **10**, 105–110.

O'Leary, M.H., Madhaven, S. and Paneth, B. (1992). Physical and chemical basis of carbon isotope fractionation in plants. *Plant Cell Environ.* **15**, 1099–1104.

Palmer, J.D. and Delwiche, C.F. (1996). Second-hand chloroplasts and the case of the disappearing nucleus. *Proc. Natl. Acad. Sci. Wash.* **93**, 7432–7435.

Passioura, J.B. (1976). Translocation and the diffusion equation. *Transport and Transfer Processes in Plants* (eds I.F. Wardlaw and J.B. Passioura). Academic Press, New York, pp. 357–361.

Paul, J.H. and Pichard,S.L. (1996). Molecular approaches to studying natural communities of autotrophs. *Microbial Growth on C_1 Compounds* (eds M.E. Lidstrom and F.R. Tabita). Kluwer, Dordrecht, pp. 301–309.

Peisker, M. and Henderson, S.A. (1992). Carbon: terrestrial C_4 plants. *Plant Cell Environ.* **15**, 987–1004.

Poole. I., Weyers, J.D.B., Lawson, T. and Raven, J.A. (1996). Variations in stomatal density and index: implications for palaeoclimatic reconstructions. *Plant Cell Environ.* **19**, 705–712.

Raven, J.A. (1991). Implication of inorganic C utilization: ecology, evolution and geochemistry. *Can. J. Bot.* **69**, 908–924.

Raven, J.A. (1993). The evolution of vascular land plants in relation to quantitative function of dead water-conducting cells and of stomata. *Biol. Revs.* **68**, 337–363.

Raven, J.A. (1995). The early evolution of land plants: aquatic ancestors and atmospheric interactions. *Bot. J. Scotl.* **47**: 151–175.

Raven, J.A. (1996). Inorganic carbon assimilation by marine biota. *J. Exp. Mar. Biol. Ecol.*, **203**, 39–47.

Raven, J.A. (1997a). Inorganic carbon acquisition by marine autotrophs. *Adv. Bot. Res.*, **27**, 85–204.

Raven, J.A. (1997b). The role of marine biota in the evolution of terrestrial biota: gases and genes. *Biogeochemistry*, in press.

Raven, J.A. (1997c). The roles of vacuoles: a cost-benefit analysis. *Adv. Bot. Res.*, **25**, 59–86.

Raven, J.A. and Farquhar, G.D. (1990). The influence of N metabolism and organic acid synthesis on the natural abundance of C isotopes in plants. *New Phytol.* **116**, 525–529.

Raven, J.A. and Spicer, R.A. (1996). The evolution of CAM. *Crassulacean Acid Metabolism, Biochemistry, Ecophysiology and Evolution* (eds K. Winter and J.A.C. Smith). Springer Verlag, Heidelberg, pp. 360–385.

Raven, J.A. and Sprent, J.I. (1989). Phototrophy, diazotrophy and palaeoatmospheres: biological catalysis and the H, N and C cycles. *J. Geol. Soc. Lond.* **146**, 161–170.

Raven, J.A., Johnston, A.M. and Turpin, D.H. (1993). Influence of changes in CO_2 concentration and temperature on marine phytoplankton $^{13}C/^{12}C$ ratios: an analysis of possible mechanisms. *Global Planetary Change* **8**, 1–12.

Raven, J.A., Johnston, A.M., Newman, J.R. and Scrimgeour, C.M. (1994). Inorganic carbon acquisition by aquatic photolithotrophs of the Dighty Burn, Angus, UK: uses and limitations of natural abundance measurements of carbon isotopes. *New Phytol.* **127**, 271–286.

Raven, J.A., Walker, D.I., Johnston, A.M., Handley, L.L. and Kübler, J.E. (1995). Implications of ^{13}C natural abundance measurements for photosynthetic performance by marine macrophytes in their natural environment. *Mar. Ecol. Progr. Ser.* **123**, 193–205.

Richardson, T.L. and Cullen, J.J. (1995). Changes in buoyancy and chemical composition during growth of a coastal marine diatom: ecological and biogeochemical consequences. *Mar. Ecol. Progr. Ser.* **128**, 77–90.

Richardson, T.L., Ciott, A.M., Cullen, J.J. and Villareal, T.A. (1996). Physiological and optical properties of *Rhizosoenia formosa* (Bacillariophyceae) in the context of open-sea vertical migration. *J. Phycol.* **32**, 741–757.

Robinson, J.J. and Cavanaugh, C.M. (1995). Expression of land form II Rubisco in chemoautotrophic symbioses: implications for interpretation of stable carbon isotope value. *Limnol. Oceanogr.* **40**, 1496–1502.

Robinson, J.M. (1994). Speculations on carbon dioxide starvation, Late Tertiary evolution of stomatal regulation and floristic modernisation. *Plant Cell Environ.* **17**, 345–354.

Romanek, C.S., Grady, M.M., Wright, I.P., Mittletehidt, D.W., Sockl, P.A., Pillinger, C.T. and Gibson, E.K. Jr (1994). Record of fluid-rock interactions on Mars from the meteorite ALH84001. *Nature* **372**, 655–657.

Rowan, R., Whitney, S.M., Fowler, A. and Yellowlees, D. (1996). Rubisco in marine symbiotic dinoflagellates: form II enzymes in eukaryotic oxygenic phototrophs encoded by a nuclear multigene family. *Plant Cell* **8**, 539–553.

Rozema, J., van de Staaij, J., Björn, L.O. and Caldwell, M. (1997). UV-B as an environmental factor in plant life: stress and regulation. *Trends Ecol. Evoln.* **12**, 22–28.

Rye, R., Kuo, P.H. and Holland, H.D. (1995). Atmospheric carbon dioxide concentrations before 2.2 billion years ago. *Nature* **378**, 603–605.

Schäfer, G., Purschke, W. and Schmidt, C.L. (1996). On the origin of respiration: electron transport proteins from archaea to man. *FEMS Microbiol. Revs.* **18**, 173–188.

Schidlowski, M. (1988). A 3,800 million-year isotopic record of life from carbon in sedimentary rocks. *Nature* **333**, 313–318.

Schmidt, H-L., Kexel, H., Butzeulecher, M., Schwarz, S., Gleixner, G., Thimet, J., Werner, R.A. and Gensler, M. (1995). Non-statistical isotope distribution in natural compounds: mirror of their biosynthesis and key for their origin assignment. *Stable Isotopes in the Biosphere* (eds E. Wada, T. Yoneyama, M. Minagawa, T. Ando and B.D. Fry). Kyoto University Press, Kyoto, pp. 17–35.

Schopf, J.W. (1968). Microflora of the bitter springs formation, late Precambrian, Central Australia. *J. Palaeontol.* **42**, 651–688.

Schopf, J.W. (1993). Microfossils of the early Archaean Apex Chert: new evidence for the antiquity of life. *Science* **260**, 95–111.

Shixing, Z. and Huineng, C. (1995). Megascopic multicellular organisms from the 1700-million-year-old Tuanchazi Formation in the Jixian Area, North China. *Science* **270**, 620–622.

Sinna, N.R. and Kellog, E.A. (1996). Parallelism and diversity in multiple origins of C_4 photosynthesis in the grass family. *Am. J. Bot.* **83**, 1458–1470.

Sirevåg, R. (1995). Carbon metabolism in green bacteria. *Anoxygenic Photosynthetic Bacteria* (eds R.E. Blankenship, M.T. Madigan and C.E. Bayer). Kluwer, Netherlands, pp. 871–883.

Smith, E.C. and Griffiths, H. (1996a). The occurrence of the chloroplast pyrenoid is correlated with the activity of a CO_2-concentrating mechanism and carbon isotope discrimination in lichens and liverworts. *Planta* **198**, 6–16.

Smith, E.C. and Griffiths, H. (1996b). A pyrenoid-based carbon-concentrating mechanism is present in terrestrial bryophytes of the class Anthocerotae. *Planta* **200**, 203–212.

Sültemeyer, D. and Rinast, K.A. (1996). The CO_2 permeability of plasma membrane from *Chlamydomonas reinhardtii*: mass spectrometric ^{18}O exchange measurements from ^{13}C $^{18}O_2$ in suspensions of plasma membrane vesicles loaded with carbonic anhydrase. *Planta* **200**, 258–268.

Summons, R.E. and Powell, T.G. (1986). Chlorobiaceae in Palaeozoic seas revealed by biological markers, isotopes and geology. *Nature* **319**, 763–765.

Tabita, F.R. (1995). The biochemistry and metabolic regulation of carbon metabolism and CO_2 fixation in purple bacteria. *Anoxygenic Photosynthetic Bacteria* (eds R.E. Blankenship, M.T. Madigan and C.E. Bayer). Kluwer, Netherlands, pp. 885–914.

Taylor, W.A. (1995a). The case for hepatics on land in the Ordovician. *Am J. Bot.* **82** (**suppl**), 92.

Taylor, W.A. (1995b). Spores in earliest land plants. *Nature* **373**, 391–392.

Tidwell, W.D. and Nambudiri, E.M.V. (1989). *Tomlinsonia thomassonii*, Gen et Sp Nov, a permineralised grass from the Upper Miocene Riccardo Formation, California. *Rev. Palaeobot. Palynol.* **60**, 165–177.

Tidwell, W.D. and Nambudiri, E.M.V. (1990). *Tomlinsonia stichkania* sp. nov., a permineralised grass from the Pliocene to (?) Pleistocene China Ranch Beds in Sperry Wash, California. *Bot. Gaz.* **151**, 263–274.

Troughton, J.H., Stout, J.H. and Rafter, T. (1974). Long-term stability of plant communities. *Ann. Rep. Dir. Carnegie Inst. Wash.* **73**, 838–845.

Uemura, K., Suzuki, V., Shikanai, T., Wadano, A., Jensen, R.G., Chmara, W. and Yokoto, A. (1996). A rapid and sensitive method for determination of relative specificity of RuBisCO from various species by anion-exchange chromatography. *Plant Cell Physiol.* **37**, 325–331.

Visscher, H., Brinkhuis, H., Dilcher, D.L., Elsik, W.G., Looy, C.V., Rampino, M.R. and Traverse, A. (1996). The terminal Palaeozoic fungal event: evidence of terrestrial ecosystem destabilization and collapse. *Proc. Natl. Acad. Sci. Wash.* **93**, 2155–2158.

von Caemmerer, S. (1992). Carbon isotope discrimination in C_3-C_4 intermediates. *Plant Cell Environ.* **15**, 1063–1072.

Wedin, D.A., Tieszen, L.L., Dewey, B., Pastor, J. (1995). Carbon isotope dynamics during grass decomposition and soil organic matter formation. *Ecology* **76**, 1383–1392.

Whitney, S.M., Yellowlees, D. (1995). Preliminary investigations into the structure and activity of ribulose bisphosphate carboxylase from two photosynthetic dinoflagellates. *J. Phycol.* **31**, 138–146.

Wignall, B. and Twitchett, R.J. (1996). Oceanic anoxia and the end Permian Mass extinction. *Science* **272**, 115–1158.

Willis, K.J. (1996). Plant evolution and the ancient greenhouse effect. *Trends Ecol. Evoln.* **11**, 277–278.

Winter, K. and Smith, J.A.C. (eds.) (1996). *Crassulacean Acid Metabolism; Biochemistry, Ecophysiology and Evolution*. Springer Verlag, Heidelberg.

Wright, V.P., Vanstone, S.D. (1991). Assessing the carbon dioxide content of ancient atmospheres using paleocretes: theoretical empirical constraints. *J. Geol. Soc. Lond.* **148**, 945–947.

Yapp, C.J. and Poths, H. (1994). Productivity of pre-vascular continental biota inferred from the $Fe(CO_3)OH$ content of goethite. *Nature* **368**, 49–51.

Ziegler, H. (1995). Stable isotopes in plant physiology and ecology. *Progr. Bot.* **56**, 1–24.

Modelling changes in land plant function over the Phanerozoic

D.J. Beerling and F.I. Woodward

20.1 Introduction

The history of the Earth since its formation as a planet has been characterised by substantial fluctuations in climate, atmospheric composition and the plate positions of the major land masses (Crowley and North, 1991; Schopf and Klein, 1992). In terms of the evolution of complex biotic systems, the past 600 Ma, known as the Phanerozoic, is the most relevant period (Graham *et al.* 1996). Massive changes in the global environment occurring over this time have been reasonably well documented in the geological and fossil records and would be expected to have had major impacts on the functioning of living organisms with the implications for evolution of aquatic and terrestrial photosynthetic systems considered by Raven in Chapter 19. This chapter considers the possible impact of some of these environmental changes on the functioning of terrestrial vegetation. We consider these impacts in two ways. Firstly, the effects of changes in atmospheric composition and temperature on leaf-scale CO_2 fixation and water vapour loss over the past 400 Ma of the Phanerozoic is considered to provide a process-based perspective on the temporal dimension of plant responses over this timescale. The spatial dimension is then expanded to the global scale by modelling the effects of a Carboniferous climate, derived by a general circulation model, on ecosystem processes using a coupled vegetation–biogeochemistry model. Because the models are process based, they allow us to develop predictions of the likely geographical variation in the stable carbon isotope composition of organic matter for the Carboniferous. This map should be of use for comparison with new measurements made on Carboniferous fossil plant materials as a check on our modelling approach.

20.2. Atmospheric evolution

Models of changes in the global atmospheric composition during the Phanerozoic predict major fluctuations in the key biologically active gases CO_2 (Figure 20.1(a))

Stable Isotopes, edited by H. Griffiths.
© 1998 BIOS Scientific Publishers Ltd, Oxford.

(Berner, 1994) and O_2 (Figure 20.1(b)) (Berner and Canfield, 1989), based on estimates of long-term weathering of Mg-Si rocks and on carbon burial rates, respectively.
The atmospheric CO_2 reconstruction shows very clearly that for the majority of the
Phanerozoic the Earth experienced CO_2 concentrations above those of the present-
day (see also Chapter 19).

The large draw-down of CO_2 (Figure 20.1(a)) occurs concurrently with the rise of
terrestrial vegetation from the Devonian into the Carboniferous and there is an associated increase in atmospheric O_2 content (Figure 20.1b) as a result of increased
organic carbon burial. The upper estimate of 35% O_2 in the atmosphere is rather high
to be considered compatible with the occurrence of terrestrial vegetation (Chaloner,
1989) but is here accepted at face value. The atmospheric CO_2 curve has been broadly
validated by isotopic studies on pedogenic carbonates (Mora *et al.*, 1996), and the O_2

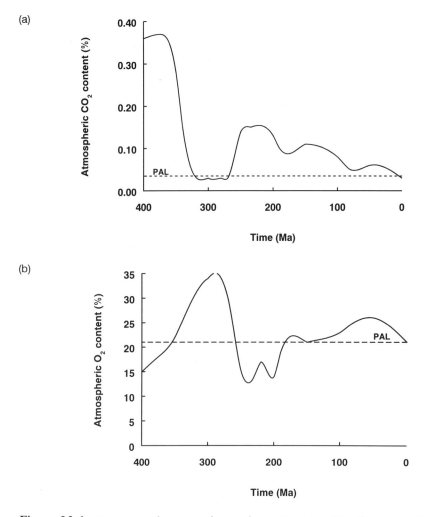

Figure 20.1. *Reconstructed patterns of atmospheric CO_2 (a) and O_2 (b) over the Phanerozoic.
Data from Berner (1994) and Berner and Canfield (1989) respectively. PAL = present-
atmospheric level.*

curve is consistent with the sedimentary record of fossil charcoal (Robinson, 1989) and both are currently accepted as the 'best guess' estimates. The impact of these large changes in atmospheric composition, and associated changes in temperature, calculated from a zero dimension global climate model (Crowley and North, 1991), on plant function at the scale of individual leaves has been investigated in the next section through a combination of modelling and some observations from the fossil record.

20.3. Phanerozoic changes in leaf function

In the short-term (centuries to millennia) the stomatal density of plant leaves has been found to be inversely related to the concentration of atmospheric CO_2 (Beerling et al., 1993; Van de Water et al., 1994; Woodward, 1987). Experimental data suggest this sensitivity can be limited at CO_2 concentrations above the PAL of 350 ppm (0.035%) for some current genotypes of trees and shrubs (Woodward, 1987). However, evidence from fossils of Carboniferous and Devonian age indicate that in the distant past plants were sensitive to changes above the PAL (McElwain and Chaloner, 1995) and these data are supported by new observations on plants growing around geothermal springs with a long history of exposure to a high CO_2 environment (I. Bettarini, personal communication).

Based on these observations therefore, we may expect that the large Phanerozoic fluctuations in atmospheric CO_2 (Figure 20.1a) were associated with large changes in leaf stomatal densities – a feature influencing the maximum rates of photosynthesis and stomatal conductance of leaves (Eamus et al., 1993) since stomata provide a major limitation to CO_2 diffusion to the chloroplasts. From analyses of recent and fossil stomatal densities, the long-term relation between stomatal density and atmospheric CO_2 concentration can be predicted by (Beerling and Woodward, 1996, 1997) :

$$sd = sd_{max} \times e^{-0.03 \times Ca} + 10 \times 10^{-6} \qquad (20.1)$$

where sd is stomatal density (m^{-2}), sd_{max} is the maximum observed stomatal density (taken as 500×10^{-6}) and the minimum value of 10×10^{-6} is the mean of the lowest stomatal densities observed on fossil material (Edwards, 1996) and C_a is the atmospheric CO_2 concentration (Pa).

The modelled trends in stomatal density (Figure 20.2a) over the Phanerozic, based on Equation 20.1 and Berner's (1994) CO_2 curve (Figure 20.1a), shows generally low values in the Devonian, very high values in the Carboniferous and low but rising values for the remaining 200 Ma. This response is consistent with observations on Carboniferous and Devonian fossils (McElwain and Chaloner, 1995), and some additional observations reported in the palaeobotanical literature (Ash, 1977; Daghlian and Person, 1977; Srinivasan, 1995). Nevertheless, further measurements from more leaves with greater temporal resolution are required to test the predicted stomatal density response in more detail. In the absence of these data we have used the modelled stomatal density trends as a starting point for reconstructing Phanerozoic changes in leaf gas exchange.

Besides the central role played by stomata in regulating leaf gas exchange, the action of changing CO_2 and O_2 concentrations on ribulose–1,5–bisphosphate carboxylase oxygenase (RUBISCO) kinetics, the primary C-fixing enzyme in C_3 terrestrial plants, also need to be considered to understand the effects of long-term changes in the environment on plant function. Here the aim is to construct a 400 Ma continuous record of CO_2 fixation and H_2O loss by leaves based on stomatal and RUBISCO kinetic considerations. The

CO_2 fixation calculation assumes no major changes in RUBISCO kinetics, an assumption supported by carbon isotope data (see Raven, Chapter 19). Therefore, the model of Beerling and Woodward (1997) has been used which combines the influence of stomatal characteristics on stomatal conductance, with a consideration of the important aspects of leaf energy balance and transpiration in regulating leaf temperature, and a mechanistic model of photosynthesis (Farquhar et al., 1980). The fully coupled model allows the prediction of photosynthesis and stomatal conductance from stomatal density data and information on irradiance, relative humidity, vapour pressure and temperature. The model requires characteristic values of stomatal length and maximum aperture width. Extensive analyses of fossil floras indicate a clear shift in dominance of different plant groups over the past 300 Ma (Knoll and Niklas, 1987) and so we have used previously assembled information on stomatal dimensions (Beerling and Woodward, 1997) taken from the palaeobotanical literature for the current simulations to reflect this shift in dominance.

The modelled changes in photosynthesis (Figure 20.2b) and stomatal conductance (Figure 20.2c) over the past 400 Ma show large and marked fluctuations in concert with the large changes in palaeoenvironment.

The large drop in atmospheric CO_2 concentration during the Carboniferous was associated with an increase in stomatal density and so higher stomatal conductances, but with no large changes in leaf photosynthetic rates. Despite the release of leaves from the diffusional constraints of having a few stomata, such as typically experienced by Devonian plants (Edwards and Axe, 1982), photosynthetic rates were severely curtailed in a high O_2 and low CO_2 environment through increased photorespiration. The continued decline in atmospheric CO_2 concentration from the middle Jurassic (about 200 Ma) through to the present day is predicted to have been associated with a gradual increase in stomatal conductance (Figure 20.2c), through the CO_2 effects on stomatal density and stomatal closure. This effect, together with a gradual increase in global temperature, contributed to a steady decrease in leaf photosynthetic rates (Figure 20.2b), in accordance with previous modelling studies of this sort (Robinson, 1994; Beerling, 1994). There is also the additional consideration that changes in mesophyll conductance might have changed in tandem with stomatal conductance over the past 400 Ma. A decrease in mesophyll conductance would lower the CO_2 concentration in the immediate RUBISCO environment, presumably lowering the potential photosynthetic rates. Such detailed considerations of the influence of leaf anatomy on leaf CO_2 fixation processes remain important but are difficult to include in a quantitative manner at present (see Chapter 8).

20.4. Terrestrial productivity in the Carboniferous

The previous section considered the temporal dimension of plant responses to Phanerozoic global environmental change and illustrated the marked influence of climate on vegetation function, particularly in the Carboniferous. In this section we consider in more detail the effects of a Carboniferous atmosphere on rates of CO_2 fixation and water loss by leaves and then extend the spatial dimension by considering the influence of the Carboniferous climate on plant and ecosystem function at the global scale.

Figure 20.2. Modelled changes in leaf (a) stomatal density predicted from Eq. 1., • indicate points from the fossil record (see text for sources), (b) predicted rates of photosynthesis for well-lit leaves (800 µmol m^{-2} s^{-1}) and (c) stomatal conductance to water vapour. Gas exchange predictions were made using the model of Beerling and Woodward (1997).

(a)

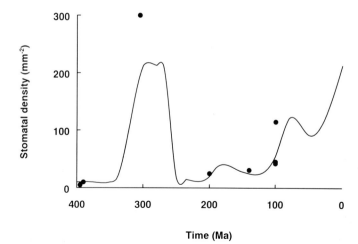

(b)

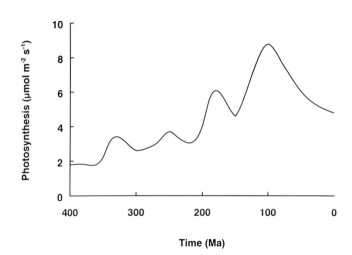

(c)

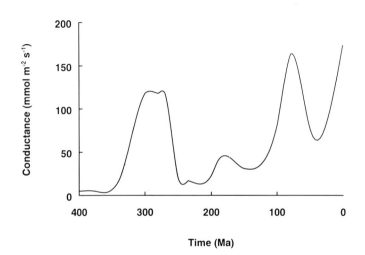

20.4.1 *Impacts on photosynthesis*

According to the atmospheric modelling of Berner (1994) and Berner and Canfield (1989) the Carboniferous era experienced high concentrations of atmospheric O_2 (ca. 35%) with low concentrations of CO_2 (0.03%). Both O_2 and CO_2 compete for the binding sites on RUBISCO, and so changes in their ratio influence the extent of CO_2 fixation (photosynthesis) and CO_2 release (photorespiration; see Chapter 8). The impact of the Carboniferous O_2 concentration on leaf gas exchange processes can by illustrated with the use of the Farquhar *et al.* (1980) mechanistic model of leaf assimilation to construct the photosynthesis versus intercellular CO_2 concentration (A/C_i) response curves. Plants photosynthesising in an atmosphere of 35% O_2 show a depressed A/C_i relationship (Figure 20.3) relative to that for plants growing at 21% O_2. In contrast, and for a comparison, plants operating in an atmospheric composition typical of the mid-Cretaceous, show a slightly depressed A/C_i response, but because the atmospheric CO_2 concentration was 3–4 times the pre-industrial value, leaf intercellular CO_2 concentrations were high leading to higher than current rates of leaf photosynthesis (Figure 20.3). This simple modelling exercise suggests that, assuming these leaf-scale effects were translated into growth rates (Raven *et al.*, 1994), the high O_2 content of the Carboniferous would have markedly reduced the productivity of terrestrial vegetation. Experimental data support this suggestion, with the growth responses of plants to O_2 increases 2–3 times the present day broadly in line with the predictions of the action of high O_2 on RUBISCO (see studies summarised by Raven *et al.*, 1994).

20.4.2 *Impacts on leaf carbon and water economies*

Photosynthesis represents the demand by the leaf for CO_2, this demand must be

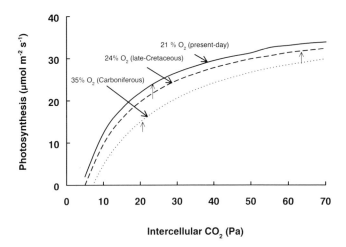

Figure 20.3. Modelled relationship between leaf photosynthesis and intercellular CO_2 concentration to changes in the atmospheric O_2 content representative of the Carboniferous and late-Cretaceous eras. The vertical arrows indicate the predicted intercellular CO_2 concentrations for the atmospheric CO_2 concentration of Berner (1994) corresponding to each era. Curves were calculated from the Farquhar et al. (1980) model of CO_2 assimilation with values of V_{max} and J_{max} of 210 and 105 mmol m^{-2} s^{-1} respectively at a temperature of 25 °C and a saturating irradiance of 1000 μmol m^{-2} s^{-1}.

balanced by the supply of CO_2 determined by stomatal conductance. Both processes need to be considered together in order to interpret the effects of the high O_2 episode of the Carboniferous on plants. Available data from the fossil record of stomatal densities, the primary limitations to maximal rates of stomatal conductance, of Carboniferous plants typically indicate rather high values for gymnosperms (Zodrow and Cleal, 1993) and pteridiosperms (Cleal and Zodrow, 1989), both dominant components of Carboniferous vegetation, see further data reviewed by Beerling and Woodward (1997). Therefore, this section investigates the functional significance of this aspect of leaf morphology on the gas exchange processes of photosynthesis and stomatal conductance. Using the model of Beerling and Woodward (1997) leaf photosynthetic rates have been simulated across a range of atmospheric CO_2 concentrations at 35% O_2 for hypothetical leaves possessing stomatal densities of 400, 200 and 100 mm^{-2}. The effect of reducing leaf stomatal density at 35% O_2 is to reduce the rate of net photosynthesis at any given atmospheric CO_2 concentration (Figure 20.4a). An important and additional effect is to increase the intercellular CO_2 concentration at a Carboniferous atmospheric CO_2 value of 30 Pa (as shown by the vertical arrows on Figure 20.4 a,b).

Higher intercellular CO_2 concentrations would help to reduce photorespiration – an important consideration for plants operating at in atmosphere with 35% O_2 (Raven et al., 1994). Leaves with a high stomatal density have higher stomatal conductances at any given CO_2 concentration (Figure 20.4(b)). Therefore, taking the photosynthetic and conductance responses together, the combined influence of an increase in leaf stomatal density is to lower considerably the instantaneous water use efficiency (WUE$_i$) of leaves. By developing leaves with a high stomatal density, Carboniferous plants increased the intercellular CO_2 concentration of the leaf but at the expense of water economy. It is interesting to note that in spite of growing in swamp conditions, aborescent lycopods of the Carboniferous tended to have xeromorphic leaves (thick cuticles, stomata in sunken pits) which, although in part may have been due to the limitations of the vascular system (Spicer, 1989), provides support for the idea that plants endured water shortage. Higher mesophyll conductances of these lycopods may also have played a role in reducing leaf water loss and detailed examination of well-preserved Carboniferous fossils might be one approach to quantifying this possibility further (see also Chapters 8 and 9).

20.4.3 *Impacts on terrestrial productivity the global scale*

The two previous sections dealt with the effects of O_2 and stomata on leaf carbon and water balances for a given specific set of environmental conditions. In this section, the leaf-scale approach has been scaled up to consider the net primary productivity of vegetation and extended geographically to the global scale. This has been achieved by driving a coupled vegetation–biogeochemistry model with a general circulation model (GCM) palaeoclimate simulation of the Carboniferous. Application of this type of modelling to palaeoecology has been described elsewhere by Beerling et al. (1996). In brief, the net primary productivity (NPP) model of Woodward et al. (1995), based on the Farquhar et al. (1980) model of photosynthesis, is used to estimate NPP from climatic and edaphic information which is then used in the biogeochemistry decomposition model CENTURY (Parton et al., 1993). CENTURY simulates the C and N dynamics of vegetation and soils. The newly-derived soil N

(a)

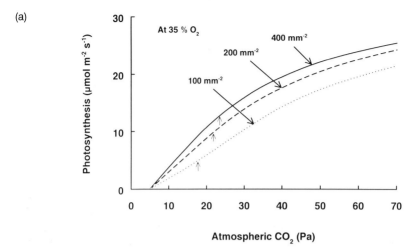

(b)

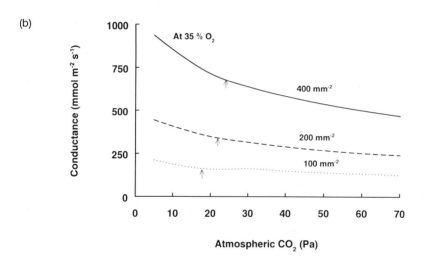

(c)

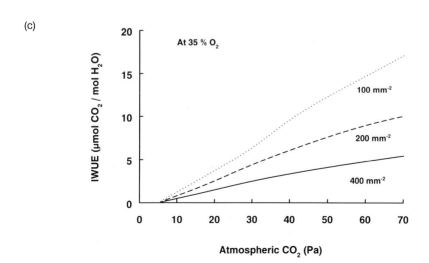

status is then used to drive the vegetation model and the iteration continued until equilibrium values of NPP are obtained (Beerling *et al.*, 1996). For the Carboniferous, the Universities Global Atmospheric Modelling Programme (UGAMP) GCM was employed to simulate the global climate dataset (monthly values of near-surface winds, precipitation, solar radiation, temperature and relative humidity) at a resolution of 3.75 x 3.75° using the land cover mask of Scotese and McKerrow (1990). The climate simulation with the UGAMP GCM for the Carboniferous was similar to that of Crowley and Baum (1994) (P. J. Valdes, personal communication). Details of the adequacies of the vegetation–biogeochemistry model in representing the responses of vegetation to changing CO_2 concentrations are given in VEMAP (1995) and for the global climate model by Valdes (1993).

The global pattern of modelled NPP for the Carboniferous and the resulting soil C concentrations derived by the biogeochemistry model are given in Figure 20.5(a,b) respectively. The area of highest NPP is concentrated in a central band through the middle of Gondawana with other areas evident on the north and south eastern edges of the super continent. Predicted soil C concentrations show similar patterns but with the extensive development of high C soils in the southern hemisphere. Overall the global terrestrial productivity of the Earth at this time is estimated to be 37.8 GT C year^{-1}, and compared with the value modelled for the present-day (45.0 GT C year^{-1}) and the mid-Cretaceous 'greenhouse' Earth (96.7 GT C year^{-1}) (Beerling *et al.*, 1996) this represents a rather low net primary productivity value. The low productivity reflects the combined effects of the cool wet Carboniferous climate and the impact of high O_2 low CO_2 on photosynthesis. The model suggests that the accumulation of the Carboniferous coal measures occurred through water logging rather than being a result of highly productive ecosystems.

The global NPP and soil C maps have been compared with the geological evidence in an effort to validate the global projections. Crowley and North (1991) summarised the available geological evidence, including the global distribution of coal, evaporites and tillites for the Carboniferous and these data are reproduced in Figure 20.6. A comparison of Figure 20.5 and Figure 20.6 shows that the band of high NPP and high soil C concentration correspond directly with the distribution of Carboniferous coals. In addition, the location of evaporites above and below this band correspond to the areas of low NPP. The presence of high soil C values in the southern hemisphere does not translate directly into the presence of coals. This is most readily explained by the presence of tillites in these areas (Figure 20.6). As consolidated deposits of glacial till, tillites are indicative of ice sheet activity suggesting that any potential for coal formation in this region of Gondawana was considerable reduced by the activity of the ice sheets eroding any organic deposits. Overall, we suggest based on this limited comparison with the geological data that the global maps of NPP and soil C represent a reasonable first attempt at reconstructing the action of Carboniferous palaeoenvironments on ecosystem processes.

Figure 20.4. *The impact of a reduction in stomatal density on (a) net rate of photosynthesis, (b) stomatal conductance and (c) instantaneous water use efficiency (photosynthesis / stomatal conductance) for a range of atmospheric CO_2 concentrations. The photosynthetic and environmental conditions for the simulations were as given in Figure 20.3, but with a relative humidity of 50 % and assuming a stomatal pore length of 10 μm and a maximum aperture width of 5 μm. Simulations were made using the model of Beerling and Woodward (1997).*

(a)

NPP (t C ha⁻¹ y⁻¹)

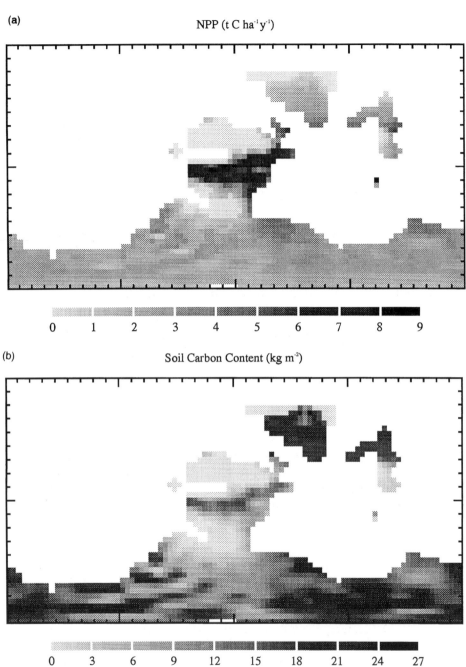

(b)

Soil Carbon Content (kg m⁻²)

Figure 20.5. Global map of Carboniferous (a) net primary productivity and (b) soil carbon concentration derived from driving a coupled vegetation-biogeochemistry model with a GCM simulation of Carboniferous climate. For (a) the scale bar is in tonnes of C ha⁻¹ year⁻¹ and for (b) soil carbon concentration g m⁻².

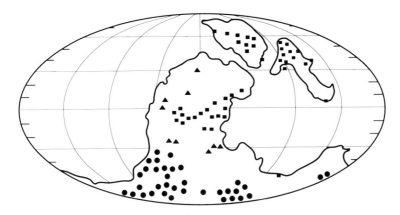

Figure 20.6. *Global distribution of geological data for the Carboniferous in the Westphalian (ca. 305 Ma).* ● = *coals,* ■ = *tillites,* ▲ = *evaporites. (Redrawn from Crowley and North, 1991)*

20.5 Global patterns of leaf carbon isotope composition

One approach to testing the predictions of vegetation function in the Carboniferous has been to use the modelled changes in the C_i/C_a ratio, and historical estimates of $\delta^{13}C_a$, to predict the likely isotopic composition of leaves, $\delta^{13}C_p$ from:

$$\delta^{13}C_p = \delta^{13}C_a - a - (b-a) \times \frac{C_i}{C_a} \qquad (20.2)$$

where $\delta^{13}C_a$ is the isotopic composition of the atmospheric CO_2 utilised by the plant, a is the discrimination against $^{13}CO_2$ by diffusion in free air and through the stomatal pores (4.4‰), b is the fractionation associated with RUBISCO (27‰) and C_i/C_a is the ratio of CO_2 concentration inside the leaf (C_i) relative to that in the atmosphere (C_a). ^{13}C discrimination can be calculated after removing the effects of $\delta^{13}C_a$ and focuses on the effects of leaf gas exchange on $\delta^{13}C_p$:

$$\Delta^{13}C = a + (b-a) \times \frac{C_i}{C_a} \qquad (20.3)$$

At the scale of individual leaves, the approach of modelled versus measured $\delta^{13}C_p$ values gave general agreement over the past 400 Ma suggesting our current understanding of RUBISCO kinetics is adequate for extrapolating photosynthetic process back in geological time (Beerling, 1997; Beerling and Woodward, 1997). More accurate predictions of historical leaf $\delta^{13}C_p$ however require $\delta^{13}C_p$ to be weighted according to the productivity of the system (c.f. Lloyd and Farquhar, 1994). This requirement arises because during conditions favourable for photosynthesis, more C will be fixed with an the isotopic value distinctive to the stomatal and mesophyll function at that time, whereas during conditions unfavourable for plant growth, e.g. drought, stomatal changes imprint a different isotopic signature on newly fixed carbon of which less is produced.

Global scale predictions of leaf $\delta^{13}C_p$ for the Carboniferous have been made (Figure 20.7) by calculating the average C_i value through the entire canopy and weighting it

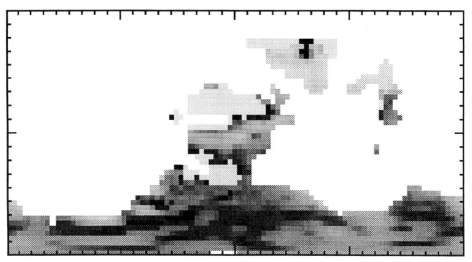

C isotope composition

| -27 | -26 | -25 | -24 | -23 | -22 | -21 | -20 | -19 | -18 | -17 | -16 | -15 |

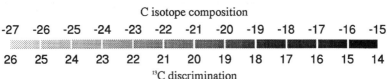

| 26 | 25 | 24 | 23 | 22 | 21 | 20 | 19 | 18 | 17 | 16 | 15 | 14 |

^{13}C discrimination

Figure 20.7. Global distribution of net primary productivity weighted leaf $\delta^{13}C$ and ^{13}C discrimination values predicted for the Carboniferous.

according to the monthly net primary productivity calculated by the vegetation model. The resulting C_i/C_a ratio is then used in equations 20.2 and 20.3 to calculate $\delta^{13}C_p$ and ^{13}C discrimination respectively. Equation 20.3 requires an estimate of $\delta^{13}C_a$ of the atmospheric CO_2 utilised by the plants in the Carboniferous. Here, we have used a recently reported value of -1‰, derived from marine inorganic carbon (Mora *et al.*, 1996).

The resulting predictions of the global pattern of NPP-weighted leaf ^{13}C discrimination show (Figure 20.7) that the highest values occur in the northern extension of Gondwana where NPP is also highest. This occurs because in these regions conditions are relatively favourable for growth (warm temperatures and abundant precipitation) so stomatal control on diffusion, and therefore photosynthesis, is not strongly limiting allowing RUBISCO to discriminate maximally against $^{13}CO_2$. Conversely, in the majority of the remaining land surface, ^{13}C values are rather low (Figure 20.7) because of low soil moisture and low temperatures. This map of stable carbon isotope composition can, in theory, be tested against observations on fossilised plant leaves and bulk organic matter (coals) (c.f. Beerling and Woodward, 1997). According to Figure 20.7, $\delta^{13}C_p$ values of organic matter as low as -18‰ should be relatively common, assuming no diagenetic alteration. Indeed, this prediction is in accord with recent $\delta^{13}C_p$ measurements made on lower Carboniferous fossilised plant material from Ireland which showed the range of $\delta^{13}C_p$ values from -25.1 to -16.5‰ (Jones, 1994). Clearly more detailed measurements are needed to test rigorously Figure 20.7 but these available data provide support for the modelling approach set out in this Chapter.

20.6 Concluding remarks

Whole scale changes in global climate and atmospheric composition occurred during the evolution of terrestrial vegetation. These changes are predicted to have affected both the nature of the physiology of plants evolving under these varying conditions and to have impacted directly on plant function. The modelling described here has shown that the influence of changes in CO_2 over the past 400 Ma on leaf morphology could have been extensive and changes in the atmospheric CO_2: O_2 ratio must have considerably affected the efficiency of photosynthesis. According to the results of gas exchange modelling on individual leaves, high atmospheric O_2 and low CO_2 concentrations would have severely curtailed terrestrial productivity – a result in line with carbon isotope measurements. Global-scale modelling confirms the leaf-scale studies and reproduces a pattern of productivity in agreement with the geological data. Calculation of NPP-weighted stable carbon isotope composition and its comparison with measurements made on fossil carbon provide some support for the approach outlined here. Future studies on the interaction between Carboniferous climate and vegetation will require a consideration of the importance of fire in regulating ecosystems dynamics.

20.7 Acknowledgements

We thank Paul Valdes (University of Reading) for providing the Carboniferous palaeoclimate dataset. DJB gratefully acknowledges funding of this work through the Royal Society.

References

Ash, S.R. (1977) An unusual Bennettitalean leaf from the Upper Triassic of the South-Western United States. *Palaeontology* 20, 641–659.

Beerling, D.J. (1994) Modelling palaeophotosynthesis: late Cretaceous to present. *Phil. Trans. Roy. Soc.* B346, 421–432.

Beerling, D.J. (1997) Interpreting environmental and biological signals from the stable carbon isotope composition of fossilized organic and inorganic carbon. *J. Geol. Soc. (Lond.)* 154, 303–306.

Beerling, D.J. and Woodward, F.I., (1996) Palaeo-ecophysiological perspectives on plant responses to global change. *Trends Ecol. Evol.* 11, 20–23.

Beerling, D.J. and Woodward, F.I. (1997) Changes in land plant function over the Phanerozoic: reconstructions based on the fossil record. *Bot. J. Linn. Soc.* 124, 137–153.

Beerling, D.J, Chaloner, W.G., Huntley, B., Pearson, A. and Tooley, M.J. (1993) Stomatal density responds to the glacial cycle of environmental change. *Proc. Roy. Soc.* B251, 133–138.

Beerling, D.J., Woodward, F.I. and Valdes, P.J. (1996) Global terrestrial productivity in the mid Cretaceous (100 Ma) : model simulations and data. *Geol. Soc. Am. Spec. Pap.* (in press).

Berner, R.A. (1994) 3GEOCARB II : a revised model of atmospheric CO_2 over Phanerozoic time. *Am. J. Sci.* 294, 56–91.

Berner, R.A. and Canfield, P. (1989) A new model for atmospheric oxygen over Phanerozoic time. *Am. J. Sci.* 289, 333–361.

Chaloner, W.G. (1989) Fossil charcoal as an indicator of palaeoatmospheric oxygen level. *J. Geol. Soc. (Lond.)* 146, 171–174.

Cleal, C.J. and Zodrow, E.I. (1989) Epidermal structure of some *Medullosan Neuropteris* foliage from the middle and upper Carboniferous of Canada and Germany. *Palaeontology* 32, 837–882.

Crowley, T.J. and Baum, S.K. (1994) General circulation model study of late Carboniferous interglacial climates. *Palaeoclimates* **1**, 3–21.

Crowley, T.J. and North, G.R. (1991) *Paleoclimatology*. Oxford University Press, Oxford.

Daghlian, C.P. and Person, C.P. (1977) The cuticular anatomy of *Fenelopsis varians* from the lower Cretaceous of central Texas. *Am. J. Bot.* **64**, 564–569.

Edwards, D. (1996) New insights into early land ecosystems : a glimpse of a Lilliputian world. *Rev. Pal. Pal.* **90**, 159–174.

Edwards, D. and Axe, L. (1992) Stomata and mechanics of stomatal functioning in some early land plants. *Cour. Forsch. Seck.* **147**, 59–73.

Eamus, D., Berryman C.A. and Duff, G.A. (1993) Assimilation, stomatal conductance, specific leaf area and chlorophyll responses to elevated CO_2 of *Maranthes corymbosa*, a tropical monsoon rain forest species. *Aust. J. Plant Physiol.* **20**, 741–755.

Farquhar, G.D., Cammerer, S. von. and Berry, J.A. (1980) A biochemical model of photosynthetic CO_2 assimilation in leaves of C_3 species. *Planta* **149**, 78–90.

Graham, J.B., Dudley, R., Aguilar, N.M. and Gans, C. (1996) Implications of late Palaeozoic oxygen pulse for physiology and evolution. *Nature* **375**, 117–120.

Jones, T.P. (1994) ^{13}C enriched lower Carboniferous fossil plants from Donegal, Ireland: carbon isotope constraints on taphonomy, diagenesis and palaeoenvironment. *Rev. Pal. Pal.* **81**, 462–464.

Knoll, A.H. and Niklas, K.J. (1987) Adaptation, plant evolution, and the fossil record. *Rev. Pal. Pal.* **50**, 127–149.

Lloyd, J. and Farquhar, G.D. (1994) ^{13}C discrimination during CO_2 assimilation by the terrestrial biosphere. *Oecologia*, **99**, 201–215.

McElwain, J.C. and Chaloner, W.G. (1995) Stomatal density and index of fossil plants track atmospheric carbon dioxide in the Palaeozoic. *Ann. Bot.* **76**, 389–395.

Mora, C.I., Driese, S.G. and Colarusso, L.A. (1996) Middle to late Paleozoic atmospheric CO_2 levels from soil carbonate and organic matter. *Science*, **271**, 1105–1107.

Parton, W.J. *et al.* (1993) Observations and modeling of biomass and soil organic carbon matter dynamics for the grassland biome worldwide. *Global Biogeochem. Cycles* **7**, 785–809.

Raven, J.A., Johnston, A.M., Parsons, R. and Kübler, J. (1994) The influence of natural and experimental high O_2 concentrations on O_2-evolving phototrophs. *Biol. Rev.* **69**, 61–94.

Robinson, J.M. (1989) Phanerozoic O_2 variation, fire and terrestrial ecology. *Palaeo. Palaeo. Palaeo. (Global and Planetary Change Section)* **75**, 223–240.

Robinson, J.M. (1994) Speculations on carbon dioxide starvation, late Tertiary evolution of stomatal regulation and floristic modernisation. *Plant, Cell Env.*, **17**, 345–354.

Schopf, J.W. and Klein, C. (eds.) (1992) *The Proterozoic biosphere. A multidisciplinary study*. Cambridge University Press, Cambridge.

Scotese, C.R. and McKerrow, W.S. (1990) Revised world maps and introduction. In: *Palaeozoic Palaeogeography and Biogeography* (eds W.S. McKerrow and C.R. Scotese). *Geol. Soc. Mem.* **12**, 1–21.

Spicer, R.A. (1989) Physiological characteristics of land plants in relation to environment through time. *Trans. Roy. Soc. Edin. (Earth Sciences)* **80**, 321–329.

Srinivasan, V. (1995) Conifers from the Puddledock locality (Potomac Group, Early Cretaceous) in eastern North America. *Rev. Pal. Pal.* **89**, 257–286.

Van de Water, P.K., Leavitt, S.W. and Betancourt, J.L. (1994) Trends in stomatal density and $^{13}C/^{12}C$ ratios of *Pinus flexilis* needles during the last glacial-interglacial cycle. *Science* **264**, 239–242.

Valdes, P. (1993) Atmospheric general circulation models of the Jurassic. *Phil. Trans. R. Soc.* **341**, 317–326.

VEMAP members. (1995) Vegetation/ecosystem modeling and analysis project : comparing biogeography and biogeochemistry models in a continental-scale study of terrestrial ecosystem responses to climate change and a CO_2 doubling. *Global Biogeochem. Cycles* **9**, 407–437.

Woodward, F.I. (1987) Stomatal numbers are sensitive to increases in CO_2 from pre-industrial levels. *Nature* **327**, 617–618.

Woodward, F.I., Smith, T.M. and Emanuel, W.R. (1995) A global land primary productivity and phytogeography model. *Global Biogeochem. Cycles* **9**, 471–490.

Zodrow, E.L. and Cleal, C.J. (1993) The epidermal structure of the Carboniferous gymnosperm from *Reticulopteris*. *Palaeontology* **36**, 65–79.

Carbon isotopes, diets of North American equids, and the evolution of North American C$_4$ grasslands

Thure E. Cerling, John M. Harris and Bruce J. MacFadden

21.1 Introduction

The evolution of North American grasslands has traditionally thought to have been documented by the evolution of equid teeth which became higher and had more pronounced crowns (hypsodont) as grasses became abundant (Kowalevkey, 1873; MacFadden, 1992; Simpson, 1951). Recently, it has been shown that the proportion of C$_4$ biomass in diet is preserved in tooth enamel by its δ^{13}C value (Lee-Thorp and van der Merwe, 1987; Quade et al., 1992; Wang et al., 1994) and that in North America the major change in δ^{13}C values occurred in the late Miocene (Cerling et al, 1993; Wang et al., 1994) rather than the early Miocene when hypsodonty became popular in equids (Simpson, 1951; MacFadden, 1992). In this chapter we show the spatial development of C$_4$ grasses in North America and show that the late Hemphillian (4–7 ma)is the period of C$_4$ ecosystem expansion.

The diet of mammals is preserved in fossil tooth enamel which is not altered during fossilization and diagenesis (Wang and Cerling, 1994). This was previously noted for collagen (DeNiro and Epstein, 1978; van der Merwe, 1978) and has been extended to enamel studies (Cerling et al., 1993; Quade et al., 1992; Wang et al 1994). The end member value for a pure-C$_3$ diet is variable because the isotopic composition of C$_3$ plants varies by more than 10‰ because of environmental variables. Although the average δ^{13}C of C$_3$ plants is about -26 to -27‰, considerably more negative values have been observed under closed canopy conditions because the isotopic composition of the atmosphere at the forest floor is enriched in CO$_2$ due to of soil respiration and is therefore depleted by several permil, and because of ^{13}C depletion in C$_3$ plants growing under low light levels (see Chapters 8 and 9). Medina et al. (1986) and van der Merwe and Medina (1989) have recorded ^{13}C depletion in C$_3$ plants to values of about -37‰ on the Amazon forest floor. However, in arid conditions C$_3$ plants close stomata to reduce water loss, and lower values of C_i/C_a are associated with an enrichment of ^{13}C by a few permil; values up to -22‰ in C_3 plants have been observed in semi-arid regions of western North America (Ehleringer et al., 1986; Ehleringer and Cooper 1988). The isotopic enrichment ^{13}C from diet in herbivores to tooth enamel is 14‰ to 15‰ based on the estimated diets of wild animals (Cerling et al., 1997) and captive animals with a known diet (Wang et al., 1994). Taken together, the range of values for animals having pure-C_3 diets is considerable: animals living in closed canopies could have $\delta^{13}C$ values as negative as -20‰, in humid areas, open canopy values are expected to be between -10 to -13‰, whereas herbivores from arid environments could have $\delta^{13}C$ values as high as -8 to -9‰. C_4 plants however, have a much smaller range in $\delta^{13}C$ values, and the expected end member value for tooth enamel is probably between +2 and +4‰. Because C_4 plants do not thrive under closed canopy conditions, the mixing line for animals with mixed C_3 and C_4 diets is between the 'humid-C_3' and 'arid-C_3' plants and C_4 plants, as shown in Figure 21.1.

As compared to the general timing of evolution for various CO_2 concentrating mechanisms (see Chapter 19), recently Cerling et al. (1993) and Wang et al. (1994) showed that C_4 biomass increased in North America in the late Neogene, apparently at a similar time to an important C_4 biomass expansion in Pakistan (Cerling et al. 1994; Quade et al., 1989; Quade and Cerling 1995), in Africa (Leakey et al., 1996; Morgan et al., 1994), and in South America (MacFadden et al., 1996). In this paper we examine a much larger sample size to see if the apparent biomass expansion really occurred in North America in the late Neogene, compare the spatial distribution of C_4 biomass through time as compared to the modern distribution, and see which elements of the equid fauna adapted to the changing food source. We do this by examining a large number of equids from all over the North American continent from about 20° to >60° latitude. Equids were selected because they are facultative grazers. Grasses are today the dominant C_4 plants in North America; the evolutionary story of equids is based on the assumption that their tooth structure was developed to feed on grasses starting in the early Neogene (Hemingfordian to Barstovian).

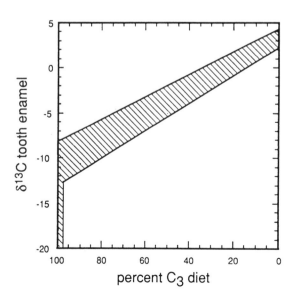

Figure 21.1. Relationship between diet and $\delta^{13}C$ of tooth enamel. The effect of a closed canopy shifts the isotopic composition of vegetation because of increased biogenic CO_2 on the forest floor; C_4 plants do not have a closed canopy effect because they do not grow under closed canopy conditions. C_3 plants are enriched in ^{13}C in arid conditions compared to more humid conditions. (Changes in the isotopic composition of the atmosphere would result in a translation of the diet field.)

21.2 Sampling protocol and sources of material

We analyzed the carbon isotopic composition of enamel from more than 300 equid individuals of Hemingfordian to Rancholabrean age (up to 20 Ma) from North America. These ranged in latitude from Alaska to central Mexico, although most were from the conterminous United States. Enamel was separated from dentine using a Dremel® tool. The enamel separate was ground to less than 100 mesh, and treated with 10‰ H_2O_2 or 1% NaOCl for sixteen hours and repeatedly rinsed in distilled water and centrifuged, followed by a sixteen hour treatment with 1 M acetic acid with following rinses with distilled water and ethanol. The purified powder was reacted with 100% H_3PO_4 at 25° C for about 40 hours and the resultant gas was cryogenically purified. Samples were treated with Ag metal to remove SO_2. We have found that high temperature treatment (8 hours with Ag metal at 450° C) or longer low temperature treatment (several days with Ag wool at 50° C) removes the contaminant SO_2, which is recognized by its pink color at -180 °C and its detection on the mass spectrometer at mass/charge ratio of 64. Results are reported in the standard permil (‰) notation where $\delta^{13}C = ((R_{sample} - R_{standard}) - 1)^* 1000$ where R_{sample} and $R_{standard}$ are the $^{13}C/^{12}C$ ratios in the sample and standard, respectively.

Samples were selected primarily from museum collections, although a few were collected in the field. These samples were from the American Museum of Natural History, the Florida Museum of Natural History, Hagerman National Monument, the Idaho Museum of Natural History, the Natural History Museum of Los Angeles County, the Nebraska State Museum, Southern Methodist University, the Texas Memorial Museum, the University of Arizona, the U. S. Geological Survey, the University of California Berkeley, the University of California Riverside, and the Utah Museum of Natural History. Samples were chosen to represent a wide range of geographic localities in North America and a fairly complete coverage of time from the early Miocene to present. In general, we analyzed a single tooth from each individual (generally P2–4 or M2–3, occasionally m3). In some cases is was not possible to determine which molar or premolar tooth was sampled. Analyses of different teeth from the same individual (P²-M³) generally do not show significant differences between the M¹ and other teeth, or between the deciduous teeth and other teeth.

21.3 Transition to C_4 diet in equids: occurrence of C_4-dominated diets, hypsodonty, and the radiation of equid genera

$\delta^{13}C$ values of equids from a wide latitudinal (central Mexico to Alaska) and longitudinal range (Florida to the west coast of North America) show that prior to late Hemphillian all samples had $\delta^{13}C$ values between -8 and -14‰ indicating a diet dominated by C_3 vegetation (Figure 21.2). Average $\delta^{13}C$ values for the North American Land Mammal Ages (NALMAs) Hemingfordian, Barstovian, Clarendonian, and early Hemphillian range between -10‰ and -11‰ indicated a C_3 dominated diet. This confirms the observation of Cerling *et al.* (1993) and Wang *et al.* (1994) that the increase in the height of equid teeth in the early Miocene of North America does not correspond to a change in the diet from C_3 to C_4 plants.

Furthermore, the overall pattern shows that certain regions of North America were

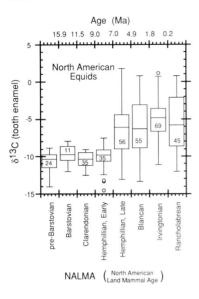

Figure 21.2. $\delta^{13}C$ *values of North American equids by North American Land Mammal Age. Graph gives the number of samples (n), the mean, 1S.D., and range for each time interval. Ages are from Woodburne and Swisher (1995), except that the early-late Hemphillian boundary is placed at 7.0 Ma based on the correlation of the Blacktail Creek Ash from the late Hemphillian age Coffee Ranch locality (M. Perkins, personal communication).*

susceptible to the expansion of C_4 grasses, whereas other regions were not. For example, the south central Great Plains became dominated by C_4 plants beginning in the late Hemphillian, whereas the Northern Rocky Mountains and Cascade regions do not show evidence for significant C_4 biomass at any time in the Neogene; nor do they have many C_4 grasses today (Figure 21.3). The rapid emplacement of the general pattern of C_4 grasses implies that the conditions that control the distribution of C_4 plants in North America, which is probably related to the atmospheric CO_2/O_2 ratio and to regional patterns of heat and moisture distribution (see Chapters 19 and 20), have been in place since the late Hemphillian. We will discuss the possible climatic and atmospheric conditions below.

The Last Chance Canyon locality within the Dove Spring Formation in the Ricardo Group of southern California is the site of the oldest known C_4 plants documented by Kranz anatomy and by stable isotopic analyses (Nambudiri *et al.*, 1987; Tidwell and Nambudiri, 1989) and is estimated to be about 12.5 Ma (Whistler and Burbank, 1992; Whistler, personal communication). We analyzed three equids from this locality between 12 and 13 Ma age and found a very limited range in $\delta^{13}C$, from -9.3‰ to -9.5‰ (Table 21.1). This implies that the proportion of C_4 plants in this locality was

Table 21.1. $\delta^{13}C$ *values for equids from the Dove Spring Formation, California, between 12 and 13 Ma*

Sample number	Species	Age (Ma)	$\delta^{13}C$
LACM 1744	*Cormohipparion occidentale*	12.0	-9.5
LACM 1741	*Cormohipparion occidentale*	12.2	-9.3
LACM 1736	*Pliohippus tantulus*	13.0	-9.5

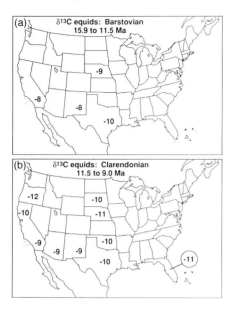

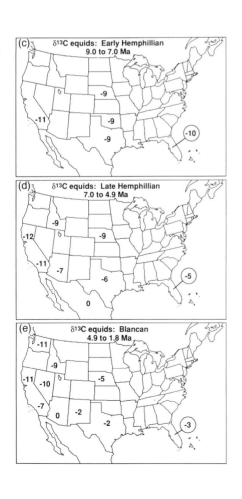

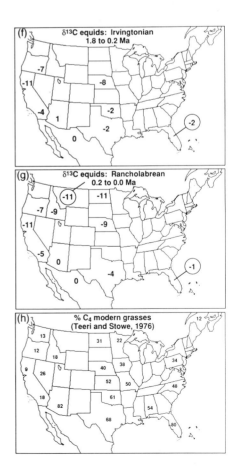

Figure 21.3. Maps of central North America showing the spatial development of C_4 ecosystems as recorded by the diet of equids and the distribution of modern C_4 grasses. Geographical distribution of $\delta^{13}C$ in fossil equid teeth from North America (A)–(G).

(A) Barstovian: 15.9 to 11.5 Ma

(B) Clarendonian: 11.5 to 9.0 Ma

(C) Early Hemphillian: 9.0 to 7.0 Ma

(D) Late Hemphillian: 7.0 to 4.9 Ma

(E) Blancan: 4.9 to 1.8 Ma

(F) Irvingtonian: 1.8 to 0.2 Ma

(G) Rancholabrean: 0.2 to 0.0 Ma

(H) Modern distribution of C_4 grasses (percent of total species). Modified from Teeri and Stowe (1976).

probably very small, representing an insignificant proportion of the biomass as far as a dietary resource was concerned. C_4 grasses in the region today tend to be associated with perennial springs. Webber (1933) and Tidwell and Nambudiri (1989) report palms, oaks, and other non-grasses as important fossil flora in the locality. Axelrod (1973) interprets the flora to indicate a thorn scrub to chaparral vegetation, similar to those characteristic of Mediterranean climates, although we note that Mediterranean climates have C_3 grasses rather than C_4 grasses.

The chronology of equid hypsodonty, radiation of equids, the first documented C_4 plants and the change to a C_4 diet in equids is somewhat puzzling. Hypsodonty, which is the great increase in the crown height of molars and premolars, began in the early to middle Miocene for equids and was well underway by 15 million years ago (MacFadden, 1988, 1992). From 17 to 12 Ma the number of equid genera greatly increased from 5 to 12. The association of hypsodonty with the great increase in the number of equid genera has been traditionally explained to be the result of the spread of savanna grasslands (e.g., Kowalevsky, 1873; MacFadden, 1992; Simpson, 1951). Our results (Yang *et al.*, 1994; and this study) show that the spread of C_4 grasslands, which is the dominant modern grass in the southern to central midwestern United States, occurred much later. The first documented occurrence of C_4 grasses is about 12.5 Ma in southern California in what is now the Mojave Desert, which ironically does not have many C_4 grasses today. The transition to a C_4 dominated diet occurs between about 7 and 6 million years ago which is the time when equid diversity began its great decline to only one genus today (see discussion below). It is likely that this time corresponds to the establishment of the North American grassland rather than savanna conditions (Axelrod, 1985; Gregory, 1971); the middle and late Miocene ecosystems, savanna or otherwise, had very little C_4 grass biomass.

This raises interesting problems in the way that one thinks about equid diversity, tooth hypsodonty, and diet. The expansion of C_4 grasses in North America started about 7 million years ago, which is about 10 million years after the apparent adaptation to grazing with the resulting hypsodonty in equid teeth. If the adaptation in the early to middle Miocene was indeed to be able to eat grasses, then the kinds of grasses must have been C_3 grasses which are not abundant today in the central and southern great plains.

21.4 Geographical and temporal distribution of C_4 grasses in North America

The modern distribution of C_4 grasses in North America is characterized by abundant C_4 grasses in regions where the growing season temperatures are high; C_4 grasses are the dominant grass in the central to southern Great Plains, the Sonoran and Chihuahuan Deserts, and Florida; C_3 grasses are dominant in Canada, the Pacific Northwest, and California (Figure 21.3H). We analyzed equids for much of the last 20 million years over most of this geographic range to establish the temporal and spatial establishment of C_4 grasses in North America. Figure 21.3 shows the average value of $\delta^{13}C$ for tooth enamel from equids from each NALMA from Barstovian to Rancholabrean (Figures 21.3A to 21.3G) for different states (except California where northern and southern California are distinguished from each other) and it also shows the modern distribution of C_4 grasses as a proportion of the extant grass flora. From this figure it is apparent that the present distribution, with a high fraction of C_4 grass in the southern United States from Arizona to Florida and southern states and Mexico, and a low fraction of C_4 grasses in California and in the Pacific Northwest, was established by Blancan time (4.3 to 1.9 Ma), with the pattern evolving in the late Hemphillian (7 to 4.3 Ma). The wide geographic coverage in the Clarendonian (Figure 21.3B), from 11.6 to 9.0 Ma, shows that C_3 biomass dominated the diets of equids in the Middle Miocene throughout North America.

21.5 Regional patterns and variations

21.5.1 *Southeastern USA*

Our knowledge of the carbon isotopic values of equids from the southeastern USA comes from the extensive fossiliferous sequence from Florida reported by MacFadden and Cerling (1996); also see preliminary data in Wang *et al.*, 1994. As indicated in Figure 21.3H modern grassland ecosystems in Florida consist predominantly (80 %) of C_4 grasses Teeri and Stowe (1976). Carbon isotopic sampling of 53 individual tooth enamel samples representing the diversity of Florida fossil equid genera (*Pseudhipparion, Neohipparion, Hipparion, Nannippus, Cormohipparion, Pliohippus, Dinohippus,* and *Equus*) indicate that during the Clarendonian and early Hemphillian, ecosystems in this region were exclusively or almost exclusively C_3-based. The middle and early part of the late Hemphillian are not represented in Florida. By the latest Hemphillian, as represented by the Bone Valley fossil deposits at about 4.5 Ma ago, Florida ecosystems had undergone a major shift so that mean carbon isotopic values of the equids are between -5 to -1‰ during the Pliocene and Pleistocene (Figure 21.3), indicating establishment of C_4 grasslands in this region during this time.

21.5.2 *Northern California*

We examined fossils from several sites in Alameda County, San Francisco Bay region, northern California. This region today has very few C_4 plants (Figure 21.3H) because of its Mediterranean climate, with the rainy season coming in the winter and the dry season in the summer. C_4 grasses make up less than 10% of the total number of grass species for the San Francisco Bay region. Late Hemphillian through to Rancholabrean sites (Pinole Formation, Doolan Canyon, and other sites in Alameda County) yielded equids with $\delta^{13}C$ values between -10 and -13‰, indicating a C_3-dominated diet, with little or no C_4 component. Five samples have average $\delta^{13}C$ values of -11.9±1.3‰ and range from -10.2 to -13.1‰, which indicates vegetation with a $\delta^{13}C$ value of about -26 to -27‰.

21.5.3 *Southern California – Mojave, Rancho La Brea*

The southern California region includes the Mojave desert which has most of its rainfall during the winter months, and therefore has little C_4 vegetation. The oldest known C_4 plants in North America are from the Dove Springs Formation (Tidwell and Nambudiri, 1989) and are about 12.5 Ma based on the stratigraphy and dating of Whistler and Burbank (1992). Equids from this locality do not show evidence for a significant component of C_4 biomass in their diet (see discussion in Section 21.3) indicating that although C_4 plants were present, their abundance in the ecosystem was minimal.

Other localities from the Mojave desert, principally in the region of Anza Borrega, show a mixture of C_3 and C_4 diet sources, with $\delta^{13}C$ values ranging from about -3‰ to -7‰ for Blancan though to Rancholabrean localities. This region today gets some spillover from the southwest monsoon, and presumably is responsible for the C_4 grasses found in the region today and in the past (see also Chapter 14). This is part of the steep gradient in $\delta^{13}C$ values from Arizona to coastal California. The samples analyzed so far, it is emphasized, are bulk samples. The detailed profiles possible using laser ablation techniques (Cerling and Sharp, 1996; Sharp and Cerling, 1996) will make it possible to determine the range in $\delta^{13}C$ in the annual diet and may make it possible to study migration patterns.

Samples from the Rancho La Brea (Los Angeles, California) excavations also show a C_4 component, ranging from about -4‰ to -8‰. Again, detailed profiles using laser ablation methods (Cerling and Sharp, 1996) and examination of non-migratory species will be necessary to determine if C_4 plants were abundant in the region or if this mixed signal is due to migration. Today there are very few C_4 plants in the Los Angeles Basin.

21.5.4 *Chihuahuan and Sonoran Deserts and Central Mexico*

The Chihuahuan and Sonoran deserts of northern Mexico and southern New Mexico and Arizona have summer monsoon rains that favour the growth of C_4 and CAM plants. Winter rains promote C_3 growth for a few months of the year, but the C_4 biomass is abundant throughout these deserts. Clarendonian samples from Arizona averaged -9.4‰. Seven Late Hemphillian samples from Arroyo de los Poños in Chihuahua, Mexico ranged from 0 to +1.7‰, and together with the samples from Ocote in Mexico are the first pure C_4 grazers. More than 20 samples from Blancan through to Rancholabrean, averaged +0.3‰ and ranged from -3.1 to +2.1‰, showing a population that has little C_3 component in its diet. This is in significant contrast to samples collected from the nearby Mojave Desert region (see Section 21.5.3 above). Figure 21.4 shows that equids from Blancan through to Rancholabrean, from the Sonoran and Chihuahuan Deserts collected in Arizona and northern Mexico, have a higher C_4 component than other geographic regions. Samples from the high elevation (> 2000 meters) site of San Josecito Cave have $\delta^{13}C$ values ranging from -8 to -11‰, which is compatible with the observation of C_3 grasses being dominant in high elevations, even at low latitude. Finally, three samples from the Ocote locality in central Mexico (late Hemphillian) showed an essentially pure C_4 diet, ranging in $\delta^{13}C$ from -0.3 to -1.1.

21.5.5 *Northern Great Basin: eastern Washington, eastern Oregon, southern Idaho, northern Nevada*

The northern Great Basin region has a semi-desert vegetation that is little influenced by the southwest monsoon. Precipitation falls primarily in the winter and early spring (December through to April), and summers are very hot accompanied by low precipitation. There are few native C_4 plants, making up less than 30% of the total number of grass species. Estimates of total C_4 biomass using carbon isotopes in modern soils indicates that C_4 plants make up between 0 and 10% of the total biomass (see Cerling and Quade, 1993). We include samples from southeastern Washington (Ringold Formation) and eastern Oregon (Owhyee River region) in the Northern Great Basin grouping

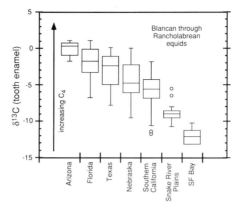

Figure 21.4. *Box-and-whisker diagram of $\delta^{13}C$ values of Blancan to Rancholabrean equids from different geographic regions of North America show differentiation in their respective diets. Equids from the San Francisco Bay region (cool, dry summers) and the Snake River Plains (hot, dry summers) have diets dominated by C_3 biomass. Equids from regions characterized by very long summers with significant precipitation (Florida, Texas, Arizona) have diets dominated by C_4 biomass. The mixed C_3/C_4 diets from Nebraska may be due to the shorter summers (compared to Texas) so that the equids ate C_3 grasses in part of the year and C_4 grasses during other parts of the year. The mixed C_3/C_4 diets of equids from southern California may be due to migration across the Sonoran/Mojave boundary.*

because their climate is similar to that in southern Idaho and northern Nevada, all of which are in the rain shadow of the Cascades or Sierra Nevada Mountains.

All our samples from this region are Clarendonian or younger, and all show a dependence on C_3 vegetation. Of the 28 samples that are Blancan or younger, two have $\delta^{13}C$ values of -5.4% and -6.4% indicating that there may have locally been some C_4 plants present; one of these is from Idaho, near Hagerman, and one from the Owyhee River region of southeastern Oregon. Figure 21.4 shows that the carbon isotopic composition for Blancan to Rancholabrean samples from southern Idaho, principally from the Hagerman region, has a narrow range; excluding the sample with a value of -6.4‰, the average $\delta^{13}C$ values for these samples from Idaho is -9.2 ± 0.6 ‰ ($n = 16$), which is compatible with a diet that averages about -23 to -24 ‰. This diet is different than for the equids of the San Francisco Bay region (see above) and indicates either a persistently small fraction (ca. 10%) of C_4 plants in the diet, or that the local C_3 plants were shifted by 2 to 3 ‰ because of heat and moisture stress (Ehleringer *et al.* 1986; Ehleringer and Cooper, 1988). We favour the latter hypothesis that heat and moisture stress in the northern Great Basin was higher than in the San Francisco Bay area, as it is today.

21.5.6 *Southern Great Plains*

The southern Great Plains region, south of 37°N, is distinct from the Northern Great Plains: today the region is characterized by winters with a few days of frost, and by very long, hot summers with rains spread throughout the year; consequently they have C_3 grasses growing in the winter and early spring, with C_4 grasses growing through the long hot season, beginning in April or May and through to September. As in other localities, samples older than early Hemphillian suggested a C_3 diet. However, by late Hemphillian there is an indication of a C_4 component in the diet and Blancan equids clearly have a C_4-dominated diet (Figure 21.3).

The first site with a significant C_4 component in the diet is at the Coffee Ranch locality estimated to be 6.8 Ma, based on the presence of the Blacktail Ash and correlation of ashes to the Rocky Mountain region (M. Perkins, personal communication). We sampled four different species of equids at Coffee Ranch (all M3s): *Astrohippus ansae, Dinohippus interpolatus, Nannippus lenticularis,* and *Neohipparion eurystyle:* (Table 21.2).

At least 3 species have $\delta^{13}C$ values greater than -6.5‰, indicating a positive component of C_4 biomass in the diet, although all species had a value as negative as -9‰ to -11‰. A detailed profile of a single tooth (UCMP 140466) using laser ablation (Sharp and Cerling, unpublished data) shows a range in $\delta^{13}C$ of 8‰ indicating that a change in diet occurred during the growth of the tooth, which took place over about 1.5 years. From the data at Coffee Ranch it is not possible to make a strong case of selective diet of one species of equids compared to another; additional detailed analyses using laser ablation may shed light on their respective selectivities of dietary resources.

For Blancan through to Rancholabrean samples, the $\delta^{13}C$ values of equids from the coastal plain region tend to have a higher C_4 component ($\delta^{13}C$ from 0 to -3‰) than from the more northerly Edwards Plateau to Texas Panhandle region ($\delta^{13}C$ from -2 to -8‰). Figure 21.4 shows that the more southerly samples from Texas have a higher C_4 dietary component in Blancan and younger samples than samples from the northern Great Plains.

21.5.7 *Northern Great Plains*

There is a clear climatic distinction between the northern and the southern Great Plains in north America. The southern Great Plains have relatively warm winters com-

Table 21.2. $\delta^{13}C$ values for equids from Coffee Ranch, Texas

Sample	Species	Age (Ma)	$\delta^{13}C$
UCMP 140465	*Astrohippus ansae*	6.8	-9.1
UCMP 140466	*Astrohippus ansae*	6.8	-6.1
UCMP 140467	*Astrohippus ansae*	6.8	-8.1
F: AM 129414	*Dinohippus interpolatus*	6.8	-6.4
UCMP 140462	*Dinohippus interpolatus*	6.8	-7.9
UCMP 140463	*Dinohippus interpolatus*	6.8	-9.7
UCMP 140464	*Dinohippus interpolatus*	6.8	-7.9
UCMP 140468	*Nannippus lenticularis*	6.8	-8.4
UCMP 140469	*Nannippus lenticularis*	6.8	-8.0
UCMP 140470	*Nannippus lenticularis*	6.8	-10.8
F: AM 129415	*Neohipparion eurystyle*	6.8	-5.8
F: AM 129416	*Neohipparion eurystyle*	6.8	-2.7
UCMP 140471	*Neohipparion eurystyle*	6.8	-5.8
UCMP 140472	*Neohipparion eurystyle*	6.8	-10.1
UCMP 140473	*Neohipparion eurystyle*	6.8	-8.4
SMU 70533	equid	6.8	-6.0

pared to the bitterly cold northern Great Plains. However, both have sufficient summer rainfall for C_4 grasses to flourish in the hot (>25 °C) summer temperatures. The northern Great Plains have a spring flush of C_3 grasses in April and May that give way to C_4 grasses in July and August (Ode *et al.*, 1980). Our data set includes samples from Nebraska through to North Dakota (40 to 49 °N). Blancan, Irvingtonian, and Rancholabrean samples from Nebraska range from about -3 to -9.5 ‰, averaging -5.9±1.9‰ ($n = 15$).

21.5.8 *Geographic Summary*

The changing distribution of C_4 grasses in North America is clearly shown in the $\delta^{13}C$ of equid teeth. Samples of early Hemphillian age and older have a C_3 diet everywhere on the continent. In the late Hemphillian the first indications of a C_4 diet are manifested in equid samples from below 37 °N in Florida, Texas, and Mexico. By Blancan times the present pattern of C_4 grasses seems to have been firmly established, with the strong C_4- to C_3-gradient from Arizona to California being recognizable, and with the northern California coastal region maintaining a virtually pure C_3 ecosystem (Figure 21.3). Figure 21.4 shows that the highest component of C_4 biomass in post-Hemphillian diets is in the southern part of North America, with the areas affected by the southwest monsoon (Sonoran and Chihuahuan Deserts) being dominated almost exclusively by C_4 grasses. In contrast, the northern Great Basin and the San Francisco Bay regions are dominated by C_3 grasses. The difference in $\delta^{13}C$ between the northern Great Basin equids compared to the San Francisco Bay equids (-9 and -12‰, respectively) may be due to differences in the isotopic composition of C_3 biomass related to water and heat stress (Ehleringer *et al.*, 1986; Ehleringer and Cooper, 1988; see Chapter 9).

21.6 History of C_4 grasses versus C_4 grasslands in North America

As described above, North American grasses using the C_4 photosynthetic pathway have been identified in the late Miocene (Tidwell and Nambudiri, 1989), which is about 12.5 Ma old at this level (Whistler and Burbank, 1992; Whistler, personal communication). The identification of *Tomlinsonia thomassonii* as a C_4 plant is unassailable as it has the characteristic Kranz anatomy and its $\delta^{13}C$ value is -13.7‰ (Nambudiri *et al.*, 1978; Tidwell and Nambudiri, 1989) which is in the observed range of modern C_4 plants. As noted above, the diets of equids from this locality do not show evidence for a significant C_4 component in the diet, suggesting that perhaps *Tomlinsonia thomassonii* was a minor component of the ecosystem.

The first clear indication of widespread C_4 biomass of interest to equids is in the late Hemphillian. By Blancan times the modern distribution of C_4 biomass was essentially in place. Thus we concur with Gregory (1971) and Axelrod (1985) who suggested that the North American steppe vegetation was established at the end of the Miocene or in the early Pliocene.

21.7 Equid diversity during the Neogene

The spread of grasslands and major adaptive radiation of equids during the middle Miocene is one of the classic examples of plant/animal co-evolution interpreted from the fossil record. At the height of equid diversity during the Barstovian and Clarendonian, a dozen genera were contemporaneous in North America and many of these apparently co-existed locally (Figure 21.5). Depending upon local conditions, this middle-Miocene equid diversity consisted of a combination of presumed browsers, mixed-feeders, and the earliest-recognized grazers. By about 10 million years ago, the short-crowned, presumed browsers (*Anchitherium*, *Hypohippus*, and *Megahippus*) had become extinct, and the remaining North American equid genera consisted of moderately to high crowned forms interpreted to have been, respectively, mixed-feeders and predominantly grazers. During the past 10 million years there has been a steady decline in North American equid generic diversity from 11 in the Clarendonian, to 10 in the early Hemphillian, 8 in the late Hemphillian, 3 in the Blancan, and only the extant genus *Equus* since the Irvingtonian (Figure 21.5; MacFadden, 1992).

The explanation for the extinction of browsing-adapted equids seems to be related to the expansion of open-country savanna/grassland habitats at the expense of woodlands during the middle Miocene. The dramatic drop in hypsodont (high-crowned) grazing equids from the early Hemphillian to the Pleistocene seems to have resulted the onset of C_4 grasslands after the late Miocene global carbon shift (Cerling *et al.*, 1993). Prior to this time, terrestrial grasslands were predominantly C_3-based and capable of supporting a relatively high diversity, as seen in some 10 genera of grazing horses (Figure 21.5; early Hemphillian). After the late Miocene carbon shift, the grasslands changed in character with an increase in C_4 grasslands, possibly of lower

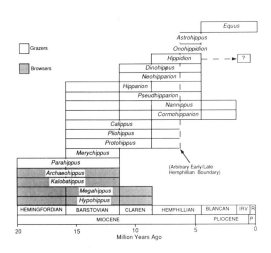

Figure 21.5. Plot of equid diversity since the middle Miocene. From MacFadden (1992) and reproduced with permission of Cambridge University Press.

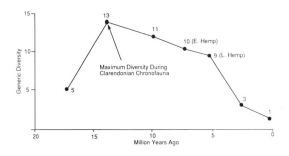

productivity as global climate became relatively more arid and with higher seasonality. These new C_4 grasslands were capable of only supporting a lowered diversity of grazing taxa, including the horses (e.g., Gregory, 1971; Webb, 1977, 1983).

What do the carbon isotopic data add to the traditional story of horse and grassland co-evolution? Prior to the late Miocene global carbon shift, all horses represented in Figure 21.5 were feeding on C_3 plant foodstuffs. From the dental crown heights, this almost certainly represents the spectrum of feeding adaptations from mixed C_3 browsing and grazing to predominantly C_3 grazing. During the late Hemphillian, the $\delta^{13}C$ values for the various equid genera span a range from -8 to -4‰ (Figure 21.6), indicating the first influence of C_4 grass in the diets of forms including *Dinohippus* and possibly *Neohipparion, Nannippus,* and *Pseudhipparion*. In contrast to the traditional text-book notion of horse lineages getting larger through time ('Cope's Rule'), *Nannippus* and *Pseudhipparion* evolved smaller body sizes, indicating independent 'dwarfing trends', although *Nannippus* attained a virtually pure C_4 diet in Arizona in the Blancan. The general reason for dwarfing in phylogenetic lineages is somewhat problematic, but in the case of these equids it may represent one kind of evolutionary strategy of feeding on lower nutritive-value C_4 grasses.

During the Blancan, two of the equid genera, *Equus* and *Nannippus*, are clearly feeding almost exclusively on C_4 grasses (Figure 21.6); they therefore are facultative grazers like extant *Equus*. *Cormohipparion*, which was widespread in North America during the Miocene but only known from a relictual species in Florida (Hulbert, 1987) during the Pliocene, has a mean $\delta^{13}C$ value of –6.5‰, indicating that this genus was an isotopically mixed feeder. Its teeth, which were only moderately high-crowned relative to contemporaneous *Nannippus* and *Equus*, also suggest a mixed diet, feeding upon some combination of browsing and grazing. During the Pleistocene, *Equus* is predominantly a grazer, but there is an interesting shift from about -1.5‰ in the

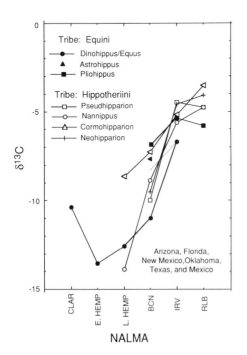

Figure 21.6. *$\delta^{13}C$ values for different equid lineages from southern North America. Samples are from <37° latitude and do not include samples from California. This figure shows that all extant equid lineages shown make the transition to a partial or complete C_4 diet.*

Irvingtonian to -4.5‰ in the Rancholabrean. A possible explanation for this shift, based on preliminary data from Florida (MacFadden and Cerling, 1996), is that the hypergrazer niche complex occupied by forms such as *Equus* and *Mammuthus* during the Irvingtonian was also occupied by *Bison*, a widespread Rancholabrean immigrant (mean $\delta^{13}C$ value of Florida fossil *Bison* is -1.0‰; MacFadden and Cerling, 1996). While the available data indicate that *Equus* has been a facultative grazer since its origin during the Pliocene (Blancan), in some ecosystems this equid may also eat other locally available and isotopically more negative foods, including forbs and shrubs (e.g., Berger, 1986), depending upon a complex array of factors, including competition from other grazers and/or seasonally available patchy food resources. This interesting speculation awaits further studies.

21.7 Global expansion of C$_4$ ecosystems

Previously, Cerling *et al* (1993) noted that the first evidence for C$_4$ diet in North America appeared to be at about the same time as the expansion of C$_4$ grasses in Pakistan, and therefore called on some global forcing. Recently, additional work in Africa (Morgan *et al.*, 1994; Cerling *et al.*, 1997) and in South America (MacFadden *et al.*, 1996) show that the first evidence for a C$_4$ diet is about 8 Ma for both continents. Thus, it appears that C$_4$ grasses expanded on four widely separated continents between 8 and 6 Ma (Cerling *et al.*, 1997). In North America, the expansion is greatest in the southern part of the continent.

C$_4$ plants are adapted to low atmospheric CO_2 concentrations under conditions of high water and heat stress (Ehleringer *et al.*, 1991). In such conditions they have an advantage over C$_3$ plants which tend to undergo photorespiration, which decreases their net photosynthesis. The pattern of C$_4$ expansion in North America is compatible with such a scenario, with the present global ecosystem, the 'C$_4$-world', being in place during the Pliocene and continuing to the present day. The previous 'C$_3$-world', which lasted for several billion years, evolved under conditions fundamentally different from the 'C$_4$-world' (see Chapters 19 and 20).

21.8 Summary and implications

We have shown that prior to about 7 million years ago C$_4$ grasses made up an insignificant fraction of the diet of North American horses, yet by 4 million years ago they were very important in certain parts of North America. This study shows that the present distribution of C$_4$ ecosystems in North America began in the late Hemphillian and may have been completed by Blancan times, or between about 7 and 4 million years ago. In the past 4 million years climates and ecosystems have shifted boundaries, but some of the distinctive features of the modern distribution of C$_4$ ecosystems (the south – north gradient of C$_4$ to C$_3$ grasses in the Great Plains, the sharp gradient from the C$_4$-dominated Sonoran Desert to the C$_3$-dominated Mojave Desert, the persistent paucity of C$_4$ grasses in the northern Great Basin) are recognizable in the Blancan (ca. 4 million years ago). As these features are today controlled by regional weather patterns it implies that these patterns, such as the southwest monsoon and the summer Great Basin high pressure system, have been in place for at least this long.

Equids in North American have long served as a classic example of adaptation to a grazing diet, beginning in the Hemingfordian and Clarendonian, 20 to 15 million years ago with the development of hypsodonty (high-crowned teeth suitable for grazing), and the expansion into many different families. However, the stable carbon isotope composition of equid enamel indicates that during this time of the early development of hypsodonty and expansion of equid diversity the diet of equids was solely C_3 plants. If these C_3 plants were grasses, then conditions favouring C_3 grass must have been different than that today in North America. When C_4 plants do become an important part of the diet of equids, diversity decreases dramatically, probably because of a decrease in the diversity of habitats available to equids. Thus, the development of the extensive C_4 grasslands characteristic of North America is associated with a decrease in equid diversity. It is likely that the earlier habitats favouring hypsodonty and the expansion of diversity were not pure grasslands, although C_3 grasses may have made up an important component of the pre-7-million year old landscape in North America much as C_4 grasses make up an important component of modern savannas.

The time of the expansion of C_4 ecosystems in North America is very close to that observed in Africa, South America, and Pakistan, implying some global forcing as previously suggested (Cerling *et al.*, 1993). The modern atmosphere, one characterized by relatively low CO_2 concentrations, favours C_4 photosynthesis under high temperature conditions because C_3 plants are less efficient because of photorespiration. A possible scenario for the expansion of C_4 ecosystems may be the gradual lowering of atmospheric CO_2 to the present conditions; in so doing the atmospheric CO_2 concentration may have crossed an important threshold for C_3 photosynthesis in certain regions (Cerling *et al.*, 1997).

Acknowledgements

We thank W. Akersten, T.M. Bown, J.D. Bryant, J. Hearst, H. Hutchison, L.L. Jacobs, E.H. Lindsay, E.L. Lundilius, H. G. McDonald, M. Voorhies, D. Whistler, D. Winkler, and M.O. Woodburne for assistance in obtaining fossils from North American localities, and J.R. Ehleringer for discussions. This work was supported by NSF.

References

Axelrod, D.I. (1973) History of the Mediterranean ecosystem in California, In: *Ecological Studies. Analysis and Synthesis 7* (eds F. diCastri, and H.A. Mooney). Berlin, Springer-Verlag, pp. 225–277.

Axelrod, D.I. (1985) Rise of the grassland biome, central North America, *Bot. Rev.* 51, 163–201.

Berger, J. (1986) *Wild Horses of the Great Basin: Social Competition and Population Size*, University of Chicago Press, Chicago. 326 pp.

Cerling, T.E. and Sharp Z.D. (1996) Stable carbon and oxygen isotope analysis of fossil tooth enamel using laser ablation. *Palaeogeogr. Palaeoclim.Palaeoecol.* 126, 173–186.

Cerling, T.E., Wang, Y. and Quade, J. (1993) Expansion of C_4 ecosystems as an indicator of global ecological change in the late Miocene. *Nature* 361, 344–345.

Cerling, T.E., Harris, J.M., MacFadden, B.J., Ehleringer, J.R., Leakey, M.G., Eisenmann, V and Quade, J. (1997) Global vegetation change through the Miocene–Pliocene Boundary. *Nature* 389, 153–158.

DeNiro, M.J. and Epstein, S. (1978) Influence of diet on the distribution of carbon isotopes in animals. *Geochim. Cosmochim. Acta* 42, 495–506.

Ehleringer, J.R. and Cooper, T.A. (1988) Correlations between carbon isotope ratio and micro-habitat in desert plants. *Oecologia* **76**, 562–566.

Ehleringer, J.R., Field, C.B., Lin, Z. and Kuo, C. (1986) Leaf carbon isotope and mineral composition in subtropical plants along an irradiance cline. *Oecologia* **70**, 520–526.

Ehleringer, J.R., Sage, R.F., Flanagan, L.B. and Pearcy, R.W. (1991) Climate change and the evolution of C_4 photosynthesis. *Trends Ecol. Evol.* **6**, 95–99.

Gregory, J.T. (1971) Speculations on the significance of fossil vertebrates for the antiquity of the Great Plains of North America. *Abh. Hess. Landes. Bodenforsch.* **60**, 64–72.

Hulbert, R.C. (1987). A new *Cormohipparion* (Mammalia, Equidae) from the Pliocene (latest Hemphillian and Blancan) of Florida. *J. Vert. Paleo.* **7**, 451–468.

Kowalevsky, V. (1873), Sur l'*Anchitherium aurelianense*, Cur. et sur l'histoire paleontologique des Chevaux. *Mem. Acad. Imp. Sci. St. Petersburg,* 7th Ser. **20**, 1–73.

Leakey, M.G., Feibel, C.S., Bernor, R.L., Harris, J.M., Cerling, T.E., Stewart, K.M, Stoors, G.W., Walker, A., Werdelin, L. and Winkler, A.J. (1996), Lothagam: A record of faunal change in the Late Miocene of East Africa. *J. Vert. Paleo.* **16**, 556–570.

Lee-Thorp, J.A. and N.J. van der Merwe (1987) Carbon isotope analysis of fossil bone apatite. *S. Afr. J. Sci.* **83**, 712–715.

MacFadden, B.J. (1988) Fossil horses from 'Eohippus' (*Hyrachotherium*) to Equus, 2: rates of dental evolution revisited. *Biol. J. Linn. Soc.* **35**, 37–48.

MacFadden, B.J. (1992) *Fossil Horses: Systematics, Paleobiology, and Evolution of the Family Equidae.* Cambridge University Press, Cambridge, 369 pp.

MacFadden, B.J. and Cerling, T.E. (1996) Mammalian herbivore communities, ancient feeding ecology and carbon isotopes: A 10 million-year sequence from the Neogene of Florida. *J. Vert. Paleo.* **16**, 103–115.

MacFadden, B.J., Cerling, T.E. and Prado J. (1996) Cenozoic terrestrial ecosystem evolution in Argentina: evidence from carbon isotopes of fossil mammal teeth. *Palaios* **11**, 319–327.

Medina, E., Montes, G., Cuevas, E. and Rokzandic, Z. (1986) Profiles of CO_2 concentration and $\delta^{13}C$ values in tropical rain forests of the upper Rio Negro Basin, Venezuela. *J. Trop.Ecol.* **2**, 207–217.

Morgan, M.E., Kingston, J.D. and Marino, B.D. (1994) Carbon isotopic evidence for the emergence of C4 plants from Pakistan and Kenya. *Nature* **367**, 162–165.

Nambudiri, E.M.V., Tidwell, W.D., Smith, B.N. and Hebbert, N.P. (1987) A C_4 plant from the Pliocene. *Nature* **276**, 816–817.

Ode, D.J., Tieszen, L L. and Lerman, J.C. (1980) The seasonal contribution of C_3 and C_4 plant species to primary production. *Ecology* **61**, 1304–1311.

Quade, J. and Cerling, T.E. (1995) Expansion of C_4 grasses in the late Miocene of northern Pakistan: evidence from stable isotopes in paleosols. *Palaeogeogr. Palaeoclim. Palaeoecol.* **115**, 91–116.

Quade, J., Cerling, T.E. and Bowman, J.R. (1989) Development of Asian monsoon revealed by marked ecological shift during the latest Miocene in northern Pakistan. *Nature* **342**, 163–166.

Quade, J., Cerling, T.E. and Bowman, J.R. (1989) Systematic variations in the carbon and oxygen isotopic composition of pedogenic carbonate along elevation transects in the southern Great Basin, USA. *Geol. Soc. Am.Bull.* **101**, 464–475.

Quade, J., Cerling, T.E., Barry, J.C., Morgan, M.E., Pilbeam, D.R., Chivas, A.R., Lee-Thorp, J. A. and van der Merwe, N.J. (1992) A 16-Ma record of paleodiet using carbon and oxygen isotopes in fossil teeth from Pakistan. *Chem.Geol. (Isotope Geosci. Sect.)* **94**, 183–192.

Sharp, Z.D. and Cerling, T.E. (1996) A laser GC-IRMS technique for in situ stable isotope analyses of carbonates and phosphates: *Geochim. Cosmochim.Acta* **15**, 2909–2916.

Simpson, G.G. (1951) *Horses: The Story of the Horse Family in the Modern World and Through Sixty Million Years History.* New York, Oxford University Press, 247 pp.

Tidwell, W.D. and Nambudiri, E.M.V. (1989) *Tomlinsonia thomassonii, get. et sp. nov.*, a permineralized grass from the upper Miocene Ricardo Formation, California. *Rev. Palaeobot. Palynol.* **60**, 165–177.

van der Merwe, N.J. and Medina, E. (1989) Photosynthesis and $^{13}C/^{12}C$ ratios in Amazonian rain forests. *Geochim. Cosmochim. Acta* **53**, 1091–1094.

Wang, Y., Cerling, T.E. and MacFadden, B.J. (1994) Fossil horses and carbon isotopes: new evidence for Cenozoic dietary, habitat and ecosystem changes in North America. *Palaeogeogr. Palaeoclim. Palaeoecol.* **107**, 269–279.

Wang, Y. and Cerling, T.E. (1994) A model of fossil tooth enamel and bone diagenesis: implications for stable isotope studies and paleoenvironment reconstruction. *Palaeogeogr. Palaeoclim. Palaeoecol.* **107**, 281–289.

Webb, S.D. (1977) A history of savanna vertebrates in the New World. Part I: North America. *Ann. Rev. Ecol. Syst.* **9**, 393–426.

Webb, S.D. (1983) The rise and fall of the late Miocene ungulate fauna in North America. In: *Coevolution*, (ed M. H. Nitecki), University of Chicago Press, Chicago. pp. 267–306.

Webber, I.E. (1933) Woods from the Ricardo Pliocene of Last Chance Gulch, California. Carnegie Inst. Washington **412**, 113–134.

Whistler, D.P. and Burbank, D.W. (1992) Miocene biostratigraphy and biochronology of the Dove Spring Formation, Mojave Desert, California and characterization of the Clarendonian mammal age (late Miocene) in California. *Geol. Soc. Am. Bull.* **104**, 644–658.

Woodburne, M.O. and Swisher, C.C. (1995) Land mammal high-resolution geochronology, intercontinental overland dispersals, sea level, climate and vicariance, in *Geochronology, Time Scales and Global Stratigraphic Correlation*, (eds. W.A. Berggren, D.V. Kent, M.-A. Aubry, and J. Hardenbo). Society for Sedimentary Geology Tulsa, OK, pp. 335–364.

Carbon isotopes in lake sediments and peats of last glacial age: implications for the global carbon cycle

F.A. Street-Perrott, Y. Huang, R.A. Perrott and
G. Eglinton

22.1 Introduction

Scientific concern about potential future changes in the global carbon cycle has resulted in an upsurge of interest in past periods with carbon budgets significantly different from present, as can be seen in the three previous contributions in this volume (Chapters 19, 20 and 21). Chief among these recently (in geological timescales) has been the last glacial maximum (LGM), when the global-average climate was significantly colder, drier and windier than today (Wright et al., 1993). The atmospheric concentration of CO_2 was about 80 µmol mol^{-1} lower than its pre-industrial level of 270–280 µmol mol^{-1} (Raynaud et al., 1993). This decrease in CO_2/O_2 ratio was probably exacerbated by a slight increase in the atmospheric O_2 content, resulting from an enhanced burial rate of organic carbon in the glacial ocean (Sarnthein et al., 1988). We infer that C_3 plants such as trees suffered from a significant increase in the ratio of photorespiration to photosynthesis (Ehleringer et al., 1991; see Chapters 8, 19 and 20); in contrast, plants able to use the C_4 or CAM photosynthetic pathways would have been selectively advantaged by lower pCO_2 and widespread aridity (see Chapter 21). Other things being equal, the ecophysiological impact of these changes would have been greatest in the tropical high mountains, where ambient $[CO_2]$ at the present upper treeline declined to about 120–140 µmol mol^{-1} at the LGM (Street-Perrott, 1994). Aquatic ecosystems would also have been particularly vulnerable to carbon limitation, because CO_2 diffuses 10^4 times more slowly in water than in air (Keeley and Sandquist, 1992; see Chapter 19). Hence, tropical mountain lakes provide a useful extreme situation for studying the carbon cycle in the past.

Stable Isotopes, edited by H. Griffiths.
© 1998 BIOS Scientific Publishers Ltd, Oxford.

22.2 The carbon cycle at the LGM

22.2.1 *Biome distribution*

Vegetation reconstructions based on pollen and plant macrofossils show that there was a global reduction in forest at the LGM compared with preindustrial times. Open vegetation types including tundra, steppe and grassland expanded (CLIMAP Project Members, 1981; Wright *et al.*, 1993). The loss of land through encroachment by ice was roughly compensated by the exposure of the continental shelves due to lower sea level (Prentice *et al.*, 1993).

Significant controversy surrounds vegetation changes in specific areas, notably in the tropics, where the implications for estimates of carbon storage are particularly great. Many rainforest plants and animals have highly disjunct distributions, which were attributed by biogeographers to the restriction of tropical rainforest to isolated enclaves, or 'refugia', during glacial times (Hamilton, 1976; Simpson and Haffer, 1978). This circumstantial argument was reinforced by pollen evidence for the extension of savanna grasslands in the equatorial lowlands (Hamilton, 1982; Van der Hammen, 1974; Van der Hammen and Absy, 1994; Van der Kaars, 1991; Maley, 1991) and for a general descent of the upper treeline by 1000 – 1700m on the tropical high mountains (Flenley, 1979). Conventionally, these changes were explained by increased aridity in the lowlands and by cooler temperatures at high altitudes (Flenley, 1979; Hamilton, 1982; Van der Hammen, 1974), with only limited support from well-dated, independent palaeoclimatic evidence such as lake levels, moraines and aeolian deposits.

Recently, however, this consensus has broken down. Firstly, Colinvaux (1987) and others have criticized the biogeographical evidence for glacial refugia. Secondly, it has become clear that large parts of the equatorial lowlands were occupied not by savannas, but by forests dominated by 'montane' taxa (*Podocarpus*, *Olea* and *Ilex* in Africa; *Podocarpus* and *Araucaria* in South America; and various Podocarpaceae, *Araucaria*, oaks and southern beech in Indonesia (Colinvaux *et al.*, 1996; Elenga *et al.*, 1991; Hope and Tulip, 1994; Van der Kaars, 1991; Maley, 1991; Servant *et al.*, 1993; Stuijts *et al.*, 1988). This has led to speculation that lower temperatures were the main factor determining forest composition, even in the lowland tropics (Colinvaux *et al.*, 1996; Servant *et al.*, 1993; Stuijts *et al.*, 1988). However, as noted by Van der Kaars (1991), many of these ancient gymnosperms and montane angiosperms are also drought-tolerant, giving them a competitive advantage at times of lower rainfall and/or physiological drought due to lower CO_2 concentration (Idso, 1989).

Thirdly, it has been demonstrated that the substantial decrease in temperatures (5–12 °C) implied by a literal translation of treeline lowering into glacial cooling is incompatible both with the CLIMAP (1981) sea-surface temperature (SST) estimates and with the results of most climate-model simulations for the LGM (Rind, 1990; Rind and Peteet, 1985; Webster and Streten, 1978; Wright *et al.*, 1993). Note, however, that some recent empirical and modelling studies indicate a significant reduction in tropical SSTs (Charles, 1997; Guilderson *et al.*, 1994). It appears to be much easier to explain the ice-age distribution of both tropical lowland and highland forests if the direct physiological effects of lower $[CO_2]$ and lower CO_2/O_2 are taken into account alongside the impact of climatic change (Street-Perrott, 1994; Jolly and Haxeltine, 1997).

22.2.2 *Carbon storage on land*

The combination of generally harsher climate and lower atmospheric CO_2 resulted in net decreases in primary production and in carbon storage in terrestrial vegetation and soils. Since the total atmospheric CO_2 content was also lower, the carbon lost from the continents must have been transferred to the oceans. Published estimates of this land–ocean transfer (Figure 22.1) are grossly discrepant, ranging from near zero to 1830 Gt C: a huge range compared with the preindustrial carbon content of vegetation and soils (ca. 2100 – 2200 Gt: Bird *et al.*, 1994; Sundquist, 1993). In general, palaeoecological data have provided the largest estimates whilst biome-modelling studies, the smallest. Most recent assessments converge on a range of 300 – 700 Gt C, with the exception of those by Adams and colleagues (Adams, 1995; Maslin *et al.*, 1995). The latter are based on present 'potential' rather than actual vegetation. They also employ maximal estimates of forest loss in tropical and northern latitudes at the LGM. Such large uncertainties severely hamper understanding of the ice-age carbon cycle as a whole.

22.2.3 *Global carbon-isotope mass budget*

The global carbon-isotope mass balance provides a crucial constraint on the magnitude of the carbon loss from the continents to the oceans, as shown by Bird *et al.* (1994). Their simple three-box model is illustrated in Figure 22.2. In their LGM scenarios, the atmospheric carbon content was evaluated from ice-core measurements (Raynaud *et al.*, 1993), and its $^{13}C/^{12}C$ ratio from data in Leuenberger *et al.* (1992) and Marino *et al.* (1992). Carbon stripped from the continents and the atmosphere was

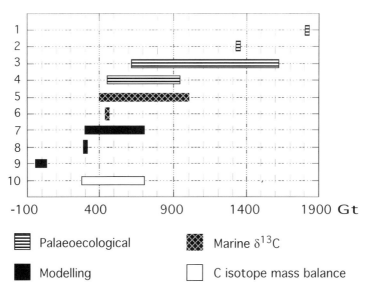

Figure 22.1. *Published estimates of land–ocean carbon transfer at the LGM. The basis of each estimate is indicated by shading. References: 1. Adams, 1995; 2. Adams et al., 1990; 3. Maslin et al., 1995; 4. Van Campo et al., 1993; 5. Maslin et al., 1995, 6. Crowley, 1991; 7. Prentice et al., 1993; 8. Friedlingstein et al., 1992; 9. Prentice and Fung, 1990; 10. Bird et al., 1994.*

simply added to the oceanic reservoir, neglecting any losses to storage in continental and marine sediments (Sundquist, 1993). The whole-ocean carbon isotope ratio ($\delta^{13}C$ = -0.32‰ PDB) was based on analyses of foraminifera in deep-sea cores (Duplessy *et al.*, 1988). Given the large size of the oceanic carbon stock, the model is very sensitive to this parameter, so Bird *et al.* (1994) employed a range of estimates from -0.3‰ to -0.4‰. They also used two independent estimates of the weighted-mean isotopic composition of pre-industrial land biomass (-25‰ and -22‰). The former is selected here for illustrative purposes.

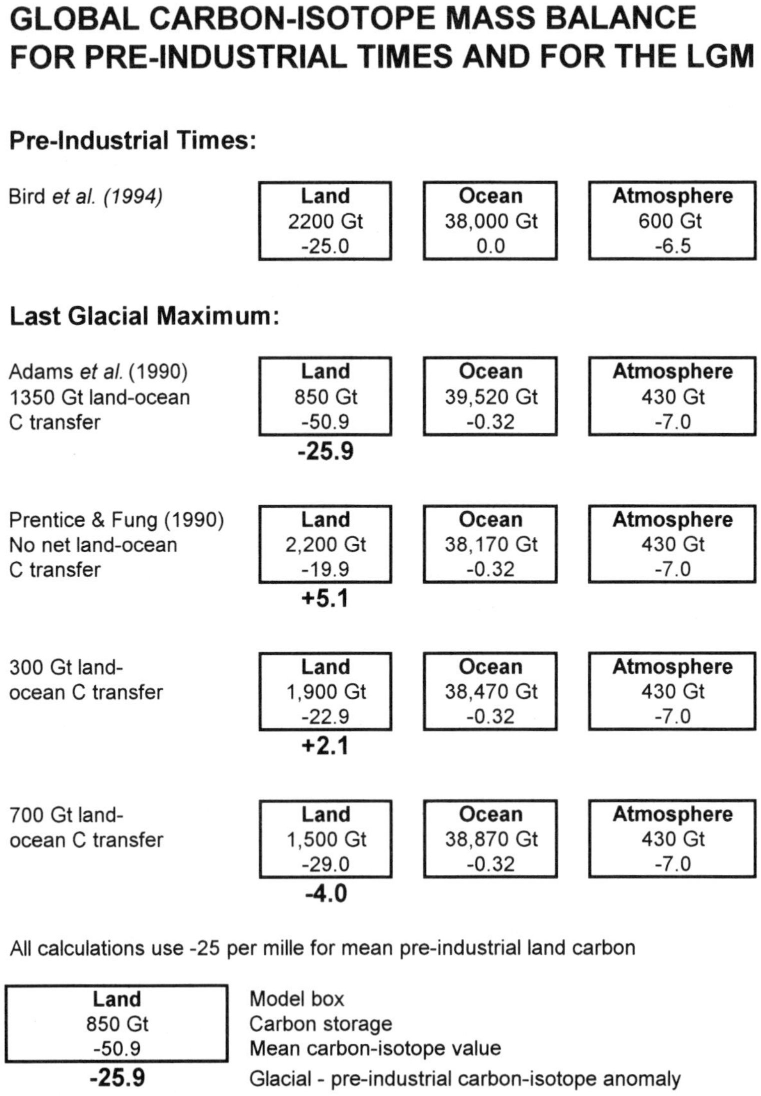

GLOBAL CARBON-ISOTOPE MASS BALANCE FOR PRE-INDUSTRIAL TIMES AND FOR THE LGM

Pre-Industrial Times:

Bird *et al. (1994)*

Land	Ocean	Atmosphere
2200 Gt	38,000 Gt	600 Gt
-25.0	0.0	-6.5

Last Glacial Maximum:

Adams *et al.* (1990)
1350 Gt land-ocean
C transfer

Land	Ocean	Atmosphere
850 Gt	39,520 Gt	430 Gt
-50.9	-0.32	-7.0

-25.9

Prentice & Fung (1990)
No net land-ocean
C transfer

Land	Ocean	Atmosphere
2,200 Gt	38,170 Gt	430 Gt
-19.9	-0.32	-7.0

+5.1

300 Gt land-
ocean C transfer

Land	Ocean	Atmosphere
1,900 Gt	38,470 Gt	430 Gt
-22.9	-0.32	-7.0

+2.1

700 Gt land-
ocean C transfer

Land	Ocean	Atmosphere
1,500 Gt	38,870 Gt	430 Gt
-29.0	-0.32	-7.0

-4.0

All calculations use -25 per mille for mean pre-industrial land carbon

Land	Model box
850 Gt	Carbon storage
-50.9	Mean carbon-isotope value
-25.9	Glacial - pre-industrial carbon-isotope anomaly

Figure 22.2. *Sample calculations with the Bird* et al. *(1994) global carbon-isotope mass-balance model. For full explanation see text. Each box gives the size and mean bulk $\delta^{13}C$ value of the respective reservoir. Δ values below each box are derived as G (glacial minus preindustrial) isotope anomalies.*

Four ice-age scenarios with widely differing estimates of carbon transfer from the continents are shown in Figure 22.2. These are, respectively, the largest plausible palaeoecological estimate (Adams *et al.*, 1990), the smallest model-based estimate (Prentice and Fung, 1990), and the lower and upper bounds of the most probable range (300 – 700 Gt). The first scenario implies that ice-age land carbon had a mean isotope value of -50.9‰, which is highly unlikely. In contrast, the zero-change scenario of Prentice and Fung (1990) results in a positive isotopic shift of 5.1‰ on land, which is necessary in order to balance the negative isotopic shifts in the ocean and atmosphere. The two 'consensus' scenarios imply that carbon-isotope value of continental organic matter changed by +2.1‰ to -4.0‰ with respect to its pre-industrial mean. In other words, it is not even certain whether the net isotopic shift was positive or negative. Better estimates of the bulk $\delta^{13}C$ composition of land carbon are clearly essential to constrain carbon-cycle models of the LGM.

The $\delta^{13}C$ values of C_3 plants became less negative by 1.5‰ to 4‰ (fossil wood and pine needles) during glacial times as reported by Krishnamurthy and Epstein (1990), Leavitt and Danzer (1992) and Van der Water *et al.* (1994). In contrast, fossil leaves of the C_4 desert shrub *Atriplex confertifolia* showed slightly more negative $\delta^{13}C$ values (-0.7‰ to -1.0‰), probably reflecting a slight decrease in the $\delta^{13}C$ value of atmospheric CO_2 (Leuenberger *et al.*, 1992; Marino *et al.*, 1992). It should be noted, however, that environmental conditions (such as changes in VPD) induce a shift in $\delta^{13}C$ in the opposite direction for C_4 as opposed to C_3 plants (see Chapter 8). Since C_3 biomass is preponderant on a global scale, the net isotopic change would appear to have been positive, placing an upper limit of 430 – 665 Gt on the amount of carbon lost from the continents (calculated here from a sensitivity analysis with the Bird *et al.* (1994) model, using the range of parameters specified above). However, the existing macrofossil data are few and biased towards western North America. Freshwater lake sediments and peats have the potential to provide a more representative sample of land carbon, provided that the effects of diagenesis on their bulk carbon-isotope composition can be reliably assessed. The next section summarizes recent bulk-isotope data from freshwater sediments of last glacial age.

22.3 Carbon-isotope data on total organic matter from lake sediments and peats

A positive isotopic shift during the last glaciation (i.e. greater enrichment in ^{13}C) has been observed in TOC from lake sediments in Europe, Africa, Asia and Australasia. The sites listed in Table 22.1 span a latitudinal range from Scotland to New Zealand and a range of altitudes from sea level to over 4 200 m a.s.l. The modern natural vegetation at all the sites with these enriched signals is dominated by C_3 plants, commonly trees, although two are located in forest/grassland ecotones and one in subtropical desert.

At the majority of sites Δ_G varies from 1‰ to 27‰ (Table 22.1). All the sites recording a glacial/interglacial amplitude of >15‰ are small lakes (meteoritic or volcanic craters) with a small carbon pool. However, not all small lakes show large shifts, especially if they are well buffered (Håkansson, 1986). With the exception of very large lakes such as Tanganyika and Victoria (M.R. Talbot, unpublished data), the tropical sequences display glacial/interglacial amplitude of 8‰ or more. In contrast, a survey of radiocarbon-dated samples from 90 Swedish sites concluded that the average

Table 22.1. *Glacial/interglacial variations in carbon-isotope ratios of total organic carbon (TOC) from freshwater lake sediments and peats*

Site	Altitude (m)	Natural vegetation	Δ_G (Glacial minus interglacial change in $\delta^{13}C$) and range (‰)	Ref.
Dead Sea, Israel	-406	Desert	-4 (-25 to -21)*	1
Lundin Palaeolake, Scotland	24	CTF	12 (-23 to -35)*	2
Vanstads mosse, Sweden	74	CTF	9 (-22 to -31)*	3
Lake Maratoto, New Zealand	<100	WTF	12 (-23 to -35)	4
Lake Rotomanuka, New Zealand	<100	WTF	3 (-28 to -31)*	4
Lake Ngaroto, New Zealand	<100	WTF	9 (-24 to -33)	4
Lake Rotorua, New Zealand	<100	WTF	7 (-25 to -32)	4
Lake Mangakaware, New Zealand	<100	WTF	7 (-26 to -33)	4
Lake Bosumtwi, Ghana	100	LRF	27 (-4 to -31)	5
Körslättamossen, Sweden	118	CTF	7 (-23 to -30)*	6
Lake Barombi Mbo, Cameroon	300	LRF	12 (-22 to -34)	7
Meerfelder Maar, Germany	337	CTF	20 (-14 to -34)	8
Searles Lake, California	493	Desert	7 (-19 to -26)	9
Lago Grande de Monticchio, Italy	656	WTF	3 (-22 to -25)	10
Lake Tanganyika, Tanzania	773	FG	7 (-22 to -29)	11
Lake Biwa, Japan	<1000	WTF	6 (-20 to -26)	12
Lake Trummen, Sweden	<1000	CTF	10 (-20 to -30)*	13
Lake Kvarnsjön, Sweden	<1000	CTF	6 (-18 to -25)*	13
Lake Akerhultagöl, Sweden	<1000	CTF	7 (-23 to -30)*	13
Håkulls mosse, Sweden	<1000	CTF	5 (-24 to -29)*	13
Björkeröds mosse, Sweden	<1000	CTF	7 (-22 to -29)*	13
Lake Asbotorpssjön, Sweden	<1000	CTF	6 (-24 to -30)*	13
Lake Flarken, Sweden	<1000	CTF	12 (-17 to -29)*	13
Ranviken Bay, Sweden	<1000	CTF	8 (-23 to -31)*	13
Lake Striern, Sweden	<1000	CTF	1 (-25 to -26)*	13
Lac du Bouchet, France	1207	CTF	2 (-26 to -28)	14
Lake Sokorte Dika, Kenya	1500	MRF	12 (-14 to -26)*	15
Lake St. Moritz, Switzerland	1768	Alpine	5 (-24 to -29)*	16
Lake Tritivakely, Madagascar	1800	MRF	31 (+4 to -27)**	17
Nilgiri Hills peats, India	>2000	FG	11 (-13 to -24)	18
Kashiru peatbog, Burundi	2240	MRF	13 (-16 to -29)	19
Sacred Lake, Kenya	2350	MRF	17 (-14 to -31)	20
Lake Fuquene, Colombia	2580	MRF	-11 (-25 to -14)	21
Lake Rutundu, Kenya	3078	Ericaceous	8 (-19 to -27)	22
Lake Kimilili, Kenya	4150	Afroalpine	14 (-7 to -21)*	23
Small Hall Tarn, Kenya	4282	Afroalpine	9 (-14 to -23)*	22

* Incomplete sequence giving minimum estimate of full glacial–interglacial range
** Range may be exaggerated due to presence of siderite in sediments
Present natural vegetation: CTF, cool temperate forest; WTF, warm temperate forest; LRF, lowland rainforest; MRF, montane rainforest; F/G, forest–grassland mosaic.
References: 1. Margaritz *et al.*, 1991; 2. Whittington *et al.*, 1996; 3. Hammarlund and Keen, 1994; 4. McCabe, 1985; 5. Talbot and Johannessen, 1992; 6. Hammarlund and Lemdahl, 1994; 7. Giresse *et al.*, 1994; 8. Brown *et al.*, 1991; 9. Stuiver, 1964; 10. Robinson, 1993, Zolitschka and Negendank, 1996; 11. Hillaire-Marcel *et al.*, 1989; 12. Meyers and Horie, 1993; 13. Håkansson, 1985; 14. Truze and Kelts, 1993; 15. Harkness and Street-Perrott, unpublished data; 16. Ariztegui and McKenzie, 1995; 17. Gibert *et al.*, 1993; 18. Sukumar *et al.*, 1993; 19. Aucour *et al.*, 1993; 20. Huang *et al.*, 1995b; 21. Pawellek and Street-Perrott, unpublished data; 22. Ficken and Street-Perrott, unpublished data; 23. Hamilton, 1982.

glacial/interglacial shift was 5‰, with the most rapid change centred on the beginning of the Holocene, around 10 000 yr BP (^{14}C years before present) (Hammarlund, 1993). Some temperate and tropical sequences show a distinct increase in ^{13}C composition during the late-glacial cold stage (the Younger Dryas stadial, 11,000 – 10,000 yr BP) as well as during the full-glacial (Giresse *et al.*, 1994; Hammarlund and Keen, 1994; Huang *et al.*, 1995b; Robinson, 1993; Talbot and Johannesson, 1992; Zolitschka and Negendank, 1996).

So far, we have only found two lakes which became more depleted in [13]C (i.e. Δ_G became negative) during the last glacial. The Dead Sea is presently surrounded by desert with a high proportion of C_4 plants. Its climate was wetter during the last ice age, which may have boosted the incidence of C_3 plants despite lower $[CO_2]$. The other site, Lake Fuquene, lies in the Andean montane rainforest belt. Its isotopic variations are puzzling but appear to correlate with the abundance of green algae in the sediments (H. Hooghiemstra, personal communication).

Most published studies have attributed the positive shift in carbon-isotope values in tropical lake sediments of last-glacial age to the spread of C_4 plants, notably savanna grasses and sedges. These would have been selectively advantaged compared with C_3 plants by drier climates and lower $[CO_2]$ (Aucour *et al.*, 1993; Giresse *et al.*, 1994; Hillaire-Marcel *et al.*,1989; Jolly and Haxeltine, 1997; Street-Perrott, 1994; Talbot and Johannessen, 1992). However, complications arise in the interpretation of the carbon-isotope record of bulk organic matter in lake sediments, since it may be derived not only from terrestrial higher plants, but also from aquatic plants and micro-organisms. Organic matter contributed by algae and microbes to lake sediments can blur or even distort the palaeovegetational signal. For instance, the exceptionally heavy $\delta^{13}C$ values (up to -4.4‰ PDB) encountered in Lake Bosumtwi, Ghana, were attributed to algae by Talbot and Johannessen (1992), though without direct evidence. Fortunately, the recently-developed technique of compound-specific carbon-isotope analysis now permits the measurement of $\delta^{13}C$ values for individual organic compounds. By determining the carbon-isotope compositions of compounds derived from specific biological sources (biomarkers), it is possible to produce independent carbon-isotope stratigraphies for terrestrial and aquatic organisms or communities. This provides a new and powerful way to investigate past changes in the carbon cycle.

22.4 Molecular-isotopic analyses: a case study from Sacred Lake, Mt. Kenya

Sacred Lake is a small (0.51 km²), shallow (5 m), oligotrophic crater lake situated at 2350 m a.s.l. on the slopes of Mt. Kenya, East Africa, in humid montane rainforest. Sediment cores from the lake exhibit extremely large variations in the $\delta^{13}C$ values of TOC: more than 17‰ (from -31.5‰ to -14.1‰: Huang *et al.*, 1995b). [14]C and U/Th dating shows that heavy values occurred during cold stages such as the LGM and light ones during warm intervals. The heaviest $\delta^{13}C$ values (most enriched in ^{13}C) in the cores coincided with maximum levels of *Gramineae* (grass) pollen as reported by Coetzee Q3 (1967). Based on the isotopic differences between higher plants using the C_3 and C_4 photosynthetic pathways, it was hypothesized that the lake catchment was occupied during the last glacial by open moorland or afroalpine vegetation rich in C_4 tussock grasses (Street-Perrott, 1994). This is not as unlikely as it seems, since C_4 grasses still survive in dry, high-altitude sites above treeline (Young and Young, 1983).

Changes in the abundance of sedges and other families containing C_4 or CAM taxa were minor (Coetzee, 1967).

We have applied the combined approach using biomarker distributions and compound-specific carbon-isotope analysis to the sediments of Sacred Lake and have obtained the first direct evidence for a substantial algal contribution to the bulk carbon-isotope record (Huang and Murray, 1995; Huang et al., 1995a,b, 1996). Specific biomarkers, 1,6,17,21-octahydro-botryococcene and sacredicenes, have been identified in high concentrations (up to 90% of the total aliphatic hydrocarbons) throughout two parallel sediment cores, SL1 and SL2. Structural analogies with isoprenoid hydrocarbons produced by the planktonic green alga *Botryococcus braunii* (Maxwell et al., 1968; Metzger et al., 1985) led us to attribute them to *Botryococcus* sp. (Huang and Murray, 1995; Huang et al., 1995a,b, 1996). The presence of fossil *Botryococcus* colonies has been confirmed by scanning electron microscopy (Huang et al., unpublished data).

More importantly, the variations in the $\delta^{13}C$ values of octahydro-botryococcene and sacredicenes with depth, parallel the bulk $\delta^{13}C$ curve throughout core SL1. The $\delta^{13}C$ values of octahydro-botryococcene in glacial-age sediments range from -5.1‰ to -11.4‰, corresponding to the $\delta^{13}C$ maximum in TOC. The highest value recorded (-5.1‰) was found in a horizon laid down at the LGM (ca. 18,200 BP): (Figure 22.3). The abundances and $\delta^{13}C$ values of these algal biomarkers in the sediments clearly indicate a major algal contribution to the sedimentary bulk carbon-isotope record.

However, the C_{23} to C_{33} n-alkanes in the Sacred Lake sediments show strong odd/even predominance, characteristic of a higher-plant leaf-wax origin (Eglinton and Hamilton, 1963). The $\delta^{13}C$ values of these land-plant biomarkers vary from -18.5‰ in the last glacial to -32.1‰ in the Holocene, confirming that a switch from C_4 to C_3 plants occurred between the last glacial and the present interglacial, as suggested by Street-Perrott (1994). Further details can be found in Street-Perrott et al. (1997).

22.5 Influence of source carbon, fractionation and transformation on the ^{13}C record of the last glaciation

The large positive shift in the $\delta^{13}C$ values of both terrestrial and lacustrine organic matter during the last glacial (Table 22.1 and Section 22.4) implies significant changes in either the isotopic signature of the source carbon, or fractionation, or both.

Considering first the source carbon, it is clear that the small (<1‰) negative shift in the $\delta^{13}C$ value of atmospheric CO_2 (Leuenberger et al., 1992) was outweighed by other factors, such as the positive isotopic shift (<6‰) caused by decreased trapping of biogenic CO_2 in a generally lower and sparser vegetation canopy (cf. Van der Merwe and Medina, 1989). Reduced inputs of biogenic CO_2 due to decreased biomass and slower decomposition rates would cause a positive shift in the source carbon for submerged plants and algae (Håkansson, 1985), as would the enhancement of reservoir effects under a generally more arid climate (Talbot, 1990). Recycling of terrestrial C_4 plant detritus within a waterbody would have the same effect.

Fractionation by C_3 plants would have been affected in a complex way by the generally prevailing combination of colder temperatures, reduced precipitation and lower $[CO_2]$. Light intensity may have increased due to diminished cloud and more open vegetation. An increase in stomatal density has been observed in fossil leaves of full-glacial age from both Europe and North America (Beerling et al., 1993; Van der Water

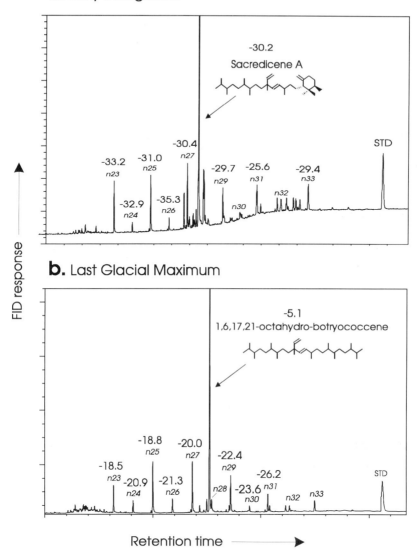

Figure 22.3. *Gas-chromatograph traces for the aliphatic hydrocarbon fraction of a) early interglacial and b) full-glacial sediment samples from Sacred Lake, Mt. Kenya, showing the distributions and $\delta^{13}C$ values of n-alkanes, sacredicene and 1,6,17,21-octahydro-botryococcene. The interpolated ages of these layers are 10 300 and 18 200 ^{14}C yr BP, respectively.*

et al., 1994). Empirical data show that the net effect of these opposing factors on trees was an enrichment in ^{13}C of approximately 1.5‰ to 4‰ (Krishnamurthy and Epstein, 1990; Leavitt and Danzer, 1992; Van der Water *et al.,* 1994).

The evidence from Sacred Lake and other sites indicates that plants using the C_4 pathway became dominant in many tropical areas during the LGM. At present, C_4 grasses and sedges are the most likely candidates, based on pollen evidence (Giresse *et al.,* 1994;

Hillaire-Marcel et al., 1989; Huang et al., 1995b; Talbot and Johannesson, 1992). On average, C_4 plants are about 14‰ more enriched in ^{13}C than C_3 plants (O'Leary, 1981) giving a potentially large isotopic signal, as observed in the terrestrial-plant leaf waxes from Sacred Lake. So far there is little evidence for an increased abundance of CAM plants, apart from a widespread but otherwise inexplicable increase in the aquatic-CAM macrophyte *Isoetes* microspores in the glacial-age sediments of lakes in the Americas (Watts and Bradbury, 1982; Colinvaux et al., 1997; Watts and Hansen, 1994).

Explaining the enormous shift in Δ with ^{13}C relatively enriched by over 55‰ in the $\delta^{13}C$ values of algal compounds in Sacred Lake during glacial times, compared with the early Holocene, is more difficult. The impact of lower lake-water temperatures alone is not clear. Temperature affects the solubility of CO_2, the diffusion of CO_2 through water and the key enzymatic reactions in opposing ways; a recent study of particulate organic matter in Lake San Moritz, however, suggests that the temperature fractionation factor is small and in the wrong direction (+0.36‰/°C): (Ariztegui and McKenzie, 1995).

Today, weathering of carbonates is mainly caused by CO_2 originating from soil respiration, producing bicarbonate with a $\delta^{13}C$ value of around -12‰. Under very cold conditions with reduced biomass, carbonate dissolution would be effected to a greater degree than today by carbonic acid of atmospheric origin, yielding bicarbonate up to 9‰ more enriched in ^{13}C (Håkansson, 1986). In all probability, more soil carbonate was available for dissolution due to a greater incidence of immature soils. Enhanced weathering of silicate minerals during the ice age may have provided an additional bicarbonate source (Hammarlund, 1993; Sundquist, 1993). Since HCO_3^- is 7 to 11‰ more enriched in ^{13}C than $CO_{2(aq)}$ at typical lake-water temperatures, uptake of bicarbonate by algae and submerged plants tends to result in a significant positive shift in bulk tissues (Keeley and Sandquist, 1992; Goericke et al., 1994; see also Chapters 15 and 19).

A major cause of decreased fractionation in many lakes was probably carbon limitation (McCabe, 1985). This may have arisen for various reasons, including: decreased atmospheric pCO_2 (Hollander and McKenzie, 1991); elevated pH and salinity in areas with diminished rainfall (Schidlowski et al., 1985); perennial ice cover (Wand and Mühle, 1990); reduced turbulence which seems unlikely under glacial conditions (Osmond et al., 1981); the presence of biological barriers to diffusion such as oil films produced by *Botryococcus*; and enhanced aquatic productivity, triggered by increased soil erosion and possibly also by decreased cloud (Goericke et al., 1994; McCabe, 1985). Scarcity of $CO_{2(aq)}$ induces active uptake of HCO_3^- in many submerged aquatic plants and algae (Keeley and Sandquist, 1992; Sharkey and Berry, 1985). In combination, these effects can produce a positive shift of more than 20‰ (Sharkey and Berry, 1985; see Chapters 15 and 19).

Finally, microbial activity in the sediment column may have an important impact on the carbon-isotope record. Utilization of the co-genetic CO_2 produced by methanogenesis imparts a very heavy isotopic signature (Herczeg, 1988). However, the selective removal of labile compounds during early diagenesis generally results in a small negative shift of around 2‰ to 3‰ in TOC, since resistant compounds such as lipids tend to be isotopically-depleted (Tyson, 1995).

Sacred Lake is situated in an extinct volcanic crater and we have summarised the likely $\delta^{13}C$ of C pools for LGM and interglacial periods (Figure 22.4). Today, its small, steep-sided catchment is occupied by dense forest overlying carbonate-free soils. The

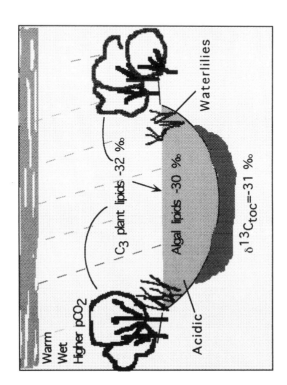

Figure 22.4. Sketch diagrams to summarize the isotopic characteristics of autochthonous and allochthonous organic matter in Sacred Lake, Mt. Kenya, under full-glacial and early interglacial conditions.

lake is acid (pH 5.0–6.1), poorly buffered and oligotrophic. During the last glacial period it was alkaline (pH 7–8.5), with a higher salinity (L. Ben Khelifa, personal communication). High concentrations of algal biomarkers in the sediments imply that the lake supported dense blooms of green algae (Figure 22.4). The *Botryococcus* cells must have operated a CO_2-concentrating mechanism involving HCO_3^- uptake, since the $\delta^{13}C$ value of 1,6,17,21-octahydro-botryococcene at 18,200 yr BP (-5.1‰; Figure 22.3) exceeded that of contemporary atmospheric CO_2 (Leuenberger *et al.*, 1992).

Our data suggest that the full-glacial superheavy $\delta^{13}C$ values of lacustrine organic matter in Sacred Lake primarily reflect carbon limitation due to lower $[CO_2]$ and higher pH, resulting in HCO_3^- use by the lake biota (Figure 22.4). The positive isotopic shift would have been exacerbated by the recycling of terrestrial C_4 rather than C_3 plant detritus, by a reduction in the amount of biogenic CO_2 and organic matter emanating from the surrounding vegetation and soils, and by an increase in reservoir effects resulting from a more arid climate. Microbially derived sedimentary hopanoids are scarce and have normal isotopic values, ruling out methanogenesis as an important factor. It is also unlikely that the isotope values of lacustrine algae were affected by perennial ice cover, since the lowest ice-covered lake in East Africa today is the Curling Pond, Mt. Kenya (4,790 m a.s.l.) (Löffler, 1964).

22.6 Conclusions

(1) The global carbon cycle at the LGM is poorly understood. Carbon-isotope data provide an important constraint on estimates of the amount of carbon transferred from the continents to the oceans.

(2) Freshwater lake sediments and peats of glacial age generally exhibit a positive shift (i.e. ^{13}C enrichment) in the $\delta^{13}C$ values of bulk organic carbon compared with interglacial samples. This shift frequently exceeds 5‰ at temperate sites and 8‰ in tropical sites.

(3) Molecular-isotopic analyses of sedimentary organic carbon from Sacred Lake, Mt. Kenya, show that a large positive shift occurred during glacial times in biomarkers of both terrestrial and aquatic origin.

(4) Phototrophs with CO_2-concentrating mechanisms (C_4 grasses and sedges on land; green algae and possibly CAM isoetids in lakes) are strongly represented in lake sediments deposited during the LGM, suggesting that lower pCO_2 had a significant impact on both terrestrial and aquatic ecosystems.

(5) Many of the existing difficulties in explaining the distribution of the major biomes, particularly tropical forests, during the LGM can be resolved if the direct physiological impact of lower pCO_2 is taken into account in addition to changes in temperature and precipitation.

References

Adams, J.M. (1995) *The Role of Terrestrial Ecosystems in Glacial-to-Interglacial Changes in the Global Carbon Cycle: An Approach Based on Reconstruction of Palaeovegetation*. D. Sc. Thesis, Université d'Aix-Marseille II, Luminy, France.

Adams, J.M., Faure, H., Faure-Denard, L., McGlade, J.M. and Woodward, F.I. (1990) Increases in terrestrial carbon storage from the Last Glacial Maximum to the present. *Nature* **348**, 711–714.

Ariztegui, D. and McKenzie, J.A. (1995) Temperature-dependent carbon-isotope fractionation of organic matter: a potential paleoclimatic indicator in Holocene lacustrine sequences. *Paläoklimaforschung* 15, 17–28.

Aucour, A.-M., Hillaire-Marcel, C. and Bonnefille, R. (1993) A 30,000 year record of ^{13}C and ^{18}O changes in organic matter from a equatorial peatbog. In: *Climate Change in Continental Isotopic Records. Geophys. Monogr.* 78, 343–351. American Geophysical Union, Washington.

Beerling, D.J., Chaloner, W.G., Huntley, B., Pearson, J.A. and Tooley, M.J. (1993) Stomatal density responds to the glacial cycle of environmental change. *Proc. Roy. Soc. London* B251, 133–138.

Bird, M.I., Lloyd, J. and Farquhar, G.D. (1994) Terrestrial carbon storage at the LGM. *Nature* 371, 566.

Brown, H., Eakin, P.A., Fallick, A.E. and Creer, K. (1991) Variations in the carbon isotopic composition of organic matter in lacustrine sediments of Meerfelder Maar. In: *Organic Geochemistry: Advances and Applications in the Natural Environment* (ed. D.A.C. Manning). Manchester University Press, Manchester, pp. 352–354.

Cepák, V. and Lukavský, J. (1994) The effect of high irradiances on growth, biosynthetic activities and the ultrastructure of the green alga *Botryococcus braunii* strain Droop 1950/807–1. *Algol. Stud.* 72, 115–131.

Charles, C. (1997) Palaeoclimatology: cool tropical punch of the ice ages. *Nature* 385, 681–683.

CLIMAP Project Members (1981) Seasonal reconstructions of the Earth's surface at the last glacial maximum. *Geol. Soc. Am. Map Chart Ser.* 36.

Coetzee, J.A. (1967) Pollen analytical studies in East and Southern Africa. *Palaeoecology of Africa* 3, 1–46.

Colinvaux, P.A. (1987) Amazon diversity in the light of the palaeoecological view. *Quat. Sci. Rev.* 6, 93–114.

Colinvaux, P.A., Bush, M.B., Steinitz-Kannan, M. and Miller, M.C. (1997) Glacial and post-glacial pollen record from the Ecuadorian Andes and Amazon. *Quat. Res.* 48, 69–78.

Colinvaux, P.A., Liu, K.B., De Oliveira, P., Bush, M.B., Miller, M.C. and Kannan, M.S. (1996) Temperature depression in the lowland tropics in glacial times. *Clim. Ch.* 32, 19–33.

Crowley, T.J. (1991) Ice age carbon. *Nature* 352, 575–576.

Duplessy, J.-C., Shackleton, N.J., Fairbanks, R.J., Labeyrie, L.D., Oppo, D. and Kallel, N. (1988) Deepwater source variations during the last climatic cycle and their impact on the global deepwater circulation. *Paleoceanography* 3, 343–360.

Eglinton, G. and Hamilton, R.J. (1967) Leaf epicuticular waxes. *Science* 156, 1322–1334.

Elenga, H., Vincens, A. and Schwartz, D. (1991) Présence d'éléments forestiers montagnards sur les Plateaux Bateke (Congo) au Pléistocène supérieur: nouvelles données palynologiques. *Palaeoecol. Afr.* 22, 239–252.

Ehleringer, J.R., Sage, R.F., Flanagan, L.B. and Pearcy, R.W. (1991) Climatic change and the evolution of C_4 synthesis. *Trends Ecol. Evol.* 6, 95–97.

Flenley, J.R. (1979) *Geological History of the Equatorial Rain Forest.* Butterworths, London.

Friedlingstein, P., Delire, C., Mueller, J.F. and Gerard, J.C. (1992) The climate induced version of the continental biosphere: a model simulation of the last glacial maximum. *Geophys. Res. Lett.* 19, 897–900.

Gibert, E., Gasse, F., Fehri, A., Ferry, L., Robinson, L., Saos, J.L. and Taieb, M. (1993) Chronologie et isotopes stables de la matière organique dans les lacs des plateaux de Madagascar. Exemples des lacs Itasy et Tritivakely. *Isotope Techniques in the Study of Past and Current Environmental Changes in the Hydrosphere and Atmosphere.* IAEA-SM-329, 90–91. International Atomic Energy Agency, Vienna.

Giresse, P., Maley, J. and Brenac, P. (1994) Late Quaternary palaeoenvironments in the Lake Barombi Mbo (West Cameroon) deduced from pollen and carbon isotopes of organic matter. *Palaeogeogr. Palaeoclimatol. Palaeoecol.* 107, 65–78.

Goericke, J.P., Montoya, J.P and Fry, B. (1994) Physiology of isotopic fractionation in algae and cyanobacteria. In: *Stable Isotopes in Ecology and Environmental Science* (K. Lajtha, and R.H. Michener, eds). Blackwell, Oxford, pp. 187–221.

Guilderson, T.P., Fairbanks, R.G. and Rubenston, J.L. (1994) Tropical temperature variations since 20,000 years ago: modulating interhemispheric climate change. *Science* **263**, 663–665.

Håkansson, S. (1986) A marked change in the stable carbon isotope ratio at the Pleistocene-Holocene boundary in southern Sweden. *Geol. För. Stockh. Förh.* **108**, 155–158.

Hamilton, A.C. (1976) The significance of patterns of distribution shown by forest plants and animals in tropical Africa for the reconstruction of Upper Pleistocene palaeoenvironments: a review. *Palaeoecol. Afr.* **9**, 63–97.

Hamilton, A.C. (1982) *Environmental History of East Africa. A Study of the Quaternary.* Academic Press, London.

Hammarlund, D. (1993) A distinct $\delta^{13}C$ decline in organic lake sediments at the Pleistocene-Holocene transition in southern Sweden. *Boreas* **22**, 236–243.

Hammarlund, D. and Keen, D.H. (1994) A Late Weichselian stable isotope and molluscan stratigraphy from southern Sweden. *Geol. För. Stockh. Förh.* **116**, 235–248.

Hammarlund, D. and Lemdahl, G. (1994) A Late Weichselian stable isotope stratigraphy compared with biostratigraphical data: a case study from southern Sweden. *J. Quat. Sci.* **9**, 13–31.

Herczeg, A.L. (1988) Early diagenesis of organic matter in lake sediments: a stable carbon isotope study of pore waters. *Chem. Geol.* **72**, 199–209.

Hillaire-Marcel, C., Aucour, A.-M., Bonnefille, R., Riollet, G., Vincens, A. and Williamson, D. (1989) ^{13}C/palynological evidence of differential residence times of organic carbon prior to its sedimentation in East African rift lakes and peat bogs. *Quat. Sci. Rev.* **8**, 207–212.

Hollander, D.J. and McKenzie, J.A. (1991) CO_2 control on carbon-isotope fractionation during aqueous photosynthesis: a paleo-pCO_2 barometer. *Geology* **19**, 929–932.

Hope, G. and Tulip, J. (1994) A long vegetation history from lowland Irian Jaya, Indonesia. *Palaeogeogr. Palaeoclimatol. Palaeoecol.* **109**, 385–398.

Huang, Y. and Murray, M. (1995) Identification of a 1,6,17,21-octahydro-botryoccene in a sediment. *J. Chem. Soc., Chem. Commun.*, 335–336.

Huang, Y., Murray, M., Eglinton, G. and Metzger, P. (1995a) Sacredicene, a novel monocyclic C_{33} hydrocarbon from sediment of Sacred Lake, a tropical freshwater lake, Mount Kenya. *Tetrahedr. Lett.* **36**, 5973–5976.

Huang, Y., Street-Perrott, F.A., Perrott, R.A. and Eglinton, G. (1995b) Molecular and carbon isotope stratigraphy of a glacial/interglacial sediment sequence from a tropical freshwater lake: Sacred Lake, Mt. Kenya. In *Organic Geochemistry: Developments and Applications to Energy, Climate, Environment and Human History* (J.O. Grimalt, and C. Dorronsorro, eds.). A.I.G.O.A., Spain, pp. 826–829.

Huang, Y., Murray, M., Metzger, P. and Eglinton, G. (1996) Novel unsaturated triterpenoid hydrocarbons from sediments of Sacred Lake, Mt. Kenya. *Tetrahedron* **52**, 6973–6982.

Idso, S.B. (1989) A problem for paleoclimatology? *Quat. Res.* **31**, 433–434.

Jolly, D. and Haxeltine, A. (1997) Effect of low glacial atmospheric CO_2 on tropical African montane vegetation. *Science* **276**, 786–787.

Keeley, J.E. and Sandquist, D.R. (1992) Carbon: freshwater plants. *Plant Cell Environ.* **15**, 1021–1035.

Krishnamurthy, R.V. and Epstein, S. (1990) Glacial-interglacial excursion in the concentration of atmospheric CO_2: effect in the ^{13}C/^{12}C ratio in wood cellulose. *Tellus* **42B**, 423–434.

Leavitt, S.W. and Danzer, S.R. (1992) $\delta^{13}C$ variations in C_3 plants over the past 50,000 years. *Radiocarbon* **34**, 783–791.

Leuenberger, M., Siegenthaler, U. and Langway, C.C. (1992) Carbon isotope composition of atmospheric CO_2 during the last ice age from an Antarctic ice core. *Nature* **357**, 488–490.

Löffler, H. (1964). The limnology of tropical high-mountain lakes. *Verh. Internat. Verein. Limnol.* **15**, 176–193.

McCabe, B. (1985) *The Dynamics of ^{13}C in Several New Zealand Lakes*. Ph.D. thesis, University of Waikato, 278 pp.

Magaritz, M., Rahner, S., Yechieli, Y. and Krishnamurthy, R.V. (1991) ^{13}C/^{12}C ratio in organic matter from the Dead Sea area: paleoclimatic interpretation. *Naturwissenschaften* **78**, 453–455.

Maley, J. (1991) The African rain forest vegetation and palaeoenvironments during late Quaternary. *Clim. Ch.* **19**, 79–98.

Marino, B.D., McElroy, M.B., Salawich, R.J., and Spaulding, W.G. (1992) Glacial-to-interglacial variations in the carbon isotopic composition of atmospheric CO_2. *Nature* **357**, 461–466.

Maslin, M.A., Adams, J., Thomas, E., Faure, H. and Haines-Young, R. (1995) Estimating the carbon transfer between the ocean, atmosphere and the terrestrial biosphere since the last glacial maximum. *Terra Nova* **7**, 358–366.

Maxwell, J.R., Douglas, A.G., Eglinton, G. and McCormick, A. (1968) The botryococcenes – hydrocarbons of novel structure from the alga *Botryococcus braunii*, Kützing. *Phytochemistry* **7**, 2157–2171.

Metzger, P., Casadevall, E., Pouet, M.J. and Pouet, Y. (1985) Structures of some Botryococcenes: branched hydrocarbons from the B-race of the green alga *Botryococcus braunii*. *Phytochemistry* **24**, 2995–3002.

Meyers, P.A. and Horie, S. (1993) An organic carbon isotope record of glacial-postglacial change in atmospheric pCO_2 in the sediments of Lake Biwa, Japan. *Palaeogeogr. Palaeoclimatol. Palaeoecol.* **105**, 171–178.

O'Leary, M.H. (1981) Carbon isotope fractionation in plants. *Phytochemistry* **20**, 553–567.

Osmond, C.B., Valaana, N., Haslam, S.M., Uotila, P. and Roksandic, Z. (1981) Comparison of $\delta^{13}C$ values of leaves of aquatic macrophytes from different habitats in Britain and Finland: some implications for photosynthetic processes in aquatic plants. *Oecologia* **50**, 117–124.

Prentice, I.C., Sykes, M.T., Lautenschlager, M., Denissenko, O. and Bartlein, P. (1993) Modelling global vegetation patterns and terrestrial carbon storage at the last glacial maximum. *Global Ecol. Biogeogr. Lett.* **3**, 67–76.

Prentice, K.C. and Fung, I. (1990) The sensitivity of terrestrial carbon storage to climate change. *Nature* **346**, 48–50.

Raynaud, D., Jouzel, J., Barnola, J.M., Chappellaz, J., Delmas, R.J. and Lorius, C. (1993) The ice record of greenhouse gases. *Science* **259**, 926–934.

Rind, D. (1990) Palaeoclimate: puzzles from the tropics. *Nature* **346**, 317–318.

Rind, D. and Peteet, D. (1985) Terrestrial conditions at the last glacial maximum and CLIMAP sea surface temperature estimates: are they consistent? *Quat. Res.* **24**, 1–22.

Robinson, C. (1993) *Lago Grande de Monticchio: A Palaeoenvironmental Reconstruction from Sediment Geochemistry*. Ph.D. Thesis, University of Edinburgh.

Sarnthein, M., Winn, K., Duplessy, J.-C. and Fontugne, M.R. (1988). Global variations of surface ocean productivity in low- and mid-latitudes: influence on CO_2 reservoirs of the deep ocean and atmosphere during the last 21,000 years. *Paleoceanography* **3**, 361–399.

Schidlowski, M., Matzigkeit, U., Mook, W.G. and Krumbein, W.E. (1985) Carbon isotope geochemistry and ^{14}C ages of microbial mats from the Gavish Sabkha and the Solar Lake. In: *Hypersaline Ecosystems* (G.M. Friedman and W.E. Krumbein eds.). Springer-Verlag, Berlin, pp. 381–401.

Servant, M., Maley, J., Turcq, B., Absy, M.-L., Brenac, P., Fournier, M. and Ledru, M.-P. (1993) Tropical forest changes during the Late Quaternary in African and South American lowlands. *Global Planet. Ch.* **7**, 25–40.

Sharkey, T.D. and Berry, J.A. (1985) Carbon isotope fractionation of algae as influenced by an inducible CO_2 concentrating mechanism. In: *Inorganic Carbon Uptake by Aquatic Photosynthetic Organisms* (W.J. Lucas and J.A. Berry eds.). Am. Soc. Plant Physiol., Rockville, 389–401.

Simpson, B.B. and Haffer, J. (1978) Speciation patterns in the Amazonian forest biota. *Ann. Rev. Ecol. System.* **9**, 497–518.

Street-Perrott, F.A. (1994) Palaeo-perspectives: changes in terrestrial ecosystems. *Ambio* **23**, 37–43.

Street-Perrott, F.A., Huang, Y., Perrott, R.A., Eglinton, G., Barker, P., Ben Khelifa, L., Harkness, D.D. and Olago, D. (1997) Impact of lower atmospheric CO_2 on tropical mountain ecosystems. *Science*, in press.

Stuijts, I., Newsome, J.C. and Flenley, J.R. (1988) Evidence for late Quaternary vegetation change in the Sumatran and Javan highlands. *Rev. Palaeobot. Palynol.* **55**, 207–216.

Stuiver. M. (1964) Carbon isotopic distribution and correlated chronology of Searles Lake sediments. *Am. J. Sci.* **262**, 377–392.

Sukumar, R., Ramesh, R., Pant, R.K. and Rajagopalan, G. (1993) A $\delta^{13}C$ record of late Quaternary climate change from tropical peats in southern India. *Nature* **364**, 703–705.

Sundquist, E.T. (1993) The global carbon dioxide budget. *Science* **259**, 934–941.

Talbot, M.R. (1990) A review of the palaeohydrological interpretation of carbon and oxygen isotopic ratios in primary lacustrine carbonates. *Chem. Geol. (Isotop. Geosci. Sect.)* **80**, 261–279.

Talbot, M.R. and Johannessen, T. (1992) A high resolution palaeoclimatic record for the last 27,500 years in tropical West Africa from the carbon and nitrogen isotopic composition of lacustrine organic matter. *Earth Planet. Sci. Lett.* **110**, 23–37.

Truze, E. and Kelts, K. (1993) Sedimentology and palaeoenvironments from the maar Lac du Bouchet for the last climatic cycle, 0 – 120,000 years (Massif Central, France). *Lect. Notes Earth Sci.* **49**, 237–275.

Tyson, R.V. (1995) *Sedimentary Organic Matter: Organic Facies and Palynofacies.* Chapman and Hall, London.

Van Campo, E., Guiot, J. and Peng, C. (1993) A data-based reappraisal of the terrestrial carbon budget at the Last Glacial Maximum. *Global Planet.Ch.* **8**, 189–201.

Van der Hammen, T. (1974) The Pleistocene changes of vegetation and climate in tropical South America. *J. Biogeogr.* **1**, 3–26.

Van der Hammen, T. and Absy, M.L. (1994) Amazonia during the last glacial. *Palaeogeogr. Palaeoclimatol. Palaeoecol.* **109**, 247–261.

Van der Kaars, W.A. (1991) Palynology of eastern Indonesian marine piston cores: a Late Quaternary vegetational and climatic record for Australasia. *Palaeogeogr. Palaeoclimatol. Palaeoecol.* **85**, 239–302.

Van der Merwe, N.J. and Medina, E. (1989) Photosynthesis and $^{13}C/^{12}C$ ratios in Amazonian rain forests. *Geochim. Cosmochim. Acta* **53**, 1091–1094.

Van der Water, P.K., Leavitt, S.W. and Betancourt, J.L. (1994) Trends in stomatal density and $^{13}C/^{12}C$ ratios of *Pinus flexilis* needles during the last glacial-interglacial cycle. *Science* **264**, 239–243.

Wand, U. and Mühle, K. (1990) Extremely ^{13}C-enriched biomass in a freshwater environment: examples from Antarctic lakes. *Geodät. geophys. Veröff.* **I 16**, 361–366.

Watts, W.A. and Bradbury, J.P. (1982) Paleoecological studies of Lake Patzcuaro on the west central Mexican plateau and at Chalco in the basin of Mexico. *Quat. Res.* **17**, 56–70.

Watts, W.A. and Hansen, B.C.S. (1994) Pre-Holocene and Holocene pollen records of vegetation history from the Florida peninsula and their climatic implications. *Palaeogeogr. Palaeoclimatol. Palaeoecol.* **109**, 163–176.

Webster, P.J. and Streten, N.A. (1978) Late Quaternary ice age climates of tropical Australasia: interpretations and reconstructions. *Quat. Res.* **10**, 279–309.

Wright, H.E. Jr., Kutzbach, J.E., Webb, T. III, Ruddiman, W.F., Street-Perrott, F.A. and Bartlein, P.J. (1993) *Global Climates Since the Last Glacial Maximum.* University of Minnesota Press, Minneapolis.

Young, H.J. and Young, T.P. (1983) Local distribution of C_3 and C_4 grasses in sites of overlap on Mt. Kenya. *Oecologia* **58**, 373–377.

Zolitschka, B. and Negendank, J.F.W. (1996) Sedimentology, dating and palaeoclimatic interpretation of a 76.3ka record from Lago Grande de Monticchio, southern Italy. *Quat. Sci. Rev.* **15**, 101–112.

Stable isotopes, the hydrological cycle and the terrestrial biosphere

Joel R. Gat

23.1 Introduction

At first sight the role of the terrestrial biosphere in the global hydrological cycle appears to be minor. The water flux to and from the continents constitutes only about 10% of the annual cyclical water flux and the amount of water held up in terrestrial plants is negligibly small compared to the major water reservoirs. A closer look, however, reveals that the biosphere plays a significant role by virtue of the recycling of water by the process of evapo-transpiration, which together with direct evaporation from surface waters more than doubles the water flux into the terrestrial atmosphere. This effect is most significant in the tropical rain forest whose very survival depends on the recycled moisture (Salati *et al.*, 1979). Moreover, the terrestrial biosphere is located at a very critical junction of the water cycle on the earth surface, which controls the partitioning of the precipitation waters into the surface or subsurface run-off systems, on the one hand, and on the other hand the return of water to the atmospheric moisture by means of evapo-transpiration (the atmospheric 'run-off'). The surface–biosphere system is shown schematically in Figure 23.1.

As is well known, the isotopic signature of waters throughout the water cycle is engendered primarily in accompaniment of the phase transition (liquid to vapour and vice versa) at the ocean/air interface, where the *d*-excess value of the atmospheric waters is established (Merlivat and Jouzel, 1979). The *d*-excess value is defined as $d=[\delta D-(8 \times \delta \,^{18}O)]$ (Dansgaard, 1964). The isotopic content of hydrogen and oxygen in the waters is measured relative to the SMOW standard as δD and $\delta^{18}O$ (Craig, 1961). Further changes occur during the rainout of the atmospheric moisture as the marine air is disconnected from its vapour source and moves onto the continents. There, as air cools down and loses more and more of its water, the air moisture and precipitation becomes depleted in the heavy isotopic species. The resultant mosaic of different isotopic

Stable Isotopes, edited by H. Griffiths.
© 1998 BIOS Scientific Publishers Ltd, Oxford.

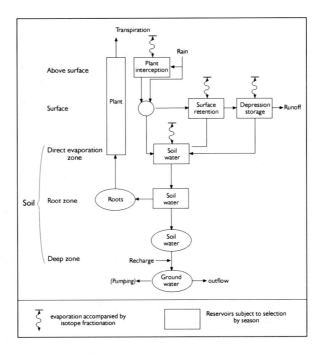

Figure 23.1. *The hydrological compartments and flow pattern at the land–surface/atmosphere interface. Adapted from Gat and Tzur (1967).*

compositions in precipitation (Yurtsever and Gat, 1981) labels the input of waters into the terrestrial reservoirs, namely the ice packs and glaciers, groundwaters and lakes. To a first approximation, the hydrological reservoirs inherit the isotopic composition of the incoming precipitation but with minor distortions (up to 1‰ in $\delta^{18}O$) imposed by the processes at the surface (Gat and Tzur, 1967). In large aquifer units and deep lakes, where the water residence time is multi-annual, the isotope composition of the input can be characterized by the long-term averaged value of precipitation, which is then either conserved in the system, as is the case of groundwaters at moderate temperatures, or later fractionated by additional processes such as exposure to the atmosphere, as occurs over large lakes (Gat, 1995).

In contrast to this, the residence time of water in the 'boxes', (i.e. the hydrological reservoirs) of the surface (Figure 23.1), is very short. As a result the isotopic rules of the game change, since they occur on a short time scale, as little as that of a single shower in some instances. The nature of isotopic change is thus shifted from that of isotope fractionation proper to the selection between waters having varied isotope contents. The subtle changes in isotope composition, which are almost negligible in the long-term averaging processes, assume great practical and diagnostic importance in considering the role played by the biosphere in the terrestrial water cycle on a short-term or seasonal basis.

23.2 Hydrological processes at the land/atmosphere interface

As rain falls onto the surface, four sites need to be considered (Figure 23.1): the canopy-intercepted waters, open water bodies on the surface, the soil waters and the water in plants. A seemingly similar process takes place in all these cases, namely the return to the atmosphere of part of the rainwater by evapo-transpiration. The total amount of re-evaporated water from all these surface reservoirs accounts in most cases for more than 50% of the incoming precipitation and approaches 100% in the arid zone. Details depend both on climate, the plant cover and surface structure and morphology.

The largest share of the evapo-transpiration flux is provided by the transpiration of the plant cover, with evaporation from water intercepted on the canopy also account-ing for a surprisingly large share (35% in the tropical rain forest according to Molion (1987) and 14.2% and 20.3% respectively from deciduous and coniferous trees in the Appalachian Mountains, USA (Kendall, 1993). Direct evaporation from surface waters is usually not very large, except under special circumstances which promote the accumulation of water in impoundment and large lakes. Direct loss of water by evaporation from the soil, which makes up the balance of the total water flux is not appreciable whenever there is an ubiquitous plant cover which contrasts with the influence of the respiratory CO_2 (see Chapters 12 and 24). The loss of water from the soil is then predominantly by uptake of water by plants. This water is then trans-ported to the atmosphere through the intermediacy of the transpiration flux, accounted for above.

Whereas the final effect of all these evapo-transpiration processes is the return of (part of) the precipitation to the atmosphere, there are important differences in details of the process mainly due to the intimacy of exposure to the atmosphere and the dif-ferent residence times of waters in the surface reservoirs (boxes) from which evapo-transpiration occurs. These range in time from a scale of minutes on the canopy to many years in large lakes. When one considers the isotopic changes associated with these processes one finds significant differences between these various pathways which are then imprinted on the resultant fluxes of water to the surface waters, groundwaters and atmospheric runoff, and which enable the quantification many of these processes and relationships.

23.3 The isotope signature imposed by processes at the surface on the hydrological cycle

23.3.1 *Open-surface waters*

Surface waters which are directly exposed to the atmosphere are one of the most widely studied systems (reviewed by Gat, 1995) and thus can be used as a model to understand the rules of the game involved in the isotopic changes imposed by the hydrologic pathways. Within the context of the present discussion, surface waters run the full gamut of scale from the limited interception volume on the canopy which is saturated even by a small rain-shower, to the continental-scale run-off systems.

The isotopic balance equation for a mixed surface water body is given by:

$$\frac{d(V\delta_L)}{dt} = F_{in} \cdot (\delta_i - F_{out}) \cdot (\delta_L - E \cdot \delta_E) \tag{23.1}$$

where V is the volume of the system; F_{in} the inflow flux including the precipitation, F_{out} and E the liquid outflow and evaporation fluxes, respectively; δ_L, δ_i and δ_E are the isotope compositions of the water, influx and evaporation, and can refer to either ^{18}O or D signatures.

The equivalent water balance equation is:

$$\frac{dV}{dt} = F_{in} - F_{out} - E \tag{23.2}$$

On substituting Equation 2 into Equation 1 one obtains:

$$\frac{d\delta_L}{dt} = \frac{F_{in}}{V}(\delta_i - \delta_L) - \frac{E}{V}(\delta_E - \delta_L) \tag{23.3}$$

Evidently the momentary change (increase) in isotope composition of the lake waters depends on the balance between the isotope fluxes of the inflow waters and the isotopic fractionation which accompanies the evaporation process. δ_E can be expressed explicitly as a function of ambient parameters (i.e. humidity and temperature) through the use of a proper evaporation model such as that of Craig and Gordon (1965).

The isotope composition of the residual waters is further and further removed from that of the influx, the larger the ratio of E/F_{in} (Figure 23.2). It is to be noted, however, that the isotopic composition of the return flux to the atmosphere (δ_E) then approaches that of the inflow (δ_{in}) and under terminal-lake conditions when $F_{in}=E$, material balance considerations dictate that $\delta_E=\delta_{in}$. Thus paradoxically, at the time that the effect on the isotopic composition of the lake water is maximal, the effect of the evaporation flux on the isotope composition of the air moisture vanishes.

For the case that δ_{in} is constant in time, or alternatively that the lake is so deep and well-mixed that the input term can be averaged over times which are long relative to the input noise, then the change in isotope composition (at hydrologically steady state) is determined exclusively by the difference between δ_{in} and δ_E. In the case of a shallow lake, however, where the water residence time is short so that the lake does not comprise a full annual cycle, as well as in a seasonally stratified lake, a selection

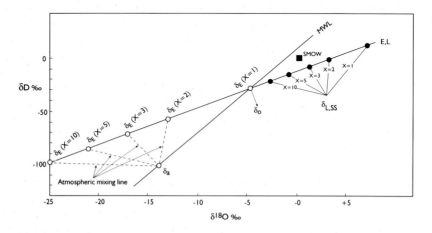

Figure 23.2. The isotopic enrichment of an evaporating water body as a function of its throughflow ratio ($X=F_{in}/E$), and the corresponding value of δ_E according to the Craig-Gordon model. Values were calculated for the arbitrary choice of humidity of 50% and $\delta^{18}O=-5‰$, with the ambient air assumed at isotopic equilibrium with the inflow waters, both $\delta_i = \delta_o$ and δ_a being situated on the Global Meteoric Water Line (GMWL).

process occurs in addition to the evaporative enrichment of the heavy isotopes in the lake waters, whereby part of the incoming waters take part in the evaporation to a larger extent than others (on a seasonal basis, for example). Such a selection process will result in an isotopic shift to the extent that the selected waters show differences in their isotopic values, as is the case for the seasonal cycle in precipitation (Yurtsever and Gat, 1981).

The waters intercepted on the canopy present an extreme case. The interception volume is smaller than that provided even by a moderate shower, so that intra-storm and inter-storm variations in isotopic composition come into effect. Since these variations are relatively large (Gat, 1980 a,b) the selection process easily overshadows the isotopic enrichment due to isotope fractionation, as is indeed evident from comparison of the isotopic composition of the throughflow precipitation to that of the total rainfall (Dewalle and Swistock, 1994; Leopoldo, 1981; Saxena 1986). The effect of the isotope fractionation can then be seen only in the downwind atmospheric waters on a large area integrated scale, through the effect on the d-excess value of the atmospheric moisture (Gat and Matsui, 1991).

23.3.2 *Soilwaters*

Waters infiltrating from the surface drain through the void spaces in the soil. In very homogeneous soils one can recognize a piston flow pattern where the infiltrate of each rain event pushes the previous one ahead, as visualized by the tritium content of the moisture column (Dincer *et al.*, 1974; Gvirtzman and Magaritz, 1990). In heterogeneous soils and fractured matrixes the pattern is more complicated.

The hold-up in the soil column and transport through it does not, in itself, affect the isotopic composition of the infiltrating waters, except as far as mixing between moving and standing water parcels occurs. However, when evaporation from within the topsoil occurs during dry intervals between rain events, this results in an enrichment of the heavy isotopes in the residual waters (Zimmerman *et al.*, 1967). At some depth beneath the surface an evaporating front develops (Menenti, 1984): above it water transport is predominantly in the gaseous phase through the air-filled intergranular space; below it laminar flow and molecular diffusion in the water-filled pore space dominate. An isotopic concentration profile develops, representing a balance between the upward convective flux and the downward diffusion of the evaporative signature (Barnes and Allison, 1988). Isotopic enrichments of ^{18}O in excess of 10‰ relative to the precipitation are found in desert soils. These enriched waters can then be flushed down by subsequent rains, imparting their evaporative isotope signature to the deeper soilwaters and groundwaters (Gat, 1988; see also Chapters 10 and 12).

Direct evaporation from a bare soil surface is a factor in the water balance of an arid environment, in particular in sandy soils with a high water table (e.g. dry lakes and diffuse groundwater discharge zones in the desert (Fontes *et al.*, 1986). In other areas it is the transpiration flux that accounts for the bulk of the evapo-transpiration flux from the soil. There is ample evidence that usually there is no fractionation between isotopes as roots take up water. However, transpiration selectively utilizes the soilwaters during the growing period in spring and summer. If, as is typically the case, the seasonal input of precipitation is not yet mixed with previous water in the soil moisture of the root zone, then this results in a selection of part of the seasonal cycle in the input to the plants, resulting usually in more depleted isotope values in the deeper soil compared to the mean annual value in precipitation. One should note that the two processes of direct evaporation from the soil and seasonal selection by transpiration have opposing isotopic signals. It is thus often difficult to use isotopic profiles in the soil as a direct measure of the evaporative history of a site.

23.3.3 *Water in plants*

As is often discussed, soilwater is taken up by plants essentially unfractionated (but see Chapter 11). This feature is being used extensively to characterize the source of water for the plants, whether soil or riverwater (Dawson and Ehleringer, 1991), snowmelt waters (Kaplan *et al.*, 1995), or to distinguish between shallow or deep soil-waters in the arid zone where marked isotope gradients exist (Adar *et al.*, 1995). Yakir and Yechieli (1995) have also exploited this feature to distinguish between saline groundwaters and flashfloods at the Dead Sea shore.

Leaves are the site of the evaporation front in the plant, specifically the cells located near the stomata. These waters are highly enriched in the heavy isotopes (Gonfiantini *et al.*, 1965) when stomata are open. By virtue of the large flux/volume ratio in the transpiring leaves, steady state can be practically achieved in relation to the evaporation process.

Allison *et al.* (1985), Yakir et al. (1990), and others who calculated the expected steady-state isotopic composition, often found the overall enrichment in the leaf-waters to fall short of the expected steady-state value. Allison interpreted such a difference to be the result of dilution by unfractionated source waters in the veins of the leaves (amounting at times to up to 30% of the total leaf-water volume (Allison *et al.*, 1985)). In the case of cotton plants however, Yakir *et al.* (1990), postulated that part of the leaf-water (in the mesophyllic cells) lags behind the steady-state enrichment due to a slow water exchange with the outer water sheath, which is assumed to be in momentary steady-state relative to the evaporation process during the day and to take up unfractionated source water at night. More complex relationships among water in different parts of the leaf are presented in Chapter 10.

Whatever the internal distribution, under steady-state conditions obviously $\delta_E = \delta_{in}$, so that the transpiration flux is unfractionated with respect to the soilwater taken up by the plants. As a result, neither the soil waters nor the atmosphere recognize the isotopic fractionation which determines the isotope composition in the leaves and, *eo-ipso*, that of the oxygen and carbon dioxide flux from the plant to the atmosphere. However, selection of waters on a seasonal or other basis is immediately translated into an isotopic shift (relative to the annually averaged isotope composition of the incoming rainwater) in the recharge flux to the groundwaters as well in the evapotranspiration flux; these two shifts obviously are in opposite directions and complement each other.

23.4 The effect of the plant cover on the hydrological cycle

The incoming precipitation is partitioned at the earth's surface into three main trajectories as described in Figure 23.1: groundwater recharge, surface run-off and re-evaporation. The plant cover is located at the crucial position where this partitioning takes place and has a large effect on the relative magnitude of these three fluxes. However the effect of the plant cover on the partitioning of the isotopic signature is less dramatic. The large enrichment of the heavy isotopes which characterizes the plant waters in the leaves, fruits etc., and which label the organic matter and the oxygen and carbon dioxide fluxes from the plants, does not really come into play in relation to the rest of the hydrological cycle, mainly because of the very limited size of the water reservoir in the plant, compared to the water fluxes involved.

The role played by the plant cover is primarily a 'mechanical' one, affecting the surface roughness on the one hand, which controls the aerodynamics of the atmospheric boundary layer (which in turn affects cloud formation and the micro-climate in general). An extreme example is provided by the tropical rain forest (Molion, 1987; Gat *et al.*, 1985). On the other hand the plant cover affects the structure of the soil surface, *inter alia* the infiltrability of the soil which is usually increased, and the surface run-off which is reduced. The most important function played by the plant cover, however, is that of a pump for transport of water from the soil back into the atmosphere. The special role of canopy interception has already been discussed.

As for the effect of the plant cover on the isotopic composition of these three water fluxes as compared to that of the incoming composition of precipitation, one finds a number of opposing effects. Since a significant part of the incoming precipitation is routed to surface run-off in the absence of a vegetation cover, especially under arid and semi-arid conditions (Gat, 1986), both processes of isotope fractionation and selection are apparent under these circumstances. The first because of the lengthy exposure of surface waters to evaporation, and also the enrichment of waters in the upper soil layers. The change in the isotope composition due to the selection process, results primarily from the dependence of the run-off generation on rain intensity (Levin *et al.*, 1980), on the one hand, and the preferential loss (by re-evaporation) of small rain amounts, on the other hand, can result in isotopic shifts in view of the 'amount effect' on the isotope composition which is a function of rain amounts (Dansgaard, 1964). Thus the surface run-off as well as the groundwaters recharged through the intermediacy of the run-off show a characteristic distortion of their isotope content relative to that of the precipitation as shown in Figure 23.3. In contrast, the directly recharged groundwaters usually show a minimal degree of change in the isotope composition.

The reduction of the surface run-off component by the presence of a plant cover then has a pronounced effect on the isotope composition of the recharge flux. The fractionating effect of the surface waters is reduced, though partially compensated for by the effect of direct evaporation from the canopy. Also the selection process is shifted in favour of the long term seasonal selection of the uptake of soil water by the plants, rather than the short-term selection during the run-off generation.

Taking all these effects together one observes that the overall shift in the isotope composition of the groundwater recharge relative to the mean value for precipitation is limited to about 1.5‰ in $\delta^{18}O$ in most climate zones. Usually the trend is one of more depleted isotopic values in the groundwaters, but somewhat shifted from the meteoric water line to isotopic compositions with a lower *d*-excess value, due to the combined effect of the selection and the evaporative enrichment along 'evaporation lines' (Figure 23.3).

The effect on the atmospheric moisture can be appreciated on a regional scale primarily. Whereas in the absence of recycling of the precipitation, the gradient of isotopic composition of precipitation from the coast inland follows a simple Rayleigh Law correlated with the temperature (Dansgaard, 1964), the effect of the transpiration flux

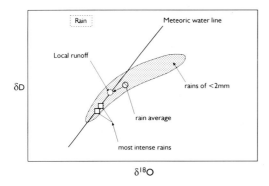

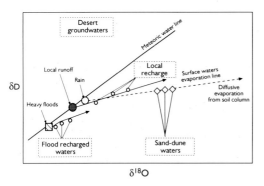

Figure 23.3. *The relationship between the isotopic composition of precipitation and the resultant recharge fluxes to the groundwater, for conditions as they apply in the arid zone (Negev, Israel). The effect of rain selection based on rain intensity, as well as the isotope fractionation due to evaporation from surface waters and soil, is shown. Adapted from Gat (1984).*

(which by and large returns the precipitation water to the atmosphere without much distortion) is to reduce such an inland gradient, as exemplified in the Amazon Basin (Salati *et al.*, 1979). The specific contribution of the fractionating evaporative flux is more readily seen in the change of the *d*-excess value of downward atmospheric waters (Gat and Matsui, 1991), as is the case under extreme conditions over large lake systems (Gat *et al.*, 1994). Figure 23.4 taken from Goodfriend *et al.* (1989) illustrates the effects of re-evaporated fractionated and non-fractionated moisture on the atmospheric waters.

23.5 Conclusions

The hydrological cycle and the terrestrial biosphere intersect at a critical point on the earth's surface, which controls both the inputs to the surface and subsurface (groundwater) run-off systems, as well as the recycling of precipitation into the terrestrial atmosphere. The modification of the isotopic composition of water at the air–land interface results from isotopic fractionation which accompanies evaporation from surface or soil water, or from the selective utilization of part of the total rainfall (on the basis of seasonality or rain intensity) in the process of run-off generation or groundwater recharge. The magnitude of such modifications is controlled by the nature of the terrain and, in particular, the vegetation cover.

Whereas the isotope enrichment of the biospheric waters (primarily in the leaves) is the result of the isotope fractionation which accompanies the evaporation of water, the effect of the biosphere on the isotopic composition of the run-off and groundwaters results primarily from the selection process and not from isotope fractionation proper. Thus, the evaporational enrichment in the leaves operates on a diurnal time scale, whereas the

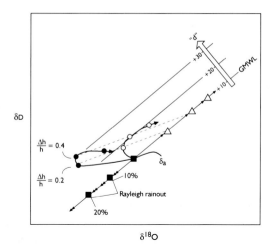

Figure 23.4. The change of the isotope composition of atmospheric moisture as a result of the addition of the evapo-transpiration flux to the atmosphere, showing the effect of isotopically unfractionated waters marked by △ (which conserves the value of the MWL) and of fractionated water from the evaporation flux marked by full circles. The addition of a 1: 1 mixture of evaporation to fractionation fluxes is shown by open circles (O). The points show the addition of 20% and 40% respectively of added moisture (△h/h) based on Goodfriend et al. (1989). Also shown is the depletion in the heavy isotopes in the air masses as expected for Rayleigh rainout of 20% and 40% respectively, of moisture at ~0°C along the Meteoric Water Line.

hydrological effect manifests itself, either, within the timescale of a single shower (primarily due to the processes occurring above ground by canopy interception) or on a seasonal timescale by the interaction of plant growth period and the soil water distribution.

The effect on the water balance of the terrestrial atmosphere depends on the nature of the process: the transpiration flux just restores the isotopic composition in the atmosphere, annulling the continental gradient of isotopes in precipitation. On the other hand, the evaporation flux, whether originating from open waters, the soil layers or the precipitation intercepted on the canopy, results in a change (increase) of the *d*-excess value in the downwind atmosphere.

References

Adar, E.M., Gev, I., Lipp, J., Yakir, D. and Gat, J.R. (1995) Utilization of oxygen–18 and deuterium in steam flow for the identification of ranspiration sources: soil water versus groundwater in sand dune terrain. In *Application of Tracers in Arid Zone Hydrology*, (eds E. Adar and C. Leibundgut), *Int. Assoc. Sci. Hydrol. Publ.* no. **238**, 329–338.

Allison, G.B., Gat, J.R. and Leaney, F.W.J. (1985) The relationship between deuterium and oxygen-18 values in leaf water. *Isotope Geosci.* **58**, 145–156.

Barnes, C.J. and Allison, G.B. (1988) Tracing of water movement in the unsaturated zone using stable isotopes of hydrogen and oxygen. *J. Hydrol.* **100**, 143–176.

Craig, H. (1961) Standards for reporting concentrations of deuterium and oxygen-18 in natural waters. *Science* **133**, 1833–1834.

Craig, H. and Gordon, L.I. (1965) Deuterium and oxygen-18 variations in the ocean and marine atmosphere. In *Stable Isotopes in Oceanographic Studies and Paleo-Temperatures*, (ed E. Tongiorgi), Lab. Geol. Nucl. Pisa, pp. 9–30.

Dansgaard, W. (1964) Stable isotopes in precipitation. *Tellus* **16**, 436–468.

Dawson, T.E. and Ehleringer, J.R. (1991) Streamside trees that do not use stream water. *Nature* **350**, 335–337.

Dewalle, D.R. and Swistock, B.E. (1994) Differences in oxygen-18 content of throughfall and rainfall in hardwood and coniferous forests. *Hydrol. Proc.* **8**, 75–82.

Dincer, T., Al-Mughrin, A. and Zimmermann, U. (1974) Study of the infiltration and recharge through the sand dunes in arid zones with special reference to the stable isotopes and thermonuclear tritium. *J. Hydrol.* **23**, 79–109.

Fontes, J.Ch., Yousfi, M. and Allison, G.B. (1986) Estimation of long term diffuse groundwater discharge in the northern Sahara using stable isotope profiles in soil water. *J. Hydrol.* **86**, 315–327.

Gat, J.R. (1980a) The relationship between surface and subsurface waters: water quality aspects in areas of low precipitation. *Hydrol. Sciences Bull.* **25**, 257–267.

Gat, J.R. (1980b) The isotopes of hydrogen and oxygen in precipitation. In *Handbook of Environmental Isotope Geochemistry*, (eds P. Fritz and J.Ch. Fontes), Volume 1, Elsevier, Amsterdam, pp. 21–47.

Gat, J.R. (1984) Role of the zone of aeration in recharge and the establishment of the chemical character of desert groundwaters. *Proc. Riza Symposium*, Munich, **2**, 487–497.

Gat, J.R. (1988) Groundwater recharge under arid conditions. In *Infiltration Principles and Practices*, (ed Yu-si Fok), Post-Conference Proceeding ICIDA, WRRC-Univ. of Hawai, pp. 245–257.

Gat, J.R. (1995) Stable isotopes of fresh and saline lakes. In *Physics and Chemistry of Lakes*, (eds A. Lerman, D. Imboden and J. Gat), Springer Verlag, New York, pp. 139–166.

Gat, J.R., Bowser, C. and Kendall, C. (1994) The contribution of evaporation from the Great Lakes to the continental atmosphere: estimate based on stable isotope data. *Geophys. Res. Lett.* **21**, 557–560.

Gat, J.R. and Matsui, E. (1991) Atmospheric water balance in the Amazon Basin: an isotopic evapotranspiration model. *J. Geophys. Res.* **96**, 13179–13188.

Gat, J.R., Matsui, E. and Salati, E. (1985). The effect of deforestation on the water cycle in the Amazon Basin: an attempt to reformulate the problem. *Acta Amazonica* **15**, 307–310.

Gat. J.R. and Tzur, Y. (1967) Modification of the isotopic composition of rainwater by processes which occur before groundwater recharge. In *Isotopes in Hydrology*, IAEA, Vienna, pp. 49–60.

Gonfiantini, R., Gratziu, S. and Tongiorgi, E. (1965) Oxygen isotope compsition of water in leaves. In *Isotopes and Radiation in Soil-Plant Nutrition Studies*, IAEA, Vienna, pp. 405–410.

Goodfriend, G.M., Magaritz, M. and Gat, J.R. (1989) Stable isotope composition of land snail body waters and its relations to environmental water and shell carbonate. *Geochim. Cosmochim. Acta* **53**, 3208–3221.

Gvirtzman, H. and Magaritz, M. (1990) Water and anion transpsort of the unsaturated zone traced by environmental tritium. In *Inorganic Contaminants in the Vadose Zone*, (eds B. Bar Yosef, N.J. Barrow and J. Goldschmidt). *Ecol. Stud.* **74**, Springer–Verlag, Berlin/Heidelberg/New York, pp. 190–198.

Kaplan, I.R., Zhang, D., Gat, J.R. (1995) The source of water used by streambed vegetation from Owen Valley in the Sierra Nevada. *EOS Supplement*, p. F214.

Kendall, C. (1993). Impact of isotopic heterogeneity in shallow systems on modeling of storm-flow generation. Ph.D. Thesis. Dept. Geology, Univ. MD, College Park. 270 pp.

Leopoldo, R.R. (1981) Aspetos hidrologicos at florista amazonica denga na regiao *de Manaus*. Ph.D. Thesis. Univ. Natl. Estado San Paulo, Botucatu, San Paulo, Brazil.

Levin, M., Gat, J.R. and Issar, A. (1980) Precipitation, flood and groundwaters of the Negev highlands: an isotopic study of desert hydrology. *In Arid Zone Hydrology: Investigation with Isotope Techniques*, Vienna: IAEA, pp. 3–22.

Menenti, M. (1984) Physical aspects and determination of evaporation in deserts. Report 10, Institute for Land and Water Management Research (ICW) Wageningen, the Netherlands.

Merlivat, L. and Jouzel, J. (1979) Global climatic interpretation of the deuterium-oxygen 18 relationship for precipitation. *J. Geophys. Res.* **84**, 5029–5033.

Molion, L.C.B. (1987) Micro-meteorology of an Amazonian rainforest. In *The Geophysiology of Amazonia*, (ed R.E. Dickinson). Wiley, New York, 285 pp.

Salati, E., Dall'ollio, A., Matsui, E. and Gat, J.R. (1979) Recycling of water in the Amazon Basin, an isotopic study. *Water Resources Res.* **15**, 1250–258.

Saxena, R.K. (1986) Estimation of canopy reservoir capacity and oxygen-18 fractionation in throughfall in a pine forest. *Nordic Hydrology* **17**, 251–260.

Yakir, D., DeNiro, M.J. and Gat, J.R. (1990) Deuterium and oxygen-18 enrichment in leaf-waters of cotton plants grown under wet and dry conditions: evidence for water compartmentation and its dynamics. *Plant, Cell and Environment* **13**, 49–56.

Yakir, D. and Yechieli, Y. (1995) Plant invasion of newly exposed hypersaline Dead Sea shores. *Nature* **374**, 803–805.

Yurtsever, Y. and Gat, J.R. (1981) Atmospheric waters. In *Stable Isotope Hydrology*, (eds J.R. Gat and R. Gonfiantini). IAEA Tech. Rep. Ser. 210, pp. 103–142.

Zimmermann, U., Ehhalt, D. and Munnich, K.O. (1967) Soil water movement and evapotranspiration: changes in the isotope composition of water. In *Isotopes in Hydrology*, IAEA, Vienna, pp. 567–584.

The $^{18}O/^{16}O$ isotope ratio of atmospheric CO_2 and its role in global carbon cycle research

Philippe Ciais and Harro A.J. Meijer

24.1 General introduction and outline

24.1.1 *The global carbon cycle*

The study of the global carbon cycle, representing the transport of carbon-containing material between the major global reservoirs, is an intriguing object of study (see e.g. Bolin *et al.* 1982; Heimann, 1993). More than ever, the field has become important in the last few decades for policy makers, due to the climatic influences that anthropogenic action is having on the global carbon cycle (Houghton *et al.*, 1996). The main causes of anthropogenic disturbance of the natural global carbon cycle is CO_2 production by the combustion of fossil fuels and changes in land use (deforestation and erosion on the one hand, intensive agriculture on the other).

Over a time scale of decades the global carbon cycle consists of three main components: oceans, the terrestrial biosphere and the atmosphere. The last named is the easiest to handle: it is relatively homogenous and well-mixed, and easily accessible for measurements. Furthermore, the atmosphere is more or less a passive carbon container, and serves as an intermediate between the other two components. It is therefore understandable that the first systematic monitoring was aimed at the CO_2 concentration of the atmosphere under background conditions (i.e. located far away from direct human or biospheric influence). Systematic research of this kind started with C.D. Keeling's monitoring network in the Pacific region, starting in 1958 with 3 stations, later extended to 12 stations, and still active to date (Keeling 1958; Keeling *et al.*, 1979, 1989, 1995).

At present there are numerous CO_2 concentration monitoring stations over the world (Francey *et al.*, 1990; Levin *et al.*, 1995; Nakazawa *et al.*, 1993; see also Boden *et al.*, 1994),

Stable Isotopes, edited by H. Griffiths.
© 1998 BIOS Scientific Publishers Ltd, Oxford.

the largest network being that of the NOAA/CMDL (Conway *et al.*, 1988, 1994; Trolier *et al.*, 1996; see also Chapter 13). Gradually a vast and reliable data set monitoring the long-term atmospheric CO_2 concentration has come into existence. Using this detailed information, it was possible to study the fate of the fossil fuel CO_2 and determine at a very coarse scale (roughly hemispheric) which region behaves as a net source or sink for carbon. Nevertheless, there was still no way to discriminate between atmosphere–ocean fluxes and atmosphere–biosphere fluxes, and observations that could make this distinction were badly needed.

Around 1977, the CO_2 concentration monitoring effort of the Scripps network was extended with systematic measurements of the $^{13}C/^{12}C$ ratio in CO_2, performed at the Centrum voor Isotopen Onderzoek (CIO) (Keeling *et al.*, 1979). This isotope ratio provides a desirable and important source of additional information: it can identify fluxes between atmosphere and ocean as distinct from those between atmosphere and biosphere. This is caused by the different carbon isotope ratios of these reservoirs. Furthermore, because the $^{13}C/^{12}C$ is different from that of the atmosphere, the annual input of fossil fuel CO_2 can be recognised and identified using the (CO_2) and carbon isotope records together. Today, many of the CO_2 networks mentioned above perform combined CO_2 concentration and $^{13}C/^{12}C$ measurements (for additional locations, see Chapter 13).

In summary, Table 24.1 gives the carbon content of the three main reservoirs mentioned above, the natural fluxes between these reservoirs, as well as the fossil fuel combustion and biomass burning inputs. As can be seen from Table 24.1, the 'one-way' annual fluxes between the reservoirs are about two orders of magnitude higher than their differences. Yet the size, direction and variability of the source and sink terms are the best indicator for the effect of human activities on the global carbon cycle, and thereby of the atmospheric 'greenhouse' effect through increasing atmospheric CO_2 concentration.

24.1.2 *The Role of Stable Isotopes, ^{13}C and ^{18}O, in CO_2*

Here we are confronted with a general measurement problem: small differences between large numbers are extremely difficult to determine precisely. One way to

Table 24.1. An overview of the main components and fluxes of the global carbon cycle (Houghton et al., 1996). The flux from atmosphere to the terrestrial biosphere is gross primary production. The net increase of the atmosphere amounts 3.3 PgC/yr (last column). The annual increase of the three reservoirs together equals the annual input of carbon due to fossil fuel combustion

Carbon reservoir	Content (Pg carbon)	Flux to atmosphere (PgC/yr)	Flux from atmosphere (PgC/yr)	Fossil fuel and Deforestation to atmosphere (PgC/yr)	Reservoir increase (PgC/yr)
Ocean	39000	90	−92		2
Land biosphere	2200	120	−121.3	1.1	0.2
Atmosphere	750	210	−213.3	5.5	3.3

tackle this problem is through 'brute force': more and more monitoring sites are coming into existence, while continuous effort is being made to increase intra- and inter-laboratory precision in measuring $[CO_2]$ concentrations and $^{13}C/^{12}C$ ratios. Using time series of $^{13}C/^{12}C$ at a few locations to make a global inventory of the ocean versus land uptake requires a precision of the order of 0.01‰. Even when maximum care is taken, very slight differences in the technique and calibration of $^{13}C/^{12}C$ atmospheric measurements may lead to large discrepancies being inferred in the global balance (see Francey et al., 1990 and Keeling et al., 1995; see also Section 24.2.1 below). Still, this approach progressively makes it possible to infer ocean- and land-fluxes at a regional scale (see Chapter 13). The effort certainly has achieved success and will be pursued in the future. At present, however, model calculations still show a spread in the source/sink sizes over some regions that are of the same magnitude as the sources/sinks themselves (Ciais et al., 1995), mainly due to the incomplete coverage of existing sampling networks over the continents. A complementary approach is to search for other measurement parameters, which may provide us with independent information. Ideally, one would like to select a means of measurement which is only sensitive to the different fluxes and pools. However, a quantity which is sensitive to the different fluxes, in a way complimentary to the current parameters, would also increase the accuracy of determination significantly. It is in this respect that the $^{18}O/^{16}O$ ratio of atmospheric CO_2 has increasingly gathered attention. Researchers became gradually aware of the fact that both soils and vegetation play an important role in controlling the global patterns in the $^{18}O/^{16}O$ ratio of atmospheric CO_2 (Ciais et al., 1996a, b; Francey and Tans, 1987; Farquhar et al., 1993; Keeling et al., 1995). On a local scale, ^{18}O in CO_2 has even been used successfully to estimate photosynthesis and respiration separately (Yakir and Wang, 1996). The critical process is the exchange of ^{18}O between CO_2 and water in plants and soils, which is considerably accelerated in presence of carbonic anhydrase (CA) (see also Chapters 10 and 12). The quantity of CO_2 that exchanges isotopes with water depends on ecosystem productivity, but the isotopic label of CO_2 emitted by the biosphere is entirely determined by the $^{18}O/^{16}O$ ratio of the water in the system that is associated with soils, plants and canopy air. Thus, the $^{18}O/^{16}O$ ratio of atmospheric CO_2 at remote background sites in the free troposphere, far from active sources, depends on: (i) the large-scale fluxes of photosynthesis as gross primary productivity (GPP) and total ecosystem respiration (R_E); (ii) the $^{18}O/^{16}O$ ratio of the specific pools of water exchanging ^{18}O with CO_2 in leaves and in soils; (iii) other non-biospheric sources of ^{18}O; (iv) large scale atmospheric mixing. In order to model the $^{18}O/^{16}O$ ratio in CO_2, one must take into account all four processes, as will be discussed later in this chapter.

The remainder of this chapter is divided into two sections: first we give a historical overview of the experimental results for ^{18}O monitoring, and the basic features of its interpretation. Then a very simple but illustrative global model is presented in Section 24.3, in which the behaviour of CO_2 and the $^{13}C/^{12}C$ ratio, is compared with that of the $^{18}O/^{16}O$ ratio. This model incorporates three of the four sources mentioned above (the non-biospheric sources being left out for simplicity), and it is meant to be an introduction to the field. Second, beginning in Section 24.4, we discuss the state-of-the art interpretation of the global $^{18}O/^{16}O$ signal, using a combination of advanced biosphere modelling and general circulation models (GCM). The results, summarized in Section 24.5, show that all the main features of the global $^{18}O/^{16}O$ signal can be described in space and time. However, the use of ^{18}O as a proxy for the terrestrial biosphere contribution to the global carbon cycle is still hindered by low accuracy in important input data such as the ^{18}O signal in meteoric water, as well as several fractionation factors (see Chapter 23).

24.2 The $^{18}O/^{16}O$ isotope ratio of atmospheric CO_2 : basic features

24.2.1 Measurements of $^{18}O/^{16}O$ in atmospheric CO_2

Measurement by an isotope ratio mass spectrometer of CO_2 yields the ratios of masses 45/44 and 46/44 and from these ratios the $^{13}C/^{12}C$ and the $^{18}O/^{16}O$ atomic isotope ratio, or rather the $\delta^{13}C$ and $\delta^{18}O$, can be determined (see Gonfiantini et al., 1995, and references therein, and Coplen, 1996). The only complication is the existence of the rare ^{17}O isotope, giving rise to mass interference on mass 45 and to a lesser extent on mass 46 (Allison et al., 1995). This means that when the systematic $\delta^{13}C$ determination of the Scripps network started at the CIO in 1977, the $\delta^{18}O$ numbers also became available. However, compared to $\delta^{13}C$, there are three complicating factors, influencing reliability and precision.

Fractionation effects. Since the mass difference between ^{18}O and ^{16}O is almost twice that between ^{13}C and ^{12}C, fractionation effects can occur in every sample (such as during incomplete collection of CO_2 from air, or fractionation during admission to the mass spectrometer) that will influence the $\delta^{18}O$ twice as much as the $\delta^{13}C$. However, since the natural fractionation effects also tend to be bigger for ^{18}O than for ^{13}C, the relative precision is not altered.

Calibration issues. The $\delta^{13}C$ so-called VPDB scale is internationally maintained using a primary reference material, NBS19 (Gonfiantini et al., 1995). This calcite must be converted to CO_2 with phosphoric acid, and the CO_2 acts as the primary standard against which all $\delta^{13}C$ measurements have to be calibrated. If treated correctly, the carbon transfer from carbonate to gas is 100%, so no extra precautions need to be taken with respect to water-free acid or reaction temperature.

For the $\delta^{18}O$ scale things are more complicated: the same NBS19 calcite acts as primary reference material for the $\delta^{18}O$ scale, but the transfer of oxygen from the calcite to the CO_2 gas is not complete since part of the calcite oxygen ends up in the water so formed. The pools of water and CO_2 gas do not have the same oxygen isotope ratio, but rather there is a temperature-dependent (and in part, reaction-kinetics dependent) fractionation between the two pools. Therefore, the CO_2 gas formed can only be used as the primary reference material for the $\delta^{18}O$ scale (called the VPDB-CO_2 scale, to indicate that the oxygen isotope ratio in the CO_2 gas is different from that in the original calcite) if the reaction circumstances are exactly prescribed in terms of the acid used and the reaction temperature. Further details can be found in Gonfiantini (1984).

As will be clear from this description, preparing the CO_2 gas from the primary reference material for calibration purposes is a much more vulnerable process for $\delta^{18}O$ than it is for $\delta^{13}C$. This is particularly the case when the calibrated $\delta^{18}O$ scale is used, not for determining the $\delta^{18}O$ of other carbonates (when errors in reference and sample preparation cancel out), but for different materials such as atmospheric CO_2 when systematic errors may commonly occur in one lab.

To date, there are still only a few labs able to maintain a $\delta^{18}O$ scale for atmospheric CO_2 that is well-calibrated against the primary reference material. Several (formal and informal) ring tests for carbonates show that, whereas the $\delta^{13}C$ determination scales are well-established around the world, with results deviating by no more than a few hundredth of a permil, the $\delta^{18}O$ determination shows differences up to 0.3‰ or more (T. Coplen, M. Verkouteren, M. Gröning, personal communication).

Fortunately, there is another primary reference material for $\delta^{18}O$ scale calibration : Vienna Standard Mean Ocean Water (VSMOW, the addition of Vienna is to indicate that this material is stored at and distributed by the IAEA located in that city). Originally solely meant for hydrological purposes, it soon became apparent that maintaining a well-calibrated $\delta^{18}O$ scale benefits very much from the use of both reference materials, and the careful comparison of their outcome. Since the relation between the two $\delta^{18}O$ scales (VPDB-CO_2 and VSMOW-CO_2) is well-known (their zero point being 0.27‰ apart, Gonfiantini et al., 1995), this has the tremendous advantage of allowing cross-calibration. In the case of the VSMOW-water preparation is via the well-defined method of isotopic equilibrium between the water and added CO_2, also involving fractionation. However, laboratory practice has shown this preparation method generally to be more reliable and precise. Moreover, finding the correct difference between the two scales is a reliable indication that both determinations are correct, since the preparation methods are entirely different.

Sensitivity to moisture. As mentioned before, the importance of the $^{18}O/^{16}O$ signal lies in the fact that it is dominated by the isotope exchange between (or rather from) water and (to) atmospheric CO_2. This implies that $\delta^{18}O$ of CO_2 will change in the presence of water, also on route between the air sampling and the final $\delta^{18}O$ measurement in the mass spectrometer. Fortunately, this isotope exchange between water and CO_2 only occurs when water is in the liquid phase, so water vapour does not influence the CO_2. Of course, the best way of sample treatment is the on-line drying of air during collection. However, sampling the air without drying and maintaining a temperature during transport such that condensation of water does not occur should function almost equally well. The latter method, however, is unreliable when the sampling site is tropical, when high temperatures, combined with high relative humidity, lead to high dew points (Gemery et al., 1996).

The effects described above have limited the usefulness of the few monitoring networks that were active in the 80s in measuring $\delta^{18}O$. In fact, for a long time, nobody thought seriously about the value of $\delta^{18}O$ records, and the $\delta^{18}O$ results were seen as a 'byproduct', and no allowances made for these problems. The Scripps-CIO cooperation more or less accidentally benefitted from the well-calibrated ^{18}O scale maintained at the CIO for hydrological research. This fortunate coincidence took care of two out of the three problems. However, the air was not dried at the sampling site, resulting in scatter clearly observable in the data. Nevertheless, an important record has been built up from the late seventies onwards with full latitudinal coverage. Figure 24.1 shows the $\delta^{18}O$ signals for three of the twelve sites: Point Barrow (Alaska), Mauna Loa (Hawaii) and the South Pole. The only other long-term record known to us is that from Cape Grim (since 1982, described in Francey et al., 1990), which is of better quality than the Scripps/CIO set, thanks to on-site drying of the air. However, the global coverage of the Scripps/CIO network since the early 80s is unsurpassed.

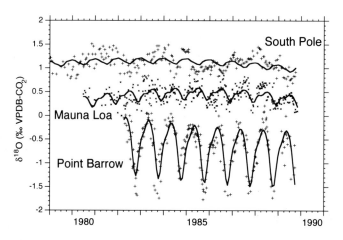

Figure 24.1. $\delta^{18}O$ *measurements of three stations of the Scripps/CIO cooperation: Point Barrow (71.3 °N), Mauna Loa (19.5 °N) and the South Pole (90.0 °S). There is a huge difference in average value visible, with a North–South gradient of almost 2‰. In spite of the visible scatter reliable seasonal signals are apparent.*

24.2.2 *Differences between $\delta^{13}C$ and $\delta^{18}O$*

The data in Figure 24.1 show striking features which were not understood for several years. Firstly, there was a huge North–South shift in the average $\delta^{18}O$ signals of almost 2‰, about ten times the change in $\delta^{13}C$, suggesting that very strong fluxes counteract any active atmospheric mixing. Secondly, the seasonal cycles are similar in magnitude to those of $\delta^{13}C$, but in the North there is a phase shift of up to several months from both $\delta^{13}C$ and the CO_2 concentration.

The principal difference in behaviour between $\delta^{13}C$ and $\delta^{18}O$ is caused by the fact that unlike $\delta^{13}C$, $\delta^{18}O$ of CO_2 is influenced by contact with water. This means:

(1) $\delta^{13}C$ is determined by fractionation effects during fluxes between reservoirs, and the $\delta^{13}C$ values of different reservoirs influence each other;
(2) under steady state conditions, $\delta^{13}C$ variations are seasonal around a global average. Observed latitudinal differences are caused mainly by the uneven distribution of fossil fuel combustion;
(3) all $\delta^{13}C$ variations are associated in some way with $[CO_2]$ variations;
(4) $\delta^{18}O$ in atmospheric CO_2 is determined by $\delta^{18}O$ in both oceanic and meteoric water through the carbonic acid equilibrium. The amount of water available is so huge that $\delta^{18}O$ in CO_2 is wholly determined by that of water, and the influence of CO_2 on H_2O is completely negligible;
(5) assuming steady state conditions, there exists a strong North–South gradient in $\delta^{18}O$. This is caused by the unequal distribution of ocean and land between the hemispheres, and by the entirely different $\delta^{18}O$ of oceanic and meteoric water;
(6) since $\delta^{18}O$ in CO_2 is determined by isotope exchange with water, not all processes that change $\delta^{18}O$ are accompanied by CO_2 concentration changes.

When the first world wide-monitoring results came into existence, it was clear that the huge gradient in $\delta^{18}O$ could only be maintained by enormous fluxes, or rather by a very large pool with a strong water equilibrium potential. The equilibrium reservoirs that were initially evident, namely oceanic exchange, plant photosynthesis and soil respiration, were far too small to account for this exchange capacity. It took several years of speculation and computation leading the exchange capacity in clouds, fog and precipitation in general, until it was learnt from plant physiologists that much more CO_2 comes into contact with leaf (chloroplast) water than actually is taken up by the plant in photosynthesis. Francey and Tans (1987) named this effect for the first time as the 'missing equilibrium flux'. In fact, for every CO_2 molecule that is taken up by photosynthesis, two others enter the leaf through the stomata. They equilibrate with the leaf water almost instantly because of CA activity and then diffuse back to the atmosphere without actually having been incorporated by the plant. This huge flux therefore only influences the $\delta^{18}O$ signature of atmospheric CO_2, and has no influence on either the $[CO_2]$ or $\delta^{13}C$ value. Some time after this discovery, a more quantitative model was presented by Farquhar et al., 1993. They combined global information on the $\delta^{18}O$ of precipitation and ground water with a description of the chloroplast water enrichment and the $\delta^{18}O$ exchange process within the leaf, and finally with a model of global biosphere. Results of this work compared favourably with the averages of the world-wide observations. Section 24.4 of this chapter, based on recently published work (Ciais et al., 1997a, b) develops this line further, and adds the influence of atmospheric circulation, making a more detailed comparison with experimental results.

In conclusion, one may say that the features of $\delta^{18}O$ in atmospheric CO_2 are now understood. The main importance of the issue from a global carbon cycle point of view is: does the $\delta^{18}O$ signal carry information that is independent of that in $[CO_2]$ and $\delta^{13}C$ signals? Obviously, the $\delta^{18}O$ signal carries information about land biospheric activity (GPP and respiration) in a way not present in the other two signals, but much more detail (fractionation effects, the $\delta^{18}O$ distribution of ground water, precipitation and water vapour over the world) is needed in order to extract this information (see Chapter 23). In Section 24.4, a first attempt at sensitivity analysis is therefore also presented.

24.3 A very simple model for the concentration, the $\delta^{13}C$ and the $\delta^{18}O$ of atmospheric CO_2

24.3.1 Model description

As an introduction to Section 24.4, and to show in what respect the $\delta^{18}O$ signal is unique, an extremely simple, 'tutorial' model is presented here. It will be shown that it is relatively easy to fit the average CO_2, $\delta^{13}C$ and $\delta^{18}O$ signal even with this very simple model, as well as the seasonal effects in CO_2 and $\delta^{13}C$, but not at all for the seasonality in $\delta^{18}O$.

In our very simple model, the earth is divided into three zones with equal air mass, North, Tropical and South. Each of these three zones contains three boxes: atmosphere; ocean; and land biosphere. The amount of carbon in the atmosphere boxes is assumed to be 250 Pg, corresponding to a CO_2 concentration of 350 ppm. The carbon contents of the ocean and terrestrial biosphere are not necessary for the model. Carbon fluxes between the different reservoirs are of four different kinds; air exchange between South and Tropical, and between Tropical and North; the atmosphere–ocean exchange; the land biosphere uptake and respiration (soil and plant respiration are taken together); finally, the exchange–diffusion flux of CO_2 into and out of the leaves and equilibrated with the leaf water. The assumed sizes of all these fluxes are given in Table 24.2.

Table 24.2. *The atmospheric carbon contents and fluxes used in the simple three zone model as exchanges between ocean (O), Biosphere (B) and atmosphere (A). The diffusion–backdiffusion is assumed to be fully proportional to photosynthesis, with coefficient K. The numbers are loosely based on Siegenthaler (1993)*

Zone	Carbon content of atmosphere (Pg C)	O ↔ A	B ↔ A Photosynthesis and (plant + soil) respiration	B ↔ A Diffusion–backdiffusion	A ↔ A' Atmospheric transport
		Fluxes (Pg C /year)			
North	250	15	55	K × 55	175 ↓
Tropical	250	30	44	K × 44	175 ↑ 175 ↓
South	250	45	1	K × 1	175 ↑

Obviously, the model is taken such that equilibrium exists as far as the carbon contents are concerned. For simplicity the influence of fossil fuel combustion is left out.

In order to describe $\delta^{13}C$ of the different reservoirs, several fractionation effects are needed. Using the existing (semi-)equilibrium situation in which atmospheric $\delta^{13}C$ is –8‰, leads to values for oceanic dissolved CO_2 at –9‰ (and bicarbonate +1‰), and the land biosphere –25‰. For $\delta^{18}O$, the isotope signature of the different components is fully determined by the water $\delta^{18}O$ values. Table 24.3 shows the assumed $\delta^{18}O$ values for the CO_2 in equilibrium with the different reservoirs, representing a combination of the $\delta^{18}O$ of the reservoir and the equilibrium temperature. The CO_2 from soil and plant respiration is assumed to have equilibrated with ground water, whereas the exchange–diffusion flux of CO_2 leads to a $\delta^{18}O$ equilibrated with the leaf water (enriched through evaporation).

24.3.2 Results for average values

There is one final parameter to determine: the ratio K between the diffusion–backdiffusion and photosynthesis (see Table 24.2). If we run the model for different values of K, we observe a tremendous influence of this ratio on the outcome of $\delta^{18}O$, both in terms of absolute value, and in the North–South gradient. Table 24.4 shows the results for values of K between 0 and 3, together with the average values observed for Point

Table 24.3. *$\delta^{18}O$ values of CO_2 equilibrated with the different reservoirs, as used in the simplified model*

zone	ocean	ground water	leaf water
	$\delta^{18}O$ (‰ VPDB-CO2)		
North	3	–12	2
Tropical	1	–5	5
South	3	–12	2

Table 24.4. *Model results for the three zones as a function of the diffusion–backdiffusion to photosynthesis ratio K. Observed values are for Point Barrow, Mauna Loa and the South Pole. All $\delta^{18}O$ values are in ‰VPDB-CO_2*

K value	North	Tropical	South
0	−4.9	−3.3	−2.1
1	−2.1	−0.7	0.0
2	−0.8	0.5	1.0
3	−0.1	1.3	1.6
observed	−0.8	0.4	1.1

Barrow ('North'), Mauna Loa ('Tropical') and South Pole ('South') for comparison with Figure 24.1. Table 24.4 shows that if one does not take the diffusion–diffusion flux into account, large discrepancies exist between the model calculations and the measurements. It was exactly this phenomenon which had puzzled the global carbon cycle community for several years. A good agreement exists between model and observations for $K = 2$ (Table 24.4). This is in agreement with what one would expect. On average, the CO_2 concentration of air inside the leaf is about 2/3 of that of the outer atmosphere. This leads to the conclusion that the amount of CO_2 diffusing out of the leaf through stomata is twice the amount actually taken up.

In a way it is striking that such a coarse model can already explain the observed average CO_2 concentration, the $\delta^{13}C$ and the $\delta^{18}O$ measurements simultaneously. However, one should remember that this model is set up using a sort of 'reverse engineering': the (roughly estimated) carbon contents and fluxes could only be used as input thanks to more sophisticated modelling/fitting of observed data. Still, very simple models such as this one serve their goal partly in an illustrative sense, but also because they provide insight into such basic items as model output sensitivity. We will return to this subject at the end of this section.

24.3.3 *The seasonal cycle*

Since the simple model performs so well in describing the average CO_2, $\delta^{13}C$ and $\delta^{18}O$ signals, the next step is to see if the seasonal cycle of these observations can also be described with the model. To this end we assume the following:

(1) there is no seasonality in fluxes for ocean exchange, or soil and plant respiration;
(2) photosynthesis in the northern and southern land model boxes is, to a high degree, seasonally dependent. This seasonal cycle is described using a simple cosine, with maxima 1.8 times the yearly average in the (northern or southern) summer, and 0.2 times the yearly average in winter. The seasonal cycle in the tropics varies moderately with the (northern) season, from 1.3 to 0.7 times the average;
(3) the diffusion–diffusion flux remains simply twice that of photosynthesis, and is thus seasonally dependent in the same way as described above.

The values of the modulations have been determined such that the results of the seasonal cycle in the CO_2 observations fitted rather well. Figure 24.2A shows the Scripps

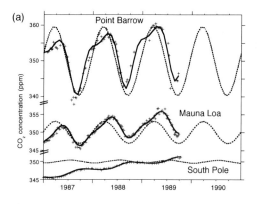

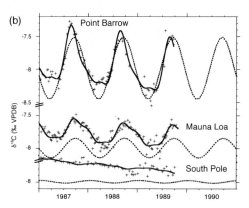

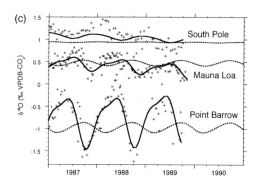

Figure 24.2. Comparison of the seasonal cycle between simple model results and observations from the Scripps/CIO cooperation. A) CO_2 concentration. The observed signals are well reproduced by the model, except for the trend caused by fossil fuel combustion, which is not described by the model. B) $\delta^{13}C$ signals. Here the simple model performs quite well in reproducing the seasonal cycle. Fossil fuel effects visible in the observations as a North–South gradient, are not described by the model. C) $\delta^{18}O$ signals. The model reproduces the average values, but the seasonal cycle is entirely wrong, both in amplitude and phase. The seasonal behaviour of $\delta^{18}O$ is also dependent on seasonal effects of the meteoric water cycle, which interact with the vegetation cycle in a complex way.

CO_2 records from Point Barrow, Mauna Loa and the South Pole. Apart from the clearly visible trend due to fossil fuel combustion, the quality and fit (in the sense of seasonal amplitude and phase) is rather good. In the Mauna Loa record the maximum of the model result is earlier than in the observed signal; this is caused by the Northern hemisphere phase assumed for the whole tropical zone. In the model South Pole signal, the Northern hemisphere phase still dominates, contrary to the experimental findings. However, the size of the amplitude is also very good for the South Pole situation. Figure 24.2B shows that the fit for the $\delta^{13}C$ signals is equally good. The South Pole and Mauna Loa records do not agree in absolute terms. This is to be expected, since the model is tuned to the average $\delta^{13}C$ value of Point Barrow, and the

North–South gradient due to fossil fuel combustion is not accounted for, and Figure 24.2B clearly shows the size of that gradient. For $\delta^{18}O$, shown in figure 24.2C, the situation is different: the phase and amplitude of the seasonal cycle are not reproduced by the model results. The main cause of this poor quality of fit is that seasonal variations in the meteoric water cycle are not taken into account, since both the ground- and leaf-water as listed in Table 24.3, should vary seasonally. However, even if one brings this extra seasonality into the model, it still fails completely to reproduce the seasonal amplitude and, especially, the phase.

24.3.4 Conclusions from the simple model calculations

The fact that our very simple model can reproduce the basic features of the CO_2 and $\delta^{13}C$ signals is both good and bad. It is satisfying that we understand the basic mechanisms that give rise to (the variations in) these signals. However, if such a coarse model already can reproduce the signals to this extent, it also means that the differences between the signals and the model carry more refined information. To put it the other way around, the CO_2 and $\delta^{13}C$ signals have only limited value as constraints for testing refined models. One can vary fluxes, reservoirs and fractionations to a considerable amount in the model and because of the agreement with CO_2 and/or $\delta^{13}C$ signals, one cannot definitely approve or reject the set of parameters. The $\delta^{18}O$ signal apparently is much more sensitive to the various input parameters. The big disadvantage of using this signal to gain additional information about the global carbon cycle is that one is also forced to describe the meteoric water cycle to a high degree of precision. Results based on the $\delta^{18}O$ signal are sensitive to many parameters outwith the direct global carbon cycle, involving the water cycle and exchange fractionation effects (see Chapter 23).

In spite of this, the next section will show that the $\delta^{18}O$ signal is a valuable additional source of information, sensitive to the gross primary production and plant respiration in a way not parallelled by other atmospheric signals.

24.4 Towards a realistic model of $\delta^{18}O$ in atmospheric CO_2: calculation of the surface fluxes

24.4.1 Carbon exchange

A description of the global exchange of CO_2 by land ecosystems requires mapping the productivity and respiratory fluxes, and their variations in time and space. In the following, 'productivity' denotes gross primary productivity (GPP) and 'respiration' the total ecosystem respiratory loss of carbon (R_E), sum of the above ground plant maintenance respiration and of the below ground respiration by roots and decomposers. For simplicity, we will neglect any net carbon storage or loss by ecosystems on an annual basis, the net ecosystem productivity (NEP) thus being equal to zero everywhere. In spite of the NEP being critical when attempting to close the budget of CO_2, it is only a small percentage of the GPP and R_E fluxes which govern the $\delta^{18}O$ of atmospheric CO_2. Global biospheric carbon models are able to calculate photosynthesis at the leaf level as controlled by climate parameters such as temperature, radiation and humidity. A wide range of models have been developed in the recent years that allow scaling up of photosynthesis from the canopy level, for different ecosystems at a spatial resolution of a few degrees ($1° \approx 120$ km) (e.g. Melillo *et al.*, 1993; Ruimy *et al.*, 1994). Alternatively, one can derive the GPP by adding the plant respiratory flux to the calculated NPP (Lloyd and Farquhar, 1994). As we have seen in the previous section, the exchange–diffusion mechanism (K) is at the heart of understanding the $\delta^{18}O$ signals. Therefore, in addition to the flux of GPP, one also needs to estimate the CO_2 concentration in the chloroplast, defined by the ratio C_c/C_a. This requires that one has to treat C_3 and C_4 species differently, because the latter are characterized by lower C_c/C_a values (see Chapter 8). In this modelling study we have used global carbon fluxes as calculated by the simple biosphere (SiB2) model, coupled to the CSU climate model (GCM). The CSU GCM is derived from the UCLA GCM, which was developed at UCLA, over a period of 20 years. Many changes have been made since the model left UCLA including revised parameterisations of solar and terrestrial radiation (Harshvardhan *et al.*, 1987), the planetary boundary layer (PBL) (Randall *et al.*, 1992), cumulus convection (Randall and Pan, 1993), cloud microphysical processes (Fowler *et al.*, 1995), and land-surface processes (Sellers *et al.*, 1986, 1992a, b, 1996a, b). Some recent results have been presented by Randall *et al.* (1989, 1991, 1996), Fowler *et al.* (1995), and Fowler and Randall (1995).

The known variables of the CSU GCM are: potential temperature; the horizontal wind components; the surface pressure; the PBL's depth and turbulence kinetic energy; the mixing ratio of three phases of water, plus rain and snow; the temperatures of the plant canopy, the ground surface, and the deep soil; the water contents of four above-ground and three below-ground moisture stores; the stomatal conductance of the plant canopy, and finally the ice temperature at land-ice and sea-ice points. The model is formulated in terms of a modified sigma co-ordinate, in which the PBL top is a coordinate surface, and the PBL itself is identified with the lowest model layer (Suarez *et al.*, 1983). The mass sources and sinks for the PBL consist of large-scale convergence or divergence, turbulent entrainment, and the cumulus mass flux. Turbulent entrainment can be driven by positive buoyancy fluxes, or by shear of the mean wind in the surface layer or at the PBL top.

For vegetated land points, the surface fluxes of sensible and latent heat, radiation, moisture, and momentum are determined using the SiB parameterisation developed by Sellers *et al.* (1986). SiB has recently undergone substantial modification (Sellers *et al.*, 1996a, b; Randall *et al.*, 1996), and is now referred to as SiB2. The number of biome-specific parameters has been reduced, and most are now derived directly from processed satellite data rather than prescribed from the literature. The vegetation canopy has been reduced to a single layer. Another major change is in the parameterisation of stomatal and canopy conductance (Collatz *et al.*, 1991, 1992; Sellers *et al.*, 1992a,b, 1996a) used in the calculation of the surface energy budget over land. This parameterisation involves the direct calculation of the rate of carbon assimilation by photosynthesis, making possible the calculation of CO_2 exchange between the atmosphere and the terrestrial biota at the dynamic time stage of the CSU GCM (Denning, 1994, Denning *et al.*, 1996a,b).

24.4.2 *Isotopic composition of water*

In our model we assume that CO_2 exchanged by leaves and by soils is in full isotopic equilibrium with soil water and that its $\delta^{18}O$ can be predicted from the $\delta^{18}O$ of water and the temperature-dependent equilibration factor (Brenninkmeijer *et al.*, 1983).

Firstly, one needs to estimate the global distribution of isotopes in meteoric water. Water vapour formed over the ocean and advected towards land undergoes a series of condensations which have the effect of preferentially removing the heavier isotopes (see Chapter 23). As a consequence, the vapour and hence the precipitation becomes more depleted in ^{18}O as one moves inland. Below 15 °C, a good correlation between $\delta^{18}O$ in precipitation and air temperature is generally observed, and a linear relationship exists between those two variables. Above 15 °C, in the tropics, the correlation with temperature is not so clear and $\delta^{18}O$ is negatively correlated to the amount of rain. In order to map $\delta^{18}O$ in precipitation at the global scale, one can use observations on the worldwide IAEA network (Rozanski et al., 1992, 1993). The network is dense over Western Europe but large gaps exist in the tropics, which are still required to allow the observed $\delta^{18}O$ in precipitation to be plotted as a function of temperature and other relevant parameters (topography, amount of rain etc.) as was done by Farquhar et al. (1993). One can also use $\delta^{18}O$ in precipitation, calculated by 3D climate models (GCM) in which the isotope hydrology is fully parameterised. Despite the fact that climate models may not have a very good degree of realism for global hydrology, they generally prove successful in predicting the worldwide distribution of isotopes in precipitation and their results compare favourably with the IAEA observations. In this work, we have followed this latter approach and used global monthly $\delta^{18}O$ in precipitation from the study of Jouzel et al. (1987).

Next, the geographical distribution of $\delta^{18}O$ in soil water is estimated across the globe, which is derived from the composition of precipitation. Soil water $\delta^{18}O$ is determined by the $\delta^{18}O$ of different precipitation events throughout the year, but moisture in the surface soil can be substantially enriched due to evaporation. Within the soil, CO_2 is expected to exchange an ^{18}O atom with water with a characteristic time $\tau = 100$ s (the rate of hydration of CO_2 over water at 25 °C) which corresponds to a path length by diffusion within the soil profile of $x = (4D\tau)^{\frac{1}{2}} = 5$ cm. This indicates that, whatever its isotopic label when produced by roots or micro-organisms, CO_2 must be in isotopic equilibrium with water within the top layers of the soil (see also Flanagan et al., 1996 and Chapter 12). Hence, the presence of an evaporating front at the surface should increase the $\delta^{18}O$ value of CO_2 escaping from the soil. It is difficult to scale up this process at the global scale since one would need to consider a succession of individual rainfall events, followed by subsequent episodes of evaporation. We have parameterised the global $\delta^{18}O$ of soil water as being equal to the monthly mean $\delta^{18}O$ of precipitation, with evaporation accounted for in a very crude manner : we calculate the isotope effect of evaporation every day by distributing evenly the monthly mean evaporative flux of the CSU climate model in each grid cell. By doing so, we ignore large shifts in $\delta^{18}O$ that may arise from short-term variability in climate. As a result, we obtain only a small enrichment of $\delta^{18}O$ in the surface soil moisture due to evaporation, of the order of 0.5‰ higher than without evaporation. Figure 24.3 shows the latitudinal influence that soil will exert on the atmospheric $\delta^{18}O$ of CO_2, under this parameterisation, the effect of soils being described by the quantity: R_E $(\delta^{18}O_S - \delta^{18}O_a - \epsilon_s)$ where $\delta^{18}O_S$ is the $\delta^{18}O$ of soil CO_2, $\delta^{18}O_a$ the $\delta^{18}O$ of atmospheric CO_2, and ϵ_s the fractionation factor for molecular diffusion of CO_2 out of the soil (8.8‰). Soils are expected to deplete the atmosphere in ^{18}O over productive ecosystems where the flux R is high corresponding to the equatorial forests and to the northern temperate and boreal forests (40° and 70°N) (see Chapter 12).

Finally, one needs to calculate the isotopic enrichment of leaf water relative to

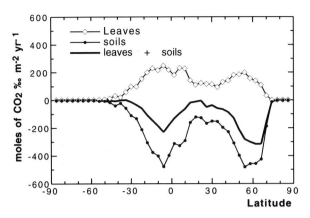

Figure 24.3. Latitudinal effect of the isotope exchange with leaf water and soil water on the $\delta^{18}O$ value of atmospheric CO_2. The quantities represented are the product of the gross CO_2 biospheric fluxes and the isotopic discriminations (see text) for, respectively, leaves and soils. The sum of leaves and soils shows the overall influence of the biosphere on the $\delta^{18}O$ of CO_2, which is to deplete the atmosphere in ^{18}O both over tropical regions and at mid-northern latitudes.

ground water accessed by roots. At steady state (but see Appendix 2 Chapter 3) and assuming constant leaf volume, one can use the Craig and Gordon equation (referred as the CG equation) (Craig and Gordon, 1965).

$$\delta_L = \epsilon_{eq} + (1-h)\cdot(\delta_i + \epsilon_k) + h\cdot\delta_E \qquad (24.1)$$

where: δ_L, leaf water (or water at point of evaporation); δ_i, ground water taken up by roots (no fractionation during root uptake); δ_E, canopy water vapour; h, relative humidity at the leaf surface; ϵ_{eq}, fractionation of ^{18}O for evaporation of water; ϵ_k, diffusive fractionation of ^{18}O in water vapour diffusing in the air.

The extrapolation of the CG equation to the globe relies on several important parameters. First is the relative humidity in the canopy close to the leaf surface. Because of transpiration, the humidity should be higher in the interior of the canopy than in the 'background' air of the PBL as measured at meteorological stations. Secondly the CG equation contains a term for the $\delta^{18}O$ of water vapour in the canopy, which is multiplied by the relative humidity. Hence, at h close to one, the effect of water vapour in determining the degree of leaf enrichment is particularly critical. Close to leaves during photosynthetic uptake, the $\delta^{18}O$ of the vapour should be enriched compared to vapour aloft, since transpired vapour is expected to have a $\delta^{18}O$ close to meteoric water. Within the canopy, the $\delta^{18}O$ of canopy vapour is expected to be a mixture between transpired vapour and vapour from outside the canopy, depending on entrainment of air from aloft to the surface (White and Gedzelman, 1984). In open areas outside the canopy, the existing surface measurements of $\delta^{18}O$ of 'background' water vapour suggest that the vapour lies close to the isotopic equilibrium with the precipitation, except in some very dry areas where raindrops can evaporate below the cloud base (Jacob and Sonntag, 1991). Isotope-specific 3D modelling studies confirm this assumption and predict that the vapour is everywhere depleted by roughly 10‰ with respect to the precipitation. We have arbitrarily neglected canopy effects, based on the idea that leaves at the top of the canopy have a larger photosynthetic uptake than leaves below. Upper canopy leaves also should have maximum photosynthesis rate during the middle of the day, when the PBL is relatively well mixed up to 1–2 km, and thus they should be in contact with water vapour that has a $\delta^{18}O$ close to that of 'background' vapour. Although this simplification needs to be tested against field measurements, we assume that the $\delta^{18}O$ of water vapour in the CG equation is equal to the 'background' $\delta^{18}O$ of vapour as calculated by Jouzel *et al.* (1987). Figure 24.3

shows the role of leaf water enrichment, mediated by the diffusion-backdiffusion flux on the $\delta^{18}O$ of atmospheric CO_2 as represented by the quantity:

$$GPP\left\{\epsilon_L + \frac{1}{1-C_c/C_a}\cdot\left(\delta^{18}O_L - \delta^{18}O_a\right)\right\} \qquad (24.2)$$

where ϵ_L is the fractionation factor for CO_2 diffusing out of the leaves, and $\delta^{18}O$ the $\delta^{18}O$ of CO_2 in the leaves (Farquhar et al., 1993). Leaves tend to enrich atmospheric CO_2 in ^{18}O over productive areas, thus having an effect opposite to soils. However, Figure 24.3 shows that in the northern hemisphere, the degree of $\delta^{18}O$ enrichment in leaves given by the CG equation does not entirely counterbalance the depletion effect of soils.

24.4.3 Non-biospheric processes

Other processes influencing the $\delta^{18}O$ of atmospheric CO_2 are the burning of fossil fuels and biomass, the isotopic enrichment of CO_2 resulting from the photolysis of O_3 in the stratosphere (Yung et al., 1991), and air–sea exchange (only this last one has been taken into account in the simple model in Section 24.3). Each one of these individual processes have a smaller influence on $\delta^{18}O$ compared to the exchange by leaves and by soils. For fossil fuels and biomass burning, CO_2 produced by combustion is assigned the same isotopic composition as the one of atmospheric O_2, i.e. $-17\permil$ VPDB-CO_2 ($+23.5\permil$ VSMOW, Kroopnick and Craig, 1972). Little is known about the isotopic composition of CO_2 in the stratosphere (Thiemens et al., 1991), but measurements over Japan by Gamo et al. (1989) indicate that there is an enrichment of roughly $4\permil$ in CO_2 above the tropopause (20 km). Lacking a clear description of fractionation leading to the enrichment of CO_2 in the stratosphere, we treat this effect by placing an arbitrary source of pure $C^{18}O^{16}O$ above 21 km in the atmospheric transport model (see below), the intensity of which is multiplied by one global tuning coefficient to reproduce the vertical profile in $\delta^{18}O$ observed by Gamo et al. (1989).

The last process to consider is the air–sea flux of CO_2. The hydration time of CO_2 being longer than the time necessary to cross the air–sea boundary by diffusion, one can assume that only CO_2 which is dissolved in sea water, in full isotopic equilibrium with H_2O, is exchanged with the atmosphere. The $\delta^{18}O$ of CO_2 in the surface ocean can thus be inferred from the $\delta^{18}O$ of sea water, close to the VSMOW value everywhere, except in the areas where fresh water enters the ocean: the main river estuaries and the Antarctic and Greenland ice sheets (Farquhar et al., 1993; Ciais et al., 1997a).

24.5 Results and discussion of 3D model

24.5.1 Atmospheric transport

We have incorporated the fluxes as described in Section 24.4 into a 3D model of atmospheric transport. The modelling of global 3D atmospheric transport has been applied to several atmospheric constituents, including CO_2. Small but persistent concentration gradients in the atmosphere are interpreted through transport models in terms of sources and sinks, to establish the budget of CO_2. Including the ^{18}O isotope in CO_2 potentially offers an independent, although indirect, constraint on the carbon fluxes. The atmospheric transport model that is used here is the TM2 tracer model developed by M. Heimann at the Max-Planck Institute of Hamburg (Heimann, 1995; Heimann and Keeling, 1989). The model is an off-line transport scheme based on meteorological wind fields analysed by the ECMWF centre. Wind fields at one location may vary from one year to the other, but the large scale patterns of atmospheric transport are conservative, and hence the simulated tracer concentration in the atmosphere does not vary from year to year because of inter-annual wind variations. The atmosphere is represented in the TM2 model by 9 vertical levels, equally spaced in sigma coordinates ($\sigma(Z)$ = pressure at altitude Z / surface pressure). As an example, the two bottom vertical layers are respectively 400 m and 800 m deep. The horizontal resolution is of 7.5 ° (48 longitudes by 25 latitudes). The time step of the model is 6 hours, but the surface fluxes are monthly averages, for each surface grid cell. Thus, the model does not resolve diurnal variations in CO_2 concentration over vegetated areas, due to the coupling between photosynthesis/respiration and the atmosphere's dynamics near the surface (e.g. nocturnal PBL thinning). The model is run for a "spin up" of 3 years, repeating the transport of year 1990, so that steady state is reached when the simulated atmospheric mean spatial gradients of concentration are constant over time. Concentration fields are archived during the fourth model year, which corresponds to atmosphere transport during 1991. Both CO_2 and $C^{18}O^{16}O$ are transported separately in the tracer model and the field of $\delta^{18}O$ in the atmosphere is calculated in the end from the ratio of the concentrations of both species.

24.5.2 *Results*

The specific $\delta^{18}O$ field relative to one process, say global exchange by leaves, can be calculated as well as the composite $\delta^{18}O$ including two or more components. The δ fields are not additive, but one can combine in an additive manner the δ-anomalies (Heimann *et al.*, 1989) (δ^*) defined as:

$$\delta^*_i = \frac{C_i(\delta^{18}O_i - \delta^{18}O_o)}{\sum_i C_i} \tag{24.3}$$

where:
$\delta^{18}O_o$, 'background' atmospheric δ set up to a constant value of 0.77 ‰; $\delta^{18}O_i$, simulated delta field corresponding to process number i; C_i, concentration field corresponding to process number i.

The simulated atmospheric $\delta^{18}O$ anomalies, zonally averaged at ground level, are shown in Figure 24.4. Leaves produce an overall increase in $\delta^{18}O$ over productive regions, which is more pronounced in the northern hemisphere than in the south because of the larger land area. Soils contribute an effect of the opposite sign. Because soil CO_2 is relatively more depleted than leaf CO_2 compared to the background air (Figure 24.3), even with identical carbon fluxes on a mean annual basis (GPP = R_E), soils predominate over leaves in determining the latitudinal distribution of atmospheric $\delta^{18}O$. Non-biospheric components are less important than the exchanges with soils or with leaves, although the emission of fossil fuels over North America, Europe and North-East Asia depletes the zonal $\delta^{18}O$ field at northern mid-latitudes by 0.3‰ compared to the South Pole. The overall $\delta^{18}O$ curve, combining all processes, indicates

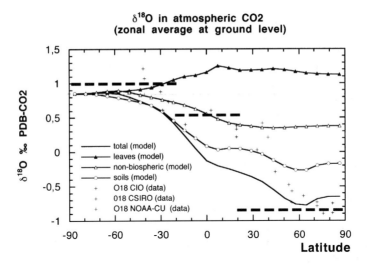

δ¹⁸O in atmospheric CO2
(zonal average at ground level)

Figure 24.4. Simulated δ¹⁸O of atmospheric CO₂ at the surface using the 3D transport model (zonal average). The different components correspond to the separate δ¹⁸O implied respectively by leaf exchange and non-biospheric sources. The overall modelled δ¹⁸O is the thin curve, which can be compared to the atmospheric obsrvations (black dots). The results of the simple model (Section 24.3) over three wide zones are represented in the straight dashed lines (K=2).

a permanent negative gradient between the northern and the southern hemisphere, which fits reasonably well the observations. However, in the tropics, the simulated δ¹⁸O is much more negative than the observed one. We have no clear explanation for this discrepancy between model and data in the tropics: one possibility is that the δ¹⁸O of ground water, and by way of consequence the δ¹⁸O of leaf water, that we assume from Jouzel *et al.* (1987), Figure 24.4, is too negative compared to the real world. Another possibility is that the prescribed GPP and R_E fluxes in the tropics are larger than the observed ones. In Figure 24.4 the results for the three zones from the simple model are shown as straight dashed lines. As pointed out in Section 24.3, the simple model proves capable of producing reasonable results.

Another valuable test of the model, where the simple model fails, is the seasonal cycle of δ¹⁸O at high northern latitudes (data from Trolier *et al.* (1996) for Barrow, Alaska), which is observed to lag behind the CO_2 seasonal cycle by 3–4 months. One can see in Figure 24.5 that the seasonal cycle of CO_2 resulting from the superimposed monthly variations in GPP and R_E fluxes is correctly represented. The seasonal cycle of δ¹⁸O also is in good agreement with that observed. Surprisingly, the effect of leaves on δ¹⁸O of CO_2 is relatively constant throughout the year, indicating that the strongly seasonal oscillation present in GPP and in ground water δ¹⁸O over boreal regions is attenuated by an opposite variation in the degree of enrichment of leaf water (inferred by the CG equation) and the temperature-controlled CO_2–H_2O isotopic equilibration factor (see also Chapter 12). However, the seasonal cycle of δ¹⁸O at high northern latitudes is governed in phase and in amplitude by the exchange with soils (Figure 24.5). Further south, it is found that both leaves and soils contribute to the phase of the δ¹⁸O seasonal cycle. For instance, at 19°N (Mauna Loa, Hawaii) roughly 0.1‰ of peak to peak amplitude comes from leaves, as opposed to 0.5‰ from soils. In the

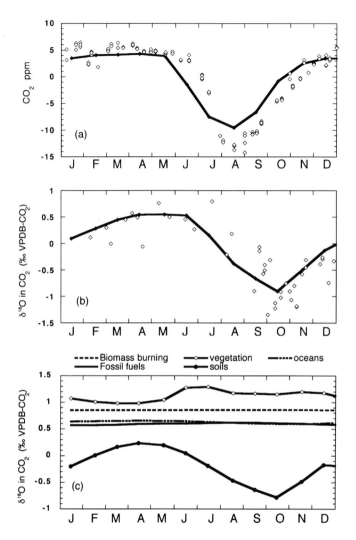

Figure 24.5. *Simulated (only monthly means are shown) and measured seasonal variations of CO_2 (A) and $\delta^{18}O$ (B) at Point Barrow, Alaska, 74°N for years 1990 and 1991. Only the atmospheric transport is changed from one year to the other, the surface fluxes remaining unchanged. The seasonal cycle in the observations has been detrended. Separation of the modelled $\delta^{18}O$ into its different components (C), respectively leaves, soils, stratospheric exchange, ocean exchange and fossil fuel combustion. (P. Peylin personal communication).*

southern hemisphere, far from active ecosystems, the role of the ocean also turns out to be important. At 40°S (data from Francey *et al.*, (1990) at Cape Grim, Tasmania) the ocean contributes 0.1‰ to the peak to peak amplitude of the seasonal cycle, against 0.1‰ from leaves and 0.25‰ from soils.

24.5.3 *Discussion and crude sensitivity test*

This first modelling study of $\delta^{18}O$ in CO_2 indicates that it is possible to capture most of the observed spatial and temporal variations in $\delta^{18}O$ by considering the processes of exchange–diffusion by leaves and the respiratory emissions of CO_2 by soils. Although the biosphere plays a dominant role, taking account of the burning of fossil fuel and the air-sea exchange of CO_2 forces results from the model to give a closer agreement with the atmospheric observations for the latitudinal gradient in $\delta^{18}O$. Respiration appears to determine the negative sign of the latitudinal gradient, as it overwhelms the effect of enrichment due to leaves. The seasonal oscillation in $\delta^{18}O$ at high northern latitudes is also controlled by the effect of soil respired CO_2 (see Chapter 12). It is clear that the $\delta^{18}O$ of atmospheric CO_2 is sensitive both to the exchange of carbon by land ecosystems but also to the $\delta^{18}O$ of water in leaves and in soils. As a crude sensitivity test, we increased the gross exchange of carbon on land, by 10% simultaneously altering GPP and R_E to preserve $NEP = 0$. The result of this test shown in Figure 24.6 indicates a small negative shift of the biospheric effect, due to the fact that the soils have a higher 'isotopic-exchange efficiency' than leaves. In another test, we maintained GPP and R_E equal to their control value, but we increased the $\delta^{18}O$ of precipitation by 1‰ everywhere, thus shifting up the $\delta^{18}O$ of leaf water and of soil water simultaneously. This test experiment, shown in Figure 24.6, indicates a reduction of the biospheric effect in the tropics compared to the control run, and virtually no changes at northern mid-latitudes. By comparing these two simple sensitivity tests, one can see that a shift of 1‰ of the $\delta^{18}O$ of the water exchanging with CO_2 has an effect comparable to a 10% change in GPP and R_E. The sensitivity to the $\delta^{18}O$ of water is important in the tropics and this may bring up uncertainties in the use of $\delta^{18}O$ in atmospheric CO_2 to constrain indirectly the carbon fluxes over tropical ecosystems. On the other hand, at northern mid-latitudes, the $\delta^{18}O$ of CO_2 appears more sensitive to the exchange of carbon to the $\delta^{18}O$ value of the water, which is encouraging. In any case, this global study points out to the need for a better characterization of the $\delta^{18}O$ of the specific pools of water in vegetation and in soils (see Chapters 11 and 12).

24.6 Conclusions

Understanding $\delta^{18}O$ in CO_2 requires a hierarchy of models to be developed which are valuable for evaluating the contribution of each process to the overall $\delta^{18}O$ atmospheric signal, and to explore the sensitivity of the simulated $\delta^{18}O$ to flux parameterisation. It

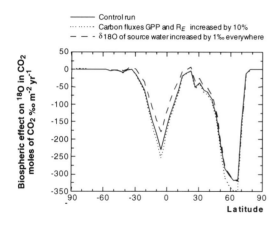

Figure 24.6. A simple illustration of the sensitivity of the $\delta^{18}O$ of atmospheric CO_2 to the biospheric exchange. Solid line is the control run identical to Figure 24.3. Dotted line represents the biospheric effect of soils and leaves when the gross fluxes of respiration and primary productivity are increased everywhere by 10%. Dashed line corresponds to an increase in $\delta^{18}O$ of precipitation of 1‰ everywhere, keeping the CO_2 fluxes at their control value.

is very informative to try and develop a very simple model, that incorporates the main effects of the study, in parallel to the more sophisticated modelling work. This serves two goals: it gives a more intuitively understandable picture of what actually is going on, and secondly, the quality of its outcome forces one to invest time in sensitivity and error analysis of a more sophisticated model. A simple model following the ideas formulated by Francey and Tans (1987) and Farquhar *et al.* (1993) and treating the atmosphere as three well mixed boxes proved successful in explaining qualitatively the latitudinal patterns of δ^{18}O in CO$_2$. However, that simple model failed to reproduce the seasonal cycle of δ^{18}O which is due to a complex interaction between seasonally varying CO$_2$ fluxes, δ^{18}O of water and temperature. A more complex 3D model accounting for the atmosphere's dynamics and for the geographical distribution of surface fluxes is further discussed, based on a recent study by Ciais *et al.* (1997a,b), but with an improved treatment of the stratospheric ^{18}O exchange. The 3D model captures the patterns of the atmospheric δ^{18}O observations (both the latitudinal gradient and the seasonal cycle) reasonably well.

Our main conclusion is that soil respiration is the dominant factor that controls the variations in δ^{18}O of CO$_2$. As the annual integral of total ecosystem respiration equals gross primary productivity, this reinforces our confidence that the signal of δ^{18}O in CO$_2$ contains an original and very valuable information on the gross carbon exchange by land ecosystems. Reducing the uncertainties can be achieved by (i) testing the sensitivity of the model to its parameterisations, to make the part between the influences of isotopic hydrology and carbon fluxes; (ii) by determining more realistic values for the isotopic fractionations, especially the fractionation of CO$_2$ diffusing out of the soil; (iii) by assessing the validity at the ecosystem level of the CG equation used to calculate the effect of leaves; and (iv) by better characterising the isotopic value of the soil moisture that exchanges isotopically with soil CO$_2$. This is clearly a long-term effort, involving appropriate experimental and field work together with the development of more process-oriented models treating the isotopic exchange of CO$_2$ and water at a finer scale.

References

Allison, C.E., Francey, R.J. and Meijer, H.A.J. (1995) Recommendations for the reporting of stable isotope measurements of carbon and oxygen in CO$_2$ gas. In: *Reference and intercomparison materials for stable isotopes of light elements*, IAEA-TECDOC 825, IAEA, Vienna, pp. 155–162.

Boden, T.A., Kaiser, D.P., Sepanski, R.J. and Stoss, F.W. (1994) *Trends '93: A Compendium of Data on Global Change.* CDIAC Oak Ridge ESD publication 4195

Bolin, B., Degens, E.T., Kempe, S. and Ketner, P. (Eds) (1982) *The global carbon cycle*, SCOPE vol. 13, John Wiley, Chichester.

Brenninkmeijer, C.A.M., Kraft, P. and Mook, W.G. (1983) Oxygen isotope fractionation Between CO$_2$ and H$_2$O, *Isotope Geoscience* 1, 181–190.

Ciais, P., Tans, P.P., White, J.W.C., Trolier, M., Francey, R.J., Berry, J.A., Randall, D.R., Sellers, P.J., Collatz, J.G. and Schimel, D.S. (1995) Partitioning of ocean and land uptake of CO$_2$ as inferred by δ^{13}C measurements from the NOAA Climate Monitoring and Diagnostics Laboratory global air sampling network. *J. Geophys. Res.* 100, 5051–5070.

Ciais, P., Denning, S.,Tans, P.P., Berry, J.A., Randall, D., Collatz, J.G., Sellers, P.J., White, J.W., Trolier, M., Meijer, H.A.J., Francey, R., Monfray, P. and Heimann, M. (1997a) A three dimensional synthesis study of δ^{18}O in atmospheric CO$_2$. Part 1 Surface fluxes,*J. Geophys Res.*, 102, 5857–5872

Ciais, P., Tans, P.P., Denning, S., Francey, R., Trolier, M., Meijer, H.A.J., White, J.W., Berry, J.A., Randall, D., Collatz, J.G., Sellers, P.J., Monfray, P. and Heimann, M. (1997b) A three dimensional synthesis study of $\delta^{18}O$ in atmospheric CO_2. Part 2 Simulations with the TM2 transport model, *J. Geophys Res.*, pp. 5873–5883.

Collatz, G.J., Ball, J.T., Grivet, C. and Berry, J.A. (1991) Physiological and environmental regulation of stomatal conductance, photosynthesis, and transpiration: a model that includes a laminar boundary layer. *Agric. and Forest Meteorol.* **54**, 107–136.

Collatz, G.J., Ribas-Carbo, M. and Berry, J.A. (1992) Coupled photosynthesis–stomatal conductance model for leaves of C4 plants, *Aust. J. Plant Physiol.* **19**, 519–538.

Conway, T.J., Tans, P.P., Waterman, L.S., Thoning, K.W., Masarie, K.A. and Gammon, R.H. (1988) Atmospheric carbon dioxide measurements in the remote global troposphere, 1981–1984. *Tellus* **40**, 81–115.

Conway, T.J., Tans, P.P., Waterman, L.S., Thoning, K.W., Kitzis, D.R., Masarie, K.A. and Zhang, N. (1994) Evidence for interannual variability of the carbon cycle from the NOAA/CMDL global air sampling network. *J. Geophys. Res.* **99**, 22831–22855.

Coplen, T. (1996) New guidelines for reporting stable hydrogen, carbon, and oxygen isotope-ratio data. *Geochim. Cosmochin. Acta* **60**, 3359–3360.

Craig, H, Gordon, A. (1965) Deuterium and Oxygen-18 variations in the ocean and the marine atmosphere. In *Stable Isotopes in Oceanographic Studies and Paleo-Temperatures*, (ed E. Tongiori) pp. 9–130.

Denning A.S. (1994) Investigations of the transport, sources, and sinks of atmospheric CO_2 using a general circulation model. *Atmospheric Science Paper No. 564, Colorado State University.*

Denning, A.S., Collatz, J.G., Zhang, C., Randall, D.A., Berry, J.A., Sellers, P.J. and Wofsy, S.C. (1997) Simulations of terrestrial carbon metabolism and atmospheric CO_2 in a general circulation model. Part 1: Surface carbon fluxes. *Tellus*, in press.

Denning, A.S., Randall, D.A. (1996b) Simulations of terrestrial carbon metabolism and atmospheric CO_2 in a general circulation model. Part 2: Simulated CO_2 concentrations. Submitted to *Tellus*.

Farquhar, G.D., Lloyd, J., Taylor, J.A., Flanagan, L.B., Syvertsen, J.P., Hubick, K.T., Wong, S.C. and Ehleringer, R. (1993) Vegetation effects on the isotope composition of oxygen in atmospheric CO_2. *Nature* **363**, 439–443.

Flanagan, L.B., Brooks, J.R., Varney, G.T. and Ehleringer, J.R. (1997) Discrimination against $C^{18}O^{16}O$ during photosynthesis and the oxygen isotope ratio of respired CO_2 in boreal forest ecosystems. *Global Biogeochem Cycles*, **11**, 83–88.

Fowler, L.A., Randall, D.A., Rutledge, S.A. (1995) Liquid and ice cloud microphysics in the CSU General Circulation Model. Part I: Model description and simulated microphysical processes. *J. Clim.* **9**, 486–529.

Fowler, L.A., Randall, D.A. (1995) Liquid and ice cloud microphysics in the CSU General Circulation Model. Part II: Impact on cloudiness, the Earth's radiation budget, and the general circulation of the atmosphere. *J. Clim.* **9**, 530–560.

Francey, R.J., Tans, P.P. (1987) Latitudinal variation in oxygen-18 of atmospheric CO_2. *Nature* **327**, 495–497.

Francey, R.J., Robbins, F.J., Allison, C.E. and Richards, N.G. (1995) The CSIRO global survey of CO_2 stable isotopes. In *Baseline 1988*, ed W. S.R., 1990.

Gamo, T., Tsutsumi, M., Sakai, H., Nakazawa, T., Tanaka, M., Honda, H., Kubo, H. and Itoh, T. (1989) Carbon and oxygen isotopic ratios of carbon dioxide of a stratospheric profile over Japan. *Tellus*, **41B**, 127–133.

Gemery, P.A., Trolier, M., White, W.C. (1996) Oxygen isotope exchange between carbon dioxide and water following atmospheric sampling using glass flasks. J. Geophys. Res. **101**, 14415–14420.

Gonfiantini, R. (1984) Advisory group Meeting on stable isotope reference samples for geochemical and hydrological investigations. *Report to the Director General, IAEA, Vienna 1984.*

Gonfiantini, R., Stichler, W. and Rozanski, K. (1995) Standards and intercomparison materials distributed by the International Atomic Energy Agency for stable isotope measurements. In *Reference and intercomparison materials for stable isotopes of light elements*, IAEA-TECDOC 825, IAEA, Vienna, 155–162.

Harshvardhan, Davies, R., Randall, D.A.,Corsetti, D., Corsetti, T.G. (1987) A fast radiation parameterisation for general circulation models. *J. Geophys. Res.* **92**, 1009–1016.

Heimann, M. and Keeling, C.D. (1989) A three-dimensional model of atmospheric CO_2 transport based on observed winds : 2. Model description and simulated tracer experiments, In *Aspects of climate variability in the Pacific and the Western Americas*, Geophysical monograph 55, (ed D.H. Peterson), AGU, Washington (USA), pp. 237–27 .

Heimann, M. (Ed.) (1993) The global carbon cycle, NATO ASI series I: Global Environmental Change, Vol 15, Springer Verlag, Berlin.

Heimann, M. (1995) The global Atmospheric tracer model TM2, Max-Planck-Institut fur Meteorologie, Bundesstrasse 55, D-20146, Hamburg, Germany August 1995. Technical Report 10.

Houghton, J.T., Meira Filho, L.G., Callander, B.A., Harris, N., Kattenberg, A. and Maskell, K. (1996) *Climate Change 1995, Intergovernmental Panel on Climate Change*, Cambridge University Press, Cambridge.

Jacob, H., Sonntag, C. (1991) An 8-year record of the seasonal variation of ^2H and ^{18}O in atmospheric water vapor and precipitation at Heidelberg, Germany. *Tellus* **43**, 291–300.

Jouzel, J., Russell, G.L., Suozzo, R.J., Koster, R.D., White, J.W.C., Broecker, W.S. (1987) Simulations of the HDO and $H_2^{18}O$ atmospheric cycles using the NASA/GISS general circulation model : The seasonal cycle for present-day conditions *J. Geophys. Res.* **92**, 14739–14760.

Keeling, C.D. (1958) The concentration and isotopic abundances of atmospheric carbon dioxide in rural areas. *Geochim. Cosmochim. Acta* **13**, 322–334.

Keeling, C.D, Mook, W.G. and Tans P.P. (1979). Recent trends in the $^{13}C/^{12}C$ ratio of atmospheric carbon dioxide. *Nature* **277**, 121–123.

Keeling, C.D, Bacastow, R.B., Carter, A.F., Piper, S.C., Whorf, T.P., Heimann, M., Mook, W.G. and Roeloffzen, J.C. (1989) .A three-dimensional model of atmospheric CO_2 transport based on observed winds: I. Analysis of observational data. In *Aspects of climate variability in the Pacific and the Western Americas*, Geophysical monograph 55, (ed D.H. Peterson), AGU, Washington (USA), pp. 165–236.

Keeling, C.D., Whorf, T.P., Wahlen, M. and van der Plicht, J. (1995) Interannual extremes in the rate of rise of atmospheric carbon dioxide since 1980. *Nature* **375**, 666–670.

Keeling, R. (1995) The atmospheric oxygen cycle: the oxygen isotopes of atmospheric CO_2 and O_2 and the O_2/N_2 ratio. *Rev. Geophys.* 1253–1262.

Kroopnick, P.M., Craig, H. (1972) Atmospheric oxygen: Isotopic composition and solubility fractionation. *Science* **175**, 54–55.

Levin, I.,Graul, R. and Trivett, N.B.A. (1995) Long-term observations of atmospheric CO_2 and carbon isotopes at continental sites in Germany. *Tellus* **47B**, 23–34.

Lloyd, J. and Farquhar, G.D. (1994) ^{13}C discrimination during CO_2 assimilation by the terrestrial biosphere. *Oecologia* **99**, 201–215.

Melillo, J.M., McGuire, A.D., Kicklighter, D.W., Moore III, B., Vororsmarty, C.J. and Schloss, A.L. (1993) Global climate change and terrestrial net primary production. *Nature* **363**, 234–240.

Nakazawa, T., Morimoto, S., Aoki, S. and Tanaka, M. (1993) Time and space variations of the carbon isotopic ratio of tropospheric carbon dioxide over Japan. *Tellus* **45B**, 258–274.

Randall, D.A., Harshvardhan and Dazlich, D.A., Corsetti, T.G. (1989) Interactions Among Radiation, Convection, and Large-Scale Dynamics in a General Circulation. *Model. J. Atmos. Sci.* **46**, 1943–1970.

Randall, D.A., Harshvardhan and Dazlich, D.A. (1991) Diurnal variability of the hydrologic cycle in a general circulation model. *J. Atmos. Sci.* **48**, 40–62.

Randall, D.A., Shao, Q., Moeng, C-H. (1992) A second-order bulk boundary-layer model. *J. Atmos. Sci.* **49**, 1903–1923.

Randall, D.A. and Pan, D-M. (1993) Implementation of the Arakawa–Schubert parameterisation with a prognostic closure. In: *The Representation of Cumulus Convection in Numerical Models* (eds. K. Emanuel and D. Raymond) American Meteorological Society, Boston, pp. 137–144.

Randall, D.A., Sellers, P.J., Berry, J.A., Dazlich, D.A., Zhang, C., Collatz, J.A., Denning, A.S., Los, S.O., Field, C.B., Fung, I., Justice, C.O. and Tucker, C.J. (1996) A revised land-surface parameterization (SiB2) for GCMs. Part 3: The greening of the Colorado State University General Circulation Model. *J. Clim.* **9**, 738–763.

Rozanski, K., Araguas-Araguas, L., Gonfiantini, R. (1992) Relation between long-term trends of oxygen-18 isotope composition of precipitation and climate, *Science* **258**, 981–985.

Rozanski, K., Araguas-Araguas, L. and Gonfiantini, R. (1993) Isotopic patterns in modern global precipitation, In: *Climate change in continental isotopic records*, (eds P.K Swart, K.C. Lohmann and J.MacKenzie Geophysical monograph 78, AGU, Washington (USA), pp. 1–37.

Ruimy, A., Saugier, B., Dedieu, G. (1994) Methodology for the estimation of terrestrial net primary production from remotely sensed data. *J. Geophys. Res.* **99**, 5263–5283.

Sellers, P.J., Mintz, Y., Sud, Y.C., Dalcher, A. (1986) A simple biosphere model (SiB) for use within general circulation models. *J. Atmos. Sci.* **43**, 505–531.

Sellers, P.J., Berry, J.A., Collatz, G.J., Field, C.B. and Hall, F.G. (1992a) Canopy reflectance, photosynthesis, and transpiration. III. A reanalysis using enzyme kinetics – electron transfer models of leaf physiology. *Remote Sens. Environ.* **42**.

Sellers, P.J., Heiser, M.D. and Hall, F.G. (1992b) Relations between surface conductance and spectral vegetation indices at intermediate (100 m^2 to 15 km^2) length scales. *J. Geophys. Res.* **97**, 19033–19059.

Sellers, P.J., Randall, D.R., Collatz, J.A., Berry, J.A., Field, C.B., Dazlich, D.A., Zhang, C. and Bounoua, L. (1996a) A revised land-surface parameterization (SiB2) for atmospheric GCMs. Part 1: Model formulation. *J. Clim* **9**, 676–705.

Sellers, P.J., Los, S.O., Tucker, C.J., Justice, C.O., Dazlich, D..A, Collatz, G.J., Randall D.A. (1996b) A revised land surface parameterization (SiB2) for atmospheric GCMs. Part 2: The generation of global fields of terrestrial biophysical parameters from satellite data. *J. Clim.* **9**, 706–737.

Siegenthaler, U. (1993) Modelling the present-day oceanic carbon cycle. In: *The global carbon cycle*, NATO ASI series I: Global Environmental Change, Vol. 15, Springer Verlag, Berlin.

Suarez, M.J., Arakawa, A. and Randall, D.A. (1983) Parameterization of the planetary boundary layer in the UCLA general circulation model: Formulation and results. *Mon. Wea. Rev* **111**, 2224–2243.

Thiemens, M.H., Jackson, T., Mauersberger, K., Schueler, B. and Morton, J. (1991) Oxygen isotope fractionation in stratospheric CO_2, *Geophys. Res. Lett.* **18**, 669–672.

Trolier, M. (1997) Monitoring the isotopic composition of atmospheric CO_2 : measurements from the NOAA global air sampling network. *J. Geophys. Res.* in press.

White, J.W.C. and Gedzelman, S.D. (1984) The isotopic composition of atmospheric water vapor and the concurrent meteorological conditions, *J. Geophys. Res.* **89**, 4937–4939.

Yakir, D., Wang, X.F. (1996) Fluxes of CO_2 and water between terrestrial vegetation and the atmosphere estimated from isotope measurements, *Nature* **380**, 515–517.

Yung, Y.L., DeMore, W.B., Pinto, J.P. (1991) Isotopic exchange between carbon dioxide and Ozone via O(1D) in the stratosphere, *Geophys. Res. Lett.* **18**, 13–16.

Index